S0-BAC-948

# Professional
# Microsoft® Robotics
# Developer Studio

(Continued)

# Part IV: Robotics Hardware

# Professional
# Microsoft® Robotics
# Developer Studio

Kyle Johns

Trevor Taylor

**WILEY**

Wiley Publishing, Inc.

# Professional Microsoft® Robotics Developer Studio

Published by
**Wiley Publishing, Inc.**
10475 Crosspoint Boulevard
Indianapolis, IN 46256
www.wiley.com

Copyright © 2008 by Wiley Publishing, Inc., Indianapolis, Indiana

Published simultaneously in Canada

ISBN: 978-0-470-14107-6

Manufactured in the United States of America

10 9 8 7 6 5 4 3 2 1

**Library of Congress Cataloging-in-Publication Data:**

Johns, Kyle, 1965-
    Professional Microsoft robotics developer studio / Kyle Johns, Trevor Taylor.
        p.   cm.
    Includes index.
    ISBN 978-0-470-14107-6 (paper/website)
        1.  Robotics.   I.  Taylor, Trevor, 1955-   II.  Title.
    TJ211.J55   2008
    629.8'9—dc22

                                                            2008014648

No part of this publication may be reproduced, stored in a retrieval system or transmitted in any form or by any means, electronic, mechanical, photocopying, recording, scanning or otherwise, except as permitted under Sections 107 or 108 of the 1976 United States Copyright Act, without either the prior written permission of the Publisher, or authorization through payment of the appropriate per-copy fee to the Copyright Clearance Center, 222 Rosewood Drive, Danvers, MA 01923, (978) 750-8400, fax (978) 646-8600. Requests to the Publisher for permission should be addressed to the Legal Department, Wiley Publishing, Inc., 10475 Crosspoint Blvd., Indianapolis, IN 46256, (317) 572-3447, fax (317) 572-4355, or online at http://www.wiley.com/go/permissions.

**Limit of Liability/Disclaimer of Warranty:** The publisher and the author make no representations or warranties with respect to the accuracy or completeness of the contents of this work and specifically disclaim all warranties, including without limitation warranties of fitness for a particular purpose. No warranty may be created or extended by sales or promotional materials. The advice and strategies contained herein may not be suitable for every situation. This work is sold with the understanding that the publisher is not engaged in rendering legal, accounting, or other professional services. If professional assistance is required, the services of a competent professional person should be sought. Neither the publisher nor the author shall be liable for damages arising herefrom. The fact that an organization or Website is referred to in this work as a citation and/or a potential source of further information does not mean that the author or the publisher endorses the information the organization or Website may provide or recommendations it may make. Further, readers should be aware that Internet Websites listed in this work may have changed or disappeared between when this work was written and when it is read.

For general information on our other products and services please contact our Customer Care Department within the United States at (800) 762-2974, outside the United States at (317) 572-3993 or fax (317) 572-4002.

**Trademarks:** Wiley, the Wiley logo, Wrox, the Wrox logo, Programmer to Programmer, and related trade dress are trademarks or registered trademarks of John Wiley & Sons, Inc. and/or its affiliates, in the United States and other countries, and may not be used without written permission. Microsoft is a registered trademark of Microsoft Corporation in the United States and/or other countries. All other trademarks are the property of their respective owners. Wiley Publishing, Inc., is not associated with any product or vendor mentioned in this book.

Wiley also publishes its books in a variety of electronic formats. Some content that appears in print may not be available in electronic books.

*To Ryan, Tanner, Kaitlyn, Kelsey, Colin, and little Abby; but most of all to Marie.*
**—Kyle**

*To Denise, whose support enabled me to undertake writing this book.*
**—Trevor**

# About the Authors

**Kyle Johns** is a principal software developer at Microsoft, where he is currently a member of the Microsoft Robotics Developer Studio Team. After receiving a master's degree in computer science from the University of Utah, he designed 3D graphics hardware for flight simulators at Evans and Sutherland. He joined Microsoft as one of the original members of the DirectX Team and then went on to help develop the graphics system software in the early days of the Xbox project. Recently he has been enjoying the opportunity to apply his 20 years of 3D graphics experience to the field of robotics by developing the Robotics Developer Studio Simulation Environment.

**Trevor Taylor** is a consultant in the field of robotics education. After 20 years in the IT industry, including co-founding a consulting company that became a Microsoft Solution Provider Partner, he moved to the Queensland University of Technology in 2002. For six years he taught a variety of subjects, including Visual Basic and Web development using ASP.NET. During this period he also worked part-time on a doctorate in computer vision and robotics. In early 2008, Trevor left QUT to concentrate on developing course materials for teaching robotics and to finish writing his thesis. Trevor has worked with MRDS since the very first Community Technology Preview in June 2006 and is an active and well-known contributor to the community.

# Credits

**Executive Editor**
Chris Webb

**Development Editor**
Maureen Spears

**Technical Editors**
David E. Buckley
Steve Tudor

**Production Editor**
William A. Barton

**Copy Editor**
Luann Rouff

**Editorial Manager**
Mary Beth Wakefield

**Production Manager**
Tim Tate

**Vice President and Executive Group Publisher**
Richard Swadley

**Vice President and Executive Publisher**
Joseph B. Wikert

**Project Coordinator, Cover**
Lynsey Stanford

**Proofreaders**
Jeremy Bagai,
David Fine,
Paul Sagan,
Word One

**Indexer**
Robert Swanson

# Acknowledgments

Various organizations have assisted with the production of this book by making software services for their robots available for inclusion in the book examples and/or providing images for use in the figures. The authors wish to acknowledge their support. These organizations include (in alphabetical order) the following:

CoroWare, Inc.: www.coroware.com

Institute for Personal Robots in Education (IPRE): www.ipre.org

LEGO: www.lego.com

Lynxmotion, Inc.: www.lynxmotion.com

Parallax, Inc.: www.parallax.com

Picblok Corporation Pty Ltd: www.picblokcorporation.com

RoboticsConnection (Summerour Robotics Corporation): www.roboticsconnection.com

Surveyor Corporation: www.surveyor.com

Throughout the preparation of the book, the Microsoft Robotics Developer Studio Team have been most helpful in resolving problems and answering questions. To them we extend a big "thank-you."

# Contents

# Contents

# Contents

# Contents

# Contents

# Contents

# Foreword

I am pleased to have the opportunity to introduce this book and Microsoft Robotics Developer Studio to its audience. I hope that the combination helps foster the creation of interesting applications for robots because it is clear that the key to the success of this emerging new industry is software that delivers value.

Anyone who has been around for a while recognizes that the current robotics industry has many similarities to the early PC industry. Both are characterized by both the tremendous passion and anticipation of the early pioneers and questions about the value of the technology. With the advent of word processing, spreadsheets, and thousands of other applications, no one asks me any longer why I own a PC. In fact, most of them also own PCs. Similarly, the personal robotics market that is just emerging has the same or even greater potential if the creativity of its community can be unlocked.

It is the vision and energy of the existing robotics community that directly contributed to Microsoft investing in the creation of Microsoft Robotics Developer Studio. While many employees at Microsoft have had a personal interest in robotics, it was not until the leadership from diverse parts of the robotics community invited Microsoft to participate that the company decided to get started. Until that time, robotics developers were considered part of an audience consisting of casual, academic, and professional developers who might use Visual Studio and our embedded operating systems.

In 2004, things began to change. I happened to be in the right place at the right time to help make that happen. At the time, I was working in the offices of Microsoft's chief software architect, aka Bill Gates. Part of my job was to promote synergies across the company, as well as to be an extra set of eyes and ears for Bill. That enabled me to attend a number of meetings with notable members of the robotics community whose projects ranged from educational robotics to academic research and from consumer robotics to industrial automation. Despite the diversity of the audience, these people all reflected a common message — something significant was happening; and they encouraged Microsoft to participate and apply its software expertise just as it had in the past to PCs and the Web.

I began a dialogue with Bill about how Microsoft might respond to this invitation. The discussion was very timely. Bill, having recently visited a number of universities to speak on the importance of computer science to future technology, was treated to several campus tours that included some innovative work in robotics. This enabled him to see firsthand the tremendous progress and investments being made in the robotics community. As a result, he asked me to spend the next few months gathering more data and formulating a specific proposal about what we might do.

I then talked with even more people in the robotics community, including Red Whittaker from CMU, Sebastian Thrun from Stanford, and Rod Brooks from MIT, to name just a few. All of them confirmed my earlier impressions that the market for robotics was primed to move forward in a new, more personal dimension. However, they also indicated that significant challenges were holding back its progress. Specifically, every robot was an island unto itself, resulting in limited sharing of technology. Developing software for robots has historically required a great deal of technical expertise and resources because the tools and technologies needed have been typically tied to specific hardware platforms. Therefore, despite a growing interest, it was hard and often expensive for anyone to participate.

I compiled my research into a 60-page proposal outlining how Microsoft could help address these challenges, which I sent to Bill, who requested that I review it with Craig Mundie, who serves as chief technology officer focused on forward-looking strategies for the company, and Rick Rashid, Microsoft's senior vice president of research, to get their insights and feedback from the perspectives of the academic and research communities. Both executives were very supportive. Craig offered some advanced technologies he had been incubating and Rick offered to host the new incubation.

From there it took about nine months to prototype what would become Microsoft Robotics Developer Studio. After another executive review, the prototype was approved, leading to a new product and business for Microsoft. In June 2006, the product was formally previewed; it was released in December that same year.

What was delivered is an impressive collection of software. It starts with the runtime, which provides a very simple, but flexible model for development. The Concurrency and Coordination Runtime (CCR) and Decentralized Software Services (DSS) libraries were originally intended as a future programming model to enable developers to write applications for the increasingly prevalent multi-processor, distributed scenarios into which computing technology is evolving. CCR enables developers to move beyond the limitations of single-threaded applications without the conventional complexities of programming locks and semaphores. DSS takes that simple programming model and applies it across the network, providing a simple yet elegant, flexible model that can run on a single PC or across networked servers. Modeled on conventional Web-style interfaces such as HTTP and SOAP, it extends the so-called Representational State Transfer (REST) model by providing a powerful notification (publish-subscribe) framework that makes the system very efficient, yet applicable to both Windows and web applications.

Furthermore, the architectural framework provides a better model for resiliency and software maintenance. Software modules, which we refer to as services, operate in isolation from each other, both from a data sharing model as well as execution standpoint. This means that failures (bugs) in one service are unlikely to corrupt the data or execution of other services. In addition, individual services can be shut down, restarted, or replaced dynamically, meaning software maintenance can be performed while an application is running. Because data sharing is done through message-passing and is represented as documents, it is an easy model for maintaining the integrity of the data shared between models. This code-data separation also makes it possible to easily create different views or user interfaces of the data without affecting the code creating the data. Therefore, it is possible to have many different ways to visualize the data. In fact, one of the easiest ways to view the state of a service is to simply use a web browser because each service state has a URL.

Finally, the CCR/DSS programming model provides for rich *composability* — that is, applications can be composed of functions provided by other modules. This is reflected in the number of generic services that are included with Microsoft Robotics Developer Studio. It enables specialization of services at one level, such as a specific motor, while enabling more advanced services, such as a drive service, to define partner contracts that enable it to operate with a variety of different motor hardware. This capability to apply series across different robots, results in not only greater opportunity for accessing enabling technologies, but also a larger market for those creating those technologies.

In addition to the programming model afforded by the CCR and DSS runtimes, we added tools to make it easier to develop applications. While supporting a wide variety of programming languages, including C#, C++, Visual Basic, and Python, we added the Visual Programming Language (VPL), which enables an easy drag-and-drop approach to creating applications. This tool not only provides a somewhat easier way to create applications, but also facilitates rapid prototyping because it offers the optional generation of human-readable C# code. Moreover, because of the common services-oriented programming model, the language is infinitely extensible.

The Visual Simulation Environment provided also makes development easier. Its 3D graphical presentation is augmented with a software physics engine for a realistic emulation of robots and their environment. Because it uses the same services framework, it is possible to develop applications in simulation and easily move them to the real hardware, which often saves time and resources. Services can be created that run without regard to whether they apply to the simulation or the real world. The software physics engine means that interaction between simulated entities is freely available. For example, when two entities collide, the simulation automatically applies the result both operationally and visually.

Finally, we added a great deal of sample code — in terms of robot models, enabling technologies, and tutorials — to the toolkit. The objective is to help people get started. Most of these samples are provided in both compiled and source form, so every sample can serve as a learning experience. Of course, with so many samples and tutorials, getting started may sometimes be overwhelming.

That's where this book comes in. It provides a number of additional ways to explore beyond the samples in the toolkit and therefore is a good companion for developers just beginning. It also illustrates creative ways to use the tools and libraries in Microsoft Robotics Developer Studio to build innovative new robotics applications.

I believe that by enabling a wider audience of contributors personal robotics has the potential to become as much a reality as personal computing has become today. I hope that Microsoft Robotics Developer Studio and this book provide important catalysts toward that end.

Tandy Trower
General Manager
Microsoft Robotics Group

# Introduction

Microsoft Robotics Developer Studio (MRDS) introduces a new way to program robots in the Windows environment. It attempts to bring some order to the chaos that has marred the field of robotics, at least as far as Windows-based applications are concerned. If it meets its objectives, you will see the emergence of a common standard for robotics software in the next few years. This is potentially very significant, and it is a great time to get in on the ground floor.

MRDS was developed over a relatively short period of time by a very small team. It is not your average Microsoft product. In an unusual move for Microsoft, key portions of the code for MRDS are available in source form. This makes it readily extensible and offers many opportunities for programmers, whether you are a hobbyist, a research student, or an employee of a large corporation, to write new services that integrate directly into the system and to easily share these services with others.

The speed of development and the limited resources of the MRDS team has meant that documentation has tended to lag behind. This book provides a comprehensive introduction to MRDS for experienced C# programmers. It contains a wealth of examples. We did not set out to duplicate the existing MRDS documentation, but instead to complement it.

Because programming with MRDS is so fundamentally different from what you might have been used to in the past, getting started can be difficult. This is where a good textbook with well-constructed examples can be useful. When this book was first proposed, no such textbook was available.

In the spirit of sharing, all of the source code for this book is available online. The examples address many of the key issues involved in building robotics applications, and we hope that they will act as launching pads for many more new and exciting services. In the years to come, we hope to look back at this book and laugh at how primitive the examples were.

Kyle has extensive experience in simulation and games development, and he is the MRDS team member responsible for the MRDS simulator. Trevor has worked with robots and computer vision for many years and has been active in the MRDS community since the very first Community Technology Preview in 2006. We have a passion for robotics and we have endeavored to pass on our knowledge through this book.

Many readers may be wondering why the examples are only in C#. The answer is simple: Weigh this book. Adding VB would double that, adding C++ would triple it, and so on. It was simply not feasible to offer a multi-language version of the book in its first edition.

We welcome feedback on the book and suggestions for further examples. The book's website (www.wrox .com) provides a feedback mechanism. You can also get updates at www.proMRDS.com. VS 2008 versions of the code will be posted to this site, as well as updates when MRDS V2.0 is officially released. In addition, the MRDS Discussion Forum hosts an active community that is constantly exchanging ideas and code, so you are encouraged to participate in the Forum.

If we compare the current state of development in robotics to the release of the IBM PC in the 1980s, then we have an exciting and challenging time ahead. Welcome to the field of robotics!

# Who Is This Book For?

This book is for programmers who want to learn about the rapidly growing field of robotics. The primary focus of the book is on developing software to control robots, and many of the examples are designed to work with the MRDS simulator. However, you can't call yourself a roboticist unless you have worked with some real robots, so also included are examples using a wide variety of different robot platforms.

We assume that you have prior programming experience with C#, the preferred language for MRDS. Although you can use other .NET languages with MRDS, all of the examples in this book are in C#. If you do not know C#, don't let this deter you, especially if you have experience in Java or C++.

Although the book is intended to be used by both professional programmers and undergraduate students, it is not aimed at beginning programmers.

# What Is This Book's Focus?

MRDS is a framework for developing software to control robots. Therefore, the book emphasizes the programming techniques and patterns that you use to write MRDS services.

A major advantage of the approach taken in the book is that you can get started without buying any hardware because MRDS includes an excellent simulator. The simulation examples, and the examples for VPL (Visual Programming Language), can be run without real robots.

The book is not intended to replace the large amount of documentation that is already available for MRDS. Instead, it attempts to lead you through various aspects of MRDS by way of examples.

Both of the authors have extensive experience in using MRDS. All of the sample services are complete and based on the authors' experience.

# What Does This Book Cover?

The book is divided into four parts. You should work through Part I first because it contains all the background information and concepts that are crucial to understand MRDS. After that, you can jump to any of the remaining parts because they are designed to be relatively independent of each other.

## Part I: Robotics Developer Studio Fundamentals

The book begins with a brief overview of MRDS in Chapter 1. It includes instructions for installing MRDS and setting up the sample code from the book that is available for download from www.wrox .com. Some quick examples enable you to verify that the code is working, and you get a first taste of MRDS. You will find references to a variety of resources at the end of the chapter that will be helpful as you learn MRDS.

Chapter 2 launches into MRDS concepts with a range of examples demonstrating how to program in the multi-threaded environment provided by the Concurrency and Coordination Runtime (CCR). The CCR presents a new programming paradigm, but it is key to the power of MRDS. Concurrency is essential to robotics, where often many tasks must be performed simultaneously. Consider a robot driving around on its own — sensors must be read to detect obstacles; steering decisions have to be made in real time; and motors need to be controlled. Chapter 2 sticks to basic CCR constructs and you do not need a robot to follow along. The chapter concludes with a list of fundamental code patterns that can be used in MRDS programming.

MRDS programs are written as services, which run under the control of Decentralized Software Services (DSS), so Chapter 3 explains the important features of services and how to write them. This chapter may present a lot of information that is new to you. Several topics are addressed, such as starting and stopping services; configuring services; and how to package your services for deployment. The examples demonstrate how two services can communicate with each other, and again no robots are required.

In Chapter 4, you work through examples of common tasks, such as subscribing to other services to receive notifications and adding a user interface using either Windows Forms or Web Forms. This chapter also discusses abstract services, known in MRDS as *generic contracts*. The sample services, Dashboard and TeleOperation, enable you to control a variety of robots, either real or simulated. These examples include a wide range of useful code patterns, including how to use a web camera as an input device. In this chapter, you finally use a real robot!

## Part II: Simulations

In Chapter 5, the MRDS simulator is introduced with an example that shows off the physics features of the simulator. The various features of the simulator are demonstrated, including the Simulation Editor. The robot models and other environmental entities provided with the simulator are also presented. A good simulator is invaluable for prototyping services, and it is well worth spending time becoming familiar with the simulator. The built-in robot simulations enable you to test ideas quickly.

Chapter 6 moves on to developing new simulation entities. A four-wheeled robot, based on the Corobot from CoroWare, is created, along with a simulated infrared range sensor. In addition to the simulation, you learn how to build a new service that complies with an existing generic contract — the Quad Differential Drive is compatible with the generic (two-wheeled) Differential Drive. Even if you do not intend to build your own simulated robot, the new simulation entities in this chapter might be useful in your own projects.

Based on the Corobot in Chapter 6, the code in Chapter 7 implements a simulated Robo-Magellan competition similar to the one run by the Seattle Robotics Society. The robot must navigate around a course using a simulated GPS to locate a set of markers. It coordinates with a referee service that times how long it takes the robot to locate the targets. The SimMagellan service combines computer vision with a wandering behavior and obstacle avoidance to successfully navigate the course and identify the markers. This is a great example to use as a jumping-off point for more complex and, it is hoped, smarter behaviors.

Up to this point, the book concentrates on wheeled robots. In Chapter 8, a simulated robotic arm is built that is based on the Lynxmotion L6 arm, which has six degrees of freedom. Later, in Chapter 15, a real robotic arm is used and the two match closely enough that the same services can be used with either. In building the simulated arm, basic concepts are covered, including joints, coordinate frames, and inverse kinematics. Some pre-defined procedures, such as stacking dominos, put the arm through its paces and are fascinating to watch.

Chapter 9 brings together a variety of different simulations, including robot soccer; a hexapod (six-legged walking robot); a Maze Simulator; and ExplorerSim. The ExplorerSim service uses a simulated Pioneer 3DX to explore a maze using a laser range finder and to build a map as it goes. The services in this chapter are not only good examples of how to write simulations, they are also useful in designing and testing your own applications. For example, the Maze Simulator can be used with your own services so that you can test algorithms for exploration, maze solving, and other applications of artificial intelligence.

## Part III: Visual Programming Language (VPL)

The basics of the Visual Programming Language (VPL) are covered in Chapter 10. It begins by outlining the concepts behind a data flow language and builds the obligatory "Hello World" example, but with a twist — this one talks to you. Then it moves on to discuss variables, loops, custom activities (the rough equivalent of subroutines), and some simple debugging. The chapter is rounded out with information on using lists and the Switch activity.

In Chapter 11, more advanced topics are discussed such as processing sensor data and controlling robots. If you have a real robot, such as the LEGO NXT, you can use it for some of the examples. However, the examples were written using the MRDS simulator so that you do not need a robot. Instructions are also provided for creating C# code from a VPL diagram and compiling it into a service. This facility is handy if you want to prototype in VPL and then hand-optimize the code later or add features that are easier to implement directly in C#. The last part of the chapter touches on controlling multiple robots with a couple of different examples.

Chapter 12 provides several VPL examples. The VPLExplorer service performs a similar function to the ExplorerSim service introduced in Chapter 9. The robotic arm makes a reappearance with a service that can be used to control the arm using an Xbox controller. With only minor changes, the VPL diagram can be made to work with either a simulated arm or a real robotic arm. Line-following sensors are constructed in the simulator to enable an iRobot Create robot to follow a line on the ground. Lastly, simple computer vision is used to enable a simulated Corobot to follow a ball.

## Part IV: Robotics Hardware

The variety of different robots that are supported under MRDS is outlined in Chapter 13. Some advice is given on selecting a robot if you plan to buy one. This type of information is a moving target — it changes on an almost daily basis. You should check the Microsoft MRDS website and the book's website (www.proMRDS.com) for updates. The chapter then explains some basic terminology and provides background information on robotics. This coverage is intended to ensure that everyone is up to speed with all the buzzwords before continuing with the rest of the chapters. No new services are introduced.

Chapter 14 builds services for the LEGO NXT and the Parallax Boe-Bot. Updated services for the Boe-Bot are included in this chapter that provide additional functionality to the services supplied by Parallax. The discussion in this chapter focuses on remotely controlled robots, so Bluetooth figures prominently. A simple "dance" program exercises the robots, and just by changing the manifest, the same program is run using simulated robots. An extended online version of this service is available that handles multiple robots simultaneously. Wandering behavior using sensors is developed in the last part of the chapter and implemented on both of the real robots using their specific capabilities, rather than generic contracts.

The Lynx 6 arm makes a comeback in Chapter 15, and forward and inverse kinematics, which control the arm motions, are discussed. Revised versions of the Lynxmotion services are provided that incorporate new kinematics algorithms. A service to record arm motions is developed as an example of the two fundamentally different approaches to controlling an arm — moving individual joints or setting the position of the end effector (the tip of the arm) using 3D coordinates. In this case, the reverse process now applies and the simulated arm can be used with the service developed for the real arm.

In Chapter 16, the wonderful world of autonomous robots is introduced. Up to this point in the book, the robots are all controlled remotely. Now the "brains" are moved onto the robots. Two different approaches are explored: an embedded PC and a PDA (personal digital assistant). These devices run a slimmed-down version of Windows known as Windows CE or Windows Mobile. In this environment you have to use the .NET Compact Framework (CF), which necessitates some changes to how you write services. A Stinger robot is set up to wander autonomously using first a PDA mounted on top of the robot and then an eBox 2300 embedded PC.

Finally, in Chapter 17, entirely new services are built from scratch. A new generic contract is defined for a generic brick (robot brain) that encapsulates the common elements of many different low-end robots as far as sensors and actuators are concerned. A test service is built to enable easy testing of services based on the generic brick. Using this new contract, services are implemented for the Picblok Integrator and the K-Team Hemisson robots. The onboard monitor program (firmware) for the Integrator's PICAXE chip is developed in this chapter as well as the MRDS services. Calibration of sensors and drive motors is covered using the Hemisson robot as an example.

# What You Need to Run the Examples

As pointed out earlier, you do not need a real robot to use much of the code in this book because you can use the simulator in MRDS. The book has been deliberately designed so that people on a tight budget, such as students, can learn the fundamentals of MRDS without real robot hardware.

If you do want to work with a real robot, you should read Chapter 13, which discusses the available robots. The robots selected for use in this book cost only a couple of hundred dollars and are well within the reach of the average professional programmer. They are small enough to use in an office environment and are readily available — most can be ordered online. If you have a bit more money to spend, the principles in the book apply equally to robots costing many thousands of dollars.

To run the examples, you should remember the following:

❑   The Microsoft Robotics Developer Studio development environment runs on Windows XP (Service Pack 2) or Vista. It requires the .NET Framework V2.0 Service Pack 1. The simulator requires a machine with a graphics card that supports DirectX 9 with support for 2.0 Pixel and Vertex Shader programs.

❑   All the code in this book was written using the Version 1.5 Refresh of MRDS, released in late December 2007. The MRDS software can be downloaded from the Microsoft website and is free of charge for noncommercial use. The MRDS kit automatically installs .NET 3.0, DirectX 9.0, XNA 1.0, and the AGEIA PhysX engine 2.7, which the MRDS simulator uses.

❑ You need Visual Studio 2005 to compile code for this release of MRDS — either VS2005 Professional or the Express Edition of C#, available as a free download from the Microsoft website. It is advisable to apply Visual Studio Service Pack 1. Visual Studio 2005 automatically installs .NET V2.0, but you must apply the .NET Service Pack 1.

*The code has not been tested using Visual Studio 2008, which was released while the book was being written. Visit the book's website (www.proMRDS.com) for the latest information on using Visual Studio 2008 with MRDS and the book examples.*

❑ Up to four gigabytes of disk space is required to install the full Visual Studio 2005 Professional with the MSDN Library. The Express Edition of C# does not require as much space, but it cannot be used to develop mobile (CF) applications such as those in Chapter 16.

❑ It is recommended that you have at least 512MB of memory on your PC. If you intend to use the Visual Programming Language or the simulator, you should have a minimum of 1GB of memory or your system will page heavily and run very slowly.

❑ Services can be built that target Windows Mobile or Windows CE devices such as personal digital assistants or embedded PCs. For the examples in Chapter 16 you need a mobile device.

❑ Rather than provide a CD with the book that would quickly become outdated, the source code for all of the examples in the book is available for download from the Wrox website (www.wrox.com), as described in the Source Code section that follows. As new versions of MRDS are released by Microsoft, you will also be able to download updated versions of the sample code from www.proMRDS.com.

*After the 1.5 Refresh version of Robotics Developer Studio was released, Microsoft changed the name of the development kit to Microsoft Robotics Developer Studio. We have used this new name and the abbreviation MRDS in most places in the book.*

# Conventions

To help you get the most from the text and keep track of what's happening, we've used a number of conventions throughout the book.

Code has several styles. When we're talking about a word in the text — for example, when discussing the `helloForm` — it appears in `this font`.

```
This text is used to show regular code blocks.
In code examples we highlight changes to previous examples or important code with a
gray background.
```

Sometimes you will see a mixture of styles. For example, where a section of code shows commands that you should type, the code you type is presented in **bold**:

```
C:\Microsoft Robotics Studio (1.5)>dssnewservice
```

> **Boxes like this one hold important, not-to-be forgotten information that is directly relevant to the surrounding text.**

*Tips, hints, tricks, and asides to the current discussion are offset and placed in italics like this.*

As for styles in the text:

❑    We *highlight* new terms and important words when we introduce them.

❑    Keyboard strokes are shown like this: Ctrl+A.

❑    We show filenames, URLs, and code within the text like so: `index.html`

❑    We indicate a break in a line of code with the ↵ symbol.

# Source Code

As you work through the examples in this book, you may choose either to type in all the code manually or to use the source code files that accompany the book. All of the source code used in this book is available for download at `www.wrox.com`. Once at the site, simply locate the book's title (either by using the Search box or by using one of the title lists) and click the Download Code link on the book's detail page to obtain all the source code for the book.

*Because many books have similar titles, you may find it easiest to search by ISBN; this book's ISBN is 978-0-470-14107-6.*

The book code is supplied in DssDeploy format, which is a self-expanding executable. Alternately, you can go to the main Wrox code download page at `www.wrox.com/dynamic/books/download.aspx` to see the code available for this book and all other Wrox books.

# Errata

We make every effort to ensure that there are no errors in the text or in the code. However, no one is perfect and mistakes do occur. If you find an error in one of our books, such as a spelling mistake or a faulty piece of code, we would be very grateful for your feedback. By sending in errata, you may save another reader hours of frustration and at the same time you will be helping us provide even higher quality information.

To find the errata page for this book, go to `www.wrox.com` and locate the title using the Search box or one of the title lists. Then, on the book details page, click the Book Errata link. On this page you can view all errata that have been submitted for this book and posted by Wrox editors. A complete book list, including links to each book's errata, is also available at `www.wrox.com/misc-pages/booklist.shtml`.

If you don't spot "your" error on the Book Errata page, go to `www.wrox.com/contact/techsupport.shtml` and complete the form there to send us the error you have found. We'll check the information and, if appropriate, post a message to the book's errata page and fix the problem in subsequent editions of the book.

# p2p.wrox.com

For author and peer discussion, join the P2P forums at p2p.wrox.com. The forums are a web-based system for you to post messages relating to Wrox books and related technologies and interact with other readers and technology users. The forums offer a subscription feature to e-mail you topics of interest of your choosing when new posts are made to the forums. Wrox authors, editors, other industry experts, and your fellow readers are present on these forums.

At p2p.wrox.com you will find a number of different forums that will help you not only as you read this book, but also as you develop your own applications. To join the forums, just follow these steps:

1. Go to p2p.wrox.com and click the Register link.

2. Read the terms of use and click Agree.

3. Complete the required information to join as well as any optional information you wish to provide and click Submit.

4. You will receive an e-mail with information describing how to verify your account and complete the joining process.

> *You can read messages in the forums without joining P2P but in order to post your own messages you must join.*

Once you join, you can post new messages and respond to messages other users post. You can read messages at any time on the Web. If you would like to have new messages from a particular forum e-mailed to you, click the Subscribe to this Forum icon by the forum name in the forum listing.

For more information about how to use the Wrox P2P, be sure to read the P2P FAQs for answers to questions about how the forum software works as well as many common questions specific to P2P and Wrox books. To read the FAQs, click the FAQ link on any P2P page.

# Part I

# Robotics Developer Studio Fundamentals

# 1

# Exploring Microsoft Robotics Developer Studio

Welcome to the world of robotics software! If you are just entering the world of robotics, a warning is in order. Whether you are a hobbyist, a student, or a professional, working with robots can be addictive, and if you are not careful, it can crowd out other worthy pursuits such as your career, social life, and personal hygiene. If you are willing to take that risk, keep reading.

This book is not going to tell you how to build your own robot to fetch beer from the refrigerator or pick up your roommate's socks off the floor. Instead, you will learn how to use an innovative new software development kit (SDK) called Microsoft Robotics Developer Studio (MRDS) to build software for robots already on the market, as well as custom robots and robots yet to be built.

The best feature of MRDS is that it's free for personal, academic, and development use. If you end up shipping an application that uses MRDS, there is a modest licensing fee.

If you have a LEGO NXT or an iRobot Create, or one of several other robots, you can start controlling your robot with MRDS right away. If you have a robot that doesn't yet have MRDS support, this book will show you how to write your own custom robotics services.

You don't have a robot yet? Don't worry about it. MRDS includes a sophisticated 3D simulation environment for virtual robots. You can be driving robots through a virtual environment just minutes after you download the free SDK, and the majority of the examples in the book are designed so that they will work with the simulator.

Don't be deceived by the "price," though. MRDS includes a sophisticated run-time environment that makes it easy to write powerful asynchronous and distributed applications. Several large corporations are using this environment in non-robotics applications to handle robust serving of web pages and financial transactions. In its short life, MRDS has already eaten up the road in the DARPA Urban Challenge and swum underwater in a robotic submarine.

If we're ever going to get to the point where robots overthrow our way of life and threaten the very existence of humanity, we all need to get busy writing robotics software. It's time to download Microsoft Robotics Developer Studio and become a part of the robotics revolution.

# Microsoft Does Robots?

Most people are surprised to hear that Microsoft has a robotics team and that they are actively working on software for robots. This section describes how this team came about and what Microsoft's goal is in this area of software development.

It was in late 2003 that robotics started showing up on the Microsoft radar. Tandy Trower, who was working directly for Bill Gates at the time, began to notice that a number of robotics companies and universities with robotics programs were contacting Microsoft. He wrote a "think week" paper for Bill exploring the possibility of developing a product in this space. At the same time, Bill visited several universities and saw that each had at least one robotics project or program that it was eager to show. It was after noticing a lot of activity and a nearly worldwide enthusiasm for this technology that he directed Tandy to conduct a five-month study to determine what Microsoft should do. Tandy determined that there was a business opportunity for Microsoft and recommended that he be the one to form a team to pursue it. The Microsoft robotics team was born (see the Foreword by Tandy Trower at the front of the book for more details).

Tandy drew developers from many parts of Microsoft with various areas of expertise. They worked together to integrate and develop a number of new technologies to support robotics development. Version 1.0 of the Microsoft Robotics Studio SDK was released in December of 2006, and version 1.5 followed in July of 2007, with a refresh in December 2007 that was used as the basis for this book.

According to Tandy, Microsoft sees parallels between the state of the robotics industry now and the state of the personal computer industry 25 years ago. At that time, software written for one computer would not necessarily work with another computer; as a result, there was a very limited market for software development. With the introduction of the IBM PC, computer hardware became more capable and standardized. DOS, and later Windows, served as a software development environment that enabled programs written on one machine to work on all machines.

Microsoft Robotics Developer Studio is an attempt to do something similar with the robotics industry. It provides a software platform and development environment that enable software written for one robot to also work with another robot with similar capabilities. Microsoft hopes that this development environment will help the robotics industry move forward by helping companies to invest more in robot behavior and algorithms than has been possible in the past.

This is an appropriate point to talk about versions and names. In April 2008, right in the middle of the final stages of preparing this book, Microsoft released a CTP (Community Technology Preview) of V2.0 and renamed Microsoft Robotics Studio as Microsoft Robotics Developer Studio. This is the type of event that causes authors to wake up in a cold sweat at night. After much discussion, it was decided to change references in the book from MSRS to MRDS, except where they clearly had to remain the same — i.e., the main installation directory for V1.5 is `C:\Microsoft Robotics Studio (1.5)`. It is hoped that this is not too confusing.

The code on the book's website will be updated as soon as the final version of 2.0 ships. There is no point in chasing CTPs because they are moving targets. In the meantime, MSRS — sorry, MRDS — has been designed to allow multiple versions to run side-by-side, so you can continue to work with V1.5 even while you are using a trial CTP of 2.0.

# Microsoft Robotics Developer Studio Components

The Microsoft Robotics Developer Studio SDK consists of a number of components. The Concurrency and Coordination Runtime (CCR) and Decentralized Software Services (DSS) comprise the run-time environment. They are both managed libraries, so the robotics services that operate within their environments are also implemented using managed code. The Visual Simulation Environment is a 3D simulator with full physics simulation that can be used to prototype new algorithms or robots. The Visual Programming Language (VPL) is a graphical programming environment that can be used to implement robotics services. In addition to all of these components, the MRDS team has implemented numerous samples and complete applications to provide programming examples and building blocks for user applications. The following sections provide a brief overview of each of these components, and they are covered in more detail in later chapters.

## Concurrency and Coordination Runtime (CCR)

The CCR is a managed library that provides classes and methods to help with concurrency, coordination, and failure handling. The CCR makes it possible to write segments of code that operate independently. It communicates when necessary by passing *messages*. When a message is received, it is placed in a queue, called a *port*, until it can be processed by the *receiver*.

Each code segment can run concurrently and asynchronously, and there is often no need to synchronize them because of the message queues. When it is necessary to wait until two or more operations have completed, the CCR library provides the necessary constructs.

The CCR provides one or more *dispatchers,* which determine what segment of code is currently running. The total number of threads is usually set according to the number of independent processors on the system, and code segments are scheduled when a thread becomes available.

One element of asynchronous programming that is often overlooked is error handling. It is possible to use exception handling to isolate failures within a single code segment, but the data that caused the error was likely passed to the segment via a message from another executing segment and the exception handler code has no way of knowing where the operation originated or where the error should be reported.

To solve this problem, the CCR provides causalities. A *causality* holds a reference to a port that is used to report errors. The causality is associated with a message and it follows that message as it is passed to another code segment and then again to another segment. At any time, if an error occurs, a message is passed to the error port associated with the causality. In this way, errors are reported back to the segment of code that initiated the operation.

*A complete description of the CCR as well as example code can be found in Chapter 2. You need a good understanding of the CCR, so be sure to work through Chapter 2 carefully.*

# Decentralized Software Services (DSS)

The CCR enables segments of code to pass messages and run in parallel within a single process. The Decentralized Software Services (DSS) library extends this concept across processes and even across machines.

An application built with DSS consists of multiple independent *services* running in parallel. Each service has a *state* associated with it and certain types of messages that it receives called *operations*. When a service receives a message, it may change its state and then send additional messages and *notifications* to other services.

The state of a service can be retrieved programmatically by sending a Get message to the service or it can be retrieved and displayed using a web browser. Services may *subscribe* to be notified when the state of a service changes or when other events occur. Services may also *partner* with other services so that they can send messages to those services and receive responses.

It requires a different mindset to write software using CCR and DSS. You must think in terms of independent chunks of code that handle messages asynchronously, rather than function calls and threads. Chapters 3 and 4 provide additional details about the DSS library, including code examples, and are important chapters for your understanding of MRDS.

# Visual Simulation Environment

MicrosoftRobotics Developer Studio includes a full-featured 3D simulation environment complete with physics simulation. A scene from Simulation Tutorial 5 is shown in Figure 1-1.

Figure 1-1

The simulation environment supports both indoor and outdoor scenes and comes with a variety of simulated robots. It is also extensible so that you can add custom robots, such as the CoroBot in Chapter 6 and the Lynx 6 robotic arm in Chapter 8, and other objects such as walls, furniture, and so on. It has a built-in editor so that you can add and move objects as the simulator is running.

The simulator is useful for prototyping new robotic algorithms prior to running them on the actual robot hardware. It is much cheaper to destroy a virtual robot with a programming error than a real one. In some cases, the robot you want to program may not even exist yet. In this case, the simulator can provide a good testing ground for a model of the robot.

The simulation environment contains models of several robots, including a LEGO NXT, an iRobot Create, a MobileRobots Pioneer 3DX, and a KUKA LBR3 robotic arm. The models for these robots are shown in Figure 1-2.

**Figure 1-2**

A number of other simulation entities such as cameras, sky, ground, and various other objects are also provided.

*The simulation environment is covered in Chapters 5 through 9.*

# Visual Programming Language (VPL)

In an MRDS application, there are usually several services running. Some of these are low-level services that interface directly with hardware. Others might be simulation services that connect directly to the simulation engine. Usually one or more top-level services control the behavior of one or more robots. These top-level services typically interface only with other services and are called *orchestration services*. Sometimes it is possible to implement an orchestration service without writing a single line of code, but instead by defining how data is transferred between services using a graphical diagram in the Visual Programming Language environment.

A typical screenshot of VPL is shown in Figure 1-3. In VPL this is called a *diagram*.

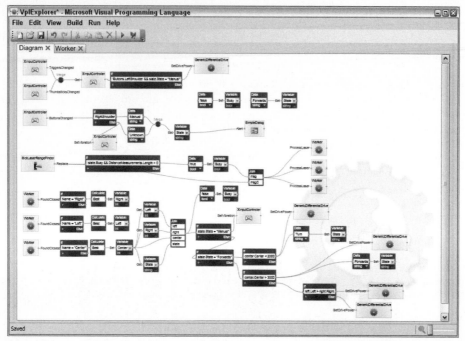

**Figure 1-3**

Each block represents a service, a calculation, a condition, or a nested diagram, and these are called *activities*. The lines between the blocks represent messages flowing from one service to another. The data flow model works well to represent the messages that are passed in a typical DSS application. VPL diagrams can be arbitrarily complex because each block can represent another nested diagram, which may in turn have other nested diagrams.

*VPL is covered in detail in Chapters 10 through 12.*

## Robotics Samples

In addition to the system components previously mentioned, MRDS also includes a number of services tailored to work with specific robots, including the LEGO NXT, the iRobot Create and Roomba, the Kondo KHR-1, the MobileRobots Pioneer 3DX, and the Parallax Boe-Bot. Various FischerTechnik robot models are also supported.

The Parallax Boe-Bot, the iRobot Create, and the Pioneer 3DX are shown in Figure 1-4 (not to scale).

**Figure 1-4**

Services for various other sensors such as a laser range finder (the large object on top of the Pioneer 3DX in Figure 1-4 on the right), infrared sensors (sticking out at the front of the Boe-Bot on the left), a camera, and a game controller are also included.

# Packages

In addition to the major 1.0 and 1.5 releases, the MRDS team has also released packages of code targeted to specific scenarios. Each of the three packages is described in the following sections. Instructions for installing the packages are included in the next section, which describes how to install MRDS. All of these packages are available from the MRDS Downloads page at `http://msdn2.microsoft.com/en-us/robotics/aa731520.aspx`.

## The Sumo Package

The sumo competition is a classic robotics contest. The MRDS team created a sumo competition for the 2007 MEDC conference in Las Vegas, Nevada, using iRobot Create robots.

A sumo competition takes place between two robots inside a circular ring. The objective is for each robot to push its opponent out of the ring without going outside itself. Each robot has sensors on its underside to detect a colored region around the outer edge of the ring. Each also has front-left and front-right bumpers and a camera.

As developers arrived at the conference, they were encouraged to implement a service to control a sumo player. The service was qualified against another basic sumo player service in the simulator and developers that passed the qualification went on to port their services to the actual hardware to compete in a real sumo ring.

It sounds like a simple contest to run but no one anticipated how much work it would be to assemble nearly 50 robots on the day before the conference. Figure 1-5 shows two of the completed sumobots at the end of that very long day.

Figure 1-5

The custom-built packages on top of the robots contain embedded PCs running Windows CE with MRDS and attached web cameras so that the robots can find each other. The built-in sensors on the Creates are used to detect the edge of the sumo ring.

The sumo package contains a simulated referee service and all of the simulation services needed to run the simulated sumo competition. The sample sumo player service is a good example of an orchestration service that interacts with other MRDS services to read sensor data and control the behavior of the robot. The referee is a different type of orchestration service that starts the matches, runs a timer, and determines when one of the sumobots has won.

The sumo package also includes code that enables the sumo player services to run on the actual hardware. Instructions for building the sumo hardware can be found at `http://msdn2.microsoft.com/en-us/robotics/bb403184.aspx`.

*Chapter 9 discusses how to simulate a sumo competition.*

## The Soccer Package

The soccer package was developed for RoboCup 2007 in Atlanta. The stated goal of RoboCup is "By the year 2050, develop a team of fully autonomous humanoid robots that can win against the human world soccer champion team." You can find more information about RoboCup at `www.robocup.org`.

The soccer package is a simulation-only competition. It includes a soccer referee service as well as a soccer field simulation environment and a sample soccer player service. Each soccer team consists of a field player and a goalkeeper. The LEGO NXT is used for the robots in the MRDS package but the environment is designed to allow other robots. In fact, the robot models used in the RoboCup competition were RobuDog models provided by RoboSoft. These articulated robot dogs were challenging to program but the resulting matches were very entertaining. Figure 1-6 shows the soccer simulation running with the RobuDog models.

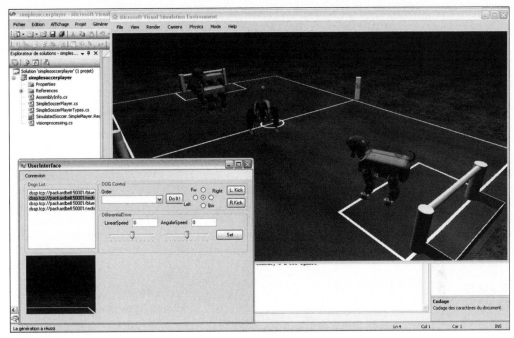

**Figure 1-6**

The RobuDog model and software are available from RoboSoft at www.robubox.com/robosoft/data/ Robocup/RobocupHome.jsp.

*Chapter 9 demonstrates how to use the soccer simulation with the simulated Corobot robot developed in Chapter 6.*

## The Courseware Package

The courseware package contains a set of labs that can be used as part of an introductory course on robotics. They illustrate common robotics problems such as reading sensors and controlling motors. The package includes the following labs:

❑ Lab Tutorial 1 (VPL): Joystick a Robot in Simulation and in Hardware

❑ Lab Tutorial 2 (VPL): Advanced Motion

❑ Lab Tutorial 3 (VPL): Sensing and Simple Behaviors

❑ Lab Tutorial 4 Part 1 (VPL/C#): Play Mastermind with the Robot via Bi-directional Speech

❑ Lab Tutorial 4 Part 2 (VPL/C#): Play Mastermind with the Robot via Vision and Text-To-Speech

❑ Lab Tutorial 5: Using Vision to Estimate the Distance to an Object

❑ Lab Tutorial 6 (VPL): Task Learning via Human-Robot Interaction

❑ Lab Tutorial 7 (C#): Multirobot Coordination

In addition to providing several good examples of MRDS services, the courseware package includes a number of services that may be useful in your own robotics projects.

# Support for MRDS

Many companies and organizations have done some very interesting things with MRDS. The following sections describe a few examples of projects and products that support MRDS.

## *The SubjuGator*

In July of 2007, students from the University of Florida Machine Intelligence Laboratory entered an autonomous robotic submarine called *SubjuGator* in the AUVSI/ONR 10th International Autonomous Underwater Vehicle Competition, where they won first place. This was the sixth version of the SubjuGator built by the university and this one had more powerful sensors and a faster computer system than previous versions. In addition, the team used Microsoft Robotics Studio for control and simulation of the submarine.

The SubjuGator was designed to operate underwater at depths of up to 100 feet. A single-board Intel Core 2 Duo-based computer running Windows XP provided the computational horsepower for monitoring and controlling all systems. The mission behavior of the submarine was controlled with Microsoft Robotics Studio communicating with a network of intelligent sensors such as cameras, hydrophones, a Doppler Velocity Log, a digital compass, an altimeter, and internal environment monitor sensors.

The SubjuGator is shown in Figure 1-7. More information about this project can be found at http:// subjugator.org.

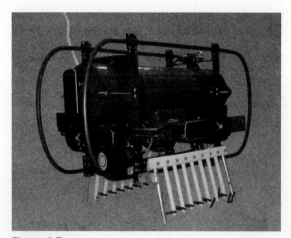

Figure 1-7

## Prospect 12

Another example of a university project using Microsoft Robotics Studio in a competition is the Princeton Autonomous Vehicle Engineering (PAVE) entry in the 2007 DARPA Urban Challenge. PAVE participated in the Urban Challenge with Prospect 12, a modified Ford Escape Hybrid fitted with several stereo and monocular cameras. PAVE was one of the 35 teams invited to participate in the National Qualification Event, but their vehicle did not compete in the final Urban Challenge race.

According to the PAVE team, their use of Microsoft Robotics Studio had over 25 services running across five dual-core servers. Prospect 12 is shown in Figure 1-8. More information about this project can be found at `http://pave.princeton.edu/main/urban-challenge`. Specific information about the Prospect 12 can be found at `http://pave.princeton.edu/main/urban-challenge/msrs`.

Figure 1-8

## The uBot-5

Speaking of university projects, one of the most interesting robotic platforms to run Microsoft Robotics Developer Studio is the uBot-5 developed at the Laboratory for Perceptual Robotics at the University of Massachusetts at Amherst. This robot is a small, lightweight research platform for mobile manipulation. It balances on two wheels using an inverse pendulum model like the Segway Scooter. The balancing is handled by a custom 12-channel FPGA-based servo controller that can update the position and velocity of all the motors at a rate of more than 2 kilohertz. An embedded PowerPC core manages low-level reflexes such as balancing and transitions between postural modes.

The uBot-5 team implemented MRDS services to control the movement of the uBot-5 and they have a simulation model that correctly balances the robot in the simulation environment. The uBot-5 is shown in Figure 1-9. More information is available at www-robotics.cs.umass.edu/Robots/UBot-5.

Figure 1-9

## *KUKA Educational Framework*

KUKA is a company well known for its high-end industrial robots, but they have also been very active in their support for Microsoft Robotics Developer Studio. The first robotic arm implemented in the MRDS Simulator, the KUKA LBR3, was developed with assistance from KUKA.

They have released their own package for MRDS called the KUKA Educational Framework. The framework includes four arm tutorials that teach about point-to-point motions and the orchestration of linked motions using the simulated LBR3. The three mobile tutorials demonstrate how to control a robotic arm attached to a mobile platform. It culminates in a task tutorial that shows how to use the mobile arm equipped with a laser range finder to pick up a box on a table and drop it into a garbage can.

These tutorials can be downloaded at no charge from www.kuka.com/usa/en/products/software/educational_framework. A scene from one of the mobile tutorials is shown in Figure 1-10.

**Figure 1-10**

Many other companies provide MRDS support for their robots and even general-purpose services. More MRDS partners are listed on the Microsoft Robotics Community page at `http://msdn2.microsoft` `.com/en-us/robotics/aa731519.aspx`.

# Setting Up Your System

This section contains instructions for preparing your system for MRDS and installing the SDK, including all the optional packages. Once you have installed all the software, follow the instructions in the last section to verify that it is working properly.

## System Requirements

According to the MRDS Data Sheet, application development for MRDS is supported on the following operating systems:

❑   Windows Vista

❑   Windows XP

❑   Windows Server 2003 R2 (32-bit x86)

❑   Windows XP 64-bit Windows Server 2003 R2 x64 editions

Note that the simulation environment is not supported on 64-bit Windows but all of the other components will work properly.

The run-time and other MRDS services can be deployed and executed on the following operating systems:

❑   Windows Vista

❑   Windows XP

- ❑ Windows XP Embedded

- ❑ Windows Embedded CE 6.0

- ❑ Windows Mobile 6.0

- ❑ Windows Server 2003 R2 (32-bit x86)

- ❑ Windows Server 2003 R2 x64 editions

- ❑ Windows XP 64-bit

*Neither Linux nor any other non-Windows operating system is supported at this time.*

The simulation environment has more rigid hardware requirements because it uses 3D graphics extensively. Several developers have had frustrating experiences because they attempted to run the simulator on systems with outdated or underpowered graphics cards.

The "Hardware Requirements" section in Chapter 5 goes into more detail about the graphics card requirements, but the bottom line is that your graphics card must support DirectX 9 graphics using Vertex and Pixel Shader Model 2.0 or higher. Most systems sold since 2006 with reasonable 3D graphics support meet these requirements. There are some lower-end systems with integrated graphics chips, such as the Intel 865G chipset family, that do not meet these requirements. This is one of the most frequently asked questions in the MRDS simulation forum, so it is a good idea to check your graphics card capabilities on the manufacturer's website if you are having trouble running simulation.

## Prerequisites

The most important software prerequisite is an appropriate version of Microsoft Visual Studio or Microsoft Visual C# Express. If you use Visual Studio, you should have at least Visual Studio 2005 with Service Pack 1 installed.

If you don't already have a copy of Visual Studio, you can download Visual C# Express at no charge from `www.microsoft.com/express/vcsharp/Default.aspx`. This development environment contains all of the components necessary to build MRDS services for all operating systems except Windows CE and Windows Mobile.

Another useful utility to install is the .NET Reflector, which can be downloaded from `www.aisto.com/roeder/dotnet`. Reflector enables you to easily view, navigate, search, decompile, and analyze .NET assemblies.

## Installing MRDS 1.5

Microsoft Robotics Developer Studio version 1.5 (December 2007 Refresh) is available as a web download from `www.microsoft.com/robotics` at the time of writing. It is highly likely that a new version will be available as this book goes to press, but you should stick with V1.5 Refresh, which is an official release, not a beta.

As noted in the introduction, this book was written using version 1.5. Most of the services will probably work with MRDS V2.0 when it is released, provided that you first convert them using the `DssProjectMigration` tool. However, the authors cannot make any guarantees about this, so you must install MRDS V1.5 to ensure that the examples work. Updates on migration issues will be posted to the book's website.

This is a critical point because services written to work with version 1.5 of the SDK will not work with later versions. This is not because the MRDS team has broken backward compatibility but because MRDS services use strong-name signing, and when a service has a reference to another service it will not run if the second service has a different version than when the first service was built. This can usually be fixed by rebuilding any services that have references to run-time services that have changed in the new release, but obviously this isn't possible if you don't have the source code.

This means that the sample code for this book will not automatically work with later releases of MRDS. The authors intend to provide updated sample code for the book as Microsoft releases new versions of MRDS. You can find updated sample code at www.wrox.com or at the authors' website at www.proMRDS .com. Keep watching the website for revised versions of the examples soon after MRDS V2.0 ships.

---

### Strong-Name Signing Difficulties

The problem of signed assemblies has also caused some difficulties for developers as they are exploring MRDS for the first time. Consider Simulation Tutorial 2, for example. It uses the SimpleDashboard service to drive the robots around in the simulation environment. The SimpleDashboard service has a reference to the SickLRF service. If you happen to rebuild the SickLRF service, which is also provided as a sample, the SimpleDashboard service won't run because the SickLRF service has now been signed with a different key than when it was originally built by Microsoft and the SimpleDashboard service won't load it due to security concerns. This can be quickly fixed by rebuilding the SimpleDashboard service, but then any other services that use the SimpleDashboard service must also be rebuilt. Fortunately, there are no other Microsoft services that reference the SimpleDashboard service because it is a top-level orchestration service.

The robotics team has tried to minimize this problem by building solutions for each of the tutorials, including all of the services that the tutorial depends on. Simply rebuilding the entire solution generally fixes the problem. One bug in the 1.5 release is that the SimpleDashboard service was not included in the solution for Simulation Tutorial 2.

---

In December 2007, the MRDS team released a refresh of the 1.5 release with bug fixes and some minor new functionality. If you downloaded version 1.5 of the SDK prior to December 2007, you should uninstall MRDS and then download and reinstall the refreshed kit. When you uninstall MRDS, it only removes the files that it installed. If you have added any of your own services to the Microsoft Robotics Studio directory structure, they will not be deleted.

If you have version 1.0 or any of the Community Technology Preview releases of 1.5 installed, it is not necessary to uninstall them. They can coexist alongside the 1.5 release. However, you should probably remove them because they are obsolete.

Using your web browser, navigate to www.microsoft.com/robotics and select the Downloads link from the right side of the page. All of the currently available downloads for MRDS are listed on this page. Click Microsoft Robotics Studio (1.5) Refresh and click Download on the next page that appears. The size of this download is 87.5MB.

It is recommended that you save the file to your hard disk and then run it after it has been downloaded. When you run the executable, the MRDS installer will run as shown in Figure 1-11.

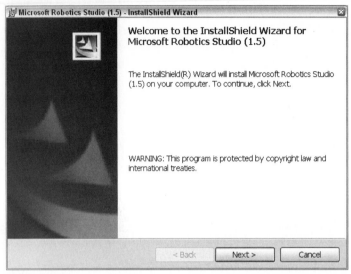

Figure 1-11

After you accept the licensing agreement (notice that MRDS is free for noncommercial use but must be licensed for commercial use), the MRDS files are installed. After the InstallShield Wizard has completed, the installer gives you options for installing additional components. Be sure to check each of these options and install them as shown in Figure 1-12.

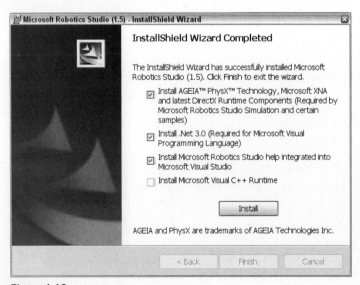

Figure 1-12

DirectX, XNA, and the AGEIA PhysX engine are required to run the simulation environment, and .NET 3.0 is required to run VPL. It is also nice to have the MRDS help files integrated into Visual Studio so that help for these APIs is visible there. Be sure that you have an active Internet connection when you run these installers because the DirectX and XNA installers get the latest installation kits from the Web.

## *Robotics Studio 1.5 Directory Structure*

By default, the MRDS 1.5 SDK is installed to `C:\Microsoft Robotics Studio (1.5)`. Some of the more important directories under this main directory are described in this section.

All the user and run-time service assemblies are stored in the `bin` directory. This directory should be in your DOS path when you open an MRDS command window. Assemblies that target the .NET Compact Framework (CF) are in the CF subdirectory.

As you might expect, the `documentation` directory contains all of the SDK documentation. The `external` directory contains installer programs for other software that MRDS depends on. This software should be installed when you install MRDS, but in case there was a problem, you can find the installers here.

The `store` directory contains content and data files used by MRDS. The contract directory cache is stored in this directory. The `store\media` directory contains all of the textures and meshes used by the simulator.

Most of the remaining files in the SDK are in the `samples` directory. The major subdirectories are as follows:

❑ `Common`: This directory contains the source files for the RoboticsCommon assembly, which contains generic contracts for many types of devices.

❑ `Config`: This directory contains manifests and configuration files for all of the samples.

❑ `Diagrams`: This directory contains sample VPL diagrams.

❑ `Misc`: This directory contains source code for services such as the SimpleDashboard service, the GameController service, and many others.

❑ `Platforms`: This directory contains subdirectories for each of the supported platforms: FischerTechnik, iRobot, Kondo, LEGO, MobileRobots, and Parallax.

❑ `Sensors`: This directory contains subdirectories for sensor services, among which are the SickLRF and the Webcam services.

❑ `Simulation`: This directory contains subdirectories for all of the simulation-specific services, such as the SimulatedDifferentialDrive, simulated sensors, the source code for all of the simulation entities, and the ArticulatedArms services.

❑ `Technologies`: This directory contains subdirectories that have source code for services that encapsulate specific technologies such as Speech.

❑ `Test`: This directory contains all of the test sample code demonstrating how to write service tests for your services.

❑ `HostingTutorials`, `RoboticsTutorials`, `ServiceTutorials`, `SimulationTutorials`, and `VPLTutorials`: These directories contain the source code for all of the tutorials described in the MRDS documentation.

An unusual feature of the SDK is that all services and files must be contained in this directory structure. It is recommended that you create a directory under the root directory with your company or organization name to hold your projects. Some developers have made the mistake of creating their own projects under the samples directory because that is where the source code for all of the other services is kept.

## Installing the Packages

As long as you are at the MRDS Downloads page, you may as well download and install the sumo package, the soccer package, and the courseware package for MRDS 1.5.

Packages do not contain a full installer. They are generated by a DSS utility program call DssDeploy. This utility gathers all of the files needed to run a particular *manifest* (a list of all the required services and their partnerships), along with any additional files specified, compresses them, and packs them into an executable. When the package is executed, it may display an EULA (end-user license agreement) that must be accepted, after which it installs the files. Finally, it displays an optional ReadMe web page that provides further information about the package.

It is recommended that you install all three packages. Some of the examples in this book assume that this code is installed. Figure 1-13 shows the sumo package being installed.

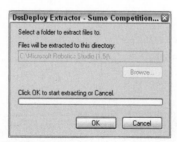

**Figure 1-13**

Packages are usually targeted for a specific release of MRDS. A package that is targeted to the 1.5 SDK will fail to install if that version of the SDK is not also installed.

When you have installed a package, an HTML file is placed in the packages directory, which contains a list of all the files that were installed as part of the package.

# Installing the Sample Code for This Book

Unless you enjoy typing in hundreds of lines of code, download the sample code package associated with this book. All of the source code used in this book, along with other necessary media files, is available for download at www.wrox.com. At the site, simply locate the book's title (either by using the Search box or by using one of the title lists) and click the Download Code link on the book's detail page to obtain all the source code for the book.

The downloaded file is a DssDeploy package just like the MRDS packages. Install it the same way that you installed the other packages. The ReadMe page that is displayed at the end of the installation provides more information about the source code, along with links to pages for each chapter.

The source code is installed in a subdirectory called ProMRDS that is placed in the MRDS top-level directory. Each chapter has its own subdirectory with more subdirectories for each project within the chapter. Binaries for the samples services, along with command files to run the samples, are installed in the bin directory. Media files required by the simulation samples are installed in the store\media directory.

# Verifying the Installation

Here are a couple of quick checks you can use to verify that you have installed everything correctly and the software is working properly on your machine.

## Runtime

Open a MRDS Command Prompt window by clicking Start ⇨ All Programs ⇨ Microsoft Robotics Studio (1.5) ⇨ Command Prompt. This brings up a DOS window that has the Path environment variable set up properly for running applications. Type the following command, as shown in Figure 1-14, and you should see a similar output: **dsshost -p:50000 -t:50001**.

Figure 1-14

This starts up a DSS node. Verify that the node is running properly by pointing a web browser to `http://localhost:50000`. Click Service Directory in the left panel and verify that the services running are the same as those shown in Figure 1-15.

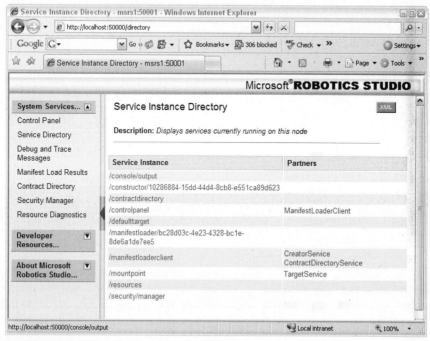

**Figure 1-15**

This verifies that the DSS runtime is properly installed. You can terminate the DSS node by pressing Ctrl+C in the Command Prompt window.

## Simulation

To verify that the simulation environment is working properly on your machine, first make sure that you have stopped the DSS node that was started in the previous section. This test runs a DSS node using the same ports, so it won't run if the previous node is still running. Click Start ⇨ All Programs ⇨ Microsoft Robotics Studio (1.5) ⇨ Visual Simulation Environment ⇨ Basic Simulation Environment.

After a few moments, a window similar to the one shown in Figure 1-16 should appear. Click View ⇨ Select Status Bar to display the frame rate. The frame rate should be somewhere near the refresh rate for your display. If it is less than 10 frames per second (FPS), you need a more powerful graphics card.

Figure 1-16

If your window is red and contains an error message, shows a solid light-blue color, or looks substantially different from the scene shown in Figure 1-16, check the capabilities of your graphics card.

## ProMRDS Sample Code

A very simple little sample program is provided to verify that you have properly installed the sample code associated with this book. Make sure that all DSS nodes have been terminated and then open an MRDS Command Prompt window and type **Welcome,** as shown in Figure 1-17.

Figure 1-17

You should see the output shown in Figure 1-17, and after a moment or two a dialog box similar to the one shown in Figure 1-18 should appear. If your speaker volume is on and speech services are correctly configured on your machine, you should hear the words in the dialog box spoken by the computer.

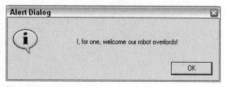

**Figure 1-18**

This verifies that the source code for the book was properly installed. Incidentally, that very simple service was generated with the VPL diagram shown in Figure 1-19. More great VPL examples can be found in Chapters 10 through 12. You can terminate the DSS node by pressing Ctrl+C in the Command Prompt window.

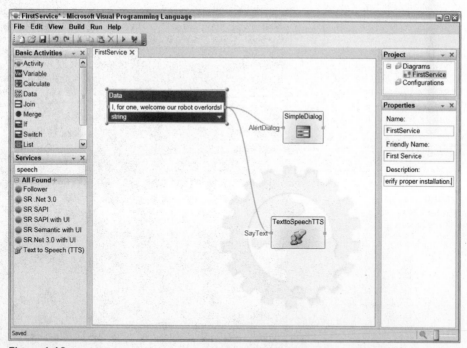

**Figure 1-19**

For an example that shows off the power of MRDS, try a different service. Follow these steps:

**1.** In Windows Explorer, navigate to the MRDS installation point and then to `ProMRDS\Chapter9\ExplorerSim`.

2.  Open `ExplorerSim.sln` in Visual Studio by double-clicking it.

3.  Once it is open, run the service with the debugger by pressing F5 or clicking Debug ⇨ Start Debugging. After a little while, a DOS Command Prompt window appears, followed by a simulation window, shown in Figure 1-20.

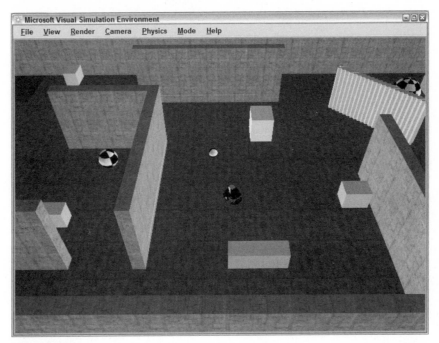

Figure 1-20

The simulated Pioneer 3DX robot wanders around building a map using information from the simulated laser range finder. You can see the map in the Map window. Figure 1-21 shows a map after the robot has done a fair amount of exploring.

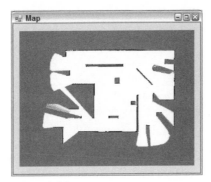

Figure 1-21

A Dashboard window also appears when you run the service.

**4.** Enter **localhost** as the Machine and **50001** as the Port and click Connect. Three services are displayed in the service directory list. Double-click simulatedlrf to select it. Figure 1-22 shows the Dashboard with the Laser Range Finder information displayed.

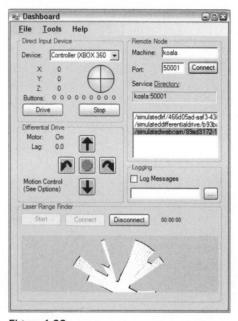

Figure 1-22

**5.** Double-click the simulatedwebcam service. A webcam window pops up to show the robot's view of the world (see Figure 1-23).

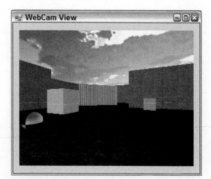

Figure 1-23

The ExplorerSim service illustrates many different aspects of MRDS and it is interesting to watch for a while. When you get tired of it, just stop the program in Visual Studio.

# Additional Resources and Support

It should be clear at this point that there is a great deal to learn about Microsoft Robotics Developer Studio. This book gives you a great start, but it is best to take advantage of all the resources available. The following sections point out a few of these.

## www.Microsoft.com/Robotics

This is Microsoft Robotics Developer Studio central. You can find information about the latest releases on this page along with pointers to many other articles and blogs written by other users. The Community link has links to partner websites where MRDS-related products are offered. There are also links to MRDS-related user groups and discussion boards.

On the main MRDS page, you can also find a link to the MRDS wiki, which contains hints, bug workarounds, and additional information about using MRDS.

Add this page to your Favorites list, as you will likely refer back to it often.

## Tutorials and User Guides

Following a long Microsoft tradition, the early documentation provided with Microsoft Robotics Developer Studio was a little thin. It is rapidly becoming more complete and now includes several informative user guides along with the tutorials.

The MRDS documentation can be accessed from a link on the Start menu. The Samples and Tutorials Overview at the beginning of the documentation provides a good roadmap for exploring the tutorials. It is strongly recommended that you do all of the tutorials. Start with the DSS tutorials to learn service programming by example. These tutorials are good companion material to Chapters 2–4 in this book. Likewise, the Simulation tutorials provide a different view of the material in Chapters 5–9. The VPL tutorials go with the material in Chapters 10–12, and the Robotics tutorials supplement the information in Chapters 13–17.

In addition to the samples and tutorials, CCR and DSS each have a good User Guide that is available in the Microsoft Robotics Developer Studio Runtime section in the MRDS documentation.

Additional sections of interest include the Courseware section and the Technology Samples section.

## Online Forums

Sometimes you might run into a problem that just can't be answered by this book or the tutorials and samples. In this case, you should turn to the online forums. There is a link on the main MRDS page to get to the forums, or you can go directly to `http://forums.microsoft.com/msdn/default.aspx?forumgroupid=383&siteid=1`.

The forums provide an opportunity to post questions in the following six categories:

❑ **Community:** Samples, feedback, and general discussion

❑ **Concurrency and Coordination Runtime:** Everything about the CCR

❑ **Decentralized Software Services:** Everything about DSS

❑ **Simulation:** Samples, entities, and graphics/physics discussion

❑ **Visual Programming Language:** Everything about VPL

❑ **Hardware Configuration and Troubleshooting:** How to set up your hardware and help with porting code to MRDS

Members of the MRDS team spend considerable time answering questions on the forums, and other users often contribute answers. It is worthwhile to search through past threads before posting a new question on the forums.

*If you are posting code to the forum, try to condense it down to the shortest piece of code that illustrates the problem. Copy and paste it into the posting window, and then select it and mark it as code by clicking the Mark Code Block button in the editing toolbar. This prevents your code from appearing in the forum with a bunch of smiley faces in it.*

## Channel 9

Channel 9 is a Microsoft-sponsored developer community with many interesting links to other developers' MRDS projects. The Channel 9 home page is `http://channel9.msdn.com` and from there you can click the Search link to search for videos, projects, and posts about MRDS. There are also links to several classic Channel 9 videos, available directly from the MRDS home page in the Featured Videos section.

Take a look at some of the projects presented and be sure to post your own project in the Sandbox when it is ready for the world to see.

# Summary

So that's Microsoft Robotics Developer Studio in a nutshell. Now that you've seen the opportunities that await you, it's time to warm up your compiler and get coding. First one to write a sentient service wins!

# 2

# Concurrency and Coordination Runtime (CCR)

Microsoft Robotics Developers Studio (MRDS) is built on two basic components: the Concurrency and Coordination Runtime (CCR) and the Decentralized Software Services (DSS). This chapter covers many of the concepts of the CCR; the next chapter discusses DSS.

To quickly clarify the differences between the CCR and DSS: The CCR is a programming model for handling multi-threading and inter-task synchronization, whereas DSS is used for building applications based on a loosely coupled service model. Services can run anywhere on the network, so DSS provides a communications infrastructure that enables services to transparently run on different nodes using all of the same CCR constructs that they would use if they were running locally.

Although you can use the CCR on its own, completely outside MRDS, this is not how you use it for creating robotics applications. Consequently, there is some overlap with DSS in this chapter because it provides an environment that makes CCR easier to use. In fact, there was some discussion during the writing of this book regarding whether CCR or DSS should be covered first. You can take a peek at the next chapter and decide for yourself in which order you want to read them.

A large amount of documentation is supplied with MRDS and the authors do not intend to reproduce it all here. You should read the online CCR User Guide (`http://msdn2.microsoft.com/en-us/library/bb905447.aspx`), which is also in the documentation that comes with MRDS. In particular, you will find the MRDS Tutorials to be an invaluable source of information.

The objective in this book is to give you a brief introduction to the important concepts and then get into coding as quickly as possible. Along the way, many new applications supplement the examples that are included in the MRDS distribution.

You will probably find that the MRDS environment is quite different from anything that you have programmed with in the past, which means that it can involve a steep learning curve. Luckily, Microsoft recognized this and built the Visual Programming Language (VPL) tool to hide many of the details about how MRDS services work. However, as a professional programmer, you need to understand what is happening "under the hood," so the book begins with the basics, and VPL is covered later in the book.

# Overview of the MRDS Framework

MRDS provides a framework for developing robotics applications. At the lowest level it is conceptually similar to the device drivers or BIOS (basic input/output system) on a PC that provide the interface to the computer hardware.

As part of MRDS, Microsoft supplies a variety of different samples that support readily available robots, but it is up to the robot manufacturers to develop and support their own code. However, MRDS provides more than just the equivalent of device drivers.

At a higher level, MRDS is similar to an operating system for robots. It is not a true operating system because MRDS services must be hosted on a Windows platform with .NET installed. In many cases this means that MRDS runs on a PC and communicates with the robot via a wireless connection. Alternatively, an embedded PC, laptop, or PDA can be mounted on the robot to run MRDS. This is an important point: You do not compile MRDS code and load it directly into a robot — there must be a Windows device somewhere, either on the robot or connected to it through a communications link.

## The Need for Concurrency

Anyone who has worked with robots knows that several things are often happening at the same time. For example, while your program is driving a robot around, it must also be listening to information from the sensors so that the robot does not bump into anything.

In the past, to handle this robotic multi-tasking, you would have had to write a multi-threaded application using Windows threads. This was a complex task. For example, you would have used mutual exclusion, or mutexes, to prevent two threads from simultaneously attempting to update the same variable. Semaphores and critical sections were used to control access to information about the current *state* of the robot so that the state did not become inconsistent. However, with mutexes comes the possibility of deadlocks. Debugging in a real-time environment to find race conditions that intermittently cause deadlocks was a nightmare.

The CCR eliminates many of the issues related to multi-threading. (However, you can still create deadlocks if you use the CCR inappropriately. You'll see an example of this later in the chapter.) It uses its own threading mechanism, which is much more efficient than the Windows threading model.

Another issue with robots is that events happen asynchronously. You don't want your service to be constantly polling the robot for sensor information. In reality, some services have to do this because the robot itself does not spontaneously send sensor updates. However, your code does not need to know about this because the details are buried inside another service that you partner with.

Similarly, you cannot afford to write timing loops that tie up the CPU. This is a common practice on cheap hobby robots because the onboard computer is quite primitive and doesn't support multi-threading. The problem with such a "busy wait" is that the robot is effectively blind while the CPU is executing a time delay. Provided that the delays are quite short, the robot will appear to be executing multiple operations simultaneously. On a Windows system, however, you need a mechanism to wait for a period of time that does not involve running the CPU at 100 percent in a tight loop.

## Services — The Basic Building Blocks

The CCR supplies the underlying infrastructure that enables multiple tasks to execute concurrently on a single computer. DSS adds another layer for combining CCR applications, called *services*, and at the same time it enables these services to run on completely separate computers and communicate via the network.

Microsoft has defined a set of *generic contracts* that describe commonly used robotics services. These contracts specify the APIs that must be used to communicate with robot components such as motors, sonar sensors, and even webcams. By standardizing these interfaces, the robotics community can share code more easily.

For example, the generic differential drive contract enables a single program called the Dashboard (see Chapter 4) to drive around any robot that has two wheels regardless of the underlying hardware, and to do this using a mouse, a joystick, or even an Xbox controller without having to configure anything.

Services are discussed in more detail in Chapter 3, but they are so fundamental to MRDS that they need to be introduced here.

## Orchestration — Putting Services Together

Every MRDS application that you build will contain one or more services. Combining these services and passing messages between them, whether they are located on the same or different computers, is one of the tasks of DSS.

Figure 2-1 shows an example of how the services might be *orchestrated* to control a robot. Combining services, a process called *partnering*, is one of the topics in the next chapter. It is the job of the orchestration service to implement high-level behaviors, such as following a line or solving a maze.

You might wonder why DSS is not shown sitting on top of the CCR in the Runtime Environment in Figure 2-1. The reason is very simple: Many services make direct calls to the CCR. Clearly, the CCR is too low level to provide orchestration; it just provides the machinery to make it all happen.

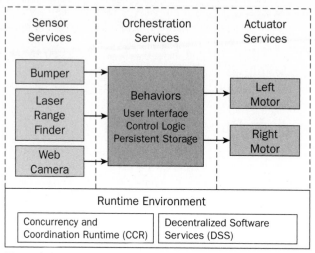

**Figure 2-1**

# Setting Up for This Chapter

Although MRDS is based on the CCR, it is possible to use the CCR on its own. However, you need the combination of the CCR and DSS in order to write MRDS services. Therefore, the examples in this chapter are in the context of a DSS service. (DSS services are covered in Chapter 3.)

You have two options at this stage:

❑ You can simply open the CCRExamples solution in the ProMRDS\Chapter2 folder.

❑ If you learn best by doing things and you like typing, then you can enter the code in this chapter into an empty service and then build and run it.

If you have successfully installed the software for this book, then you already have a folder called ProMRDS under your MRDS installation (by default it is in C:\Microsoft Robotics Studio (1.5)). If you cannot find this folder, then please go back to Chapter 1 and follow the installation instructions.

Throughout this book, you will continually see references to the ProMRDS folder. Remember that this is under the MRDS installation point, or what is called the *root* or the *mountpoint* in MRDS terminology.

We don't really believe that people like typing in code from a textbook, but even if you do, you might want to check out the sample code first, as explained below. It can also act as a reference in case you can't get your code to work due to some small typing error.

**An Important Note about Versions**

At the time of writing, Microsoft had just released Visual Studio 2008. However, all of the example code for this book was developed using Visual Studio 2005. Both of these versions can coexist on the same computer, but you should use VS 2005 if possible.

In addition, the code in this book is based on MRDS Version 1.5 with the December 2007 Refresh applied. Planning was already underway for Version 2.0 as this book was being written; a new Community Technology Preview (CTP) of 2.0 will be available by the time this book is published. At the same time, Microsoft plans to rename the product to Microsoft Robotics Developer Studio — an author's nightmare in the final stages of writing a book! Therefore, you see references to MRDS in this book, instead of MSRS. Consider them to be the same product.

You should download and install MRDS V1.5 even if a later CTP version of MRDS is available. Multiple versions can be installed on the same computer and can operate side-by-side.

Visit the website for this book for updates at www.proMRDS.com. VS 2008 versions of the code will be posted to this site as well as updates when MRDS V2.0 is officially released. You can also get downloads and errata at www.wrox.com.

## Using the CCRExamples Project

Navigate to the `ProMRDS\Chapter2\CCRExamples` folder using Windows Explorer. Double-click on `CCRExamples.sln` to open the solution in Visual Studio.

*If you are using the Express Edition of Visual C# and you have other Express products installed, then you might have to select C# from the pop-up dialog. The Express Edition does not properly recognize the solution file.*

Once you have the solution open in Visual Studio, open the file `CCRExamples.cs` and locate the method called `Start`. All DSS services have a method with this name and it is called during the initialization of the service.

Note that the code uses a lot of `Console.WriteLine` statements to display information in the MRDS DOS command window that is created automatically when you run the project. Using `Console.WriteLine` is not recommended because MRDS has other ways of displaying output. However, for these examples it is easier to just use the console.

The code in the `Start` method runs in a loop that displays a menu and asks you to enter a number for one of the examples. Then it executes the selected example code. If you enter a value of zero, then the service exits. There is no need to discuss this part of the code because it has nothing to do with the CCR or DSS.

To try out the examples, select Start Without Debugging from the Debug menu or press Ctrl+F5. You need to run without the debugger for the Causality example to work correctly, as explained below. All the other examples can be run with the debugger.

The screenshot in Figure 2-2 shows the initial output from the service. Note that the messages at the top of the window are from DSS as it starts the service. The first message from the program is "Main Thread 8," which indicates that it is running. The text menu at the bottom of the screen provides a list of the available examples. These cover the major features of the CCR and are explained in the rest of this chapter.

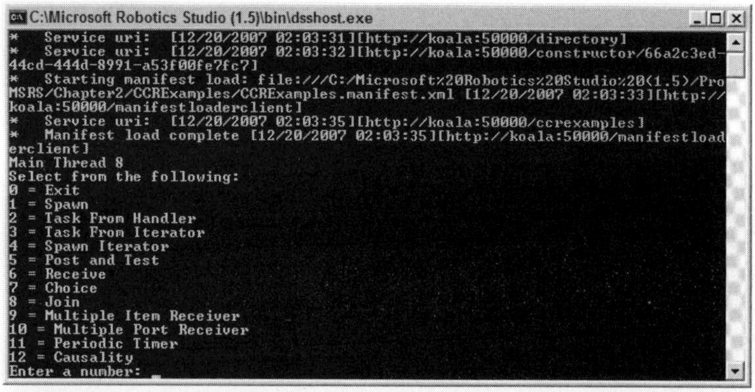

Figure 2-2

## Entering the Code Manually

If you want to enter the code yourself, then start by opening Visual Studio and creating a new DSS service. MRDS installs templates so that you can easily create new projects.

**1.** Start Visual Studio 2005 and then select File ⇨ New ⇨ Project.

The New Project dialog appears, as shown in Figure 2-3. Notice that the Robotics templates are selected from the Project Types under Visual C# and that Simple DSS Service (1.5) is highlighted.

*If you don't see the Robotics folder under Project Types, then it is likely that you installed Visual Studio after MRDS. In that case, you should reinstall MRDS.*

You can create the service wherever you like, but it is a good idea to keep your code separate from the MRDS distribution. In Figure 2-3, the location is set to the Projects folder under the MRDS mountpoint.

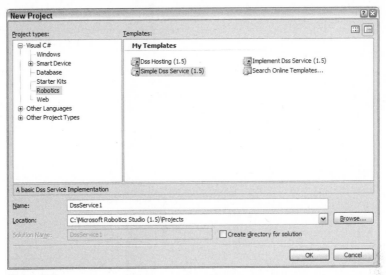

Figure 2-3

**2.** Give the project a name. It doesn't matter what you call it, but leave it as DssService1 for now.

*If you are using Visual C# Express Edition, then the dialog is slightly different and you are not asked for the location. The project is automatically created in the Projects folder under MRDS. The dialog for the Express Edition is shown in Figure 2-4.*

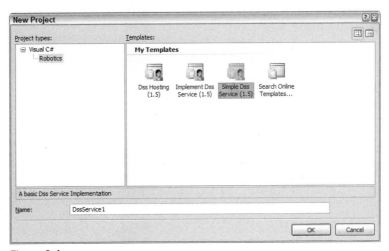

Figure 2-4

**3.** Examine the contents of the solution in the Solution Explorer. Once the project has been created, you should see the files shown in Figure 2-5 in the Solution Explorer. Note that the References are expanded to show that `Ccr.Core`, `DssBase` and `DssRuntime` were added automatically. These references must always be included for MRDS service projects.

**Figure 2-5**

**4.** Add a new `using` statement. To do so, open the main service source file, which is `DssService1.cs`. At the top of the file you will find several `using` statements. Add one for `System.Threading` as shown here:

```
using ...
using dssservice1 = Robotics.DssService1;
using System.Threading;
```

**5.** Scroll through the code to update the `Start` method. It should look like the following:

```
/// <summary>
/// Service Start
/// </summary>
protected override void Start()
{
    base.Start();
    // Add service specific initialization here.
}
```

`Start` is called when the service is started by DSS. This is where you would normally place code to perform any initialization that your service requires, but you are going to use it to execute a series of examples.

Note that in the examples that follow, each example is shown as a separate procedure. You can add each procedure to the service one at a time and put a call to it in the `Start` method. This helps to keep the code compartmentalized and easier to manage. Eventually, you will end up with a service that runs all of the examples. (The CCRExamples code also has regions around each of the examples as a further way of managing the code.)

To reiterate, you need to enter the code for each of the example functions after the end of the `Start` method and then insert an appropriate call to the example function just below `base.Start` where the comment tells you to enter your own initialization code, like so:

```
protected override void Start()
{
    base.Start();
    // Add service specific initialization here.

    // Wait until the service has started up completely.
    Wait(2000);
    Console.WriteLine("Main Thread {0}", Thread.CurrentThread.ManagedThreadId);

    // Call the Example
    ExampleFunction();
}

void ExampleFunction()
{
    // Example code goes here
}
```

After you insert the preceding code, compile the solution and run it in the debugger to ensure that everything is OK. Of course, it is much easier if you just open the CCRExamples solution.

## Tips for Coding with MRDS

Most programs are not written from scratch. Programmers often find a similar piece of code and then copy, paste, and modify it as required. The same is true of MRDS. There are literally dozens of examples provided in the MRDS samples folder, and of course there are the examples from the website for this book (which you installed into the ProMRDS folder). Novel examples are popping up on the Web all the time, such as driving your robot using a Nintendo Wii remote controller with the .NET managed library for the Wiimote as it is called (available from www.codeplex.com/WiimoteLib). Do a Google search for "MRDS wiimote" for more information.

The following are some tips for coding with MRDS:

❑ You should definitely read the documentation available online at http://msdn2.microsoft .com/en-us/library/bb881626.aspx. This documentation is also supplied as a compiled help file, called MSRSUserGuide.chm, in the MRDS documentation directory. (The online version might be more current.)

❑ If you thought that that .NET Framework Class Library was huge, now you have to add to that the CCR and DSS libraries. You should make extensive use of IntelliSense and the Object Browser in Visual Studio to find out what properties and methods are available for the various classes. You can also right-click a class name in the code and select Go To Definition to see the properties and methods of a class via reflection.

❑ No attempt is made in this book to cover all of the available classes because it would be a waste of paper and you would not want to read it anyway. All the class information is at your fingertips. An online class reference for MRDS is accessible from Visual Studio by pressing F1 while you have the cursor on a CCR or DSS class in the editor. It is also available through the online documentation page. The class reference was released while this book was being written and it was still very basic, with very few examples.

# Coordination and Concurrency Runtime

As previously mentioned, the CCR is a lightweight library that is supplied as a .NET DLL. It is designed to handle asynchronous communication between loosely coupled services that are running concurrently. This chapter contains many concepts that may be new to you, so you might want to skim through it on the first reading and then come back later to reread portions of it more carefully. It is difficult to cover all of the material sequentially because various concepts are interrelated. Figure 2-6 summarizes the CCR architecture. The following list briefly outlines how the CCR works. Then it is described in more detail throughout the rest of the chapter. You can refer back to Figure 2-6 as you read through the rest of the sections.

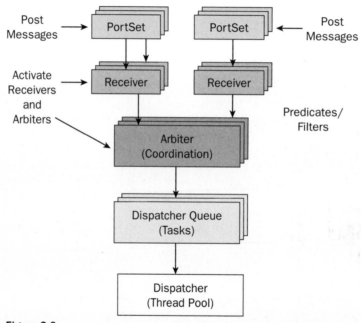

**Figure 2-6**

❑ The CCR uses an asynchronous programming model based on *message passing* using a structure called a *port*. A *message* is an instance of any data type that the CLR can handle. When you write DSS services, you declare your own classes for message types. If messages have to be sent across the network, then they are first serialized into XML and then deserialized on the destination node.

❑ Ports are message queues that only accept messages of the same data type as the one in the port declaration. This "type safety" is important for ensuring that obvious mistakes are picked up at compile time.

❑ In CCR terminology, you place messages into a port by *posting* to the port. Messages remain in ports until they are dequeued by a *receiver*. Activation conditions can be set on receivers to create complex logical expressions, such as a *Join* between two ports (two messages must arrive, which is effectively a logical AND) or a *Choice* between two ports (a message can arrive on either port, creating a logical OR). Evaluating activation conditions is the job of *arbiters*.

❑ The coordination primitives, implemented through arbiters, are used to synchronize and control *tasks*. Once a receiver's conditions have been met, a task is queued to a *dispatcher queue* and then passed to a *dispatcher* for execution.

❑ When a task is scheduled to run, a *handler* is executed. Handlers are pieces of code that run in a fully multi-threaded environment. They can run on their own if you simply request the execution of a task, or more commonly they consume messages as a consequence of being connected to a receiver. Handlers often generate new messages, and the cycle continues.

❑ There are mechanisms within the CCR to handle failures, also called *faults*, in relation to web requests or *exceptions* in the usual .NET CLR sense. A structured approach called *causalities* is analogous to try/catch.

Most of the time, the CCR handles all this transparently, and there is no need for you to explicitly define mutexes or write callback procedures, which you might be used to using for multi-threaded programming in the past. The standard Windows synchronization primitives still work, but they are not usually needed.

The DSS is based on the CCR and does not require any other components. It provides a hosting environment for running services and making them accessible via a web browser. You cannot separate DSS from the CCR. When you run the examples in this chapter, you are running a DSS node and executing a service inside this node.

## Concurrent Execution

Obviously, a key feature of the CCR is that it supports multi-threading, or *concurrency*. Understanding how concurrency works in the CCR environment is your first challenge as you learn about MRDS.

However, before examining the components of Figure 2-6, you need to understand a few key concepts: tasks, delegates and iterators. However, in order to do this, you need to use some CCR APIs that have not been discussed yet — a classic "chicken or the egg" situation.

### Tasks

At the bottom of Figure 2-6, you can see that dispatchers schedule tasks by allocating them to threads from a pool. A task contains a reference to a handler that is a piece of code you want executed.

In the CCR, tasks implement the `ITask` interface. This is useful because it gives tasks a data structure, enabling them to be passed around and queued. However, it also means that you must wrap a piece of code as an `ITask` before you can submit it to a dispatcher queue.

Running a task is relatively easy. Consider the following simple handler that loops around counting from 1 to 10, with a delay in between printing the numbers:

```
// Simple Handlers do not take parameters and do not return anything
void SimpleHandler()
{
    // Handler does not really do anything
    Console.WriteLine("Simple Handler Thread {0}",
            Thread.CurrentThread.ManagedThreadId);
```

*(continued)*

*(continued)*

```
for (int i = 0; i < 10; i++)
{
    Wait(100);
    Console.Write(i + " ");
}

// Try adding PressEnter() and see what happens.
// It gets complicated!
//PressEnter();

Console.WriteLine("Finished Simple Handler Thread {0}",
        Thread.CurrentThread.ManagedThreadId);
}
```

This is not very exciting on its own, but things get more interesting when you run this code several times concurrently.

The DsspServiceBase class (part of DSS) provides a wrapper to execute a handler as a task called Spawn. If you select the Spawn option from the menu in CCRExamples, it executes the following code:

```
Spawn(SimpleHandler);
Spawn(SimpleHandler);
Spawn(SimpleHandler);
Spawn(SimpleHandler);
Console.WriteLine("Spawns executed ...");
```

*If you are creating your own service, then place this code into the* Start *method and add the* SimpleHandler *method to your service.*

The output should look something like the following:

```
Simple Handler Thread 8
Spawns executed ...
0 1 2 3 Select from the following:
0 = Exit
1 = Spawn
2 = Task From Handler
3 = Task From Iterator
4 = SpawnIterator
5 = Post and Test
6 = Receive
7 = Choice
8 = Join
9 = MultipleItemReceive
10 = MultiplePortReceive
11 = Periodic Timer
12 = Causality
Enter a number: 4 5 6 7 8 9 Finished Simple Handler Thread 8
Simple Handler Thread 8
0 1 2 3 4 5 6 7 8 9 Finished Simple Handler Thread 8
Simple Handler Thread 8
```

```
0 1 2 3 4 5 6 7 8 9 Finished Simple Handler Thread 8
Simple Handler Thread 8
0 1 2 3 4 5 6 7 8 9 Finished Simple Handler Thread 8
```

The SimpleHandler displays its first message before all of the spawns have completed. Then it starts to count from 1 to 10, but the main thread displays the menu again so the two are interspersed in the console output. Eventually the first task completes, and then the handler runs three more times. Meanwhile, the main thread is quietly waiting in the background for you to select another menu option, effectively tying up a thread.

Notice in this case that all of the spawned threads are on thread number 8. How can this be in a multi-threaded environment? The answer is very simple: The default number of threads in the pool in this case is only two. The first thread executes the Spawn statements and displays the menu, so it is tied up. That leaves only one thread to execute the spawned tasks, which therefore have to execute sequentially.

This behavior is not something that you can count on. If the computer had a four-core processor, then DSS would have automatically created a dispatcher with four threads and all of the tasks would have executed simultaneously. Try the following example to prove this.

Assuming that you saw the same thread used repeatedly, you can change the attributes on the service so that more threads are created in the initial pool. (If you did *not* see this behavior, then you probably have a four-core processor, or two dual-core processors, etc.) This is really a DSS issue because you are changing attributes on a DsspServiceBase class, but it makes sense to discuss it here.

At the top of the code, locate the service class definition and add the ActivationSettings attribute. Set the parameters to false for ShareDispatcher and 6 for ExecutionUnitsPerDispatcher as shown:

```
/// <summary>
/// Implementation class for CCRExamples
/// </summary>
[DisplayName("CCRExamples")]
[Description("The CCRExamples Service")]
// Add the following attribute to the service to create six threads
// instead of the default, which is probably only two
[ActivationSettings(ShareDispatcher=false, ExecutionUnitsPerDispatcher=6)]
[Contract(Contract.Identifier)]
public class CCRExamplesService : DsspServiceBase
{
    ...
```

Recompile and run the service again. Select the Spawn example and see what happens. The output should be as follows:

```
Spawns executed ...
Simple Handler Thread 11
Simple Handler Thread 13
Simple Handler Thread 14
Simple Handler Thread 15
0 0 0 0 1 1 1 1 2 2 2 2 3 3 3 3 Select from the following:
0 = Exit
```

*(continued)*

*(continued)*

```
 1 = Spawn
 2 = Task From Handler
 3 = Task From Iterator
 4 = SpawnIterator
 5 = Post and Test
 6 = Receive
 7 = Choice
 8 = Join
 9 = MultipleItemReceive
10 = MultiplePortReceive
11 = Periodic Timer
12 = Causality
Enter a number: 4 4 4 4 5 5 5 5 6 6 6 6 7 7 7 7 8 8 8 8 9 Finished Simple ↵
Handler Thread 14
9 Finished Simple Handler Thread 13
9 Finished Simple Handler Thread 11
9 Finished Simple Handler Thread 15
```

All of the spawns execute simultaneously, and the menu still runs too! You can see this because each number appears four times in a row.

This is an important point: The default DSS dispatcher thread pool only has two threads for a single CPU machine (even if it is dual-core). The minimum number of threads is two.

*Before you continue, go back and comment out the* service *attribute. If you don't do this, then some of the remaining examples will have different output from what is shown in the book.*

Another way to get overlapped behavior is to create your own dispatcher so you can specify the number of threads explicitly. Add a `RunFromHandler` function as shown in the following example. It creates a new dispatcher with three threads in the pool. It then starts four tasks using the same `SimpleHandler` as above.

At this stage you don't need to know much about the dispatcher and dispatcher queue, but notice the use of `Arbiter.Activate` to activate the tasks and `Arbiter.FromHandler` to create the necessary task instances.

In the CCR you need to *activate* a task to get it to execute, which means enqueuing it to a dispatcher queue. Without `Arbiter.Activate`, the `Arbiter.FromHandler` would do nothing — it would simply create an instance of an `ITask`, but not schedule it for execution. This is a potential trap for novices.

```
// Run a Task from a Handler
void RunFromHandler()
{
    // Explicitly create a Dispatcher so we can control the pool size
    Dispatcher d = new Dispatcher(3, "Test Pool");
    DispatcherQueue q = new DispatcherQueue("Test Queue", d);

    // Activate FOUR tasks with a pool of THREE threads
```

```
Arbiter.Activate(q,
    Arbiter.FromHandler(SimpleHandler),
    Arbiter.FromHandler(SimpleHandler),
    Arbiter.FromHandler(SimpleHandler),
    Arbiter.FromHandler(SimpleHandler));
}
```

When you run this code (the menu option in CCRExamples is Task From Handler), the output is quite different from the example with `Spawn`:

```
Simple Handler Thread 18
Simple Handler Thread 19
Simple Handler Thread 17
0 0 0 1 1 1 2 2 2 3 3 3 Select from the following:
0 = Exit
1 = Spawn
2 = Task From Handler
3 = Task From Iterator
4 = SpawnIterator
5 = Post and Test
6 = Receive
7 = Choice
8 = Join
9 = MultipleItemReceive
10 = MultiplePortReceive
11 = Periodic Timer
12 = Causality
Enter a number: 4 4 4 5 5 5 6 6 6 7 7 7 8 8 8 9 Finished Simple Handler Thread 19
Simple Handler Thread 19
9 Finished Simple Handler Thread 17
9 Finished Simple Handler Thread 18
0 1 2 3 4 5 6 7 8 9 Finished Simple Handler Thread 19
```

Notice this time that three different threads start up straight away and announce themselves. They begin counting together, with groups of three numbers appearing on the screen. Then the menu is redisplayed by the main thread, which is not part of the new dispatcher's thread pool. Eventually thread 19 finishes, although it was not the first to start, and it starts executing the fourth task while threads 17 and 18 finish off.

This whole process is nondeterministic. Try running it several times and see which thread starts first, and which one finishes first. You cannot predict it. All you can say is that eventually all four tasks will be completed.

> *If you need even more proof that the CCR is multi-tasking, try the periodic timer example later in this chapter. You can actually select another menu option while the timer is running and you will see that "Tick" messages are injected into the output from the other example.*

## Delegates

Quite often a task is just a few lines of code. The simplest way to implement this code fragment is using a C# *delegate* to construct an anonymous method. A delegate is similar to a normal procedure declaration except that the procedure name is the keyword `delegate`. (Delegates do not exist in other .NET languages and you have to define a normal procedure instead.)

If you have not seen delegates before, here is a code fragment that illustrates the use of delegates. In this example, a message is sent to move a robotic arm and the Choice arbiter waits for either a response message to say that the move was successful or a Fault message indicating that there was some problem. (Don't worry about the rest of the syntax; Choice is discussed later in the chapter.)

```
_mainPort.Post(moveCommand);

bool success = true;

yield return Arbiter.Choice(
    moveCommand.ResponsePort,
    delegate(SSC32ResponseType response)
    {
    },
    delegate(Fault f)
    {
        LogError(f);
        LogError("Servos cannot be initialized.");
        success = false;
    }
);
```

In the preceding example, the first delegate has no code — it is just a placeholder. You don't need it for any purpose other than to be there when the move is successful. However, in general, there will be some code in this delegate. The second delegate, however, logs the error and sets a flag to say that the move failed.

Look carefully at the code. Notice that the success variable is defined outside of the delegates. However, the second delegate can still access it and update its value. This is a powerful feature of delegates when used in this way, but one that is only available in C#. (Other languages have to use global variables to achieve the same effect.)

*Technically, this is what is called a* lexical closure. *If you have programmed in an asynchronous environment before, then an I/O completion routine is another form of closure. It has been common in Lisp-based languages for some time and there is a similar construct in Ruby. However, it is relatively new in imperative languages. Interestingly, this concept is on the drawing board for the next version of Java.*

Think about it for a moment. The Choice arbiter has executed a delegate, probably on a different thread, and it has updated the local variable called success in the current procedure. That's cool stuff, but potentially dangerous.

Usually, a delegate contains only a small amount of code, but it can call other procedures and be quite complex. A key point in multi-threaded design, though, is to try to keep your tasks short and ensure that they do not block.

Another concept illustrated in the preceding code that might be new to you is yield return. This is called a *continuation*. The code is suspended at this point, and some time later it resumes execution with the next statement after the yield return. The next section discusses this further.

## Iterators

A key concept for the CCR is the use of an *iterator*, which allows sequential execution of code but without blocking the execution thread when it needs to wait for a message. When an operation must be performed that will take some unknown amount of time to execute, the iterator effectively remembers the current location in the code and then relinquishes control until a message is received. At some later point in time, the message arrives and the code resumes execution from the point where it left off.

Iterators were introduced to C# in version 2.0 to support `foreach` iteration without having to implement the entire `IEnumerable` interface. Each time that an iterator is called, it is supposed to return the next value in the sequence using `yield return`, or terminate the iteration using `yield break`. Other .NET languages do not have this support (yet) and so it is currently much easier to program with the CCR in C#.

The CCR uses iterators based on a sequence of tasks to enable you to write code that looks like it executes sequentially, but in reality it is a series of asynchronous steps. An iterator is declared as a method of type `IEnumerator<ITask>`, which means it will iterate over tasks. The syntax `<ITask>` is known as a generic and is discussed next. The compiler also recognizes that this type of method can contain `yield return` and `yield break` statements, which are not valid in normal code.

---

### Generics

C# uses generic classes, which are denoted by the syntax `<type>`. These are similar to templates in C++ if you are familiar with them.

In the case of `IEnumerator<ITask>`, the function is declared to be an enumerator that returns objects of type `ITask`. Other enumerators can also be built that return different types, but you only need to use tasks for MRDS. Also notice the syntax for declaring a `Port` that receives integer values as messages:

`Port<int>`

You can use any .NET CLR data type, including user-defined classes, as the message type for a `Port`.

---

Consider the following routine, which waits on messages using two different ports. The waiting is done using `yield return`:

```
private static IEnumerator<ITask> IteratorExample()
{
    Port<int> p1 = new Port<int>();
    Port<int> p2 = new Port<int>();

    p1.Post(0);

    bool done = false;

    while (!done)
```

*(continued)*

*(continued)*

```
    {
        yield return Arbiter.Receive(false, p1,
            delegate(int i)
            {
                Console.WriteLine("P1 Thread {0}: {1}", ↵
Thread.CurrentThread.ManagedThreadId, i);
                p2.Post(i + 1);
            }
        );

        yield return Arbiter.Receive(false, p2,
            delegate(int i)
            {
                Console.WriteLine("P2 Thread {0}: {1}", ↵
Thread.CurrentThread.ManagedThreadId, i);
                if (i >= 10)
                    done = true;
                else
                    p1.Post(i + 1);
            }
        );
    }

    yield break;
}
```

This looks like normal sequential code, but each time `yield return` is executed it returns a `Task` instance and remembers the location it reached in the code so that execution can resume from the following statement the next time the iterator is called (hence the name "continuation"). The CCR keeps calling the iterator, executing the sequence of tasks it returns, until eventually `yield break` is executed and the iteration stops.

Notice that the last line of the routine is `yield break`. Because this is the end of the routine, there is really no reason to add this statement. However, sometimes the compiler cannot determine whether you have finished the code or not, and it complains with an error that "Not all code paths return a value."

In this case, the returned tasks are *receivers*. (Receivers are discussed in the section "Receivers and Arbiters" later in the chapter.) For now, all you need to know is that a receiver waits for a message to arrive on a port. While the receiver is waiting, the operating system thread is freed to do other things (or the CPU will be idle if no tasks are waiting to execute).

---

### Some Key Points about Iterators

One of the major concerns in multi-threaded programming is the blocking of threads. This is highly undesirable because it takes the thread out of circulation for the period that it is blocked. In the `IteratorExample` code, it appears that the `Arbiter.Receive` blocks execution, but in fact the thread is not blocked at all: It is returned to the thread pool until the receiver executes.

---

Second, and this is again jumping ahead a little here, the receivers in the example are nonpersistent. (The first parameter in the `Arbiter.Receive` is `false`.) This is essential for an iterator to work because the `yield return` must wait for completion. If you specify a CCR arbiter that is persistent, its work is never finished because it remains available to process new messages. In that case, the `yield return` will never resume execution of the procedure.

Under the covers, the C# compiler rewrites an iterator into an invisible class with several methods. One of these methods is `MoveNext`, which is part of the machinery that handles advancing to the next code segment after a `yield return`. You don't need to know about this to use an iterator, but it is interesting.

Note that iterators have to be executed as tasks. If you simply insert a call to the `IteratorExample` method in your code, nothing will happen. There are no compilation errors, but the code is not executed.

The `yield` statement cannot be used inside a delegate, inside a `catch` block, or in a `try` if it has an associated `catch`. In fact, it cannot be used anywhere except inside an iterator. It causes compilation errors when used in normal code.

Iterators cannot return values in `out` or `ref` parameters so you have to use global variables if you want to return values to the caller.

In CCRExamples, the preceding `IteratorExample` code can be executed by selecting the Task From Iterator example, which executes the following code:

```
void RunFromIterator()
{
    Dispatcher d = new Dispatcher(4, "Test Pool");
    DispatcherQueue taskQ = new DispatcherQueue("Test Queue", d);

    Console.WriteLine("Before Iterator submitted - thread {0}",
Thread.CurrentThread.ManagedThreadId);

    Arbiter.Activate(taskQ,
        Arbiter.FromIteratorHandler(IteratorExample),
        Arbiter.FromIteratorHandler(IteratorExample));

    Console.WriteLine("After Iterator submitted - thread {0}",
Thread.CurrentThread.ManagedThreadId);

}
```

This code creates a new dispatcher with four threads and then activates the iterator twice by constructing a task array using `Arbiter.FromIteratorHandler`.

The output from the example should look something like the following, but you will find that the order of the output changes every time you run it:

```
Before Iterator submitted - thread 9
After Iterator submitted - thread 9
P1 Thread 20: 0
P2 Thread 22: 1
P1 Thread 21: 0
P2 Thread 21: 1
P1 Thread 21: 2
P2 Thread 21: 3
P1 Thread 21: 4
P2 Thread 21: 5
P1 Thread 21: 6
P2 Thread 21: 7
P1 Thread 20: 8
P1 Thread 23: 2
P2 Thread 23: 3
P1 Thread 23: 4
P2 Thread 23: 5
P1 Thread 23: 6
P2 Thread 23: 7
P1 Thread 23: 8
P2 Thread 23: 9
P1 Thread 23: 10
P2 Thread 21: 9
P1 Thread 21: 10
P2 Thread 21: 11
P2 Thread 23: 11
```

In the output, you can see that the iterators are submitted on thread 9. However, threads 20 to 23 are used to do the work of the iterators because a new dispatcher is specified in the `Arbiter.Activate`. The iterators post messages backward and forward between their two ports, p1 and p2, causing the execution to bounce between two threads.

The threads run as fast as their little feet will carry them (threads have much smaller feet than Hobbits). However, they do not all run at the same speed. They start out counting in synch, 0, 1, 0, 1, but then thread 21 gets a burst of speed, almost finishes the iterations, and unfortunately it hands the baton to thread 20, who fumbles it. Thread 23 jumps in and tries to finish the race, but is eventually beaten by thread 21. That's about as exciting as the CCR gets.

Similar to what you saw earlier with `Spawn`, there is also a `SpawnIterator` method provided by DSS. An example is included in CCRExamples as follows:

```
// A slightly different iterator with a parameter for SpawnIterator
private static IEnumerator<ITask> SpawnIteratorExample(int n)
{
    Port<int> p1 = new Port<int>();
    Port<int> p2 = new Port<int>();

    p1.Post(n);

    bool done = false;
```

```
        while (!done)
        {
            yield return Arbiter.Receive(false, p1, delegate(int i)
            {
                    Console.WriteLine("P1 Thread {0}: {1}", ↵
Thread.CurrentThread.ManagedThreadId, i);
                    p2.Post(i + 1);
            });

            yield return Arbiter.Receive(false, p2, delegate(int i)
            {
                Console.WriteLine("P2 Thread {0}: {1}", ↵
Thread.CurrentThread.ManagedThreadId, i);
                    if (i >= n+10)
                        done = true;
                    else
                        p1.Post(i + 1);
            });
        }

        yield break;
}
```

This code differs slightly from the previous code because the iterator accepts a parameter, which is the number to start counting from.

The corresponding code to launch the iterator is as follows:

```
SpawnIterator<int>(5, SpawnIteratorExample);
Console.WriteLine("Spawn Iterator executed ...");
```

Notice that the call to `SpawnIterator` has two parameters, and that the type of the first parameter is declared using a generic `<int>`. `Spawn` and `SpawnIterator` each have three overloads that accept one, two, or three parameters, which are passed to the handler. There are several similar overloads for common `Arbiter` APIs so that you can pass parameters to handlers.

When the code is executed (select the SpawnIterator example in CCRExamples), the output should be as follows. Note that again a single thread is used in all cases because `SpawnIterator` uses the default DSS dispatcher and there are only two threads in the pool:

```
Spawn Iterator executed ...
P1 Thread 8: 5
P2 Thread 8: 6
P1 Thread 8: 7
P2 Thread 8: 8
P1 Thread 8: 9
P2 Thread 8: 10
P1 Thread 8: 11
P2 Thread 8: 12
```

*(continued)*

**49**

*(continued)*

```
P1 Thread 8: 13
P2 Thread 8: 14
P1 Thread 8: 15
P2 Thread 8: 16
```

You need to become familiar with iterators because they are used extensively in the MRDS sample code.

## Ports and Messages

The most important class in the CCR is the Port, which is basically a First-In-First-Out (FIFO) queue for *messages*.

Messages, also referred to as *requests* or *responses*, are just objects of a specified type. You can create your own classes and send instances of these classes as messages. (In fact, this is how services work, and their APIs are defined in terms of the *operations* they can perform, which are implemented using different classes of messages. There is more on this in the next chapter.)

Consider the following simple port, for example:

```
Port<int> intPort = new Port<int>();
intPort.Post(42);
```

In the preceding code snippet, the Port class creates a new port that accepts only integers. Because Port is a generic class, you can just as easily create a port for strings, doubles, or even more complex objects such as lists or classes that you define yourself.

This code creates a new port called intPort, and sends the value 42 to it as a message using the Post method. Because intPort has an explicit type of int, you can only *post* (enqueue) integer values to it. If you attempt to post a string, then you will get a compilation error. This type-checking helps to eliminate subtle bugs.

The posted value of 42 remains in the port queue until it is dequeued either by being explicitly read or by a *receiver*. (Receivers are covered in the next section.)

If messages are never removed from the port, then they just keep accumulating, which poses a potential memory leak. You can empty a port by calling its Clear method. It is also possible to set policies that prevent the build-up of messages in a port (as explained later).

---

### Caution on Modifying Messages

You might assume that once you post a message, you can forget about it and dispose of the original variable or reuse it. That would be a bad mistake! The memory for the message is still in use in the port queue until the message has been processed.

When posting messages in a loop, you have to create a new instance of the message each time around the loop. If you are concerned about memory usage, then you can dispose of the message in the handler once you have finished with it. Otherwise, the CLR

---

> garbage collector will eventually determine that the message has no remaining references and reclaim the memory.
>
> Another common mistake is modifying the contents of a message object (usually a class instance) before it has been received. This affects the message that has already been sent. The simple solution is to always create new messages.

The major advantage of ports is that if you have a handle to a port — for example, if it is a global variable — then you can post messages to it from any thread and it will always be a safe operation. Furthermore, if all the receivers are busy, the message simply waits until it can be processed. The sender of the message does not have to wait because posting does not block (except for certain policies).

## Reading from a Port

To see what the port contains (for debugging purposes), you can use the `ToString` method:

```
// Display the state of the port
Console.WriteLine(intPort.ToString());
```

A port has an `ItemCount` property that can be used to determine how many items are queued. However, if you want to see whether an item is available and retrieve it (if there is one), you can use the `Test` method, which returns the item as an object, or `null` if there is no item available. An overloaded version returns a Boolean result and passes back the item (or `null` if there is no item) as an `out` parameter. Note that `Test` removes the item atomically, i.e., in a thread-safe manner, so that multiple threads cannot inadvertently remove the same item.

```
int j;
// Use Test to read from the port (or fail)
if (intPort.Test(out j))
    Console.WriteLine("Got value: " + j);
else
    Console.WriteLine("No value!");
```

This is a straightforward way to use a port as a FIFO queue that can be fed data from anywhere in the application. However, your code would have to continually poll the port to determine whether new items were available — for example, by using timer events. This is not the way that ports are usually used; it is more common to set up receivers that execute tasks automatically when messages arrive.

If you want to flush a port, you can execute a loop that keeps removing items using `Test` until no more are available. However, it is much more efficient just to use `Clear`.

Here is a complete example that uses a port and the `Test` method. You should be able to follow it based on the discussion so far. Enter the code into your service just below the `Start` method if you are building the code yourself, and then put a call to `PostAndTest` inside the `Start` method. If you are using the CCRExamples service, you can run the service and select the Post and Test item from the menu (refer to Figure 2-2).

```
// Post messages and use Test to retrieve them
void PostAndTest()
{
    int i, j;
    // Create a new integer port
    Port<int> intPort = new Port<int>();
    // Post the answer to life the universe and everything
    intPort.Post(42);
    Console.WriteLine("Posted a value");

    // Display the state of the port
    Console.WriteLine(intPort.ToString());

    // Use Test to read from the port (or fail)
    if (intPort.Test(out j))
        Console.WriteLine("Got value: " + j);
    else
        Console.WriteLine("No value!");

    // View the port status again
    Console.WriteLine(intPort.ToString());
    // Try Test a second time with nothing to read
    if (intPort.Test(out j))
        Console.WriteLine("Got value: " + j);
    else
        Console.WriteLine("No value!");
    // Final status
    Console.WriteLine(intPort.ToString());

    // Pause for user to read the output
    PressEnter();
}

void PressEnter()
{
    Console.ForegroundColor = ConsoleColor.Yellow;
    Console.Write("Press Enter: ");
    Console.ResetColor();
    Console.ReadLine();
}
```

The `PressEnter` function is just a convenience to make the code pause. However, due to the asynchronous nature of the CCR, it does not always function as expected! In some examples, the "Press Enter" text will be interleaved with other text on the screen because the code is multi-threaded. (This is why it is displayed in yellow on the screen — so it stands out.)

When you run this code, you should see output similar to the following in the command window:

```
Posted a value
Port Summary:
    Hash:1424
    Type:System.Int32
    Elements:1
    ReceiveThunks:0
```

```
Receive Arbiter Hierarchy:

Got value: 42
Port Summary:
    Hash:1424
    Type:System.Int32
    Elements:0
    ReceiveThunks:0
Receive Arbiter Hierarchy:

No value!
Port Summary:
    Hash:1424
    Type:System.Int32
    Elements:0
    ReceiveThunks:0
Receive Arbiter Hierarchy:

Press Enter:
```

Notice that after posting to the port, the number of elements is 1 under `Port Summary`. After `Test` has executed, the number of elements is 0. On the second attempt to use `Test`, the code displays the message "No value!"

It is worth pointing out here that the number of `Receive Thunks` (a thunk is a chunk of receiver code — that is, the handler in a task) is zero in all cases, and nothing is listed under `Receive Arbiter Hierarchy`. This is because no receivers are specified in this example.

## PortSets

A port can process only a single data type. Another class, called a `PortSet`, aggregates several different types of messages. In effect, it is a bunch of message queues that can all be treated as a single entity.

If you open `CCRExamplesTypes.cs`, you will find the definition of the *operations port* for the service, which contains three types of messages: `DsspDefaultLookup`, `DsspDefaultDrop`, and `Get`. (This is the minimum set of messages required for a service, explained in the next chapter.)

```
/// <summary>
/// CCRExamples Main Operations Port
/// </summary>
[ServicePort()]
public class CCRExamplesOperations : PortSet<DsspDefaultLookup, ↩
    DsspDefaultDrop, Get>
{
}
```

Multiple messages of different types can be posted to a `PortSet`. Each message type can have a different handler, or you can create complex combinations of messages that must arrive to trigger a particular handler.

Note that it doesn't make sense to declare a `PortSet` with the same data type appearing in the list twice. If you have two possible sets of integer values, for example, you should use different enums or wrap them in different classes.

---

### Limits on Declaring Portsets

Note that there are some inherent limitations to creating a `PortSet` declaratively. The CCR only allows up to 20 types in a `PortSet` for what is called the *desktop CLR*, i.e., the usual .NET environment. For the .NET Compact Framework (CF) environment, this is restricted to only eight because of a limitation in the Just-In-Time (JIT) compiler.

Novices sometimes run up against these limits because they add new operations to an existing service, thereby breaking it. It is particularly easy to exceed the limit for CF services.

However, there is a solution: Create the `PortSet` programmatically using `typeof`.

---

Looking again at the preceding example, the code can be rewritten as follows so that the `PortSet` is actually created at runtime as part of the class constructor:

```
/// <summary>
/// CCRExamples Main Operations Port
/// </summary>
[ServicePort()]
public class CCRExamplesOperations : PortSet
{
    public CCRExamplesOperations()
        : base(
            typeof(DsspDefaultLookup),
            typeof(DsspDefaultDrop),
            typeof(Get)
        )
    { }
}
```

However, there is still a problem with this approach. By creating the `PortSet` dynamically, there is no way to have the strong type-checking that is usually applied at compile-time when you try to post to a port.

The workaround is quite simple: Define overloads for the `Post` method for each of the different types (inside the `CCRExamplesOperations` class definition):

```
public void Post(DsspDefaultLookup msg)
{
    base.PostUnknownType(msg);
}
public void Post(DsspDefaultDrop msg)
{
    base.PostUnknownType(msg);
}
public void Post(Get msg)
{
    base.PostUnknownType(msg);
}
```

Notice that these overloads use the `PostUnknownType` method. This method looks up the list of acceptable types at runtime and posts to the associated port in the `PortSet`. If there is no match, then it throws an exception. You can also use `TryPostUnknownType` in your code, which returns a Boolean value indicating whether the post was successful or not, but doesn't cause an exception.

Having seen how posting messages works, it is now time to look at the other side — how messages are usually removed from a port.

## Receivers and Arbiters

Strictly speaking, a port has a second queue, which is a list of *receivers* to be executed to retrieve and process messages. These receivers are sometimes called *continuations* and are conceptually similar to *callback procedures* in that they execute asynchronously when some event occurs — in this case the arrival of a message (or multiple messages). Much of the coding in MRDS services involves building receivers and the *handlers* that execute when messages are received.

You have already seen how messages are posted and can be extracted using `Test`. Now consider the case of a receiver. To use a receiver, you must construct it and then attach it to the port. This can be done most easily using wrappers that DSS and CCR provide:

```
// Create a Receiver and place it in the Dispatcher Queue
// This one is NOT persistent so it only fires once
Activate(
    Arbiter.Receive(false, intPort,
        delegate(int n)
            { Console.WriteLine("Receiver 1: " + n.ToString()); }
    )
);
```

The `Activate` method is defined in `DsspServiceBase` and wraps an `Arbiter.Activate` to use the default DSS dispatcher queue, which is `Environment.TaskQueue`. You can also create your own dispatcher if you want.

The `Arbiter.Receive` constructs a receiver using the following parameters:

❑   A persistent flag, which is `false` in the preceding code, i.e., non-persistent — it only receives once and then is removed

❑   A port called `intPort` to read from

❑   A delegate that is executed in-line

If the first parameter is `true`, then the receiver is persistent and will remain in the port's receiver list after processing a message (unless it is explicitly removed later). Many of the receivers that you create to control a robot need to be persistent.

The `intPort` is the same one declared in the section "Ports and Messages" earlier in the chapter, and it accepts integers as messages. Because of this, the delegate must expect to receive an integer as a parameter, which in this case is declared as n.

You should realize by now that immediately after calling `Activate` execution continues with the next statement. The receiver quietly waits until an integer is posted to the `intPort`, or, if one has already been posted, then the receiver will fire straight away and process the value. Either way, it is quite likely that the receiver will execute on a different thread from the one that executed the `Activate`. This is an important concept, and you should be sure that you understand it.

Putting this in the context of a complete example, here is the code for the Receive example in CCRExamples:

```
// Receive a message on a port
void Receive()
{
    // Create a port
    Port<int> intPort = new Port<int>();

    // Create a Receiver and place it in the Dispatcher Queue
    // This one is NOT persistent so it only fires once
    Activate(
        Arbiter.Receive(false, intPort,
            delegate(int n)
                { Console.WriteLine("Receiver 1: " + n.ToString()); }
        )
    );

    // Add another receiver, but make it persistent
    Activate(
        Arbiter.Receive(true, intPort,
            delegate(int n)
                { Console.WriteLine("Receiver 2: " + n.ToString()); }
        )
    );

    Wait(100);
    Console.WriteLine("Receivers Activated:\n" + intPort.ToString());

    // Post a message
    // We could do this even before the receivers were
    // activated because it will just stay in the queue.
    // However, we want to see the effect on the receiver
    // queue in the port.
    intPort.Post(10);
    Wait(100);
    Console.WriteLine("After First Post:\n" + intPort.ToString());

    // Wait a while then post another message
    Wait(100);
    intPort.Post(-2);
    // The Receiver is still active so wait for it to process
    // the second message as well
    Wait(100);
    // And finally a third message ...
    intPort.Post(123);
    Wait(100);
}
```

The preceding example creates two receivers: the first is nonpersistent and the second is persistent. Then it posts three messages to the port. It shows what happens to the state of the port by displaying port summaries. The output is shown here:

```
Receivers Activated:
Port Summary:
    Hash:1424
    Type:System.Int32
    Elements:0
    ReceiveThunks:2
Receive Arbiter Hierarchy:
Receiver`1(Onetime) with method unknown:<Receive>b__10 nested under
    none
Receiver`1(Persistent) with method unknown:<Receive>b__11 nested under
    none

Receiver 1: 10
After First Post:
Port Summary:
    Hash:1424
    Type:System.Int32
    Elements:0
    ReceiveThunks:1
Receive Arbiter Hierarchy:
Receiver`1(Persistent) with method unknown:<Receive>b__11 nested under
    none

Receiver 2: -2
Receiver 2: 123
```

In the first port summary, there are two receivers in the receiver list. After Receiver 1 fires, there is only one receiver left. This second receiver is persistent, as you can see from the two messages that it displays in response to integers being posted to the port.

Persisted handlers are vital to service-oriented applications. Consider a web server. It waits for HTTP requests to come in on a port (this is TCP/IP port 80, which is not the same as a CCR port) and then in response to this request it sends back a web page. This cycle of request/response (possibly with an error) is fundamental to services. When you learn about the DSS in the next chapter, you will see that this is the pattern of what is called a *service operation*.

## Choice

The Choice arbiter, which you can think of as a logical OR, waits on two receivers until one of them fires. It then shuts down the unused receiver.

One common use of Choice is to handle responses from a service. The response can be either the requested data or an exception, which is returned as a SOAP Fault. In this case, the Choice has two receivers that correspond to success or failure of the operation.

The following example from CCRExamples shows how `Choice` works. You can run it by selecting the `Choice` example from the menu.

```
// Choice -- Choose the first message to arrive (Logical OR)
void Choice()
{
    // Create a PortSet that takes two different data types
    PortSet<bool, int> ps = new PortSet<bool, int>();

    // If you post the messages BEFORE setting up the Choice,
    // you might see a different result compared to posting
    // them AFTER it has been created
    //ps.Post(1000000);
    //ps.Post(true);

    // Create the Choice and activate it
    Activate(
        Arbiter.Choice<bool, int>(ps,
        // Create delegates for each type in the PortSet
            delegate(bool b)
            { Console.WriteLine("Choice: " + b.ToString()); },
            delegate(int n)
            { Console.WriteLine("Choice: " + n.ToString()); }
        )
    );

    // Post two messages - Only one will be selected
    // NOTE: If you run this often enough, you might see
    // cases where either of these is displayed because the
    // result is indeterminate if there are messages of both
    // types in the port when the Choice executes
    ps.Post(1000000);
    ps.Post(true);
}
```

When you run this code, one value displays. As the preceding code shows, the usual result is 1000000, but it is possible for a value of `true` to be displayed depending on subtle timing variations in the CCR. If you uncomment the `Post` instructions at the top of the routine and remove the Post calls at the bottom, you will most likely find that the behavior is reversed. `Choice` does not guarantee which of the ports it will choose from if two messages are waiting when it is evaluated.

In the V1.5 Refresh, Microsoft introduced a new concise syntax for using `Choice` that can be used inside iterators. As you should understand by now, the `yield return` statement can wait on a message to be received. Many of the operations you can perform on services are declared so that they return a `PortSet`. For example, saving the state of a service (covered in Chapter 3) can now be done as follows:

```
yield return (Choice)SaveState(_state);
```

The `SaveState` method returns a `PortSet` that can receive either a Default Replace Response or a `Fault`. You should really check to ensure that it was successful — i.e., did not return a `Fault` — it is unusual for this to fail in a service that has been properly debugged. However, this single line of code does suspend execution until the `SaveState` has completed, which is important if you are trying to save the state as part of the shutdown process in your service's `Drop` handler, for example.

A slightly longer example shows another way to use it:

```
PortSet<SuccessResult, Exception> successResultPort = SomeFunction();

yield return (Choice)successResultPort;
SuccessResult s = successResultPort;
if (s == null)
    Console.WriteLine("Exception:" + (Exception)successResultPort);
```

Success/Failure ports are discussed later in the chapter under the section "Error Handling," but it suffices to say that SomeFunction returns a PortSet that can have two possible results: SuccessResult or Exception. Presumably the function kicks off some tasks that will eventually post a result. This use of Choice does not require delegates and is much cleaner.

The example also demonstrates how a message can be extracted from a PortSet implicitly. When the variable s is initialized, it receives a SuccessResult message if the PortSet is holding one. (You know that there is some type of message in the PortSet because of the preceding Choice.) If there is no SuccessResult, then the code grabs the Exception from the PortSet and displays it.

Note that Choice does not persist. In fact, it performs a *teardown* as soon as one of the receivers executes. This actually prevents a second message from arriving, although in the binary case with only two message types it is hard to see how both types of message could arrive unless there was a bug in the code. The Choice class can handle an arbitrary number of receivers, but you will most commonly see it with only two.

## Join

Similar to a logical AND, a JoinReceive arbiter waits for two receivers to complete before continuing. In terms of synchronization, this allows a "rendezvous" point in the code where two tasks must both complete before the rest of the code is executed.

Consider the Join example from CCRExamples:

```
// Join -- Wait for two messages (Logical AND)
void Join()
{
    // Set up three different ports
    Port<bool> p1 = new Port<bool>();
    Port<int> p2 = new Port<int>();
    Port<string> p3 = new Port<string>();

    // Join on ports p1 and p2
    Arbiter.Activate(Environment.TaskQueue,
        Arbiter.JoinedReceive(
            false, p1, p2,
            delegate(bool b, int i)
            {
                Console.WriteLine("Join 1: {0} {1}", b, i);
                // Now post to p2 so the other Join can complete
```

*(continued)*

*(continued)*

```
                        p2.Post(i+1);
                }
            )
        );

        // Join on ports p2 and p3
        Arbiter.Activate(Environment.TaskQueue,
            Arbiter.JoinedReceive(
                false, p2, p3,
                delegate(int i, string s)
                {
                    Console.WriteLine("Join 2: {0} {1}", i, s);
                    // Now post to p2 so the other Join can complete
                    p2.Post(i-1);
                }
            )
        );

        // Now post to the ports
        // NOTE: It is not possible to tell which Join will be
        // executed because it depends how quickly the messages
        // arrive. In general, all three messages would not be
        // sent at the "same time" as in the code below.

        p1.Post(true);
        p3.Post("hello");
        p2.Post(99);
    }
```

When you run this example, the output might look like the following:

```
Join 2: 99 hello
Join 1: True 98
```

Alternately, the result might be as follows:

```
Join 1: True 99
Join 2: 100 hello
```

Try executing the example repeatedly; you should see both behaviors. It all depends on where the CCR is up to in its cycle of evaluating receivers for execution (because three messages arrive almost simultaneously); it is therefore nondeterministic.

Notice here that there is contention for messages on port p2. However, both of the Joins eventually run as long as enough messages are posted.

## Combining Arbiters

Arbiters can be nested. This means that you can place a Choice inside a Join, or vice versa, which enables you to create arbitrarily complex logic. This chapter does not provide any examples of this, but you can play around with it on your own by combining the simple examples shown previously.

Bear in mind that a `Choice` atomically removes all nested arbiters from their ports once it is satisfied. Consider this when designing your logic.

## Receiving Multiple Messages

A common requirement is to accumulate a number of messages before proceeding to execute the next step in the code. The CCR provides two different ways to do this:

❑ Multiple Item Receive

❑ Multiple Port Receive

The `Join` discussed previously is considered *static* because it only takes two ports defined at compile-time. You can also create a *dynamic* `Join` by waiting for multiple messages to arrive on a port.

The following example shows how to read six messages from a port in one atomic operation. This is a persistent receiver, so it groups messages together in clumps of six items and displays them:

```
// Receive multiple messages of the same type
void MultipleItemReceive()
{
    Port<int> p = new Port<int>();

    Arbiter.Activate(Environment.TaskQueue,
        Arbiter.MultipleItemReceive(
            true, p, 6,
            delegate(int[] array)
            {
                string s = "";
                for (int i = 0; i < array.Length; i++)
                    s += array[i].ToString() + " ";
                Console.WriteLine("{0} Items: {1}", array.Length, s);
            }
        )
    );

    for (int i = 0; i < 4; ++i)
        p.Post(i + 1);
    Wait(100);

    Console.WriteLine(p.ToString());

    for (int i = 0; i < 4; ++i)
        p.Post(i + 1);
    Wait(100);

    Console.WriteLine(p.ToString());

    for (int i = 0; i < 4; ++i)
        p.Post(i + 1);
    Wait(100);

    Console.WriteLine(p.ToString());
}
```

Notice that the code posts messages in three batches of four at a time (for a total of 12 messages), so the receiver fires partway through the second batch. The output looks like the following if you run the MultipleItemReceive example in CCRExamples:

```
Port Summary:
    Hash:1424
    Type:System.Int32
    Elements:4
    ReceiveThunks:1
Receive Arbiter Hierarchy:
JoinSinglePortReceiver(Persistent) with method unknown: ↵
<MultipleItemReceive>b__1c nested under
    none

6 Items: 1 2 3 4 1 2
Port Summary:
    Hash:1424
    Type:System.Int32
    Elements:2
    ReceiveThunks:1
Receive Arbiter Hierarchy:
JoinSinglePortReceiver(Persistent) with method unknown: ↵
<MultipleItemReceive>b__1c nested under
    none

6 Items: 3 4 1 2 3 4
Port Summary:
    Hash:1424
    Type:System.Int32
    Elements:0
    ReceiveThunks:1
Receive Arbiter Hierarchy:
JoinSinglePortReceiver(Persistent) with method unknown: ↵
<MultipleItemReceive>b__1c nested under
    none
```

Three port summaries are displayed so that you can see what is happening. The first summary shows that there are four messages (elements) in the port and one receiver (thunk), which is a Multiple Item Receiver. Another four messages are posted and, before the second port summary can be displayed, the receiver outputs the values of six messages. Two items are left when the second summary is displayed. The last four messages are posted, and the receiver picks them up for output, and then the port summary shows that there are no messages remaining to be processed.

The MultipleItemReceive can only receive from a single port, unlike the JoinReceive shown earlier. If you want to use different ports, then you need to use a Multiple Port Receiver instead.

An example of the Multiple Port Receiver is also included in CCRExamples. It looks like this:

```
// Receive multiple messages of the different types
void MultiplePortReceive()
{
    // Create a port set that accepts two different data types
```

```
PortSet<int, double> pSet = new PortSet<int, double>();

//Arbiter.Activate(Environment.TaskQueue,
Activate(
    Arbiter.MultipleItemReceive(
        pSet, 10,
        delegate(ICollection<int> colInts, ICollection<double> colDoubles)
        {
            Console.Write("Ints: ");
            foreach (int i in colInts)
                Console.Write(i + " ");
            Console.WriteLine();
            Console.Write("Doubles: ");
            foreach (double d in colDoubles)
                Console.Write("{0:F2} ", d);
            Console.WriteLine();
        }
    )
);

// Generate some random numbers and post them using
// different message types depending on their values
Random rnd = new Random();
for (int i = 0; i < 10; ++i)
{
    double num = rnd.NextDouble();
    if (num < 0.5)
        pSet.Post((int)(num * 100));
    else
        pSet.Post(num);
}
}
```

Although this is referred to as a Multiple Port Receiver, it actually uses the `MultipleItemReceive` but with a `PortSet`. This particular `PortSet` accepts integers and doubles. Notice that the delegate has two parameters, which are collections of the corresponding types. The piece of code at the bottom of the example randomly generates a set of integers and doubles.

Every time you run the example, you will get a different result. Two examples of the output are shown here:

```
Ints: 36 25 15
Doubles: 0.74 0.71 0.87 0.81 0.51 0.51 0.79
Ints: 23 17 34 18 45 40
Doubles: 0.53 0.56 0.96 0.59
```

Note that there are different numbers of items in each of the collections, but the total number of items is ten in both cases.

## Interleave

Services typically offer a large number of different operations. As you will see in Chapter 3, these are defined as a set of classes, called *operations*, which are the message types you can send to a service. Each of these message types needs to have a corresponding receiver.

Setting up multiple receivers is the job of an `Interleave`. It also provides important facilities for protecting resources and housekeeping by using three groups of receivers: Tear Down, Exclusive and Concurrent:

❑ The Tear Down group contains receivers that should be called when the `Interleave` is supposed to shut down. The processing of messages stops as soon as a Tear Down receiver is executed; and once it completes, the entire `Interleave` is disposed, effectively cancelling all of the receivers. Teardown messages take priority.

❑ The Concurrent receiver group is easy to understand: It operates as if you had simply set up a bunch of receivers yourself. They can all run in parallel (if there are enough threads).

❑ The Exclusive receiver group is the key, however. It only allows a single receiver to execute at a time. It waits for an executing Concurrent receiver to finish before starting an Exclusive receiver. If more Exclusive receivers are triggered before the first one finishes, then they are queued and executed one after another. Concurrent receivers cannot execute while an Exclusive receiver is running.

There is a subtle issue here in that the actual messages are held by the `Interleave`, which is said to be *guarding* the ports. You will not see the messages queued in the individual ports in a `PortSet` if it is controlled by an `Interleave`. They are released one at a time if they are Exclusive and remain in the pending exclusive requests queue in the `Interleave`.

Exclusivity is an important concept for services. In particular, services usually contain an internal *state*. This state contains information that controls the operation of the service, and it must remain consistent. If several threads updated the state at the same time, there is no telling what might happen. Therefore, any handler that wants to update the state should be Exclusive.

Reading the state, however, can be done concurrently because a Concurrent handler can never run at the same time as an Exclusive handler, and it will therefore always see a consistent view of the state.

An `Interleave` persists in the sense that it contains a set of receivers. However, the individual receivers in an `Interleave` can be persistent or nonpersistent. In general they are persistent. Notice in the following example that the `DropHandler` in the `TeardownReceiverGroup` is nonpersistent. It wouldn't make sense for this to be persistent because once you drop a service, it is gone!

```
Activate(
    Arbiter.Interleave(
        new TeardownReceiverGroup(
            Arbiter.Receive<DsspDefaultDrop>(false, _mainPort, DropHandler)
        ),
        new ExclusiveReceiverGroup(
            Arbiter.Receive<LaserRangeFinderResetUpdate>(true, _mainPort, ↵
LaserRangeFinderResetUpdateHandler),
            Arbiter.Receive<LaserRangeFinderUpdate>(true, _mainPort, ↵
LaserRangeFinderUpdateHandler),
            Arbiter.Receive<BumpersUpdate>(true, _mainPort, ↵
BumpersUpdateHandler),
            Arbiter.Receive<BumperUpdate>(true, _mainPort, ↵
BumperUpdateHandler),
            Arbiter.Receive<DriveUpdate>(true, _mainPort, ↵
DriveUpdateHandler),
```

```
                        Arbiter.Receive<WatchDogUpdate>(true, _mainPort, ←
WatchDogUpdateHandler)
                    ),
                    new ConcurrentReceiverGroup(
                        Arbiter.Receive<Get>(true, _mainPort, GetHandler),
                        Arbiter.Receive<dssp.DsspDefaultLookup>(true, _mainPort, ←
DefaultLookupHandler)
                    )
                )
            );
```

Services sometimes use one `Interleave` during service initialization, and then a different `Interleave` once they are fully up and running. In this case, the receivers in the initialization `Interleave` might not be persistent.

Alternatively, once you have set up an `Interleave`, you can add to it using its `CombineWith` method. The `DsspServiceBase` class automatically creates a `MainPortInterleave` for you. The following code shows how some receivers can be added without affecting existing receivers. Note that no receivers are specified in this case for the `TeardownReceiverGroup` or the `ConcurrentReceiverGroup` but you still have to supply a parameter in this construct.

```
        MainPortInterleave.CombineWith(
            new Interleave(
                new TeardownReceiverGroup(),
                new ExclusiveReceiverGroup(
                    Arbiter.Receive<simengine.InsertSimulationEntity>(true, ←
_notificationTarget, InsertEntityNotificationHandler),
                    Arbiter.Receive<simengine.DeleteSimulationEntity>(true, ←
_notificationTarget, DeleteEntityNotificationHandler),
                    Arbiter.Receive<FromWinformMsg>(true, _fromWinformPort, ←
OnWinformMessageHandler)
                ),
                new ConcurrentReceiverGroup()
            )
        );
```

As a final example, consider the case where you have a sequence of tasks that must be executed as a unit and not preempted. You can add them all as Exclusive receivers. The first one in this multi-step process is executed when a new message arrives. Just before it completes, it posts a message to the second one, and so on. Each task passes control to the next one. No Concurrent receivers can execute while an Exclusive receiver is ready to run. However, other Exclusive tasks can sneak into the queue.

## *Dispatchers and Dispatcher Queues*

Dispatchers and dispatcher queues enable you to create as many thread pools as you like and to assign different priorities and scheduling policies. This is in stark contrast to the .NET CLR, which operates with a single thread pool (if you use `System.Threading`).

A dispatcher manages a pool of threads, and it can have multiple dispatcher queues feeding into it. The items in a dispatcher queue are (usually) the combination of a handler and some data from a port. Although dispatchers can handle any number of dispatcher queues, quite often a dispatcher will only have one queue.

When you create a DSS service, the base class (DsspServiceBase) creates a default dispatcher and dispatcher queue for you. Consequently, you do not usually have to worry about creating these objects or managing them when you are writing a DSS service.

The number of threads managed by a dispatcher can be specified at construction time. Unlike the CLR thread pool, this pool of threads will not automatically grow or shrink during the lifetime of the dispatcher. (This helps to make the CCR more efficient but at the cost of possible bottlenecks if you execute too many long-running tasks and soak up all of the threads.)

The following code shows the constructor signatures for the dispatcher class obtained using reflection. You can do this for any of the classes in MRDS by typing the class name into your code, right-clicking it with the mouse, and selecting Go To Definition from the pop-up menu. This is a useful feature. Alternatively, you can use the Object Browser.

```
namespace Microsoft.Ccr.Core
{
    public sealed class Dispatcher : IDisposable
    {
        public Dispatcher(int threadCount, string threadPoolName);
        public Dispatcher(int threadCount, ThreadPriority priority, ↵
bool useBackgroundThreads, string threadPoolName);
        public Dispatcher(int threadCount, ThreadPriority priority, ↵
DispatcherOptions options, string threadPoolName);
        public Dispatcher(int threadCount, ThreadPriority priority, ↵
DispatcherOptions options, ApartmentState threadApartmentState, ↵
string threadPoolName);
        ...
```

By default, the number of threads is the number of CPUs, or CPU cores if it is a multi-core CPU; but the minimum number of threads is two, even for a single CPU. (You get the default number of threads when you specify zero in the constructor.)

If you have a resource that must only be accessed by one thread at a time, then you can use just a single thread in a dispatcher specifically for this purpose. Because there is only one thread, it doesn't matter how many tasks are queued — only one task can execute at a time. This is an issue, for example, if you have legacy code that uses a Single-Threaded Apartment (STA) model. Windows Forms is a case in point, and running forms requires special consideration, a subject covered in Chapter 4.

Dispatchers and dispatcher queues are relatively lightweight, so there is no real penalty in having several of them. However, it is unlikely that you will need more than one for most applications.

One scenario in which you might want multiple dispatchers is when you want to set different thread priorities. All threads in a dispatcher have the same priority, but you can have multiple dispatchers. ThreadPriority is an enum in the System.Threading namespace. You can find the values and their descriptions using IntelliSense, but for ease of reference they are listed in the following table:

## Thread Priority Values

| Value | Description |
|-------|-------------|
| Lowest | Threads can be scheduled after threads with any other priority. |
| BelowNormal | Threads can be scheduled after threads with Normal priority but before threads with Lowest priority. |
| Normal | Threads can be scheduled after threads with AboveNormal priority but before threads with BelowNormal priority. (This is the default.) |
| AboveNormal | Threads can be scheduled before threads with Normal priority but after threads with Highest priority. |
| Highest | Threads can be scheduled before threads with any other priority. |

The `threadPoolName` is useful if you use the Threads window of the Visual Studio Debugger because you can use it to identify threads from particular dispatchers.

There are a few other properties defined for dispatchers that you might use occasionally:

❑ `WorkerThreadCount` can be used to examine or change the number of threads in the pool.

❑ `ProcessedTaskCount` tells you how many tasks are queued for processing.

❑ `PendingTaskCount` keeps track of how many tasks have been executed since the dispatcher was created.

A dispatcher queue is used to queue tasks. If you are wondering why this functionality is not integrated into the dispatcher, it's because this enables queues to be defined with separate scheduling policies. These policies are defined in an `enum` called `TaskExecutionPolicy`. Descriptions of each of the policies are given in the following table and should be self-explanatory. `Unconstrained` is the default policy. The other policies involve either queue depth or scheduling rate, and there are two possibilities: discard tasks or force threads attempting to submit new tasks to wait.

## Dispatcher Task Execution Policy Values

| Value | Description |
|-------|-------------|
| Unconstrained | All tasks are queued with no constraints (the default). |
| ConstrainQueueDepthDiscardTasks | Tasks enqueued after the maximum depth are discarded. |
| ConstrainQueueDepthThrottleExecution | Maximum depth is enforced by putting posting threads to sleep until the queue depth falls below the limit. |

*Table continued on following page*

| Value | Description |
|---|---|
| ConstrainSchedulingRateDiscardTasks | Tasks enqueued while the average scheduling rate is above the limit are discarded. |
| ConstrainSchedulingRateThrottleExecution | Once the average scheduling rate exceeds the limit, posting threads are forced to sleep until the rate falls below the limit. |

# Implementing Common Control Structures

Most programmers look for patterns or templates that they can use in their coding. This section provides a number of patterns that are commonly used in MRDS programming.

## Sequential Processing

Executing a sequence of operations is quite easy with an iterator, as you have already seen. However, sometimes you might want to spin off a separate task but you need to wait for it to finish. You can take two approaches to this:

❑   Create a completion port to wait on and have the other task send a message when it is finished.

❑   Use the ExecuteToCompletion arbiter.

Using a completion port that you create yourself is quite easy and you should be able to do that at this point. It doesn't matter what type of port it is because you will only send a single message to it to indicate to the first task that the second task has finished. In fact, you can set up a string of receivers, with each of them referring to the next handler in the sequence. The handlers can "daisy chain" from one to another by sending a message to the completion port when they have finished. This multi-step process can call a mixture of simple handlers or iterators.

The ExecuteToCompletion arbiter works as shown in the next example. A parent iterator yields to the ExecuteToCompletion arbiter to run a child iterator. In the following code, a new iterator task is explicitly created. You could also use Arbiter.FromIteratorHandler, discussed previously. Notice that in this case a double is passed to the iterator.

```
private IEnumerator<ITask> Parent()
{
    // Do some work
    ...

    // Now call the child
    Console.WriteLine("Yielding to child");
    yield return Arbiter.ExecuteToCompletion(
        taskQueue,
        new IterativeTask<double>(3.1415926f, Child)
    );
```

```
        Console.WriteLine("Child completed");

        // Carry on with our own processing
        ...
}

private IEnumerator<ITask> Child(double number)
{
        // Execute some code
        ...
        // Yield now
        yield return Arbiter.xxx();

        // Execute some more code
        ...
        // Yield again,
        yield return Arbiter.xxx();

        // And so on
        ...
}
```

The Parent function calls the Child, which executes a series of steps. The CCR knows when an iterator is complete, either by reaching the end of the function or via yield break. This terminates the iteration, and the original yield return in the Parent is satisfied.

Note that the Child does not have to be an iterator — you could use Arbiter.FromHandler to create the task. In this case, the yield return in the parent is similar to a normal procedure call except that it provides an opportunity for the CCR to do some scheduling. This helps to break long-running tasks up into bite-sized pieces and share the CPU.

## Scatter/Gather

The term "Scatter/Gather" refers to sending out multiple requests and then waiting for all of the responses to arrive. Sending the requests is the easy part — you just need to use Post. Then you need to wait for the responses from the various operations. Depending on your logic, you might want to wait until one of the operations has completed, all of them have completed, or any combination in between.

The Choice arbiter can wait on several receivers. Examples typically focus on only two options because in most cases you are only interested in getting a response or a fault. This particular form of the Choice uses a PortSet that has only two possible message types. However, an alternate form of Choice accepts an array of receivers to which you can add as many receivers as you like. This enables you to wait for one of many responses.

To wait for all operations to complete, you can use a Multiple Item Receiver or a Multiple Port Receiver. It is possible, for example, to send several requests to the same port, in which case you need to receive multiple items. However, it is more likely that you will want to receive from multiple different ports.

Although it is a little advanced for this chapter, consider the following code from the Synchronized Dance service in Chapter 14 that controls two robots simultaneously:

```
        PortSet<DefaultUpdateResponseType, Fault> p = ↩
new PortSet<DefaultUpdateResponseType, Fault>();
        drive.RotateDegrees[] rotateRequests = ↩
new drive.RotateDegrees[_drivePorts.Count];

        for (i = 0; i < _drivePorts.Count; i++)
        {
            rotateRequests[i] = new drive.RotateDegrees();
            rotateRequests[i].Body.Degrees = _state.RotateAngle;
            rotateRequests[i].Body.Power = _state.RotatePower;
            rotateRequests[i].ResponsePort = p;
            _drivePorts[i].Post(rotateRequests[i]);
        }

        yield return Arbiter.MultipleItemReceive(p, drivePorts.Count, ↩
driveHandler);
```

The code creates a request for each robot and then sends it in the `for` loop. The `ResponsePort` of each request is set to the same `Portset`, `p`. Then the code waits on a response from each robot using the `MultipleItemReceive`. Note that it is waiting on a `PortSet` and the response can be either the default response or a fault for each robot, but there is only one message per robot.

As another example, if your initialization consists of three steps that can execute in parallel, you can spawn the three steps and have each step post a message to a completion port. The startup procedure must wait for all three messages to arrive on the completion port.

## State Machines

There is nothing special about finite state machines (FSMs); they are often used in robotics applications. The SimMagellan example in Chapter 7 and the ExplorerSim example in Chapter 9 use state machines. Some of the MRDS samples also use state machines. Because this is a fundamental concept in computer science, it is not discussed here.

## Last Message

Sometimes messages arrive fast and furious. You might not want to process all of these messages, especially if it takes a long time to process a single message.

You have seen that it is possible to check the `ItemCount` on a port. Later, in Chapter 9, you will encounter the ExplorerSim example, where the robot is processing laser range finder (LRF) information. If it cannot keep up with the data, it needs to throw away some readings and only process the latest message.

The following code fragment shows how ExplorerSim discards old LRF messages:

```
/// <summary>
/// Gets the most recent laser notification. Older notifications are dropped.
/// </summary>
/// <param name="laserData">last known laser data</param>
```

```
        /// <returns>most recent laser data</returns>
        private sicklrf.State GetMostRecentLaserNotification(sicklrf.State laserData)
        {
            sicklrf.Replace testReplace;
            Port<sicklrf.Replace> laserPort = _laserNotify;

            int count = laserPort.ItemCount;

            for (int i = 0; i < count; i++)
            {
                testReplace = (sicklrf.Replace)laserPort.Test();
                if ((testReplace != null) && (testReplace.Body.TimeStamp > ↩
    laserData.TimeStamp))
                {
                    laserData = testReplace.Body;
                }
            }
        ...
```

Notice that this code uses a particular syntax to obtain the `laserPort` from the `PortSet` for laser notifications. This implicit assignment operator is automatically generated by `DssProxy` when it creates the Proxy DLL, and you do not need to worry about it.

When you are using a joystick to drive a robot, there can be a large number of requests to change the drive power in a short period of time. Unfortunately, in this case, the `SetDrivePower` handler is normally set up as an Exclusive receiver as part of the `MainPortInterleave` that DSS creates for you.

This is jumping ahead to the next chapter, but be aware that this is a trap for beginners. You cannot get the count of waiting messages from the `SetDrivePower` port in this case because the messages are held up in the `Interleave` and the queue length on the `SetDrivePower` port will always appear to be zero!

In this case, you must extract messages from the `MainPortInterleave`. You can check how many messages are pending using the `PendingConcurrentCount` and `PendingExclusiveCount` properties, and then you can extract messages using the `TryDequeuePendingTask` method. Alternatively, the handler can post all incoming messages to an internal port that only processes one message at a time, and the internal handler can check the queue length on the internal port. This approach is used in Chapter 17, where services are developed for new robots.

## Time Delays

Perhaps you are familiar with `System.Threading.Thread.Sleep`. Although you can use `Thread.Sleep` in a CCR environment, this is *not* the recommended way to introduce a delay.

In the preceding sample code you might have noticed calls to `Wait` to suspend execution of the code temporarily. This is a convenience routine that is defined in the CCRExamples service. It simply calls `Thread.Sleep`.

If you look in the code, another `Wait` method is commented out that uses the `TimeoutPort`, defined in the `CcrServiceBase` class:

```
/// <summary>
/// Wait for a specified period of time
/// </summary>
/// <param name="millisec">Delay time in milliseconds</param>
/// <remarks>Suspends execution without using Thread.Sleep, but creates
a deadlock.</remarks>
void Wait(int millisec)
{
    // Create OS event used for signalling.
    // By creating a new one every time, you can call this routine
    // multiple times simultaneously because there is no shared resource.
    AutoResetEvent signal = new AutoResetEvent(false);

    // Schedule a CCR Timeout task that will execute in parallel with
    // this method and signal when the Timeout has completed.
    Activate(
        Arbiter.Receive(
            false,
            TimeoutPort(millisec),
            delegate(DateTime timeout)
            {
                // Done. Signal so that execution can continue.
                signal.Set();
            }
        )
    );

    // Block until Timeout completes
    signal.WaitOne();
}
```

The receiver waits for the specified time delay in milliseconds. (There is also an overload for `TimeoutPort` that uses a `TimeSpan`.) It is not a persistent receiver (the first parameter is `false`).

Notice that the delegate does nothing except set the signal because the objective is just to suspend execution for some length of time. The last line of the routine blocks execution and waits for the signal from the delegate.

Unfortunately, when there are only two threads, this can lead to a deadlock. To test this, uncomment the code for `Wait` and comment out the original (one-line) `Wait` method. Make sure that you have commented out the `ActivationSettings` attribute at the top of the code so you don't have six threads. Now recompile the code.

> *If you are lucky enough to have a quad-processor PC, then you should not comment out* `ActivationSettings` *but instead change the number of threads to two. Otherwise, you will get four threads by default and you won't see the deadlock.*

Run the program and select the Spawn example, which is option number 1. What happens? A start message appears and then you are left staring at a blinking cursor. Don't stare for too long because nothing is going to change.

It appears that this fancy version of Wait is not a solution to the problem of blocking a thread — it still has the same effect as Thread.Sleep. The subtle issue here is that if Wait is called while another Wait is active, then two threads are blocked and there are no threads left to execute the delegates and wake up the waiting threads. The main loop calls Wait after executing your selection, and then the Spawn example also executes Wait.

You must be very careful not to deadlock the CCR by soaking up all of the threads. If you know that one of your threads will block, and there is no easy way around it, then you can explicitly specify a separate dispatcher and the number of threads to use on your service.

*For reviewing this chapter, many thanks to George Chrysanthakopoulos, lead developer for MRDS, who says that "yield is your friend."*

Inside an iterator, it is very easy to create a delay directly by using the TimeoutPort in a yield return:

```
// Wait for some time
yield return Arbiter.Receive(
                false,
                TimeoutPort(timeDelay),
                delegate(DateTime timeout) { }
            );
```

Timers set up in this way are not very reliable because they use CLR timers. There is a fair amount of variation in the time intervals that elapse. This should be corrected with new timers in version 2.0 of MRDS. In the meantime, you can search the forum for ways to work around this problem if you need highly accurate timers or timers for very short time periods. You need to use a CcrStopwatch with V1.5. You can find some threads on this topic in the MRDS discussion forum.

## Periodic Events

Executing code periodically is simply a matter of using the TimeoutPort with the persist parameter set to true and specifying the handler to call. Note that the handler must accept a DateTime parameter. This approach results in the handler being called repeatedly. If the handler cannot perform its processing within the specified time period, then the messages back up and the handler will be flat-out running all the time.

Another approach that is probably better is for the handler to call itself back when it has finished processing. For example, the following handler activates a new receiver to call itself as the last statement in the routine, but it is a nonpersistent receiver:

```
static int TimerCounter = 0;

// A handler that calls itself periodically
void PeriodicTimerExample(DateTime dt)
{
    int timeDelay = 1000;

    TimerCounter++;
    if (TimerCounter > 10)
```

*(continued)*

*(continued)*

```
            return;

        Console.WriteLine("Tick {0} ...", TimerCounter);

        // Wait for some time
        Activate(Arbiter.Receive(
            false,
            TimeoutPort(timeDelay),
            PeriodicTimerExample));
    }
```

To get the process started, you must set up an initial receiver. In the following code, the timeout is set to only 10 milliseconds because this is just to get the process going. Note also that the `TimerCounter` is reset to zero first because the timer handler should only run 10 times:

```
// Reset the counter and kick off the timer
TimerCounter = 0;
Activate(Arbiter.Receive(false, TimeoutPort(10), PeriodicTimerExample));
```

When you run this code by selecting Periodic Timer in CCRExamples, the output might look like the following:

```
Tick 1 ...
Select from the following:
0 = Exit
1 = Spawn
2 = Task From Handler
3 = Task From Iterator
4 = SpawnIterator
5 = Post and Test
6 = Receive
7 = Choice
8 = Join
9 = MultipleItemReceive
10 = MultiplePortReceive
11 = Periodic Timer
12 = Causality
Enter a number: Tick 2 ...
Tick 3 ...
Tick 4 ...
Tick 5 ...
Tick 6 ...
Tick 7 ...
Tick 8 ...
Tick 9 ...
Tick 10 ...
```

This might not seem very exciting, but you can make another selection from the menu while the timer is ticking. See what happens then! Do you run out of threads? If so, uncomment the `ActivationSettings` to give yourself six threads and try again.

You can improve the accuracy of a periodic timer by using the parameter, which is a timestamp. If the code remembers the timestamp from the previous invocation, it can subtract the two and figure out the amount of time that actually elapsed. The new delay for the next timeout can be adjusted so that the timer fires on schedule the next time. At least that is the theory — it isn't always reliable.

## Setting Limits with Timeouts

From time to time, you might want to execute an operation that might not succeed but for which there is no indication of failure. For example, if you send a request to a service that is hung, then you will never receive a reply. In this case, you want to use a timeout so that you don't end up waiting forever.

Consider the following code fragment from the Lynx6Arm simulation in Chapter 8:

```
// Wait for notification or time out if nothing is received
yield return Arbiter.Choice(
    Arbiter.Receive(false,
        TimeoutPort(DsspOperation.DefaultLongTimeSpan),
        delegate(DateTime dt)
        { LogError("Timeout waiting for visual entity"); Shutdown(); }),
    Arbiter.Receive<simengine.InsertSimulationEntity>(false,
        notificationTarget,
        delegate(simengine.InsertSimulationEntity ins)
        {
            _entity = (Lynx6ArmEntity)ins.Body;
        })
);
```

The Choice waits on the TimeoutPort and the Simulator notification port. If the simulation entity is successfully created, there is no problem. However, if it is not, the timeout will kick in.

Note that all DSS operations have a TimeSpan property that you can set before you post a request. This specifies the maximum amount of time to wait for a reply. If there is no response within that time period, then a Fault is generated automatically by DSS. This is an alternative approach.

## Asynchronous I/O

Most of the robot implementations provided with MRDS use a serial port for communication. The .NET environment makes programming easy with the SerialPort class. However, serial ports normally operate asynchronously. It is not a good idea to tie up a CCR thread waiting for a read on a serial port, but sometimes this is your only option. (For example, look at the Hemisson robot in Chapter 17.)

Covering how to use a serial port is outside the scope of this chapter. You should look at some of the sample code in MRDS and also study Chapter 17. Be aware, however, that there are many different ways to implement serial communications.

The best example is probably in the Sick Laser Range Finder code. You can find this service in samples\Sensors\SickLRF and you should look at SerialIOManager.cs, which defines a set of operations that it can perform (Open, Close, SetRate, Send) and has a public ResponsePort to which it posts received packets. External code can wait on the ResponsePort for messages, which will be either a Packet or an Exception.

This approach of using two ports is a good way to isolate the asynchronous code. One port is used for sending commands, configuration parameters, and outgoing data, and the other port is used for receiving incoming data and errors. How the service works internally is irrelevant to the CCR.

# Blocking I/O

For long operations that involve blocking I/O requests, such as writing a large file, it is not desirable to tie up a thread. In this case you can create a separate dispatcher and submit a task to it. Alternatively, create a CLR thread and allocate it exclusively for use with the I/O operations. This removes it from the CCR scheduling.

Note that when you create a service (which is covered in Chapter 3), you can specify that you want it to have its own dispatcher, as described earlier in this chapter. This takes it out of contention for threads in the main DSS thread pool. You can do this using an attribute at the top of your service, as shown in the following example:

```
[ActivationSettings(ShareDispatcher=false, ExecutionUnitsPerDispatcher=6)]
[Contract(Contract.Identifier)]
public class CCRExamplesService : DsspServiceBase
{
    ...
```

Note that some blocking operations have Asynchronous Programming Model (APM) implementations that help with multi-threading. These are the classic begin/end operations that can be used with web requests (see the IPCamera.cs sample included in the MRDS distribution) or with file I/O. All that your code needs to do is create a port for signaling the completion of the asynchronous operation, execute the Begin method, and then post a message to this port in the End method (which runs asynchronously). Meanwhile, your CCR code can use a receiver to wait on the port.

This topic comes up again later in the section "Interoperation with Legacy Code."

# Throttling

A subtle way to enforce throttling is to set the number of threads to a small number on the dispatcher. Tasks will then pile up, waiting to execute, because not enough threads are available. Remember that you can apply policies to dispatcher queues that explicitly implement throttling strategies.

In Chapter 17, the code for the Hemisson Drive implements a throttling process for SetDrivePower requests by posting all requests to an internal port. There are several operations that you can use to change the drive power, and this "funneling" of the requests through a single port enables it to examine the queue and drop messages when required — a task that isn't possible with the main interleave. This is necessary because serial communications with the Hemisson are very slow and must be done synchronously to avoid hanging the onboard monitor program. Recall that when you use a joystick to drive a robot, hundreds of requests can arrive in a short period of time.

# Error Handling

You can, of course, still use structured exception handling with try/catch/finally statements. However, in a multi-threaded environment, it is possible for an error to occur in a different thread, and the error needs to be passed back to the original caller.

You have already seen examples of using a Choice to select either a valid response or a fault. This relies on the other service trapping any errors and generating a fault message.

If an exception occurs during the processing of a message in a DSS operation handler and you don't have a try/catch, then DSS automatically converts the Exception object to a Fault and sends it back as the response.

Note that DSS includes a set of methods for tracing. In particular, you can use the LogError method to send a message to the console output of the DSS node. This is explained in Chapter 3. It is always a good idea to put calls to LogError in places where the error might be catastrophic.

It is also possible to output information to the MRDS command window using the Console.WriteLine method. This is not a good approach, but it might be better than nothing at all. Be aware, however, that writing to the console this way is a relatively slow process; and you should not use it inside a tight loop or it will slow down your service. Microsoft discourages the use of Console.WriteLine.

## Causalities

The CCR defines an error handling approach called *causalities*. These are based on the concept that no matter how you nest operations, and regardless of whether they are asynchronous or not, there is some point in the past that is the root cause of the current execution stream.

Causality contexts are passed from a sender to a receiver and propagate from there to any further receivers that are invoked. They can even handle multiple exceptions, such as might be generated from a Join, for example. In this case, the causality will contain multiple items.

Here is the example code from CCRExamples. In order to see it working, you must start the program without the debugger; otherwise, the debugger will trap the exception and the causality won't get a chance to handle it.

```
// Example of using a Causality
//
// NOTE: You must run this example without the Debugger!
// Otherwise the Debugger will trap the exception and you
// will not see the Causality working properly.
//
private void Causality()
{
    using (Dispatcher d = new Dispatcher())
    {
```

*(continued)*

*(continued)*

```
            using (DispatcherQueue taskQ = new DispatcherQueue("Causality Queue", d))
            {
                Port<Exception> ep = new Port<Exception>();
                Port<int> p = new Port<int>();

                // Set up a causality for the current thread
                Dispatcher.AddCausality(new Causality("Test", ep));

                Console.WriteLine("Main thread: {0}", ↵
        Thread.CurrentThread.ManagedThreadId);

                // Set up a receiver that will generate an exception
                Arbiter.Activate(taskQ, Arbiter.Receive(false, p, CausalityExample));

                // Post an item and the causality goes with it
                p.Post(2);
                Wait(500);

                Exception e;
                while (ep.Test(out e))
                    Console.WriteLine("Exception: " + e.Message);
            }
        }
    }

    // Receiver to execute with causality enabled
    private void CausalityExample(int num)
    {
        Console.WriteLine("CausalityExample thread: {0}", ↵
    Thread.CurrentThread.ManagedThreadId);

        Port<int> p = new Port<int>();
        // Set up another receiver
        Arbiter.Activate(Environment.TaskQueue,
            Arbiter.Receive(false, p,
            // And now an anonymous delegate
            delegate(int n)
            {
                Console.WriteLine("Anonymous method thread: {0}", ↵
    Thread.CurrentThread.ManagedThreadId);
                // Can you spot the deliberate error?
                int i = 0; n = n / i;
            }
        ));
        // Post a message to activate the receiver
        p.Post(num);
    }
```

The output from the preceding example might look as follows:

```
Main thread: 8
CausalityExample thread: 15
Anonymous method thread: 9
Exception: Attempted to divide by zero.
```

Notice that three different threads are involved here:

❑ The main thread is the one that displays the exception message because it does not yield (it just sleeps).

❑ The second thread is the receiver called `CausalityExample`, and it is deliberately run in a different dispatcher to illustrate the point.

❑ The third thread is from the default dispatcher and is a delegate receiver inside `CausalityExample`.

One other point to note about the preceding code is the use of a `using` statement to wrap all of the code in the first routine that uses the new dispatcher and dispatcher queue. This ensures that these are disposed of as soon as the block of code finishes executing, rather than waiting for the garbage collector. The other examples in CCRExamples didn't bother with this, although they probably should have done so.

## Success/Failure Ports

The next example shows how to use a `SuccessFailurePort`, which is a `PortSet` consisting of a `SuccessResult` and an `Exception`. In this case, the handler is the `GetHandler` from the Lynx6Arm service in Chapter 15. The overall approach here is to call a method that returns a port and then branch based on the result.

```
/// <summary>
/// Get Handler
/// </summary>
/// <param name="get"></param>
/// <returns></returns>
/// <remarks>Get the state for the service</remarks>
private IEnumerator<ITask> GetHandler(Get get)
{
    yield return Arbiter.Choice(
        UpdateState(),
        delegate(SuccessResult success)
        {
            get.ResponsePort.Post(_state);
        },
        delegate(Exception ex)
        {
            get.ResponsePort.Post(Fault.FromException(ex));
        }
    );
}
```

The `GetHandler` sets up a `Choice` that calls `UpdateState`, which returns a `SuccessFailurePort`. The `Choice` arbiter listens for a message, which will be either a `SuccessResult` or an `Exception`. If a `SuccessResult` is returned, then the state is sent back on the `ResponsePort`. The state is global and is updated by `UpdateState`.

Because `Get` operations are intended to be accessible via the Web, the handler must return either the current state or a SOAP `Fault`. Therefore, if an exception occurs, the `GetHandler` has to convert it to a `Fault` before posting it back to the `ResponsePort` specified in the `Get` request. This can be done using the `FromException` static method on the `Fault` class. Note that you need to include a `using` statement for `W3C.Soap` in order to use `Fault`.

The `UpdateState` method creates a new `SuccessResultPort` and then posts either a `SuccessResult` or an `Exception` to it based on the result of a `Query` command sent to the SSC-32 servo controller. In this case, the code has to convert from a SOAP `Fault` back to an `Exception` in order to return a failure status. Note that `UpdateState` actually returns a port with a message already sitting in it. This enables a `Choice` to be used in the `GetHandler`.

```
// Synchronize the state by reading the joint positions from the controller
private SuccessFailurePort UpdateState()
{
    SuccessFailurePort resultPort = new SuccessFailurePort();

    //Create new query pulse width command
    int[] channels = new int[Lynx6ArmState.NUM_JOINTS] { 0, 1, 2, 3, 4, 5 };
    ssc32.SSC32QueryPulseWidth queryCommand = new ssc32.SSC32QueryPulseWidth();
    queryCommand.Channels = channels;

    ssc32.SendSSC32Command command = new ssc32.SendSSC32Command(queryCommand);
    _ssc32Port.Post(command);

    //Update the arm state based on the query response
    Activate(Arbiter.Choice(command.ResponsePort,
        delegate(ssc32.SSC32ResponseType response)
        {
            ssc32.SSC32PulseWidthResponse queryResponse = ↵
(ssc32.SSC32PulseWidthResponse)response;
            for (int i = 0; i < _state.Joints.Count; i++)
            {
                int jointAngle = ServoAngleToJointAngle(i, ↵
PulseWidthToAngle(queryResponse.PulseWidths[i]));
                _state.Joints[i].State.Angular.DriveTargetOrientation = ↵
AngleToOrientationQuaternion(jointAngle);
                // Remember the angle too - much easier to use!
                _state.Angles[i] = jointAngle;
            }
            _state.GripperAngle = ↵
(int)Math.Round(_state.Angles[(int)JointNumbers.Gripper]);

            resultPort.Post(new SuccessResult());
        },
        delegate(Fault fault)
        {
            resultPort.Post(new Exception(fault.Reason[0].ToString()));
        }
    ));

    return resultPort;
}
```

# Interoperation with Legacy Code

In terms of the CCR, Windows Forms (WinForms) constitute legacy code. WinForms use a STA model that does not work well under the CCR. Therefore, creating and using WinForms requires special treatment. See Chapter 3 for more information.

Using GDI graphics primitives suffers from problems similar to WinForms. This includes manipulating bitmaps. You need to be careful to ensure that you don't end up with an access violation.

The Webcam service also needs to run in STA mode in order to use COM interop with older-style webcam drivers. For the STA model, you should create a dispatcher with a single thread.

The MTA (Multi-Threaded Apartment) model is a little different. If you want, or need, a fixed set of threads, then create a dispatcher with the appropriate number of threads. This will isolate these threads from the threads in the default DSS dispatcher.

If you have a requirement to use older code that uses the APM with `Begin`/`End` operations, such as asynchronous file I/O, then you can wrap these operations in classes that use CCR ports. For further information on this topic, search the discussion forum. There is also a good MSDN article called "Concurrent Affairs" by Jeffrey Richter on the MSDN Magazine website. It includes code for an APM wrapper. The URL is `http://msdn.microsoft.com/msdnmag/issues/06/09/ConcurrentAffairs/default.aspx`.

# Traps for New Players

Newcomers to MRDS often make a few common mistakes. This section briefly describes these and how to avoid them:

❑   In conventional code — i.e., not in an iterator — if you forget to use `Activate` on a `Receive`, `Choice`, and so on, then nothing will happen. The code will compile, but it will not do what you expect.

❑   If you call an iterator method directly, then nothing will happen. You must use `SpawnIterator` or some other method that activates it as a task. Again, the code will compile, but you will spend some time tearing your hair out wondering why it doesn't do anything and breakpoints inside the method never fire.

❑   Iterator syntax is incompatible with `out` and `ref` parameters, i.e., you cannot return anything. However, you can pass parameters into an iterator. The only workaround for this is to pass in a `Port<type>` and then post back an instance of `type` (which is some appropriate data type that you choose).

One way to call an iterator with a parameter is as follows:

```
yield return CcrHelpers.FromIteratorHandler(3,TestMethod);
```

where the CCR helper function is defined as follows:

```
static class CcrHelpers
{
    public static ITask FromIteratorHandler(T0 t0, IterativeHandler handler)
    {
        yield return Arbiter.ExecuteToCompletion(
            base.TaskQueue,
            new IterativeTask<T0>(t0, handler));
    }
}
```

The `SpawnIterator` method has three overloads that take one, two, or three parameters and can be used as shown in the following example:

```
SpawnIterator<int, bool>(42, true, handler);
```

❑ `Arbiter.ExecuteToCompletion` activates the task. However, you still need to use `yield return` or the code does not wait for completion. This is a little confusing.

❑ Although this next tip really belongs in the next chapter, it is still relevant here. A service has one interleave, `MainPortInterleavePort`, which is created by DSS. If you set up other independent receivers, then they are not afforded the protection of the Exclusive Receiver Group and will run concurrently. This can result in very strange behavior because the state of the service might be updated simultaneously by two threads. The solution is to use the function `MainPortInterleave.CombineWith`.

Similarly, when you use `SpawnIterator` or `Activate`, these tasks run in parallel with tasks from the main interleave.

# Summary

The CCR is a very efficient and lightweight runtime. It allows multi-threaded programming without many of the complexities of a traditional approach. However, for many people it represents a different programming paradigm and it takes a little bit of getting used to. This chapter has introduced you to a variety of different methods in the CCR.

Although a lot of the CCR has been covered, many other APIs have not been discussed. In addition to reading the documentation, you can also use the Microsoft Robotics Developers Studio discussion forum as a resource. Use the Search function first. If you can't find an answer to your question, then post a message. The forum is at http://forums.microsoft.com/MSDN/default.aspx?ForumGroupID=383&SiteID=1.

Chapter 3 discusses Decentralized Software Services (DSS). Using DSS is essential for writing MRDS services.

# 3

# Decentralized Software
# Services (DSS)

The Decentralized Software Services (DSS) are responsible for controlling the basic functions of robotics applications. As explained in the previous chapter, DSS is built on top of the CCR. There is a DSS base class from which all services are derived, and this draws heavily on the features of the CCR.

DSS is responsible for starting and stopping services and managing the flow of messages between services via *service forwarder ports*. In fact, DSS itself is composed of several services that load service configurations; manage security; maintain a directory of running services; control access to local files and embedded resources such as icons; and provide user interfaces through web pages that are accessible using a web browser.

DSS uses a protocol called, not surprisingly, DSS Protocol (DSSP). The specification for DSSP can be used for free under the Microsoft Open Software Promise. DSSP is based on the Representational State Transfer (REST) model, which should be a familiar model for anyone who has worked in web development.

In some ways, such as handling errors using faults, DSSP is similar to SOAP (Simple Object Access Protocol), which is used by web services. However, unlike web services, DSSP separates *state* from *behavior* and represents all access to a service as *operations* on the state. In contrast, the state of a web service is hidden. Making the state of a robot fully visible, especially sensor values, is essential to writing control behaviors for the robot. A key feature of both models is asynchronous event notifications, which are important in the dynamic environment of robotics.

In this chapter, you learn how to build services, including the fundamental concepts. Like the last chapter on the CCR, this one is full of new terminology, and it may take you a little while to become completely familiar with DSS.

# Overview of DSS

As you read through this chapter, you might find that there is too much information to absorb in one go. Don't panic! Read through the chapter in its entirety, and then come back and hit the highlights later. The repetition throughout the chapter is deliberate. Over time you will begin to absorb the concepts by osmosis.

The basic building block in MRDS is a *service*. Services can be combined (or composed) as *partners* to create *applications*. This process is referred to as *orchestration*.

An example of a complete application is shown diagrammatically in Figure 3-1. This Visual Explorer application uses a camera for the robot to explore its environment. The robot hardware is accessed via three *generic contracts* that describe the application programming interfaces (APIs) for a differential drive (two independently driven wheels), a set of bumpers, and a web camera. The Visual Explorer service does not have to know anything about the hardware in order to use these services.

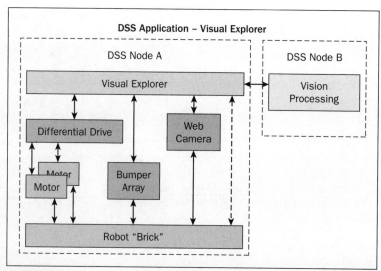

**Figure 3-1**

At the next level down, the motor service talks to the robot *brick* service, which in turn communicates directly with the robot. (A brick service is the core component for a robot.) In this example, the web camera is mounted on the robot and controlled via the brick. However, image processing, which can be a very CPU-intensive task, is carried out on a different computer (DSS Node B). Computer vision packages for MRDS such as RoboRealm (`www.roborealm.com`) or VOLTS-IQ from Braintech (`www.volts-iq.com`) could be used for this task.

The services that make up an application can be created and destroyed dynamically. However, in most cases they are specified at startup through a *manifest*. This is an XML file that lists the required services and their *partner relationships*. Figure 3-1 shows the Visual Explorer service partnered with the Vision Processing, Differential Drive, Bumper Array, and Web Camera services. The dotted connection to the

Robot "Brick" is intended to indicate that sometimes you have to bypass other services and go directly to the hardware to access devices such as LEDs, buzzers, and so on, which are not exposed through generic contracts.

One of the objectives of MRDS is to provide an environment where services are loosely coupled. In Figure 3-1, a different type of robot could be substituted for the existing one and Visual Explorer would not even know, as long as the new robot implemented the same generic contracts. Of course, in practice it is a little more complicated than this because the new robot might have motors that drive faster, a different configuration of bumpers, and a higher-resolution web camera. Nonetheless, the principle still stands.

A service consists of several components:

❑ **Contract:** This defines the messages you can send to a service, as well as a globally unique reference, called the *contract identifier*, which identifies the service and is expressed in the form of a URI (Universal Resource Identifier)

❑ **Internal state:** Information that the service maintains to control its own operation

❑ **Behaviors:** The set of *operations* that the service can perform and that are implemented by *handlers*

❑ **Execution context:** The *partnerships* that the service has with other services, and its initial state

Figure 3-2 shows all these components diagrammatically. They are discussed in more detail in the following section, and then covered again later in the chapter when you build your first service. Repetition is the key to learning!

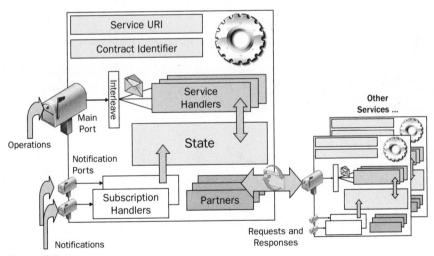

Figure 3-2

# Contracts

A *contract identifier* is used to uniquely identify different services. When you create a new service, as explained next, a contract identifier is created for you. For example, here is the contract identifier from the Dance service in Chapter 14:

```
/// <summary>
/// Dance Contract class
/// </summary>
public sealed class Contract
{
    /// <summary>
    /// The Dss Service contract
    /// </summary>
    public const String Identifier =
            "http://www.promrds.com/contracts/2007/10/dance.html";
}
```

Every service must have a class called `Contract`, and it must contain a field called `Identifier`. MRDS uses URIs in the following format:

```
http://somehost.domain/path/year/month/servicename.html
```

This looks like a valid URL, but it does not have to exist on the Internet. By default, the host name is `www.promrds.com`, but most of the Microsoft services use the host and path `schemas.microsoft.com/robotics`. The convention of using the year and month in which the service was created helps to keep the URIs unique and enables new versions of the same service to be built.

The important point to note is that the URI must be unique. You can use any host name or/path directory, that you want. This book uses a real web prefix, `www.promrds.com/contracts/`. Be aware, however, that the URI should be all lowercase or there can be problems with case sensitivity.

The programmatic interface to a service is defined as part of the *contract* via a set of classes that describe the types of messages to which a service can respond. The objective is to keep this interface clean by only using data structure definitions, and no methods. In particular, there must be at least one `PortSet` that is public and contains ports for all of the available message types. (Remember that messages are just data types, i.e., classes).

For example, here is the *main operations port* for the Dance service (although it is called a port, it is actually a `PortSet`):

```
/// <summary>
/// Dance Main Operations Port
/// </summary>
[ServicePort()]
public class DanceOperations : PortSet<DsspDefaultLookup, DsspDefaultDrop, Get>
{
}
```

Notice that this `PortSet` is "decorated" with the `[ServicePort]` attribute, which marks it is as an operations port. MRDS makes extensive use of such attributes to help build the necessary infrastructure for services declaratively. The compiler takes care of generating any code that might be necessary, so you don't have to worry about it.

`DanceOperations` contains ports for three types of messages: `DsspDefaultLookup`, `DsspDefaultDrop`, and `Get`. A `Lookup` operation enables other services to find out information about this service; `Drop` is used to shut down the service; and `Get` is used for retrieving a copy of the service's state. This is the set of message types created automatically when you create a new service.

## State

The internal *state* of a service is a class containing properties that are important to the operation of the service. Some properties are effectively constants for the duration of the service execution. This can include parameters such as the COM port that is used to communicate with a robot. By making this available externally, it is possible to reconfigure the service without recompiling it.

Other state properties change over the lifetime of the service. For example, you might implement a *finite state machine (FSM)* to control a robot as it wanders around (see Chapter 16). Then you can define the current state of the robot as Drive Straight, Turn Right, Back Up, and so on, and store this in a variable in the service state. (Do not confuse "state" in the FSM with the "state" of a service. In general, the FSM state is a small subset of the overall service state.) In this case, the state changes in response to external events, such as detecting an obstacle in the path of the robot.

Lastly, some properties in the state can represent things happening in the real world, e.g., values obtained from sensors, user input, and so on.

Here is the state from the Stinger Drive By Wire service in Chapter 16:

```
[DataContract]
public class StingerDriveByWireState
{
    // The Headless field indicates if we should run without a GUI
    private bool _headless;

    [DataMember]
    public bool Headless
    {
        get { return _headless; }
        set { _headless = value; }
    }

    // Can run with the Motor disabled for testing
    private bool _motorEnabled;

    [DataMember]
    public bool MotorEnabled
    {
        get { return _motorEnabled; }
        set { _motorEnabled = value; }
    }
```

*(continued)*

*(continued)*

```
private bool _wanderEnabled;

[DataMember]
public bool WanderEnabled
{
    get { return _wanderEnabled; }
    set { _wanderEnabled = value; }
}

private WanderModes _wanderMode;

[DataMember]
public WanderModes WanderMode
{
    get { return _wanderMode; }
    set { _wanderMode = value; }
}

private int _wanderCounter;

[DataMember]
public int WanderCounter
{
    get { return _wanderCounter; }
    set { _wanderCounter = value; }
}

private double _irLeft;

[DataMember]
public double IRLeft
{
    get { return _irLeft; }
    set { _irLeft = value; }
}

private double _irFront;

[DataMember]
public double IRFront
{
    get { return _irFront; }
    set { _irFront = value; }
}
private double _irRight;

[DataMember]
public double IRRight
{
    get { return _irRight; }
    set { _irRight = value; }
}
}
```

The `StingerDriveByWireState` class defines the internal state of the service. Notice that it is also part of the service contract because it has the `[DataContract]` attribute. When another service issues a `Get` request, it knows from the contract exactly what information it will receive in the response message.

DSS nodes implement a web server so you can use a web browser and enter the URL for a service to see its state. Requesting a web page from the service executes an `HttpGet` operation, which returns an XML file. Being able to examine the state of a service remotely using a web browser gives you some insight into the service without having to run it in the debugger.

The state can be stored in a *saved state* or *config* file, which is an XML file. This can be reloaded the next time that the service is run so it "remembers" parameter settings. An example of a config file is shown here for the Stinger Drive By Wire service:

```xml
<?xml version="1.0" encoding="utf-8"?>
<StingerDriveByWireState xmlns:s="http://www.w3.org/2003/05/soap-envelope"
xmlsn:wsa="http://schemas.xmlsoap.org/ws/2004/08/addressing"
xmlsn:d="http://schemas.microsoft.com/xw/2004/10/dssp.html"
xmlsn="http://schemas.tempuri.org/2007/11/stingerdrivebywire.html">
    <Headless>false</Headless>
    <MotorEnabled>true</MotorEnabled>
    <WanderEnabled>false</WanderEnabled>
    <WanderMode>None</WanderMode>
    <WanderCounter>0</WanderCounter>
    <IRLeft>61.3</IRLeft>
    <IRFront>80</IRFront>
    <IRRight>79.8</IRRight>
</StingerDriveByWireState>
```

The state has values that are supposed to be set by editing the saved state file: `Headless`, `MotorEnabled`, and `WanderEnabled`. It also contains internal information that provides some insight into how the Wander behavior is running: `WanderMode` and `WanderCounter`. The last three fields are the IR sensor values, obtained from other services but included here to make them easier to see.

The `WanderMode` controls the internal FSM during wandering. It uses an `enum` that is also part of the contract (because it has a `[DataContract]` attribute too):

```csharp
// Modes that the robot can be in while wandering
[DataContract]
public enum WanderModes
{
    None,
    DriveStraight,
    VeerLeft,
    VeerRight,
    TurnLeft,
    TurnRight,
    BackUp
}
```

When the state is serialized (into XML), the `WanderMode` appears as one of the names in the `enum`. There is no need to memorize "magic" numbers so you can tell what is happening when you view the state. In your code, you simply use the symbolic values, such as `WanderModes.DriveStraight`.

You can *poll* a service by requesting its state on a regular basis using a `Get` operation. This places the responsibility on your service to request the updates (and wait for them to arrive). Determining an appropriate polling interval is one of the challenges of robotics. If you poll too fast, you might overload the other service, or simply receive the same old stale information. If you don't poll often enough, you might miss important information, such as a looming wall, and then crash!

One of the key features of DSSP is that messages can be sent asynchronously, i.e., unsolicited. These are referred to as *notifications*. Services can offer *subscriptions* to their state information, so you can avoid polling. (Subscriptions are covered in Chapter 4).

For example, a laser range finder (LRF) produces range scans about once a second. Services that are interested in this data can subscribe to the LRF service and receive notifications when new range scan data becomes available. (The LRF data is part of the service's state, so the LRF service is sending state update notifications.) These updates continue to arrive until one of the services is shut down or the subscriber decides to unsubscribe.

Note that the LRF service sends the range data as part of the notification message. However, a Webcam service simply sends a message to say that a new frame is available because of the large amount of data involved. It is then up to the recipient to request the frame. If the recipient is slow and misses a couple of updates, very little bandwidth is wasted because the images are not sent in notifications anyway.

Putting this all together, the diagram shown in Figure 3-3 depicts a sequence of messages passing between hypothetical services.

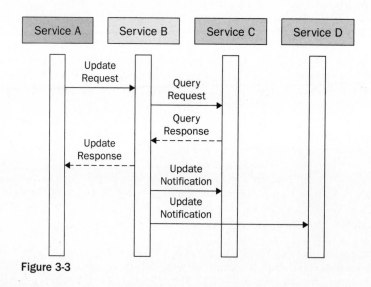

**Figure 3-3**

In Figure 3-3, note the following:

❑ Service A generates new information and sends it to Service B as an `Update` request. Perhaps Service A has obtained a new sensor reading. An `Update` only changes part of the state, as opposed to `Replace`, which overwrites the entire state.

❑ Service B realizes that in order to complete the update, it needs another piece of information from Service C, so it makes a `Query` request. (This is somewhat artificial, but just play along.) A `Query` is used to retrieve part of the state, whereas a `Get` returns the whole state.

❑ Service C responds to Service B, so then Service B can respond to Service A. The `Update` transaction is now complete. However, the state of Service B has changed and it has two subscribers: Service C and Service D. Therefore, it sends notification messages to both of these services. Note that (apart from the original subscription message) Services C and D just wait patiently and do not send any messages.

Note two distinct patterns in Figure 3-3. One is the classic request/response pattern. You have the option to wait for the response or ignore it. The response could be a `Fault` if the other service encountered an error processing the request. You can also explicitly set the request's `TimeSpan` property so that it times out if the other service doesn't respond.

The other pattern is asynchronous notifications to multiple subscribers. The subscribers first request notifications (these messages are not shown in Figure 3-3) and then they continue to receive updates whenever they become available. No acknowledgment is required for notifications.

## Behavior

Service *behavior* describes what the service actually does. Behavior includes the algorithms that a service uses to achieve its purpose, such as wandering around without bumping into obstacles. Orchestration is a form of behavior. Behavior also includes simple functions such as returning the state when it is requested.

All services implement a certain minimum set of *operations* that are defined when you create a new service (as shown earlier in "Contracts"), so even an "empty" service has some behavior. An operation is a function that a service can perform. Operations are defined by the ports that represent them, and ports use particular classes (or data types) for their messages. This is discussed in more detail below.

This terminology gets a little messy. Sometimes operations are also referred to as requests or responses — a request occurs when a message is sent to a service (using a specified data type associated with a port), and a response is sent back by the service (using another data type).

Note that operations are not required to send a response; it is up to you as the programmer to define whether or not a response is sent. At runtime, you can also set `ResponsePort` to `null` in the request — that is, you can explicitly say that you don't want a response. DSSP uses TCP/IP or HTTP, both of which are reliable transports, so there is no need to acknowledge every message.

However, if you anticipate that a request might fail and you want to know about it, then you must specify a response port (and set up a receiver to read from the port) or set up a *causality*. Causalities are like structured exception handling on steroids. They allow exceptions that occur anywhere along the execution path of a message sequence to be posted back to a single port. These exceptions might even occur concurrently due to the multi-threaded nature of the CCR.

Internally, a service uses *handlers* to process operations. Handlers operate asynchronously and possibly in parallel. Most handlers are controlled by a CCR `Interleave` (called `MainPortInterleave`), which is created automatically for you. As shown in the previous chapter, the handlers attached to an interleave can belong to the Exclusive or Concurrent groups (or the special case called TearDown, which is usually just for Drop messages).

Here is a handler for a Get operation that retrieves the state from a service:

```
/// <summary>
/// Get Handler
/// </summary>
/// <param name="get"></param>
/// <returns></returns>
[ServiceHandler(ServiceHandlerBehavior.Concurrent)]
public virtual IEnumerator<ITask> GetHandler(Get get)
{
    get.ResponsePort.Post(_state);
    yield break;
}
```

This is a very simple handler — it just sends back a copy of the service state. It is annotated with the [ServiceHandler] attribute, which specifies that it belongs to the Concurrent group. This handler is allowed to execute simultaneously with other Concurrent handlers, but it is not allowed to execute while an Exclusive handler is running.

In effect, operations are the APIs that can be used to query and manipulate a service, e.g., retrieve (Get) or change (Replace, Update) the state. In a traditional programming environment, you link your code against a library of subroutines and use *local procedure calls* (the APIs for the library) to execute the functions in the library. In MRDS, operations are executed via SOAP requests, which means that they are similar to *remote procedure calls* and can therefore be executed on another computer anywhere in the network.

If an unhandled exception occurs during the execution of a handler, then DSS automatically creates a Fault and returns it as the response. However, if you are running the code in the debugger, it catches the exception and suspends execution. You can simply continue executing the code so DSS gets a chance to handle it.

Note that the contract is embodied in a Proxy DLL that is created automatically when you compile your service by a program called DssProxy. (Because .NET assemblies are self-describing, you can find out about the contract using *reflection*, but that is another topic.) You do not make calls directly to the service DLL; instead, you call routines with the same signatures in the Proxy DLL.

> When you set up references to other services in Visual Studio, as a general rule you should always link to the Proxy DLL, rather than the service implementation DLL.

If you understand web services, then you know that parameters that are passed to and from services have to be *marshaled*, i.e., converted to a neutral format that both the client and the server understand. For web services, this format is XML. Because DSSP is based on SOAP, it also uses XML for requests over HTTP. However, for direct communication using TCP/IP, a binary serialization is used that is 10 to 20 times faster. (When you get to the "Running a DSS Node" section, you will see that two ports are specified when you start DSS: one for HTTP requests and one for direct TCP/IP connections.)

# Execution Context

The context of a service includes all of its *partners*. These are other services that it relies on to do its job. The obvious examples for a robot are the motors that drive the wheels, infrared sensors, and so on. A service can have as many partners as it likes, and the partnerships can be dynamic.

By design, partners do not have to be on the same computer. Each computer runs an instance (or possibly multiple instances) of a DSS node. A *service directory* is maintained in each DSS node. As you have already seen, one of the possible operations on a service is a lookup.

In most cases, partnerships last for the duration of the service execution. Therefore, they are often established declaratively in the code (using attributes). However, MRDS has a concept of *generic contracts*. These types of contracts specify the operations that a conforming service must support. At runtime, any service that complies with the generic contract can be used as a partner. The association between the actual service and the generic service is done using a *manifest*, which is an XML file that uses a schema defined by Microsoft.

A common example is the generic differential drive service for two-wheeled robots. The manifest can connect a service to the drive on a Boe-Bot, or a LEGO NXT Tribot, or a Stinger, and so on, and the service should have no idea what type of robot it is talking to.

One case where partnerships are established on demand is when you use the Dashboard (which is described in Chapter 4). In order to do this, the Dashboard must query the service directory to find services that implement generic contracts that it understands, such as the differential drive, game controller, laser range finder, webcam, and so on. The execution context of the Dashboard, therefore, changes over time.

The following manifest starts a Boe-Bot and the Dashboard. It runs the BASICStamp2 service, which is the "brick" or "brain" on the robot; the BSDrive service, which controls the wheels; and the Dashboard:

```xml
<?xml version="1.0" encoding="utf-8"?>
<Manifest
    xmlns="http://schemas.microsoft.com/xw/2004/10/manifest.html"
    xmlns:dssp="http://schemas.microsoft.com/xw/2004/10/dssp.html"
    >

  <CreateServiceList>

    <!--Start BasicStamp2 Brick -->
    <ServiceRecordType>
      <dssp:Contract>http://schemas.microsoft.com/robotics/2007/06/basicstamp2.html
      </dssp:Contract>
      <dssp:PartnerList>
        <!--Initial BasicStamp2 config file -->
        <dssp:Partner>
          <dssp:Service>Parallax.BoeBot.Config.xml</dssp:Service>
          <dssp:Name>dssp:StateService</dssp:Name>
        </dssp:Partner>
      </dssp:PartnerList>
    </ServiceRecordType>

    <!--Start the BoeBot drive service-->
```

*(continued)*

*(continued)*

```
      <ServiceRecordType>
        <dssp:Contract>http://schemas.microsoft.com/robotics/2007/06/bsdrive.html
        </dssp:Contract>
        <dssp:PartnerList>
          <!--Initial Drive Configuration File -->
          <dssp:Partner>
            <dssp:Service>Parallax.BoeBot.Drive.Config.xml</dssp:Service>
            <dssp:Name>dssp:StateService</dssp:Name>
          </dssp:Partner>
        </dssp:PartnerList>
      </ServiceRecordType>

      <!-- Dashboard -->
      <ServiceRecordType>
        <dssp:Contract>http://schemas.microsoft.com/robotics/2006/10/dashboard.html
  </dssp:Contract>
      </ServiceRecordType>

    </CreateServiceList>

  </Manifest>
```

You can clearly see each of the services listed inside a `ServiceRecordType`. The BASICStamp2 and BSDrive also partner with the `StateService`, which is responsible for loading the initial state from the specified config file.

Don't worry about manifests at this stage. MRDS includes a Manifest Editor that makes creating them relatively easy. This is discussed later in the chapter.

## DSSP Service Operations

The DSSP service model defines many operation classes. Luckily, most of these you will never use in an MRDS service. (If you are interested, you can read the full DSSP specification, available at `http://download.microsoft.com/download/5/6/B/56B49917-65E8-494A-BB8C-3D49850DAAC1/DSSP.pdf`).

The full list of DSSP operations is shown in the following table. You do *not* need to know all of these operations, but they are included here for completeness.

| Operation | Description |
|-----------|-------------|
| Create | Creates a new service |
| Delete | Deletes (part of) service state |
| Drop | Terminates a service (actually, requests the service to terminate itself) |
| Get | Retrieves a copy of the service state |
| Insert | Adds new information to the service state |
| Lookup | Retrieves information about the service and its context |

| Operation | Description |
| --- | --- |
| Query | Similar to Get but with additional parameters to allow structured queries. (Returns a subset of the state.) |
| Replace | Replaces the entire service state |
| Subscribe | Requests notification of all state changes |
| Submit | Special case of an Update that does not necessarily change state (typically used with Web Forms) |
| Update | Retrieves information about the service and its context |
| Upsert | Performs an Update if the state information exists, otherwise an Insert |

The only mandatory operation according to the DSSP specification is Lookup. However, for practical reasons, MRDS always creates services with Lookup, Drop, and Get. A service that only implemented Lookup would be pretty useless.

The DSS runtime provides a wrapper, CreateService, for the Create operation. Lookup operations are a little more complicated, but can be handled by requests to the Directory service.

All of the operations are based on the generic DsspOperation class:

```
public class DsspOperation<TBody, TResponse>
```

To define your service operations, you subclass one of the standard DSSP operations and supply your own body type (which is the request) and a PortSet for the response.

An example will help to clarify this. The Hemisson services in Chapter 16 have a request to read the IR sensors. It is defined as follows:

```
[DisplayName("GetSensors")]
[Description("Gets the state of the infrared sensors")]
public class QueryInfraRed : Query<SensorsRequest, PortSet<Sensors, Fault>>
{
}

[Description("Requests sensor data")]
[DataContract]
public class SensorsRequest
{
}
```

This new operation type, QueryInfraRed, is based on the Query class because it only retrieves part of the state. When a new instance is created, the Body field contains a variable of type SensorsRequest, and the ResponsePort field is a PortSet that returns either a Sensors object or a SOAP Fault. (The Sensors class is not shown here. It contains all of the IR sensor values.)

A key point to note is that the request type must be unique among all of the types in the main operations `PortSet`. If you tried to implement requests using an `int` for several of them, only the first one in the `PortSet` would ever get any messages.

This new class must be added to the main operations port for the Hemisson services, and then a handler can be written as follows:

```
///  <summary>
///  QueryInfraRed Handler
///  </summary>
///  <param name="query"></param>
///  <returns></returns>
[ServiceHandler(ServiceHandlerBehavior.Concurrent)]
public virtual IEnumerator<ITask> QueryInfraRedHandler(QueryInfraRed query)
{
    query.ResponsePort.Post(_state.Sensors);
    yield break;
}
```

Notice that this handler returns only the `Sensors` portion of the state. How this sensor information is updated is irrelevant here. It is the responsibility of the service to obtain this data from the robot and make it available.

The `[ServiceHandler]` attribute flags this method as being a handler, and the `QueryInfraRed` parameter indicates what type of operation it handles. The attribute also specifies which interleave group the handler should be in.

This behavior is in the Concurrent interleave group. This means that it is allowed to execute in parallel with other Concurrent handlers. However, it cannot execute while an Exclusive handler is running.

The third group of handlers are TearDown handlers. Usually this group only contains the `Drop` handler. TearDown handlers are not only exclusive; they also prevent further execution of any other handlers. The last thing that a service must do when it receives a `Drop` request is to post back a response.

The `QueryInfraRed` example shows a request that basically has an empty body and returns some information. The opposite is a request that sends information but does not expect any information in the response. The `PlayTone` operation from the Boe-Bot (see Chapter 14) is an example:

```
[DisplayName("PlayTone")]
[Description("Plays a tone on the speaker on the Boe-Bot.")]
public class PlayTone : Update<Tone, PortSet<DefaultUpdateResponseType, Fault>>
{
}
```

The `PlayTone` operation sends a `Tone` object as the request body, but it only receives a `DefaultUpdateResponseType` to indicate success (or a `Fault` for failure). There are pre-defined classes for the default responses to most of the operation classes so that you do not have to bother defining response types for all of your operations.

The `Tone` class is marked as part of the data contract. It also specifies two data members as the parameters for a constructor, `Frequency` and `Duration`:

```
[Description("Sound a Tone on the Speaker.")]
[DataContract]
public class Tone
{
    // Parameters
    private int _frequency;
    private int _duration;

    [DataMember, DataMemberConstructor(Order = 1)]
    [Description("Frequency in Hz (rounded to nearest 50Hz).")]
    public int Frequency
    {
        get { return this._frequency; }
        set { this._frequency = value; }
    }

    [DataMember, DataMemberConstructor(Order = 2)]
    [Description("Duration in Milliseconds (rounded to nearest 50ms).")]
    public int Duration
    {
        get { return this._duration; }
        set { this._duration = value; }
    }
}
```

As a convenience, the Proxy generator (discussed later in the section "Proxy Assemblies") will create appropriate helper methods on the proxy operations port so that you can write the following command (assuming `_stampPort` is already set up as the operations port):

```
_stampPort.PlayTone(3000, 500);
```

This constructs a new `Tone` object and sends it to the Boe-Bot.

Because the `PlayTone` operation returns a `PortSet`, you can also use it in a `Choice`:

```
yield return
    Arbiter.Choice(
        _stampPort.PlayTone(3000, 500),
        delegate(DefaultUpdateResponseType d)
        { },
        delegate(Fault f)
        {
            Console.WriteLine("Play Tone failed: " + f.Reason);
        }
    );
```

This has the added benefit that execution will not continue until the `PlayTone` request has been completed, and of course it is also defensive programming because it explicitly flags errors.

---

**A Note About Console.WriteLine**

Throughout this chapter and the previous one, you have used `Console.WriteLine` statements to output information to the console (the MRDS Command Prompt window). This is okay in simple examples, but it should not be used in real services.

You should use `LogInfo`, `LogWarning`, `LogError`, or `LogVerbose` to output diagnostic information and error messages. These messages can be read using a web browser, as you will see shortly. If you want the output to still appear on the console, you can specify where it should be directed as follows:

```
LogInfo(LogGroups.Console, "some message");
```

Log messages begin with an asterisk and have a timestamp and the service URI appended, which makes it easy to identify where they came from (see Figure 3-4 for some examples). However, this additional information clutters up the output, so for the purposes of illustration, Console.WriteLine is used in the examples. Make a mental note that this is not best practice.

---

## Generic Contracts

A generic contract is similar to an abstract class, but MRDS does not implement class inheritance in the usual sense. Generic contracts have no implementation of their own. They consist solely of the type definitions required for the contract identifier, the state, and the main operations port.

To implement a generic service, you must build a service based on the generic contract. This can be done most easily using the `/alt` qualifier to `DssNewService` to specify the generic contract that should be used. This is covered in more detail in Chapter 4.

The implementation service has two operations ports: one for the new service itself and one that receives requests based on the generic contract. In your source code for the new service, you must implement handlers for all of the operations defined in the generic service contract. There is no requirement to add more handlers to the new service (apart from `Lookup`, `Drop`, and `Get`, which all services have). The simplest implementation, therefore, consists only of handlers for the generic operations, in which case you do not need your own operations port.

The reason for defining generic contracts is to try to make MRDS services hardware-independent in the same way that device drivers hide hardware details from an operating system.

# Running a DSS Node

A program called `DssHost.exe` is responsible for implementing the run-time environment for MRDS. In the text, references to DSS usually mean the environment created by `DssHost`. In Figure 3-1, the DSS nodes are instances of `DssHost` running on a Windows-based computer (or computers). The online documentation contains a full description of `DssHost`. You can also get abbreviated help from `DssHost` itself, as described in this section, which explains how to start `DssHost` and get information from it.

`DssHost` implements a web server so that you can interrogate DSS and running services using a web browser. This is a novel approach that means you can examine and manipulate the state of services running in a DSS node across the network without installing any special client software to do it.

When you start a DSS node, several services are started automatically. You will explore most of these services in this chapter. They include the following:

- ❏ **Console Output:** Captures and filters informational and error messages
- ❏ **Constructor:** Creates new service instances
- ❏ **Control Panel:** Provides an interface to start and stop services manually
- ❏ **Manifest Loader:** Interprets manifests to set up the services comprising applications
- ❏ **Mount Service:** Provides access to the local file system (only under the MRDS folder)
- ❏ **Embedded Resource Manager:** Exposes icons, bitmaps, XSLT files, and other resources
- ❏ **Security Manager:** Manages the security on the DSS node
- ❏ **Service Directory:** Maintains a list of available services on the DSS node
- ❏ **State Partner Service:** Reads configuration files when services start

## *Directory Structure of a DSS Node*

The layout of a DSS node is as follows:

```
Root directory (or Mount point)
    Bin
    Store
        Logs
        Media
        Styles
        Transforms
```

When you deploy the MRDS runtime to another computer, this is the minimal directory structure that is created. If you have a full installation of MRDS on your computer, then there are many more directories but they are not required to run MRDS.

In a web browser, you can go to the Service Directory (as explained below) and click on the mountpoint service to see the local path where MRDS is installed, which is referred to as the *root directory* or the *mount point*. The mountpoint service can be used in your code to access local files, but it only allows you to access files below the MRDS root directory.

You can find out where your local DSS node is in your C# code using the `LayoutPaths` static class, as shown in this example:

```
string logdir = LayoutPaths.RootDir + LayoutPaths.LogDir;
```

This will give you the full path to the Logs directory if you want to write your own log file.

*File I/O operations are synchronous — that is, they block the calling thread. Therefore, you should not usually perform file I/O directly unless you know that the operations are very quick. For long operations, you might need to create a separate thread.*

## Starting DssHost

Open an MRDS Command Prompt window by clicking Start ⇨ All Programs ⇨ Microsoft Robotics Studio (1.5) ⇨ Command Prompt. Enter the following command (bold text) to see all of the available command-line options:

```
C:\Microsoft Robotics Studio (1.5)>DssHost /?
```

Now start DssHost on its own (with no services) using the following command:

```
C:\Microsoft Robotics Studio (1.5)>DssHost /p:50000 /t:50001
```

The output from DssHost should look like Figure 3-4, which shows that the Directory and Constructor services started, but no manifest was supplied (so nothing else started).

Figure 3-4

---

### Running MRDS Commands

In case you haven't realized it yet, you must run an MRDS command prompt, not a normal MS-DOS command prompt. This is necessary to establish the correct operating environment.

All MRDS tools accept parameters on the command line. You can use either a forward slash (/) or a hyphen (-) to introduce a parameter, and even mix them in the same command.

In most cases, the names of commands and the parameters are not case-sensitive. However, it looks better in print to use mixed case.

---

> Parameters have a full name, such as port, and an abbreviation, which in many cases is a single letter, such as p (for port).
>
> In addition, most commands allow you to place command-line parameters into an options file. This is a text file that contains one parameter per line. You specify an options file on the command line using an at sign (@), as in the following example:
>
> ```
> DssHost @myoptions.txt
> ```
>
> Options files are useful to avoid exceeding the maximum command-line length or when you have a lot of parameters and don't want to retype them repeatedly.

DssHost requires you to specify at least the port parameter (abbreviated to /p). In the example in Figure 3-4, the tcpport is also specified with /t. At first glance this is a little confusing. The "port" numbers you give to DssHost have nothing to do with CCR ports, and why do you need two ports?

The port parameter supplies what might more appropriately be called the HTTP port, and the tcpport parameter is the SOAP port. DssHost includes a web server that can be used to examine and control services. This web server uses the port parameter. In addition, a service receives requests (SOAP messages) through its service port, which is the tcpport parameter.

There is nothing magical about the numbers 50000 and 50001 other than the fact that they are above the range of "well-known" TCP/IP ports. In fact, in the early days of MRDS, the documentation used 40000 and 40001 in many cases.

## Exploring the DSS Web Pages

Once DssHost is running, start up a web browser and enter the following URL in the address bar: http://localhost:50000.

Because you will make frequent use of a web browser to examine services, it is a good idea to save a shortcut (or favorite) for this web page once it is displayed. Notice that the port number in the URL is the (HTTP) port that you specified on the command line when you started DssHost.

You should see something like what is shown in Figure 3-5. Notice that there are several options in the menu at the left-hand side of the window. Each of these options is explained briefly below. Try them all out for yourself. You can investigate the Developer Resources and About Microsoft Robotics Studio sections of the menu on your own.

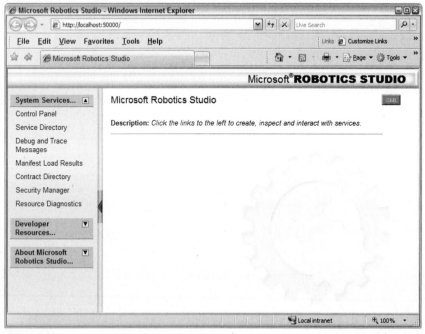

Figure 3-5

## *The Control Panel*

The Control Panel displays all of the available services. When you select the Control Panel, it might take a little while before anything is displayed because it initiates a refresh of the directory cache. Eventually, the Control Panel displays a list of all the available services. You can manually start services from this list.

You can narrow down the list of services by typing in a search string. Figure 3-6 shows the results of entering "cam," which indicates all services with "cam" in their name or description, which are obviously cameras.

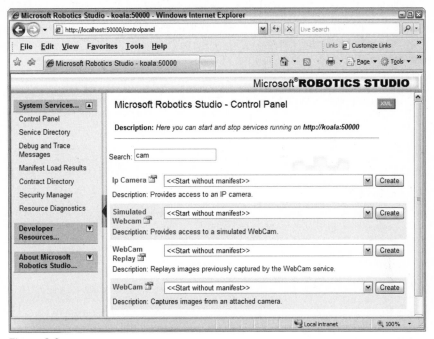

Figure 3-6

In the drop-down list beside each service, you can find a list of all of the manifests that DSS found that refer to the service. You can either select one of these manifests or just leave the drop-down set on <<Start without manifest>> and click the Create button to start a new instance of the service. If you have a web camera, make sure that it is plugged into your PC and click the Create button beside the Webcam service. (Don't select a manifest.)

## The Service Directory

A new Webcam service will be started on your computer. You can verify this by selecting Service Directory from the menu (see Figure 3-7).

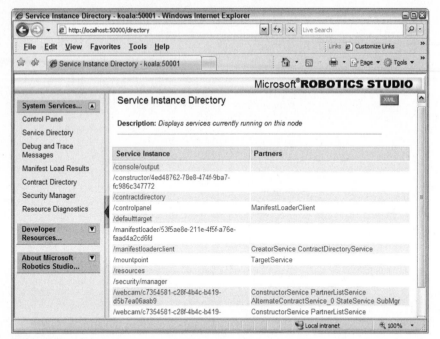

Figure 3-7

Every DSS node has a service directory that lists the currently running services. Two Webcam services are listed (the second one is cut off a little at the bottom of the window in Figure 3-7) because there is a generic Webcam contract and the service implements both the generic contract and its own contract. Notice that there are several other services as well, such as the manifest loader, console output, and the Control Panel.

Click the Webcam service in the Service Instance list to see what the output from the service looks like. An example is shown in Figure 3-8. If you look carefully in the address bar, you might be able to see that the URL is `http://localhost:50000/webcam/c7354581-c28f-4b4c-b419-d5b7ea06aab9`. Before you panic, assuming you have to know these magic numbers, it should be pointed out that they change every time you run a service — they are only there to make service URIs unique. The main point is that the Webcam service can be accessed just by appending its name to the end of the DSS node URL; the numbers are not necessary unless there are multiple instances of a service.

Figure 3-8

The Webcam service formats its output when you make a `HttpGet` request to the service and presents it as a Web Form. The form enables you to set the camera parameters, including refresh interval, display format, and capture format (image resolution). You can also select from several cameras if you have more than one connected to your PC. Finally, you can run the viewer continuously or refresh the image manually using the Start, Stop, and Refresh buttons. This is an advanced example of a web interface to a service.

## The Debug and Trace Messages Page

Select Debug and Trace Messages from the menu. You should see something like what is shown in Figure 3-9.

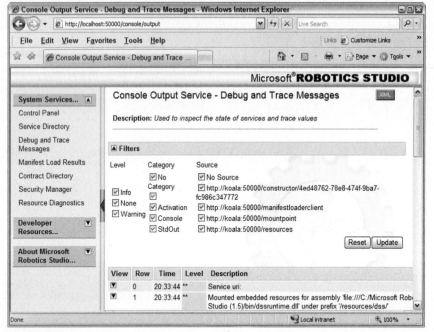

Figure 3-9

The Console Output, as it is known (the URL is `http://localhost:50000/console/output`), lists all of the informational and error messages that are generated by services running in the DSS node. You can use this to assist you with debugging.

> *Don't be confused by the name "console output" because any messages that you write using* `Console` `.WriteLine` *do* not *appear on this page.*

Notice that you can select the level of messages in the Filters section. This is one of the advantages of using `LogInfo` and the other methods: You can select at runtime what level of messages you want to see. Another advantage is that you can expand the information available by clicking the small, downward-pointing arrow in the View column beside each message. This shows you information from the stack, including the name of the source file and the line number. You might find this useful if you can't remember where a particular message comes from.

The default trace levels are set in the `DssHost` application configuration file, which is in the MRDS `bin` folder and is called `dsshost.exe.config`. (This is really a .NET feature, not MRDS.) Recall that the config file is an XML file, so you can open it in Notepad. It contains some helpful comments.

You can also enable timeout tracking on all messages in the config, so that you can detect services that are not responding to messages. You can also log all message traffic if you really want to!

The Manifest Load Results option is covered shortly because there is no manifest loaded at the moment.

## The Contract Directory

Next, click the Contract Directory menu option. This displays a screen like the one shown in Figure 3-10. This is not very exciting; it just tells you where the services reside on the local hard drive.

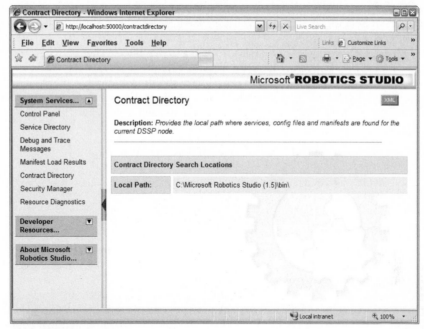

Figure 3-10

## The Security Manager Page

Select the Security Manager from the menu. This displays a Web Form. If you click the Edit button, the screen should look like the one shown in Figure 3-11.

Figure 3-11

Some users have experienced problems running DssHost due to firewall or security settings on their computers. This is most likely to happen if you are not logged in as an Administrator on your PC.

The Security Manager page shows the contents of the file store\SecuritySettings.xml, which you can edit if you wish. (If you cannot find the file, then the default security settings will be in force.) However, there is another part to the puzzle: The name of the security settings file is set in the .NET application configuration file for DssHost (which was mentioned above) — namely, bin\dsshost .exe.config. If you look in this file you should find a key called Security:

```
<!-- Comment the line below to disable security -->
<add key="Security" value="..\store\SecuritySettings.xml"/>
```

Security is discussed in more detail in the online documentation. You can also search the Discussion Forum if you have problems.

## Resource Diagnostics

The last menu option is Resource Diagnostics. This page shows information about each of the dispatchers (and their queues) running in the DSS node. You might find this information helpful when trying to diagnose a problem. An example is shown in Figure 3-12.

When you have finished exploring DSS in the web browser, enter Ctrl+C in the MRDS Command Prompt window or simply close down the window. This is a nasty way to shut down DssHost, but it works.

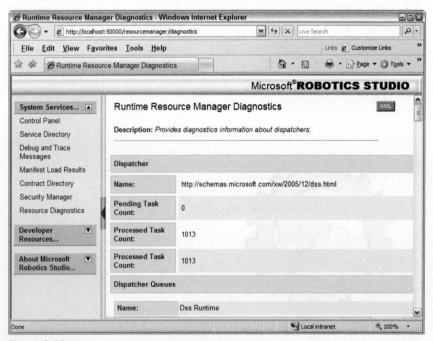

Figure 3-12

# Running a Robot Service

To run a robot service, you need to supply a manifest on the command line. Several manifests are supplied for you in the MRDS `samples\Config` folder. After you have installed the code that comes with this book, you will also have a folder called `ProMRDS\Config`.

You will no doubt find yourself typing the same `DssHost` command over and over again as you test your services. You might want to create batch files to run various services. If you place them into the MRDS `bin` folder, then they will be on the search path and will be found automatically when you type a command in an MRDS Command Prompt window. These files are fairly simple, but they save you a lot of time.

For example, here is a batch file that runs a Boe-Bot (called `RunBoeBot.cmd`):

```
@ECHO ON
REM Run a Parallax Boe-Bot
REM Type Ctrl-C in this window when you want to stop the program.
dsshost -port:50000 -tcpport:50001 ↵
-manifest:"../ProMRDS/Config/Parallax.BoeBot.manifest.xml" ↵
-manifest:"../ProMRDS/Config/Dashboard.manifest.xml"
```

*The* `DssHost` *command is all on one line in the batch file, not wrapped as it appears here in print.*

Two manifests are specified on the command line: the Boe-Bot and the Dashboard. This is because the two services are not directly related — the Dashboard connects to other services dynamically.

Notice that the paths to the manifests are relative to the location of the batch file. If you place the batch file into a different folder, i.e., not the `bin` folder, then you will need a relative path to `DssHost.exe`, and the relative paths to the manifests will be different as well.

To run this batch file, you can just double-click it in Windows Explorer. Alternatively, to run it from an MRDS Command Prompt window, enter the following command:

```
C:\Microsoft Robotics Studio (1.5) >RunBoeBot
```

## Manifest Load Results

Once you have the Boe-Bot running (or your particular robot), start a web browser again and browse to the DSS node at `http://localhost:50000`. Select Manifest Load Results in the menu; the result should look similar to what is displayed in Figure 3-13. You can expand each manifest to see additional information. Notice that the Boe-Bot manifest loaded the BASICStamp2 and BSDrive services, each with a StateService partner (an initial config file).

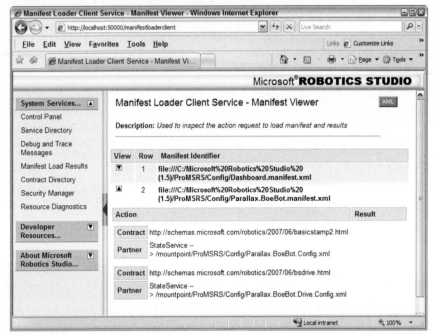

Figure 3-13

Now select the Service Instance Directory (see Figure 3-14).

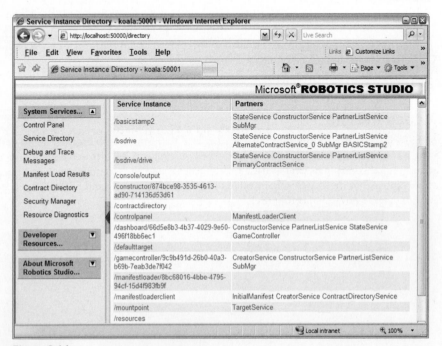

Figure 3-14

## Viewing Service State

Figure 3-14 indicates that many different services are running. Click the basicstamp2 service to view its state. This output should look like the window shown in Figure 3-15.

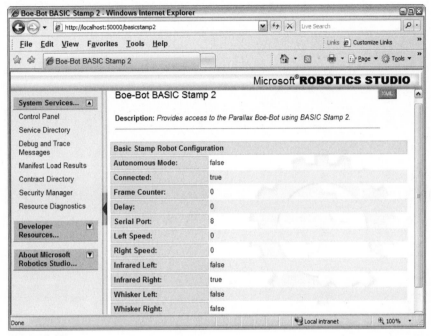

**Figure 3-15**

Notice in Figure 3-15 that the right infrared sensor specifies `true`, which indicates an obstacle in the immediate vicinity of the Boe-Bot. This page is nicely formatted because it uses an XSLT (Extensible Stylesheet Language Transform) file to format the XML output from the service.

The state information on the page does not update automatically — you need to keep clicking on the browser's refresh button.

Go back to the Service Directory and click the bsdrive service. The output should look like Figure 3-16. This is raw XML code. If you look back at Figure 3-15, you will see an XML button in the window's top-right corner. You can click this to see the raw XML behind a formatted state page.

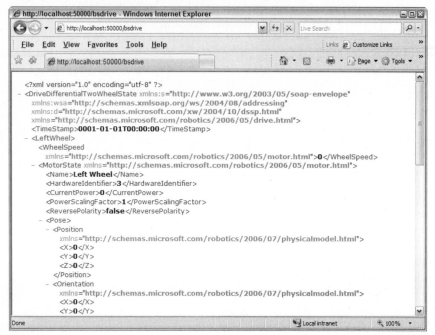

Figure 3-16

The BSDrive service does not have an associated XSLT file to format its output. However, the output is still intelligible. This is one of the advantages of XML — it is human-readable.

That completes the pictorial overview of DSS. It is hoped that the many pages it took to show the screenshots were worth more than several thousand words.

# Creating a New Service

As an MRDS programmer, services are your bread and butter. Therefore, it makes sense to begin by discussing exactly what a service is and how it operates. The best way to learn is to make a new service.

In this section you will build two services, called ServiceA and ServiceB. For your convenience, completed versions of these services are supplied in the `ProMRDS\Chapter3` folder.

ServiceA is the "master," and it partners with ServiceB so that it can get information from ServiceB. As you work through the exercises, make sure you add code to the correct service.

## *Building a Service from Scratch*

In this section you build a brand-new service. It is advisable to keep your code separate from the MRDS distribution. The Service Tutorials suggest creating new services under the `samples` folder, but this results in your code being intermixed with the Microsoft code. All of the code for the book is under the

ProMRDS folder, which makes it much easier to pick up and move elsewhere. Furthermore, there is no chance of a conflict if you update your MRDS installation, such as with the MRDS V1.5 Refresh.

If you have not done so already, create a new folder called Projects. In Windows Explorer, browse to the Microsoft Robotics Studio (1.5) folder and make a new folder under it called Projects.

Alternatively, open an MRDS Command Prompt window by running the Command Prompt option from the Microsoft Robotics Studio (1.5) folder in the Start menu. Create a new folder called Projects:

```
C:\Microsoft Robotics Studio (1.5)>md Projects
```

This is where you should create your own MRDS projects.

## Using Visual Studio

The previous chapter explained how to create a new service using Visual Studio. Now you need to create another service. The New Project dialog is shown in Figure 3-17.

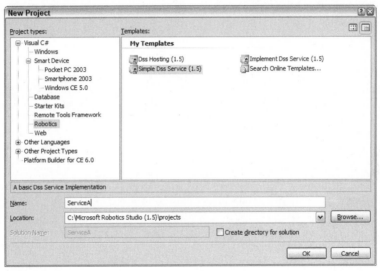

Figure 3-17

To summarize the steps involved:

1. Open Visual Studio and select File ⇨ New ⇨ Project.

2. Select the Robotics project type and the Simple Dss Service (1.5) template.

3. Make sure that the location is your Projects folder.

4. Give the service the name ServiceA.

When Visual Studio has finished creating your project, you will find the following source files listed in the Solution Explorer panel:

❑ **AssemblyInfo.cs:** This is a standard file that contains information about the output assembly (DLL). You do not need to be concerned with it here, so you can ignore it.

❑ **ServiceA.cs:** This is the main source file for the new service. Most of the work of creating a new service is done in this file.

❑ **ServiceA.manifest.xml:** This is the manifest that is used by DSS to load the service.

❑ **ServiceATypes.cs:** This contains a set of classes, also called *types,* that are used by the service and other services that wish to communicate with it.

(There are several more files that Visual Studio uses to manage the solution: `ServiceA.csproj`, `ServiceA.csproj.user`, `ServiceA.sln` and `ServiceA.suo`. You can ignore them; they don't show up in the Solution Explorer anyway.)

The source files are explained in more detail later in this chapter. In the meantime, it is instructive to look at the Visual Studio Project Properties.

Open the Project Properties and step through each of the tabs to look at the settings. The important points on each of the tabs are highlighted here:

❑ **Application tab:** The application type is Class Library (DLL). All MRDS services are dynamic link libraries (DLLs). Notice that the Assembly name is `ServiceA.Yyyyy.Mmm`, where `yyyy` is the current year and `mm` is the current month. Using the date is a simple way to create different versions of a service with the same name. Recall from earlier in the chapter that contract identifiers also contain a year and a month. The assembly name and the contract identifier should match. (It is possible to have a different assembly name, but that can become a source of confusion!)

The default namespace is `Robotics.ServiceA`. There is no need to change this unless you want to create assemblies for a particular organization.

❑ **Build tab:** The output path should be `..\..\bin`. All MRDS services are placed into the `bin` folder under the MRDS root directory.

❑ **Build Events tab:** There is a post-build event command that is only executed after successful compilation:

```
"C:\Microsoft Robotics Studio (1.5)\bin\dssproxy.exe"↵
/dll:"$(TargetPath)" /proxyprojectpath:"$(ProjectDir)Proxy "↵
/keyfile:"$(AssemblyOriginatorKeyFile)" $(ProxyDelaySign)↵
$(CompactFrameworkProxyGen) /binpath:". "↵
/referencepath:"C:\Microsoft Robotics Studio (1.5)\bin\ "↵
/referencepath:"C:\Microsoft Robotics Studio (1.5)\bin\ "
```

You do not need to know what this command does for now, and you certainly should not change it. Its purpose is to create a *proxy* for your service. If you are familiar with web services and SOAP requests, then you know what a proxy is. Otherwise, it is explained later in the section "Proxy Assemblies."

❑ **Debug tab:** When you run the debugger, it automatically starts an external program: `C:\Microsoft Robotics Studio (1.5)\bin\dsshost.exe`.

DssHost.exe runs a DSS node, which is effectively the MRDS runtime environment. Your service(s) run within this node. Although it is possible to start a DSS node from within your own application program, it isn't done that way for the examples in this book.

Notice that the working directory is the MRDS root directory and some command-line options are passed to DssHost:

```
-port:50000 -tcpport:50001
-manifest:"C:\Microsoft Robotics Studio (1.5)\projects
\ServiceA\ServiceA.manifest.xml"
```

❑ **Reference Paths tab:** There should be two directories in the list of reference paths. The first directory is where all of the service DLLs reside, as well as the components that make up MRDS. This enables you to easily reference other services. The second directory is the V2.0 .NET Framework, which is required to run MRDS V1.5.

```
c:\microsoft robotics studio (1.5)\bin
c:\windows\microsoft.net\framework\v2.0.50727
```

You can ignore the Resources, Settings, and Signing tabs because they are not important at this stage. For now, just leave Visual Studio open while you create another service from the command line as an alternative to using Visual Studio. We want two services so that they can interact with each other.

## Using the DssNewService Tool

Open an MRDS Command Prompt window from the MRDS menu by clicking Start ⇨ All Programs ⇨ Microsoft Robotics Studio (1.5). Note that you cannot use a normal MS-DOS command prompt because the MRDS Command Prompt window defines a lot of environment variables and paths to help you run MRDS commands. In addition, the MRDS command prompt automatically places you into the root directory of MRDS when it starts up.

To use the DSSNewService tool, follow these steps (what you should type appears in bold text):

1. You should already have a folder called Projects. Change to this folder:

   ```
   C:\Microsoft Robotics Studio (1.5)>cd Projects
   ```

2. Execute DssNewService to create a service for you:

   ```
   C:\Microsoft Robotics Studio (1.5)>DssNewService /service:ServiceB
   ```

   The only command-line parameter you need to provide is the service name. Note carefully that it is ServiceB. This is going to be the partner to ServiceA, which you created earlier.

3. Once the service has been created, which only takes a couple of seconds, you can open it in Visual Studio. Either locate the .sln file in Windows Explorer and double-click it, or at the command prompt enter the following:

   ```
   C:\Microsoft Robotics Studio (1.5)>cd ServiceB
   C:\Microsoft Robotics Studio (1.5)>ServiceB.sln
   ```

4.  Look around in ServiceB. It is basically the same as ServiceA. The only slight difference you might find is that it has an additional reference path:

```
c:\microsoft robotics studio (1.5)\bin\cf
```

This is for the .NET Compact Framework, which is not relevant here, but is covered in Chapter 16.

The `DssNewService` tool has several command-line qualifiers. You can read about them in the online help or enter the following command:

```
C:\Microsoft Robotics Studio (1.5)>DssNewService /?
```

We will not cover all of the options here, but a few of them are worth mentioning:

❑   You can explicitly set the URI prefix that is used in the contract identifier with /org. The default is schemas.tempuri.org.

❑   The /month and /year qualifiers affect the contract identifier and the assembly name. They default to the current month and year.

❑   You can specify the service namespace using /namespace. This might be important if you are producing code for a particular organization. You might want to coordinate this with the /org qualifier.

❑   The /clone qualifier enables you to copy an existing service. It does this using reflection, not by copying the source files. Therefore, when DssNewService has finished building your solution, all of the methods exist but there is no code. You can, of course, copy and paste the code yourself.

❑   As you become more familiar with MRDS, you will want to create new services that implement generic contracts. The /alt qualifier specifies an alternate service, i.e., the generic contract, that you wish to implement in addition to your own set of operations. This is covered in the next chapter. Note that if you specify /alt, you also need to supply an /i qualifier.

## Examining the Service Source Files

This section examines the various files that are generated when you create a new service. The source code for a new service is split into two files: the main code and the "types." You are not limited to using only these two files — you can split the code up any way you like. In fact, in the early days of MRDS, it was common to also have a "state" file that contained only the definition of the service state. This is now rolled into the types file, but you can still see many state files in the samples supplied with MRDS.

### The Types File

Begin by opening `ServiceATypes.cs` in Visual Studio. At the very top of the file you can see the contract for ServiceA:

```
namespace Robotics.ServiceA
{
    /// <summary>
    /// ServiceA Contract class
    /// </summary>
```

```
public sealed class Contract
{
    /// <summary>
    /// The Dss Service contract
    /// </summary>
    public const String Identifier = ↵
"http://schemas.tempuri.org/2008/01/servicea.html";
}
```

The namespace is based on the service name. The `Contract` class is a mandatory part of the contract, and it must contain a string called `Identifier`. (Your contract identifier will be different unless you explicitly specified `/year:2008` and `/month:01` when you created the service, which is unlikely.)

Next is the service state. All services have a state class, but at this stage it is empty. Adding fields to the service state is one of the standard steps in creating a new service:

```
/// <summary>
/// The ServiceA State
/// </summary>
[DataContract()]
public class ServiceAState
{
}
```

Notice that the service state has the `[DataContract]` attribute applied to it. This is necessary so that the class will be copied across to the Proxy DLL when the code is compiled.

The next piece of code is the main operations port, which is a `PortSet`:

```
/// <summary>
/// ServiceA Main Operations Port
/// </summary>
[ServicePort()]
public class ServiceAOperations : PortSet<DsspDefaultLookup, ↵
DsspDefaultDrop, Get>
{
}
```

A service's main operations port must have the `[ServicePort]` attribute. Note that even though it is called a port, it is really a `PortSet` that lists all of the operations supported by the service.

The default `PortSet` when you create a new service only contains `DsspDefaultLookup`, `DsspDefaultDrop`, and `Get` operations.

For an orchestration service, such as ServiceA, there is no need to add any operations because no other service will call ServiceA — it is the "master." However, ServiceB needs some additional operations (as explained later in the section "Update") because ServiceA needs to call ServiceB to perform certain functions.

The final section of the types file contains the definitions for the operation request types. In this case, there is only one for `Get`:

```
/// <summary>
/// ServiceA Get Operation
/// </summary>
public class Get : Get<GetRequestType, PortSet<ServiceAState, Fault>>
{

    /// <summary>
    /// ServiceA Get Operation
    /// </summary>
    public Get()
    {
    }

    /// <summary>
    /// ServiceA Get Operation
    /// </summary>
    public Get(Microsoft.Dss.ServiceModel.Dssp.GetRequestType body) :
            base(body)
    {
    }

    /// <summary>
    /// ServiceA Get Operation
    /// </summary>
    public Get(Microsoft.Dss.ServiceModel.Dssp.GetRequestType body, ↵
Microsoft.Ccr.Core.PortSet<ServiceAState,W3C.Soap.Fault> responsePort) :
            base(body, responsePort)
    {
    }
}
```

There are no definitions for `DsspDefaultLookup` and `DsspDefaultDrop` because these operations are implicitly handled by the `DsspServiceBase` class. You can override these operations if you wish; and you will do this for ServiceB, but in general it is not necessary — you can let the base class handle them for you.

Notice that `Get` uses two generic types (one for the request message type and the other for the response message or messages): `GetRequestType` and `PortSet<ServiceAState, Fault>`.

`GetRequestType` is known as the *body* of the message. In this case, it is a class that is already defined in the DSS service model. You can also define your own classes to use in messages, so you can determine exactly what the content of messages will be.

`Get` operations are service-specific because they return the state for the particular service. Therefore, every service needs to subclass `Microsoft.Dss.ServiceModel.Dssp.Get` and supply an appropriate `PortSet` for the response. Notice that in this case the `PortSet` contains `ServiceAState`.

The `PortSet` indicates that there are two possible responses to this message: `ServiceAState` or `Fault`. Obviously, the purpose of `Get` is to return `ServiceAState`. Because DSSP is based on SOAP, it uses the standard mechanism already defined by SOAP to return errors. If you look at the top of the file, you will see that the SOAP namespace is referenced because this is where `Fault` comes from:

```
using W3C.Soap;
```

## Main Service Implementation File

Now open the main code file for ServiceA, `ServiceA.cs`. It contains the service initialization and the service operations (or behaviors). Scroll down below the `using` statements:

```
namespace Robotics.ServiceA
{
    /// <summary>
    /// Implementation class for ServiceA
    /// </summary>
    [DisplayName("ServiceA")]
    [Description("The ServiceA Service")]
    [Contract(Contract.Identifier)]
    public class ServiceAService : DsspServiceBase
    {
```

The namespace matches `ServiceATypes.cs` as you would expect. You can change the `[DisplayName]` and `[Description]` attributes as you see fit because they are only for documentation. The contract is specified using `Contract.Identifier`, which is in the `ServiceATypes.cs` file (as you saw earlier). The new service class is a subclass of `DsspServiceBase`. This gives you access to a whole host of helper functions.

Every service must have an instance of its service state, even if the state is empty, and a main operations port:

```
        /// <summary>
        /// _state
        /// </summary>
        private ServiceAState _state = new ServiceAState();

        /// <summary>
        /// _main Port
        /// </summary>
        [ServicePort("/servicea", AllowMultipleInstances=false)]
        private ServiceAOperations _mainPort = new ServiceAOperations();
```

By convention, the service state is called _state, but you can call it anything you like. In addition, an operations port called _mainPort is created for sending messages to ServiceA. It might seem strange to send messages to yourself, but in larger services this is quite common. Again, you can use any name you like for the main port, but it is best to stick with convention.

Finally, there is the constructor for the class, which is always empty:

```
/// <summary>
/// Default Service Constructor
/// </summary>
public ServiceAService(DsspServiceCreationPort creationPort) :
        base(creationPort)
{
}
```

Next is the Start method. All services have a Start method. It is called during service creation so that the service can initialize itself:

```
/// <summary>
/// Service Start
/// </summary>
protected override void Start()
{
    base.Start();
    // Add service specific initialization here.
}
```

A call to base.Start is inserted here automatically. This is explained further in the section "Service Initialization" later in the chapter. You should not remove it.

Lastly, there is a handler for the Get operation:

```
/// <summary>
/// Get Handler
/// </summary>
/// <param name="get"></param>
/// <returns></returns>
[ServiceHandler(ServiceHandlerBehavior.Concurrent)]
public virtual IEnumerator<ITask> GetHandler(Get get)
{
    get.ResponsePort.Post(_state);
    yield break;
}
```

This handler is very simple. All it does is return a copy of the service state. It is annotated with the [ServiceHandler] attribute, which specifies that it belongs to the Concurrent group.

Notice that there are no handlers for DsspDefaultLookup and DsspDefaultDrop because these are defined in the base class and there is no need to override them. In fact, you might even wonder why there is code here for Get. The short answer is that it is one of the handlers that is commonly modified.

### The Service Manifest

The last source file to look at (we will ignore `AssemblyInfo.cs` for now) is the manifest. Open `ServiceA.manifest.xml` in Visual Studio:

```xml
<?xml version="1.0" ?>
<Manifest
    xmlns="http://schemas.microsoft.com/xw/2004/10/manifest.html"
    xmlns:dssp="http://schemas.microsoft.com/xw/2004/10/dssp.html"
    >
    <CreateServiceList>
        <ServiceRecordType>
<dssp:Contract>http://schemas.tempuri.org/2008/01/servicea.html</dssp:Contract>
        </ServiceRecordType>
    </CreateServiceList>
</Manifest>
```

When you run the debugger, this manifest is passed to `DssHost`. The manifest describes the services to be started and any partnerships. The preceding one only has a single `ServiceRecord` with the ServiceA contract identifier. Manifests are discussed in detail later in the chapter in the section "Modifying Manifests."

# Compiling and Running Services

In general, before you compile a service, you need to add references for the other services it uses and write some code to implement the service operations. However, in this case, you have a "chicken or the egg" situation because ServiceA relies on ServiceB, so you will start by compiling both services.

## Compiling a Service

You don't need to do anything special to compile a service — just select Build Solution from the Build menu. If it is not open already, open ServiceA in Visual Studio. Locate the `Start` method in `ServiceA.cs` and add a line of code to announce the startup of the service, as shown here:

```csharp
protected override void Start()
{
    base.Start();
    // Add service specific initialization here.
    Console.WriteLine("ServiceA starting");
```

Now compile ServiceA. It should compile without errors. You can run it if you want, but it just displays a message indicating that it is starting, which is not very exciting.

Open ServiceB in Visual Studio and add a `Console.WriteLine` command in `ServiceB.cs`, except obviously the message should say "ServiceB starting."

When you compile, you might notice in the Output panel that some other DLLs are created as well. These are the Proxy DLLs discussed later in the section "Proxy Assemblies."

## Setting Up References

In Visual Studio for ServiceA, expand the References in the Solution Explorer panel. References are inserted automatically for `Ccr.Core`, `DssBase`, and `DssRuntime` when the service is created. In almost all services, you will also require `RoboticsCommon.Proxy`. You can add this reference to your services if you want, but because they do not involve a real robot, it is not necessary.

You also need to add references to other services that you intend to use. In the example in this chapter, ServiceA partners with ServiceB, so ServiceA must have a reference to the ServiceB proxy. (See "Proxy Assemblies," later in the chapter). This is why you compiled both of the services in the previous section.

When you add a reference, it might take a little while before the dialog is displayed. Eventually you should see a list of DLLs. Scroll down to ServiceB, as shown in Figure 3-18. Make sure you select the Proxy assembly. (The year and month in the assembly name will be different for you.)

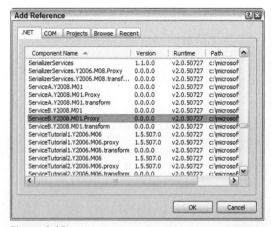

**Figure 3-18**

Note that ServiceB does *not* need a reference to ServiceA because it never calls ServiceA — it only responds to requests from ServiceA.

---

### Service Assemblies

If you were writing a service to control a real robot, you would have to find the relevant proxies and add them to your service at this point. For example, to drive a Boe-Bot you need references to `BASICStamp2.Y2007.M06.Proxy` and `BSServices .Y2007.M06.Proxy`.

The examples in this chapter use two services, each of which is in a separate project. However, there is no requirement to build every service into a separate assembly. In fact, you can combine many services into one assembly as long as each service has its own namespace and follows all the rules for service contracts.

In the case of BSServices, the project contains several different source code files. Each one of these is a separate service, but they are all combined into a single DLL. Perhaps the best example is RoboticsCommon, which combines a large number of services.

---

At the top of `ServiceA.cs` underneath the existing `using` statements, add another one:

```
using serviceb = Robotics.ServiceB.Proxy;
```

You can now recompile ServiceA with the new reference. Clearly, you have to compile ServiceB before ServiceA so that the correct reference is used.

## Proxy Assemblies

When you compile a service, a post-build event is triggered after a successful compilation. This runs `DssProxy`, which generates a Proxy DLL. You can see the command in the Project Properties on the Build Events tab, as mentioned earlier.

Service operations are executed "over the wire" by sending messages. These messages must first be serialized, i.e., converted to XML, so that they can be sent. This is necessary because the service that you are communicating with might not be on the same computer. It is not simply a matter of passing across a pointer reference to a location in memory — this will not work across the network!

This is an important point: You can't allocate a chunk of memory, e.g., create a new class instance, and send a pointer to it to another service. The contents of the area of memory have to be converted to XML, and then sent.

The Proxy DLL is responsible for performing the serialization and deserialization of messages. When you post a message, you pass a handle to the request message body (a pointer reference) to the Proxy and it creates an XML message from that. When a response message is returned, the Proxy converts the XML back into its binary representation and gives you a handle to a response message.

Therefore, the reference that you add to ServiceA is to the ServiceB proxy. You do not call ServiceB directly.

## Running Services

Now that you have compiled the services, you can try running each of them using whatever method you normally use: press F5; click the Start Debugging icon in the toolbar; or click Debug ⇨ Start Debugging.

This starts `DssHost` and runs the manifest that was generated for you when you created the service. The command that is executed can be found in the Project Properties on the Debug tab in the Start External Program textbox, and the command-line options are in the Command Line Arguments textbox.

The first time you run ServiceA, you should see something like what is shown in Figure 3-19. Notice that the message "ServiceA starting" is displayed but then nothing else happens. You must stop debugging or close the MRDS Command Prompt window to terminate the service.

Figure 3-19

Because this is a new service that has not been run before, DSS rebuilds the contract directory cache, as you can see from the message:

```
Rebuilding contract directory cache. This will take a few moments ...
Contract directory cache refresh complete
```

The contract directory cache is populated using reflection on all of the DLLs in the MRDS `bin` folder. It consists of two files in the `store` folder:

❑   `contractDirectory.state.xml`

❑   `contractDirectoryCache.xml`

You can open these files and have a look at them if you like, but you will never need to edit them. If you want to force the cache to be refreshed, you can delete these two files. The next time `DssHost` starts, it will recreate them.

Also in Figure 3-19, you can see that the Manifest Loader starts up and then loads the ServiceA manifest. When you are creating your own manifest (explained below), you might make mistakes. If you do, this is where the errors will appear.

Now add some behavior to ServiceB. Open it in Visual Studio and then open `ServiceB.cs`. Scroll down in ServiceB until you find the `Start` method. Add the following code (shown highlighted):

```
/// <summary>
/// Service Start
/// </summary>
protected override void Start()
{
    base.Start();
    // Add service specific initialization here.

    SpawnIterator(MainLoop);

}
```

```
private IEnumerator<ITask> MainLoop()
{

    Random r = new Random();
    while (true)
    {
        // Generate a random number from 0-100 and write it out
        Console.WriteLine("B: " + r.Next(100));

        // Wait a while
        yield return Arbiter.Receive(
            false,
            TimeoutPort(1000),
            delegate(DateTime time)
            { }
        );
    }
}
```

The SpawnIterator in the Start method kicks off an infinite loop called MainLoop. ServiceB then displays a random number from 0–100 on the console, waits for a second, and then repeats forever (or until you shut it down). The point of this code is that it continually generates new random numbers that are used to simulate sensor readings.

Compile ServiceB and run it. You should see output like that in Figure 3-20.

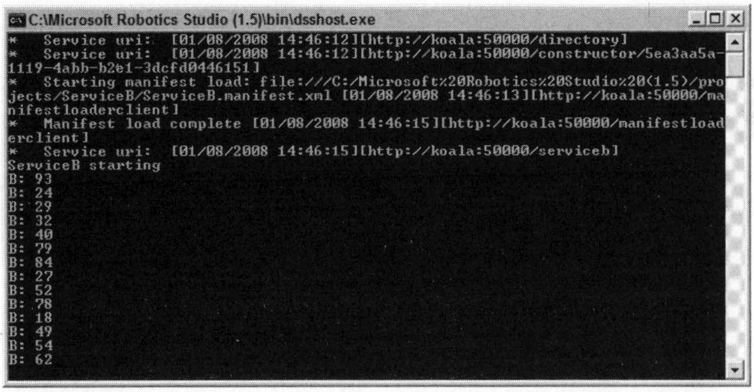

Figure 3-20

## Using the Debugger

Debugging services is just like debugging normal code — assuming that you normally debug multi-threaded code! You can set breakpoints and step through code just as usual. Try it out by setting a breakpoint inside MainLoop and running ServiceB again.

Figure 3-21 shows the debugger stopped at a breakpoint. Notice in the bottom-left corner that the Threads window is visible. Select it from the menu using Debug ⇨ Windows ⇨ Threads. This menu option is only available while the debugger is running. Two threads are assigned to the User Services Common Dispatcher. This dispatcher is set up for you by DssHost.

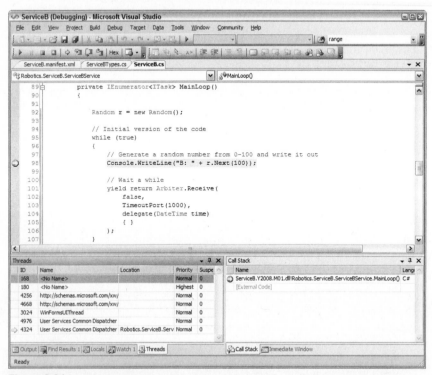

Figure 3-21

Be careful about where you set breakpoints. You can set breakpoints inside a delegate if you want to. If you try to single-step through the code, you might encounter strange behavior when you hit a yield return. Instead of stepping through a yield return, set a breakpoint on the statement after it and tell the debugger to continue execution.

> *It is important to understand that by setting up an infinite loop like this, one of the two threads created for the default dispatcher is unavailable to the CCR for scheduling service operations. In this case it is not a problem, but in general you should not tie up a thread like this, or you should specify more threads as explained in the previous chapter.*

If you are calling other services, you can debug them too. Simply open the relevant source file from another project in the current Visual Studio window. Set breakpoints in this "external" source file as required, and then run the debugger.

For example, later in the chapter you will be running ServiceA with ServiceB as a partner. You can open ServiceA in Visual Studio and set breakpoints in ServiceA.cs. Then, from the menu, click File ⇨ Open ⇨ File and browse to ServiceB.cs and open it. You can set breakpoints inside ServiceB as well. That way, as you send messages backward and forward, you stop in the code on both sides.

# Defining Service State

The concept of service state implies that it should contain all of the necessary information to enable you to save the state, restart the service some time later, and, by reloading the saved state, continue running from where the service left off. For this reason, any information that you retrieve from a service must be part of the state. Service state is exposed via the Get and HttpGet operations. The Get operation is intended to be used programmatically, whereas HttpGet is for human consumption, which enables you to observe the state using a web browser.

In the simplest form, the state is displayed as XML. Therefore, you can start working with a new service without having to write a UI. If you want to present the information in an easy-to-read form, you can supply an XSLT file and even some JavaScript code for the web page.

To define a service state, follow these steps:

**1.** Open ServiceB in Visual Studio. (Note that this is "B," not "A.") Then open the source file called `ServiceBTypes.cs`.

**2.** Scroll down through the source code to the service state, which is initially empty. Add two properties to the state, as shown here:

```
/// <summary>
/// The ServiceB State
/// </summary>
[DataContract()]
public class ServiceBState
{
    [DataMember]
    public int Interval;
    [DataMember]
    public int SensorValue;
}
```

Each of these fields, Interval and SensorValue, is declared as public and is decorated with the [DataMember] attribute. (The purpose of these two fields will become apparent shortly.) These fields are part of the service contract and will be serialized (to XML) when the state is saved or sent in a message. You can, of course, have private members in the state if you want, but they are ignored during serialization.

Purists might prefer to use a property declaration like the following, rather than a public field, although this is not strictly necessary. There is, however, a subtle difference if you want to fully document your code — the [DisplayName] attribute can only be applied to a property, not a public field.

```
private int _interval;
[DataMember]
public int Interval
{
    get { return _interval; }
    set { _interval = value; }
}
```

**3.** Now that the state for ServiceB contains some fields, go back to `ServiceB.cs` and revise the `MainLoop` function as shown:

```
// Global flag to allow terminating the Main Loop
private bool _shutdown = false;

private IEnumerator<ITask> MainLoop()
{

    Random r = new Random();

    while (!_shutdown)
    {
        // Generate a random number from 0-100 and save it in the state

        _state.SensorValue = r.Next(100);
        Console.WriteLine("B: " + _state.SensorValue);

        // Wait a while
        yield return Arbiter.Receive(
            false,
            TimeoutPort(_state.Interval),
            delegate(DateTime time)
            { }
        );
    }

}
```

The changes are as follows:

❑   A global flag called _shutdown has been added.

❑   The while loop now uses _shutdown so that it can be terminated (see below).

❑   New random values are stored in the state SensorValue property.

❑   The timer interval is read from the state Interval property.

Later in the chapter, in the section "Dropping Services," you will add a Drop handler, which can use _shutdown to terminate the MainLoop. For now, this variable has no effect on the operation of the service.

Storing the simulated sensor value into the state means that it can be accessed from other services using a Get operation.

Exposing the Interval property enables it to be changed without having to recompile the code. It is really a configuration parameter. There are two ways in which the interval can be changed: through the initial state (as explained in the next section) or via an update operation (covered later in the chapter).

# Persisting Service State

Saving state is useful for remembering the configuration of a service. MRDS provides a service called the Initial State Partner service that can be used to read a configuration file and populate your service state when the service starts up.

In the case of the Simulator, you can save the entire state of the simulation. Using this saved state, you can restart a simulation later without having to construct the environment again programmatically. You can see an example of this in the MRDS Simulation Tutorials.

## Loading State from a Configuration File

To load an initial state from a config file, you need to specify where it should come from. DSS can obtain a name for a config file in two ways:

❑   You can specify it in your code in an [InitialStatePartner] declaration.

❑   The filename can be supplied in the manifest.

### Using Initial State Partner to Load a Configuration File

To load state using a service state partner, go to the top of ServiceB.cs and locate the declaration of the state instance. Add an [InitialStatePartner] attribute as shown here:

```
/// <summary>
/// _state
/// </summary>
// Set an OPTIONAL initial state partner
// NOTE: If the config file is NOT specified in the manifest, then the
// file will be created in the MRDS root directory. If it is listed in the
// manifest, then it will be created in the same directory as the manifest.
[InitialStatePartner(Optional = true, ServiceUri = "ServiceB.Config.xml")]
private ServiceBState _state = new ServiceBState();
```

The initial state partner declaration says that the config file is optional. This is a good idea because the service won't start if the file is required and it doesn't exist, such as the first time you run the service.

Notice the ServiceUri parameter, which specifies the filename. This is not required, but if you omit it, you must specify the filename in the manifest. You should get into the habit of naming your manifests and config files in a consistent fashion. Then you can easily pick up all of the files associated with a particular service. This is why we have called the config file ServiceB.Config.xml. However, MRDS does not care what the filename is.

*If a config file is specified in a manifest without a path, then it is assumed to be in the same directory as the manifest. This is an easy way to keep the files together. However, if the config file is specified in the service code without a path (and not listed in the manifest), then it defaults to the MRDS root directory.*

The `ServiceUri` is a URI, not a directory path on your hard drive. You can, however, specify the location of a config file using the `ServicePaths` class, as shown by the following example from the Dashboard in Chapter 4:

```
    private const string InitialStateUri = ServicePaths.MountPoint + @"/ProMRDS/
Config/Dashboard.Config.xml";

    // shared access to state is protected by the interleave pattern
    // when we activate the handlers
    [InitialStatePartner(Optional = true, ServiceUri = InitialStateUri)]
    StateType _state = null;
```

`ServicePaths.MountPoint` maps to "/mountpoint," but you should not make this assumption. The mountpoint is the root directory of the MRDS installation on the machine where the DSS node is running. Directory paths are specified relative to this. The `ServicePaths` class contains a number of other URI prefixes that you can use. You can investigate it using the Object Browser or reflection (by typing `ServicePaths` in the Visual Studio editor and then right-clicking it and selecting Go To Definition).

## Specifying a Configuration File in a Manifest

If you want to specify the config file in the manifest, you must add a service state partner to ServiceB. Open `ServiceB.manifest.xml` in Visual Studio and update it by adding the following highlighted code:

```
<?xml version="1.0" ?>
<Manifest
    xmlns="http://schemas.microsoft.com/xw/2004/10/manifest.html"
    xmlns:dssp="http://schemas.microsoft.com/xw/2004/10/dssp.html"
    >
    <CreateServiceList>
        <ServiceRecordType>
    <dssp:Contract>http://schemas.tempuri.org/2008/01/serviceb.html</dssp:Contract>
            <dssp:PartnerList>
              <dssp:Partner>
                <dssp:Service>ServiceB.Config.xml</dssp:Service>
                <dssp:Name>dssp:StateService</dssp:Name>
              </dssp:Partner>
            </dssp:PartnerList>
        </ServiceRecordType>
    </CreateServiceList>
</Manifest>
```

The first time that you run ServiceB after completing the rest of the changes in this section, it will create a config file in the `Projects\ServiceB` directory because this is where the manifest is located and no path is specified in the manifest.

Your code, usually in the `Start` method, must be prepared to handle the situation where no state is defined. This might happen, for example, the first time you run a service. In this case, the global variable (usually called _state by convention) will be `null` because no config file was found.

Go to the Start method in ServiceB.cs and add the following code in the middle of the function to check the state:

```
protected override void Start()
{
    base.Start();
    // Add service specific initialization here.
    Console.WriteLine("ServiceB starting");

    // Make sure that we have an initial state!
    if (_state == null)
    {
        _state = new ServiceBState();
    }
    // Sanity check the values (or initialize them if empty)
    if (_state.Interval <= 0)
        _state.Interval = 1000;

    // Save the state now
    SaveState(_state);

    SpawnIterator(MainLoop);

}
```

As well as creating a new state if there isn't one, the code confirms that the information in the state is sensible. The value of Interval will be zero if the state has just been created. It also saves the state (config file), as explained in the next section.

## Saving the State to a Config File

Saving the current state of a service is actually quite trivial — just call the SaveState helper function in DSS. However, you need to set up the name of the config file first, as explained in the previous section.

The following code from the previous code snippet saves the state each time the service runs:

```
// Save the state now
SaveState(_state);
```

SaveState is just a helper function that posts a Replace message to the Mount service. Because this happens asynchronously, you cannot assume when it returns that the state has been saved, or even that it was successful! However, SaveState returns a PortSet, so you can use it in a Choice to determine whether it returns a Fault. More important, if you are saving the state in your Drop handler, then you can wait until the save has completed before finally shutting down the service. The simplest way to wait is as follows:

```
yield return (Choice)SaveState(_state);
```

This assumes that you are inside an iterator.

In V1.5, the default output directory for saved config files was changed to the MRDS root directory (if no path is specified in the Partner attribute). For this reason, it is preferable to supply the filename in the manifest file. Then the file will be saved to the same directory as the manifest. Alternatively, as explained in the last section, you can specify an explicit path.

Saved state files are in XML format. In the case of ServiceB, the first time the state is saved it should look like the following:

```
<?xml version="1.0" encoding="utf-8"?>
<ServiceBState xmlns:s="http://www.w3.org/2003/05/soap-envelope"
xmlns:wsa="http://schemas.xmlsoap.org/ws/2004/08/addressing"
xmlns:d="http://schemas.microsoft.com/xw/2004/10/dssp.html"
xmlns="http://schemas.tempuri.org/2008/01/serviceb.html">
  <Interval>1000</Interval>
  <SensorValue>0</SensorValue>
</ServiceBState>
```

You can edit the saved state and change the `Interval`. The file is in the MRDS root directory and is called `ServiceB.Config.xml`. Open this file in Notepad and change `Interval` to 200. Save the file.

Rerun ServiceB. The output should appear a lot faster. What you have done is to change the configuration of ServiceB without modifying the code and recompiling. Even though ServiceB rewrites the config file every time it runs, it simply propagates what is already there.

# Modifying Service State

The DSS Protocol defines several types of operations that you can use to modify the state of a service, but only two of them are commonly used in MRDS: `Replace` and `Update`.

`Replace` messages are not used very often, so some services do not implement this operation. It is easy to understand why if you consider that a sophisticated service has a lot of properties in the state and these properties might not be directly related to one another. For example, you would rarely want to set the sensor polling interval, the serial COM port for communications, and the power to the motors at the same time.

This is where `Update` comes in. The Generic Brick contract in Chapter 17 defines separate `Update` messages for several tasks, including setting the drive power, changing the brick configuration parameters, and turning on LEDs.

## Replace

A `Replace` message supplies a completely new copy of the state. Because the `Body` of the message contains an instance of the state that is specific to this service, you must define a `Replace` class in your service types file (where xxx is the service name):

```
    public class Replace : Replace<xxxState, ↵
PortSet<DefaultReplaceResponseType, Fault>>
    {
        public Replace()
        {
        }

        public Replace(xxxState body)
```

```
                    : base(body)
            {
            }
    }
```

A basic `Replace` handler might look like the following:

```
[ServiceHandler(ServiceHandlerBehavior.Exclusive)]
public IEnumerator<ITask> ReplaceHandler(Replace replace)
{
    _state  = replace.Body;
    replace.ResponsePort.Post(DefaultReplaceResponseType.Instance);
    yield break;
}
```

Notice that the handler is declared in the Exclusive group. This guarantees that the state is not updated by two different handlers at the same time, which could lead to an inconsistent state.

However, note also that the code does no checking of the incoming state. Most of the samples in the MRDS distribution are like this (if they have a `Replace` operation). It would be better to check crucial values in the state before making the replacement.

The handler sends an acknowledgment message to the `ResponsePort` when it has finished. It uses a pre-defined message called `DefaultReplaceResponseType.Instance` that makes your life easier because you don't need to create a response type. (Similar instances are available for other types of operations; use IntelliSense to look for them.) If the partner service cannot continue until the replacement is complete, then it can wait for this response message to arrive.

## Update

You use the `Update` messages when only a portion of the state needs to change. You need to define an appropriate class to hold all the information you want to update, and then write a handler to perform the update. This must be an Exclusive handler to avoid potential conflicts.

### Defining a Class for an Update Request

Go back to `ServiceBTypes.cs` in Visual Studio. At the bottom of the file below the `Get` class, add the following code:

```
/// <summary>
/// ServiceB Set Interval Operation
/// </summary>
public class SetInterval: Update<SetIntervalRequest,
            PortSet<DefaultUpdateResponseType, Fault>>
{
    public SetInterval()
    {
    }
}

/// <summary>
/// Set Interval Request
```

*(continued)*

**133**

*(continued)*

```
/// </summary>
[DataContract]
[DataMemberConstructor]
public class SetIntervalRequest
{
    [DataMember, DataMemberConstructor]
    public int Interval;
}
```

The new class, `SetInterval`, is based on `Update` and uses a request message type of `SetIntervalRequest` and returns a `DefaultUpdateResponseType` if it is successful.

The `SetIntervalRequest` class has only one member, which is the `Interval`. Notice that there is a data contract on `SetIntervalRequest` and it requests that `DssProxy` generate a helper to construct a new instance using the [DataMemberConstructor] attribute. Because of the constructor, you can execute this operation from another service using the following shorthand:

```
_servicebPort.SetInterval(1500);
```

This assumes, of course, that `_servicebPort` is a service forwarder port that points to ServiceB.

## Adding a Handler to Process Update Requests

Now you need to go to `ServiceB.cs` to add the handler for the `SetInterval` operation. At the bottom of the file, add the following handler just before the end of the service class:

```
/// <summary>
/// Set Interval Handler
/// </summary>
/// <param name="request">SetIntervalRequest</param>
/// <returns></returns>
[ServiceHandler(ServiceHandlerBehavior.Exclusive)]
public virtual IEnumerator<ITask> SetIntervalHandler(SetInterval request)
{
    if (_state == null)
        // Oops! Return a failure
        request.ResponsePort.Post(new Fault());
    else
    {
        // Set the interval
        _state.Interval = request.Body.Interval;
        // Return a success response
        request.ResponsePort.Post(DefaultUpdateResponseType.Instance);
    }
    yield break;
}
```

This handler simply updates the `_state.Interval` from the request body and then sends back an acknowledgment using one of the pre-defined default response types.

It is worth mentioning at this point that this example demonstrates defensive programming: The code checks for a null state before using it. When services start, there is no guarantee in what order they will complete their initialization. A race condition is possible if another service sends a request before the initial state of ServiceB has been loaded. ServiceB must protect itself from an access violation by confirming that the state actually exists before updating it.

Note that the ServiceB GetHandler should also be modified to determine whether the state is null:

```
/// <summary>
/// Get Handler
/// </summary>
/// <param name="get"></param>
/// <returns></returns>
[ServiceHandler(ServiceHandlerBehavior.Concurrent)]
public virtual IEnumerator<ITask> GetHandler(Get get)
{
    if (_state == null)
        // Oops! Return a failure
        // This can happen due to race conditions if another service
        // issues a Get before the Initial State has been defined
        get.ResponsePort.Post(new Fault());
    else
        // Return the state
        get.ResponsePort.Post(_state);
    yield break;
}
```

## Returning Errors

Notice in the preceding code that if there is no state, a Fault is returned. Errors are always returned this way via the response port. However, the code just creates an empty Fault object. This is not good practice, and the programmer should be reprimanded. A Fault should always include information about what caused the error.

There are several ways to create a Fault that contains useful information. Inside the catch block of a try/catch, you can use the Fault.FromException helper method to convert the Exception into a Fault.

If you want to use a string as the error message (called the *reason*), you can use the following:

```
// Create a new Fault based on the error message
Fault fault = Fault.FromCodeSubcodeReason(FaultCodes.Receiver,
        DsspFaultCodes.OperationFailed, "Some error message");
```

There are many different fault codes and DSSP fault codes to choose from. You can list them in the editor by using IntelliSense and find the ones that are most appropriate for your situation.

# Service Initialization

You have already seen the `Start` method that must be present in every service. This is where you place your initialization code. When a new service is created, it includes a call to `base.Start`, which does the following:

- ❑ Calls `ActivateDsspOperationHandlers` on all your main and alternate ports, trying to hook up handlers with the `[ServiceHandler]` attribute to ports in the operation PortSets

- ❑ Sends a `DirectoryInsert` to add your service to the Service Directory

- ❑ Does a `LogInfo` with the URI of the service

If you have a lot of initialization to do, don't call `base.Start` until *after* you have completed the initialization. This appears to be counter to the comment that is automatically inserted when you create a service, but it does not cause any problems:

```
protected override void Start()
{
    base.Start();
    // Add service specific initialization here.
}
```

In older code (from V1.0) you might see the following pattern instead of `base.Start`:

```
// Listen for each operation type and call its Service Handler
ActivateDsspOperationHandlers();
// Publish the service to the local Node Directory
DirectoryInsert();
// Display HTTP service Uri
LogInfo(LogGroups.Console, "Service uri: ");
```

Going back even further to older code, `ActivateDsspOperationHandlers` was not used; there was an explicit declaration of the main interleave — for example, if you look in `ArcosCore.cs` for the Pioneer 3DX robot:

```
Activate(Arbiter.Interleave(
    new TeardownReceiverGroup
    (
        Arbiter.Receive<DsspDefaultDrop>(false,_mainPort,DropHandler)
    ),
    new ExclusiveReceiverGroup
    (
        Arbiter.Receive<Replace>(true, _mainPort, ReplaceHandler),
Arbiter.ReceiveWithIterator<Subscribe>(true, _mainPort, SubscribeHandler),
    ),
    new ConcurrentReceiverGroup
    (
Arbiter.Receive<DsspDefaultLookup>(true,_mainPort,DefaultLookupHandler),
```

```
        Arbiter.Receive<Get>(true, _mainPort, GetHandler),
        Arbiter.Receive<HttpGet>(true,_mainPort, HttpGetHandler),
        Arbiter.Receive<HttpQuery>(true,_mainPort, HttpQueryHandler),
        Arbiter.Receive<Query>(true, _mainPort, QueryHandler),
        Arbiter.Receive<Update>(true, _mainPort, UpdateHandler),
    )
));
```

You no longer have to do all of this because DSS uses reflection to find your handlers based on the [ServiceHandler] attribute and the method signature. However, it is helpful to understand what is actually happening when your service starts.

Note that if you want to add handlers to the main interleave after the service has started, you should use the Arbiter.CombineWith method. If you create a new interleave, it runs independently of the main interleave and therefore does not guarantee exclusivity. Likewise, if you use SpawnIterator or Activate to start a task, then the task runs outside of the main interleave.

A common thread in the discussion forums is about how to introduce delays during initialization to ensure that other services have started up properly. Using Thread.Sleep is *not* recommended because it blocks a thread. If possible, check the status of other services by sending a Get request to them (and waiting for the response) or looking in the directory to see whether they are visible.

Another alternative is to use SpawnIterator to start a new thread running an iterator. Then you can use the TimeoutPort to insert delays (if you really feel that you need to), which will release the thread until the timeout occurs. With Arbiter.ExecuteToCompletion, you can even call a series of other iterators, one after another.

In summary, best practice for initialization is to use iterators, and avoid blocking threads.

# Composing and Coordinating Services

One of the benefits of writing everything as services is that you can combine, or orchestrate, services to create more complex applications. This section discusses how to start and stop services and establish partnerships.

## Starting and Stopping Services Programmatically

In most cases, you do not need to explicitly start or stop a service because all the services you require will be started when the manifest is loaded. However, this section explains the procedure. The next section ("Using the Partner Attribute") shows you an easier approach.

Remember that ServiceA is the one that is in charge. It partners with ServiceB and makes requests to ServiceB. Therefore, open ServiceA.cs in Visual Studio.

## Adding a Service Forwarder Port

You need a service forwarder port so that you can send requests to ServiceB. However, because the code will create a new instance of ServiceB, the service port must be `null` initially. Add the following code somewhere near the top of the service class:

```
// Create a port to access Service B,
// but we don't know where to send messages yet
serviceb.ServiceBOperations _servicebPort = null;
```

For the moment, ignore the question of how to establish a partnership between ServiceA and ServiceB. In the `Start` method of ServiceA, add a line of code to spawn a new task:

```
/// </summary>
protected override void Start()
{
    base.Start();
    // Add service specific initialization here.
    Console.WriteLine("ServiceA starting");

    // Start the main task on a separate thread and return
    SpawnIterator(MainTask);
}
```

This achieves the purpose of leaving a thread running but finishing the service initialization as far as DSS is concerned.

## Main ServiceA Behavior

Underneath the `Start` method, add the `MainTask`. This routine creates a new ServiceB; calls `MainLoop`, which runs until it is satisfied; and then closes down the services and the DSS node:

```
private IEnumerator<ITask> MainTask()
{
    Port<EmptyValue> done = new Port<EmptyValue>();

    SpawnIterator<Port<EmptyValue>>(done, CreatePartner);

    // Wait for a message to say that ServiceB is up and running
    yield return Arbiter.Receive(
        false,
        done,
        EmptyHandler
    );

    // Check that we have a forwarder
    if (_servicebPort == null)
    {
        LogError("There is no ServiceB");
        yield break;
    }
    else
```

```
    {
        SpawnIterator<Port<EmptyValue>>(done, MainLoop);
        yield return Arbiter.Receive(
            false,
            done,
            EmptyHandler
        );
    }

    // We no longer require ServiceB so shut it down
    // NOTE: This is not necessary -- it is just here to illustrate that
    // other services can be shut down
    LogInfo(LogGroups.Console, "Dropping ServiceB ...");
    _servicebPort.DsspDefaultDrop();

    Console.WriteLine("ServiceA finished");
    // Wait a while for ServiceB to exit
    yield return Arbiter.Receive(
        false,
        TimeoutPort(500),
        delegate(DateTime time)
        { }
    );

    // Pause for user input --
    // This is just so that the DSS node stays up for the time being
    Console.WriteLine("Press Enter to exit:");
    Console.ReadLine();
    // Shut down the DSS node
    ControlPanelPort.Post( ↩
new Microsoft.Dss.Services.ControlPanel.DropProcess());

    yield break;
}
```

## Using Completion Ports for Synchronization

As an example, MainTask creates a port called done for signaling purposes. Because no information needs to be transferred, it uses the EmptyValue class. Then it spawns the CreatePartner iterator and waits on the done port. An alternative, and much neater way to do it without signaling on a port, is to use Arbiter.ExecuteToCompletion.

The first step in CreatePartner is to call CreateService using the ServiceB contract identifier. If this is successful, then a new URI is created based on the service information. Then a ServiceForwarder is created. This enables messages to be sent to the operations port of the newly created service. (The _servicebPort was declared at the beginning of this section as a global).

```
        private IEnumerator<ITask> CreatePartner(Port<EmptyValue> p)
        {
            // Create ServiceB instance
            Console.WriteLine("Creating new ServiceB");
            yield return Arbiter.Choice(CreateService(serviceb.Contract.
    Identifier),
                delegate(CreateResponse s)
```

*(continued)*

*(continued)*

```
            {
                // Create Request succeeded.
                LogInfo(LogGroups.Console, "ServiceB created: " + s.Service);

                Uri addr;
                try
                {
                    // Create URI from service instance string
                    addr = new Uri(s.Service);

                    // Create forwarder to ServiceB
                    _servicebPort = ↵
ServiceForwarder<serviceb.ServiceBOperations>(addr);
                }
                catch (Exception ex)
                {
                    LogError(LogGroups.Console, ↵
"Could not create forwarder: " + ex.Message);
                }
            },
            delegate(W3C.Soap.Fault failure)
            {
                // Request failed
                LogError(LogGroups.Console, "Could not start ServiceB");
            }
        );

        // Signal that the create is finished
        p.Post(EmptyValue.SharedInstance);

        yield break;
    }
```

Assuming that a service forwarder port is created successfully, the `MainLoop` iterator is called (also using the `done` port to signify completion). `MainLoop` looks like this:

```
    private IEnumerator<ITask> MainLoop(Port<EmptyValue> p)
    {
        // Issue several Gets to ServiceB
        yield return Arbiter.ExecuteToCompletion(Environment.TaskQueue,
            Arbiter.FromIteratorHandler(GetData));

        // Change the update interval on ServiceB
        _servicebPort.SetInterval(1500);

        // Do some more Gets to see the effect of the changed interval
        yield return Arbiter.ExecuteToCompletion(Environment.TaskQueue,
            Arbiter.FromIteratorHandler(GetData));

        // Finally, post a message to say that we are finished
        p.Post(EmptyValue.SharedInstance);

        yield break;

    }
```

`MainLoop` in turn calls another iterator to request the sensor data from ServiceB. Then it changes the update interval in ServiceB and again reads the sensor data several times using `GetData`:

```
// Request "sensor" data from ServiceB
private IEnumerator<ITask> GetData()
{
    for (int i = 0; i < 10; i++)
    {
        // Send a Get request to ServiceB
        yield return Arbiter.Choice(_servicebPort.Get(),
            delegate(serviceb.ServiceBState s)
            {
                // Get request succeeded
                LogInfo(LogGroups.Console,  ↩
"ServiceB Sensor: " + s.SensorValue);
            },
            delegate(W3C.Soap.Fault failure)
            {
                // Get request failed
                LogError(LogGroups.Console,  ↩
"Get to ServiceB failed" + failure.Reason);
            }
        );

        // Wait for 1 second
        yield return Arbiter.Receive(
            false,
            TimeoutPort(1000),
            delegate(DateTime time)
            { }
        );

    }

    yield break;
}
```

This code uses the `LogInfo` method rather than `Console.WriteLine`. The output is therefore clearly differentiated from the `Console.WriteLine` output, as you can see in Figure 3-22.

Notice that `GetData` uses `Get` operations on ServiceB, repeated at one-second intervals. This timer interval does not change, but partway through, ServiceA changes the interval in ServiceB. You should be able to see the effects of these mismatched time intervals quite clearly in the output. This merely illustrates that sometimes the data you get might be a little stale.

### Dropping Services

When the `MainLoop` is finished, a `Drop` message is sent to ServiceB (from `MainTask`). Then the code waits for the user to press Enter. This is not something that you would normally do. In particular, `Console.ReadLine` is not CCR-friendly and should not be used. However, it enables you to poke around inside the DSS node using a web browser before the whole thing is closed down.

To stop a service, you send it a `Drop` message. By default, the DSS runtime handles `Drop` requests for you, so you don't have to write any code. However, ServiceB does something a little nasty — it runs an infinite loop on a dedicated thread. All sorts of "nasties" can arise when you are programming MRDS services, so it is worthwhile to review `Drop` handlers here.

Go to the bottom of the `ServiceB.cs` file in Visual Studio. (If you have been following the instructions, there should be a `SetIntervalHandler` there). Insert the following code:

```
/// <summary>
/// Drop Handler
/// </summary>
/// <param name="drop"></param>
/// <returns></returns>
[ServiceHandler(ServiceHandlerBehavior.Teardown)]
public virtual IEnumerator<ITask> DropHandler(DsspDefaultDrop drop)
{
    // Tell the main loop to stop
    _shutdown = true;

    Console.WriteLine("ServiceB shutting down");

    // Make sure you do this or the sender might be blocked
    // waiting for a response. The base handler will send
    // a response for us.
    base.DefaultDropHandler(drop);

    // That's all folks!
    yield break;
}
```

Make sure that the `Drop` handler is in the TearDown group. If you inadvertently place it into the Concurrent group, very strange things happen. The service will disappear from the service directory, but it continues executing!

The only function that the `DropHandler` needs to perform is to set the `_shutdown` flag so that the main loop terminates on the next iteration. Then it calls `base.DefaultDropHandler` to finish the job.

Just declaring this handler is sufficient to override the default handler defined in DSS — there is nothing else you need to do to make it work.

Now ServiceA can send a message to ServiceB to tell it to terminate itself:

```
_servicebPort.DsspDefaultDrop();
```

The last step is for ServiceA to send a `DropProcess` message to the Control Panel, which shuts down the DSS node.

## Running the Services Together

Recompile both ServiceA and ServiceB. Then run ServiceA in the debugger. If you have done everything properly, then Service B should also start up, as shown in Figure 3-22.

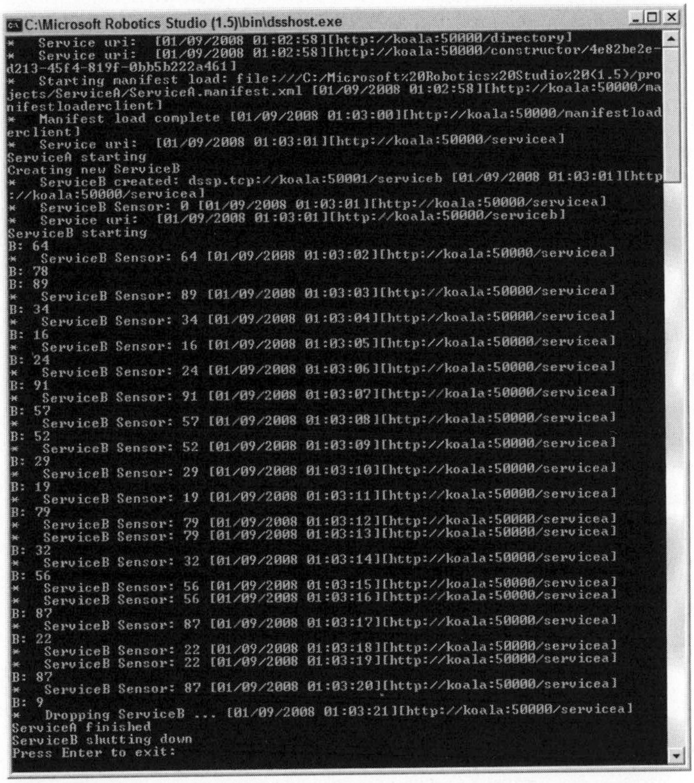

Figure 3-22

If you read through the output in Figure 3-22, you will see that ServiceB is created by ServiceA, but the first Get response from ServiceB is before it has output any "sensor" readings, and consequently the value is zero. (The first "reading" is when ServiceB displays B: 64).

After 10 iterations, the timer interval for ServiceB is changed by ServiceA. Then you see sensor readings doubling up because ServiceA continues to poll ServiceB at the same rate as before. This is now too fast for the incoming "sensor data" at the ServiceB end.

Lastly, ServiceA tells ServiceB to shut down by sending a Drop message.

# Using the Partner Attribute to Start Services

Rather than create services explicitly, you can create a service partner declaratively using the [Partner] attribute. This is the usual approach. However, sometimes you do not know ahead of time what services you will have to partner with — for example, the Dashboard dynamically partners with Differential Drive services, Laser Range Finder services, and Webcam services. When it starts, it doesn't know which of these services might be available. More to the point, it doesn't know which DSS node you are going to point it at.

To specify a partner at compile time, go back to the top of ServiceA.cs and change the declaration of _servicebPort as follows:

```
// Partner with ServiceB
[Partner("ServiceB", Contract = serviceb.Contract.Identifier,
    CreationPolicy = PartnerCreationPolicy.CreateAlways, Optional = false)]
private serviceb.ServiceBOperations _servicebPort =
    new serviceb.ServiceBOperations();
```

Notice that the contract identifier for ServiceB is specified as a parameter to the [Partner] attribute. In addition, ServiceB is *not* optional. The CreationPolicy is set to CreateAlways. This causes the constructor service to create a new instance of ServiceB for you. It doesn't matter whether you create a new instance of the service operations port or leave it null if the config file exists because the State service will take care of it.

Four service creation policies are available:

- ❑ CreateAlways
- ❑ UseExistingOrCreate
- ❑ UseExisting
- ❑ UsePartnerListEntry

The simplest approach is CreateAlways. This is equivalent to creating the service explicitly but requires only a couple of lines of code. In some cases this makes sense, such as when a service uses another service that is never called directly by anyone else, i.e., there is a one-to-one partnership.

Now that you have declared the partnership with ServiceB, there is no longer any need to call the CreatePartner routine. However, for the purposes of illustrating a point, you will insert some different code.

The FindPartner routine replaces CreatePartner:

```
private IEnumerator<ITask> MainTask()
{
    Port<EmptyValue> done = new Port<EmptyValue>();

    SpawnIterator<Port<EmptyValue>>(done, FindPartner);
```

FindPartner issues a DirectoryQuery to try to find ServiceB, but with the default long timeout:

```
private IEnumerator<ITask> FindPartner(Port<EmptyValue> p)
{
    // Find ServiceB in the local directory
    Console.WriteLine("Finding ServiceB in the Directory");
    yield return Arbiter.Choice(
        DirectoryQuery(serviceb.Contract.Identifier,
DsspOperation.DefaultLongTimeSpan),
        delegate(ServiceInfoType s)
        {
            // Request succeeded
            LogInfo(LogGroups.Console, "Found ServiceB: " + s.Service);

            Uri addr;
            try
            {
                // Create URI from service instance string
                addr = new Uri(s.Service);

                // Now create service forwarder to ServiceB
                _servicebPort =
ServiceForwarder<serviceb.ServiceBOperations>(addr);
            }
            catch (Exception ex)
            {
                LogError(LogGroups.Console,
"Could not create forwarder: " + ex.Message);
                _servicebPort = null;
            }
        },
        delegate(W3C.Soap.Fault failure)
        {
            // Request failed
            LogError(LogGroups.Console, "Could not find ServiceB");
            _servicebPort = null;
        }
    );

    // Signal that the find is finished
    p.Post(EmptyValue.SharedInstance);

    yield break;
}
```

Basically, FindPartner waits until ServiceB does a DirectoryInsert and becomes visible in the service directory. The rest of the code is the same as CreatePartner. This is one way to synchronize service startup. For it to work successfully, services must not call base.Start until they have completed all of their initialization. Otherwise, they appear in the directory prematurely. If you run this revised version of the code, there is not very much difference, so no screenshot is provided.

You can change the partner creation policy to UseExistingOrCreate, but you won't see any difference in this scenario because the "Or Create" option causes ServiceB to be created. In other words, ServiceA uses an existing instance of ServiceB if one exists; if not, it simply goes ahead and creates one anyway.

# Modifying Manifests Manually

When you create a new service, a manifest is automatically created for you. This section steps you through the process of modifying a manifest using a text editor. However, as you will see later in "Modifying Manifests Using the DSS Manifest Editor," it is preferable to use the appropriate tool for this task. The purpose of this section is to give you some understanding of what manifests look like and how they work. It is not suggested that you always edit manifests by hand.

## Use Existing Partner

In Visual Studio for ServiceA, open `ServiceA.manifest.xml`. It should have a single service record for ServiceA. Leave it alone for the moment.

Change the partner creation policy in `ServiceA.cs` to `UseExisting`:

```
// Partner with ServiceB
[Partner("ServiceB", Contract = serviceb.Contract.Identifier,
    CreationPolicy = PartnerCreationPolicy.UseExisting, Optional = false)]
    private serviceb.ServiceBOperations _servicebPort = ↵
new serviceb.ServiceBOperations();
```

This specifies that there must be a ServiceB running on the DSS node before ServiceA can start up properly. You can see from the ServiceA manifest that there is no mention of ServiceB.

Recompile the code and run ServiceA. Be patient — very patient! In fact, be patient for up to two minutes. Eventually, an error message will appear (displayed in red), as shown in Figure 3-23.

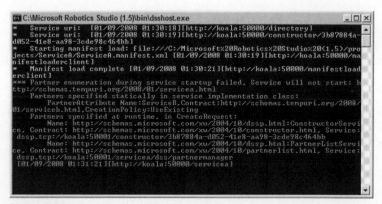

Figure 3-23

This error says, "Partner enumeration during service startup failed." Quite clearly, the error occurred because you asked for ServiceB but did not create a new instance of it. This is easy to fix.

Open the `ServiceB.manifest.xml` in Visual Studio and copy the Service Record. Then go to `ServiceA.manifest.xml` and paste it in there. The resulting ServiceA manifest should look like the following:

```xml
<?xml version="1.0" ?>
<Manifest
    xmlns="http://schemas.microsoft.com/xw/2004/10/manifest.html"
    xmlns:dssp="http://schemas.microsoft.com/xw/2004/10/dssp.html"
    >
  <CreateServiceList>
    <ServiceRecordType>
    <dssp:Contract>http://schemas.tempuri.org/2008/01/servicea.html</dssp:Contract>
    </ServiceRecordType>
    <ServiceRecordType>
    <dssp:Contract>http://schemas.tempuri.org/2008/01/serviceb.html</dssp:Contract>
      <dssp:PartnerList>
        <dssp:Partner>
          <dssp:Service>ServiceB.Config.xml</dssp:Service>
          <dssp:Name>dssp:StateService</dssp:Name>
        </dssp:Partner>
      </dssp:PartnerList>
    </ServiceRecordType>
  </CreateServiceList>
</Manifest>
```

*Be very careful when you copy and paste pieces of code between XML files. If you end up with mismatched tags, the manifest won't work at all and you will just get a syntax error. This is one reason why it is better to use the Manifest Editor (discussed in "Modifying Manifests Using the DSS Manifest Editor"), rather than edit manifests manually.*

Remember that in the section "Specifying a Configuration File in a Manifest" you added a state service partner to ServiceB. This is part of the Service Record as well. However, when you run ServiceA now, a new config file is created in the `Projects\ServiceA` folder because that is where the manifest is. It all gets a little confusing after a while.

This is the brute-force approach to partnering — it simply specifies two services in the manifest and they are both created. Save the manifest and run ServiceA again. The services should now run successfully.

## Contract Identifiers and Case Sensitivity

If you have plenty of time and nothing better to do, edit the manifest again by changing the names of the services in the contract identifiers to ServiceA and ServiceB, i.e., capitalize them so that they look pretty. Run ServiceA. What happens? Look at Figure 3-24.

Figure 3-24

First, the contract directory cache is rebuilt because you requested two services that DSS has not seen before. However, this doesn't help and you get two error messages saying that there was an error creating the service, with a Fault.Subcode of UnknownEntry.

*This is a small trap for beginners — contract identifiers are case-sensitive.*

## Use Partner List Entry

The last step is to change the partner creation policy to UsePartnerListEntry. Recompile and run ServiceA.

This time you get an error that partner enumeration failed, but ServiceB starts up and begins pumping out "sensor" data. ServiceA, however, does not start.

To fix this problem, you need to edit the manifest and explicitly make ServiceB a partner of ServiceA. The resulting manifest looks like the following, with the changes highlighted:

```xml
<?xml version="1.0" ?>
<Manifest
    xmlns="http://schemas.microsoft.com/xw/2004/10/manifest.html"
    xmlns:dssp="http://schemas.microsoft.com/xw/2004/10/dssp.html"
    xmlns:this="http://schemas.tempuri.org/2008/01/servicea.html"
    >
  <CreateServiceList>
    <ServiceRecordType>
    <dssp:Contract>http://schemas.tempuri.org/2008/01/servicea.html</dssp:Contract>
      <dssp:PartnerList>
        <dssp:Partner>
          <dssp:Name>this:ServiceB</dssp:Name>
        </dssp:Partner>
      </dssp:PartnerList>
      <Name>this:ServiceA</Name>
```

```
    </ServiceRecordType>
    <ServiceRecordType>
    <dssp:Contract>http://schemas.tempuri.org/2008/01/serviceb.html</dssp:Contract>
      <dssp:PartnerList>
        <dssp:Partner>
          <dssp:Service>ServiceB.Config.xml</dssp:Service>
          <dssp:Name>dssp:StateService</dssp:Name>
        </dssp:Partner>
      </dssp:PartnerList>

      <Name>this:ServiceB</Name>

    </ServiceRecordType>
  </CreateServiceList>
</Manifest>
```

Both services have explicit names in the manifest, and ServiceB is in the `PartnerList` for ServiceA. ServiceB has a config file specified using the `StateService` as a partner. Note that there is no path, just a filename, so the config file is assumed to be in the same directory as the manifest.

When you run ServiceA with this revised manifest, both services start up and everybody is happy.

# DSS Tools

Several tools are supplied with MRDS to assist you with various tasks. You have already seen `DssNewService`, `DssHost`, and `DssProxy`. This section briefly outlines some other tools that you should become familiar with, including the Manifest Editor, `DssInfo`, and `DssProjectMigration`. Because it is a significant tool in its own right, `DssDeploy` is covered in the following section.

## Modifying Manifests Using the DSS Manifest Editor

The DSS Manifest Editor was introduced in MRDS V1.5 as a tool to help make the creation and editing of manifests easier. We do not intend to provide a tutorial on the Manifest Editor here, but you should be aware that it exists and read up on it in the online documentation. (Actually, the Manifest Editor is not a separate program but part of VPL. A batch file called `dssme.cmd` in the MRDS `bin` directory executes VPL in Manifest mode.)

To try out the Manifest Editor, use the very last manifest from the previous section. You can start the Manifest Editor by clicking Start ➪ Microsoft Robotics Studio (1.5) ➪ Microsoft DSS Manifest Editor. It might take a little while to start, because it is listing all of the available services.

When the Manifest Editor main screen appears, click File ➪ Open and browse to the ServiceA manifest to open it. Once the manifest has been loaded, it should look like what is shown in Figure 3-25.

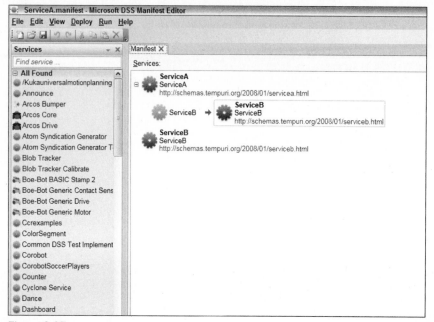

**Figure 3-25**

You can select ServiceB and add an initial configuration if you want. This will add a state partner with a nominated config file.

Apart from that, just play around in the Manifest Editor until you are comfortable with it. You will find it is much easier to use than editing XML files. It is the recommended method for editing manifests, rather than opening them in a text editor.

# DssInfo: Examining Contract Information

You can get information about contracts and the operations that a service supports using the DssInfo tool. For example, the following command displays information about the Dashboard service that is included with the code for Chapter 4:

```
C:\Microsoft Robotics Studio (1.5)>dssinfo bin\dashboard.y2007.m10.dll
```

You can control the amount of information displayed using command-line qualifiers. The default information level shows the contract, partners, the operations that are supported, and so on, as shown here:

```
Reflecting:              Dashboard.Y2007.M10.dll
DSS CONTRACT
Verbosity                ShowWarnings
Assembly:                c:\microsoft robotics studio (1.5)\bin\dashboar
                         d.y2007.m10.dll
```

```
Service:                          DashboardService
  DssContract:                    http://www.promrds.com/2007/10/dashboard.html
  Namespace:                      ProMRDS.Robotics.Services.Dashboard
  ServicePrefix:                  /dashboard
  Singleton:                      False
  Partner(s):
    [InitialStatePartner]         /mountpoint/ProMRDS/Config/Dashboard.Config.xml
    UseExistingOrCreate           http://schemas.microsoft.com/robotics/2006/09/g
                                  amecontroller.html
  ServicePort:                    DashboardOperations
    Lookup:                       DsspDefaultLookup
    Drop:                         DsspDefaultDrop
    Get:                          Get
    Replace:                      Replace
```

If you run `DssInfo` and specify `RoboticsCommon.dll`, you will see several generic contracts and some actual service implementations.

## DssProjectMigration: Migrating Services

At this stage, you do not need to migrate any services because all your code is on one computer and you have not upgraded MRDS. However, it is worth mentioning the `DssProjectMigration` tool so that you can file away it in the back of your mind for later use. This tool can be very useful in a couple of circumstances:

❑   When V2.0 of MRDS arrives, possibly around the time this book is released, you might want to take some of your existing services over to the new version. This is the primary purpose of `DssProjectMigration`, and it can migrate entire hierarchies of projects. In this case, the new (V2.0) tool will look for known changes to the code and try to fix anything that might break from V1.5.

❑   This tool ensures that a project is set up properly for its current location. If you move a project to a different folder, or if you transfer it to another computer where MRDS is installed on a different disk drive, then you need to update some of the project details. Rather than edit the `.csproj` and `.sln` files yourself, you can let `DssProjectMigration` take care of it for you.

For more information, you can refer to the online documentation when you need to use `DssProjectMigration`.

# Deploying Services

Sooner or later, you will want to distribute your services to friends, colleagues, or customers. MRDS includes a tool specifically designed for this purpose: `DssDeploy`.

There are fundamentally two different ways to deploy a service:

❑   Distribute only the service itself to a computer that already has MRDS installed, i.e., distribute the source code so other people can see it and modify it.

❑ Distribute the service plus the core components of MRDS so that the packaged service can be run on any computer (assuming the computer is running Windows XP, Vista, Mobile, or CE), i.e., distribute executables only.

For full documentation on DssDeploy, see the online help. This section simply provides two examples illustrating how to use it, but they are two commonly used examples.

## Sharing Source Code

If you are interested in sharing your code, you can package it so that other people can install it into their MRDS environment. This is what the authors have done with the ProMRDS code. In this case, the DssDeploy package contains only the source code for the relevant services and the compiled DLLs. All of the object files, debug symbol tables, and so on, are left behind because they can be recreated by compiling the source code.

To build a package for one of the chapters in this book, following these steps:

**1.** Make a listing of the files from a command prompt using the following:

```
Dir /b/o:n/s >dir.txt
```

**2.** You can then edit the output file (dir.txt) in Notepad to remove all of the object files and anything else that should not be included. (Strange files sometimes sneak into source directories when you are not looking). The list of files is turned into a DssDeploy options file, which is just a sequence of DssDeploy command-line options. You can use a hash sign (#) to insert comments.

**3.** Along with the source files, some other files should be included, such as the DLLs from the bin folder, the ReadMe file, etc. A standard template for an options file helps to ensure that you don't forget anything. The following example shows the options file for Chapter 2, which is relatively short. Called DeployChapter2Options.txt, it resides in the ProMRDS\Package folder:

```
# Chapter 2 Deploy Options File
# All Chapter files should follow the same basic format

# Include the deploy options file for later repackaging
/d:"..\Package\DeployChapter2Options.txt"

# Chapter Readme.htm and associated files
/d:"..\Chapter2\Readme.htm"
/d:"..\Chapter2\images\*"
/d:"..\Chapter2\styles\*.css"

# Add the Batch files to run the applications in this chapter
#/d:"..\BatchFiles\xxxx.cmd"

# Add the Config files for the applications (if any)
# and the Manifests
```

```
#/d:"..\Config\xxxx*"

# Include any models/meshes


# And the media files


# Now start with the source files

#/d:"..\Chapter2\Chapter2.sln"

#---------- CCR Examples ----------

# Add the Readme files for this application (if any)

# Include the binaries
/d:"..\..\bin\CCRExamples.Y2007.M12.dll"
/d:"..\..\bin\CCRExamples.Y2007.M12.proxy.dll"
/d:"..\..\bin\CCRExamples.Y2007.M12.transform.dll"

# Now the source code, but none of the compiled stuff
/d:"..\Chapter2\CCRExamples\AssemblyInfo.cs"
/d:"..\Chapter2\CCRExamples\Ccrexamples.cs"
/d:"..\Chapter2\CCRExamples\CCRExamples.csproj"
/d:"..\Chapter2\CCRExamples\CCRExamples.csproj.user"
/d:"..\Chapter2\CCRExamples\CCRExamples.manifest.xml"
/d:"..\Chapter2\CCRExamples\CCRExamples.sln"
/d:"..\Chapter2\CCRExamples\CcrexamplesTypes.cs"
```

4. To build chapter packages, you use a batch script to avoid typing long command lines. This script, called BuildChapterPackage.cmd, contains the following commands:

```
@echo off
echo Build a single chapter for ProMRDS
if "%1" == "" goto usage
if not exist DeployChapter%1Options.txt goto noopts
..\..\bin\dssdeploy /p /e+ /cv+ /s- /n:"ProMRDS Chapter %1 Package"
/r:"..\Chapter%1\Readme.htm" @DeployChapter%1Options.txt ProMRDSChapter%1.exe
copy ProMRDSChapter%1.exe c:\temp
copy /Y ProMRDSChapter%1.exe ProMRDSChapter%1.exe.safe
goto end

:usage
echo Usage: BuildChapterPackage num
echo where num is the Chapter number
:noopts
echo DeployChapternumOptions.txt file must already exist

:end
```

The /p qualifier asks DssDeploy to create a packed self-expanding executable. Deployment is restricted to only the same version of MRDS using the /cv+ qualifier, and security ACLs are turned off with /s-.

The ReadMe file for the package (/r qualifier) should be a HTML file. It is displayed automatically when the package finishes installing on the target computer. Other options, such as for including a license, signing the file, and so on, might be relevant to you if you are creating a commercial package.

The script automatically adds a name to the package, specifies the deploy options file, and creates a ".safe" version for sending via e-mail. It is not rocket science, and you might be able to come up with a more elegant solution. However, it works — as you have no doubt already seen!

5.    To deploy the package, copy it to the target PC and run it. It verifies that MRDS is installed and then creates the necessary folders and copies the files into them. Then it runs DssProjectMigration to ensure that the projects are correctly configured. Finally, it displays the ReadMe file in a web browser.

## Distributing Executables

When you get to Chapters 16 and 17, you will be deploying services to a PDA running Windows Mobile or an eBox embedded PC running Windows CE. These environments run the .NET Compact Framework, so they are referred to as CF services for short.

In this situation you do not want to install the full MRDS environment. (In fact, you can't do this anyway.) What you want is a runtime-only environment.

The batch script to create the package, called BuildStingerCFPackage.cmd, contains the following commands:

```
REM Create a DssDeploy package
setlocal
call "..\..\..\sdkenv.cmd"
cd "%~dp0"
DssDeploy /p /cf /n:"ProMRDS Stinger CF Example" /d:"../../BatchFiles/CF/*"
/m:../../Config/StingerCFDriveByWire.manifest.xml StingerCF.exe
pause
```

This DssDeploy command creates a package for the Compact Framework due to the /cf qualifier. It explicitly adds all the files from the BatchFiles\CF folder to the package, and then uses a manifest to collect all of the necessary services. DssDeploy analyzes the manifest and walks down the dependency tree looking for service DLLs. The MRDS runtime DLLs are automatically included.

Deploying a runtime version of MRDS in this fashion is not restricted to just CF platforms. You can also create a package without the /cf qualifier and deploy it to Windows XP or Vista.

There is one restriction on the runtime-only version, however, and it is related to the simulator. In order to run Simulation services, you need to have the AGEIA PhysX engine installed, as well as the latest versions of Microsoft DirectX and XNA. In short, you must have MRDS installed on the target machine if you want to deploy a simulation.

## *Viewing the Contents of a Package*

A tool called `ViewDssDeployContents`, written by Paul Roberts at Microsoft, is available from the Channel 9 website (`http://channel9.msdn.com/ShowPost.aspx?PostID=368096`), and you should definitely download a copy of it. Perhaps this tool will be included in MRDS V2.0.

This tool shows, in an Explorer-like tree, what files are included in the package. This is particularly handy if you want to check out a package before installing it. Otherwise, you might have no idea what files are going to be replaced. `DssDeploy` only lists the files that will be replaced if there are less than about 20 of them, but beyond that, it simply asks you something like "72 files will be replaced, OK?".

# Summary

This chapter addresses the basics of using DSS: state, operations, and partners. You have seen the various components of a DSS service and how services interact. There is a lot to learn and you probably recognized some overlap with the previous chapter. If you have not done so already, you should at least browse through the MRDS online documentation. In addition, go through the tutorials.

As noted at the beginning of the chapter, it takes some time to get used to the MRDS terminology and way of doing things. The best way to learn is to get your hands dirty and start hacking services.

The next chapter elaborates on writing DSS services to enable you to perform more advanced functions, including using Windows Forms and a web camera.

# Advanced Service Concepts

The Decentralized Software Services (DSS) are responsible for controlling the basic operations of robotics applications. As explained in the previous chapter, DSS is built on top of the CCR. This chapter moves on to more advanced features of services, including subscriptions and user interfaces. It is very likely that you will need to use the material in this chapter at some stage.

This chapter develops a service, called TeleOperation, that can be used to drive a robot with a web camera attached so you can see where the robot is going. The code introduces you to using Windows Forms as well as Web Forms with MRDS services, and demonstrates how to display live video. Don't worry if you don't have a robot with a camera because the faithful simulator comes to the rescue and you can still work through the code.

You might want your user interface to look like the cockpit of a 747, but that is not the objective in this chapter. It covers the basics of using Windows Forms (WinForms) and how to format web pages using XSLT. You can add fancy stuff yourself later.

These concepts are used in the Dashboard service that is included with this chapter. However, the discussion in the text uses the TeleOperation service because it contains a small subset of the Dashboard functions.

## Setting Up for This Chapter

Although MRDS is full of examples, a new service is provided in this chapter to perform *teleoperation*: driving a robot remotely using an onboard camera. This is a common scenario, used, for example, in bomb disposal robots. In this case, you are the brains of the robot. It is basically just a camera on wheels, and requires only two services: Generic Differential Drive and Webcam.

You can use the TeleOperation service in this chapter with any robot that has a Generic Differential Drive service. In addition, if you can attach a camera to the robot somehow and it is accessible via the Webcam service, then you can see what the robot sees. The simplest solution is a small wireless "spy camera" with a video receiver that can be connected to your PC via a video capture device. This setup is shown diagrammatically in Figure 4-1.

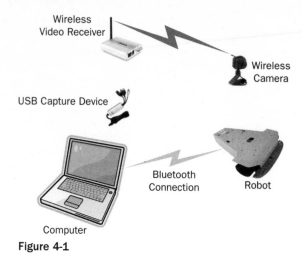

**Figure 4-1**

Figure 4-2 shows a Swann wireless camera mounted on the front of a LEGO NXT Tribot. The camera is smaller than the 9V battery that powers it. Because the camera is not a LEGO part, liberal use of Blutac is required to fasten the camera and the battery onto the Tribot.

**Figure 4-2**

Even if you do not have a wireless camera, you can still try out the service using the Maze Simulator (which is in Chapter 9) with a simulated Pioneer 3DX robot. By default, when you run the TeleOperation service from Visual Studio, it starts the Maze Simulator. To run the service with a real robot, you need to edit the manifest: `ProMRDS\Chapter4\TeleOperation\TeleOperation.manifest.xml`.

You can also edit the command-line parameters in the Debug properties for the Solution and add another manifest, but you still need to remove the Maze Simulator from the existing manifest in this case.

The design of the TeleOperation service allows you to connect to a DSS node anywhere on the network to use the robot and camera. It connects to the first Differential Drive service that it finds, and the first WebCam. This should rarely be a problem because you are unlikely to be trying to teleoperate two robots at the same time!

Figure 4-3 shows a screenshot of the TeleOperation service (left). And look, there's R2-D2!

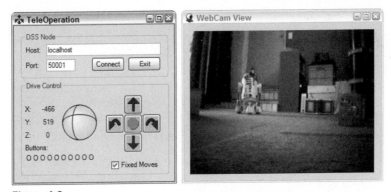

**Figure 4-3**

The TeleOperation control window is similar to the Dashboard, which is also included in the `Chapter4` folder. You can use a game controller or joystick, or use the mouse with the onscreen "trackball." You can also use the arrow keys on the keyboard to drive the robot around. The arrow buttons on the left WinForm in Figure 4-3 use the `DriveDistance` and `RotateDegrees` operations, so the robot only moves a defined distance. However, you can uncheck the Fixed Moves checkbox and then the arrow buttons just turn the motors on as appropriate and you have to use the Stop button to stop the robot. This is similar to Robotics Tutorial 4, called "Drive by Wire." This option is provided because some robot services do not implement the `DriveDistance` and `RotateDegrees` requests.

The second WinForm (on the right in Figure 4-3) is the WebCam View, which shows live video from the camera mounted on the robot. Setting up the camera can be a little bit of a compromise. It needs to be tilted down to see the floor immediately in front of the robot, but this can limit how far it can see into the distance.

## Hardware Setup

Your particular hardware might be quite different from the hardware used during the development of this chapter. For example, a Corobot has a web camera mounted on the front as a standard component, but you might have to jury-rig a camera onto your robot somehow. Regardless of your hardware setup, the key point is that the robot must have a Differential Drive service and the camera must be supported by the Webcam or IP Camera services.

The wireless camera used for testing in this chapter is a Swann Microcam with a Belkin USB Video Capture Device (VCD) to grab frames from the camera. The camera transmits in the 2.4GHz range. This is the same unregulated frequency band that is used by WiFi and Bluetooth. Consequently, there is a little interference between the camera and Bluetooth, even though Bluetooth uses frequency hopping.

Bluetooth causes streaks to appear in the video; but more important, the wireless camera causes some Bluetooth packets sent to or from the robot to be lost or corrupted. If the robot's communication protocol is not robust, then the service might hang waiting on packets that will never arrive because of "collisions." Even without direct interference, the presence of another transmitter close to the Bluetooth module on the robot "desensitizes" the Bluetooth receiver.

The Boe-Bot and LEGO NXT (shown in Figure 4-2) appear to operate OK with the Swann wireless camera attached. From time to time they miss a command, but they recover. However, the Stinger robot (in Figure 4-1) has some problems. The Stinger Serializer service hangs sometimes. This is being investigated.

The Surveyor SRV-1 has an onboard camera, but the video does not work with the TeleOperation service using the MRDS services available from Surveyor at the time of writing (version 061216) because these services do not implement the generic Webcam contract. However, a modified version of the camera service available from the book's website does implement the Webcam contract. The authors are working on services for the new Blackfin version of the SRV-1 that uses WiFi for communications and will therefore be much faster.

In addition, the SRV-1 Differential Drive service suffers from a flooding problem that makes using a gamepad almost impossible because the robot continues to move long after you release the joystick. You can drive it by carefully using the arrow keys on the keyboard. This problem has been addressed by a new Differential Drive service that is also available from the book's website (www.proMRDS.com). Unfortunately, this uncovered a different problem, which is that the robot is "deaf" while it is transmitting an image. Keep watching the book's website for updates.

## Creating the Service

By now, you should have a good idea of how to create a service, so the following instructions are very brief. In addition, the TeleOperation service contains a lot of code, so some code is not covered in the book.

You should open the TeleOperation service in Visual Studio and look at the code as you follow through the text. The completed service is available in the ProMRDS\Chapter4 folder. The Dashboard service is also included in the Chapter4 folder, but it is not discussed here.

Quite a lot of references are used by the TeleOperation service. Rather than introduce them as they are needed, the following is a complete list. Remember that whenever you add a reference, look in the properties for the reference and change the Copy Local and Specific Version properties to false. If you forget to do this you might have problems with versioning.

❑     `RoboticsCommon.Proxy`: This is almost always required for robot services, so you should get in the habit of adding it whenever you create a new service. Note that this reference is to the Proxy.

❑     `Ccr.Adapters.WinForms` and `System.Windows.Forms`: These are required to use a Windows Form in your service, which is discussed in detail later in the chapter.

❑     `GameController.Y2006.M09.Proxy`: This enables you to use a joystick or gamepad to drive the robot.

❑     `System.Drawing`: This is required so you can manipulate bitmaps from the Webcam.

You do not need to add references to the Differential Drive or Webcam services because they are included in `RoboticsCommon`. The code establishes partnerships with these services dynamically.

At the top of the main source file, `TeleOperation.cs`, are appropriate `using` statements:

```
// Added references
// For Forms
using Microsoft.Ccr.Adapters.WinForms;
using System.Windows.Forms;
// For handling webcam images
using System.Drawing;
using System.Drawing.Imaging;
using System.Runtime.InteropServices;
// Game Controller
using game = Microsoft.Robotics.Services.GameController.Proxy;
// Generic Differential Drive
using drive = Microsoft.Robotics.Services.Drive.Proxy;
// Webcam
using webcam = Microsoft.Robotics.Services.WebCam.Proxy;

// For locating services in the Directory
using ds = Microsoft.Dss.Services.Directory;

// For HttpGet
using Microsoft.Dss.Core.DsspHttp;
// For HttpStatusCode
using System.Net;
// Specifically for HttpPost handling
using System.Collections.ObjectModel;
using System.Collections.Specialized;
using Microsoft.Dss.Core.DsspHttpUtilities;
```

Knowing which references you need is something that comes with experience. Alternatively, you can look at other similar services and see what they use. In particular, you can use the tool tips (IntelliSense) in another service to find out where a certain data type comes from.

For example, the TeleOperation Windows Form (`DriveControl.cs`) also requires the following statement so that it can draw the "trackball":

```
using System.Drawing.Drawing2D;
```

Because TeleOperation is based on the Dashboard, you could figure this out by looking at the Dashboard code.

Some of the using statements that are automatically added when you create a new Windows Form are *not* required. However, they do no harm so you can leave them there. In any case, using statements are merely a convenience. Provided that you know which assembly a particular class is in (and you have added an appropriate reference), you can always specify its full name.

Throughout the rest of the chapter, various message types are mentioned. The Start method sets up the necessary receivers in the main interleave. The code is reviewed here in full, rather than line by line scattered throughout the chapter:

```
            // Hook up all of the Form events
            MainPortInterleave.CombineWith(Arbiter.Interleave(
                new TeardownReceiverGroup
                (
                ),
                new ExclusiveReceiverGroup
                (
                    // Form handling
                    Arbiter.ReceiveWithIterator<OnLoad>(true, _eventsPort,
OnDriveControlLoadHandler),
                    Arbiter.Receive<OnClosed>(true, _eventsPort,
OnDriveControlClosedHandler),
                    Arbiter.ReceiveWithIterator<OnLoad>(true,
webCamEventsPort, OnWebCamFormLoadHandler),
                    Arbiter.Receive<OnClosed>(true, _webCamEventsPort,
OnWebCamFormClosedHandler),
                    // Connection request
                    Arbiter.ReceiveWithIterator<OnConnect>(true, _eventsPort,
OnConnectHandler)
                ),
                new ConcurrentReceiverGroup
                (
                    // Game Controller-Added later
                    // Drive notifications - Not currently used

                    // Movement commands
                    Arbiter.ReceiveWithIterator<OnMove>(true,
_webCamEventsPort, OnMoveHandler),
                    Arbiter.ReceiveWithIterator<OnMove>(true, _eventsPort,
OnMoveHandler),
                    Arbiter.ReceiveWithIterator<OnMotionCommand>(true,
_eventsPort, OnMotionCommandHandler),
                    // WebCam -- Be careful because this can lock up if Exclusive
                    Arbiter.ReceiveWithIterator<webcam.UpdateFrame>(true,
_webCamNotify, CameraUpdateFrameHandler)
                )
            ));
```

Briefly, the _eventsPort and the _webCamEventsPort are used for messages from the main Form (called DriveControl) and the Webcam Form, respectively. General Windows Form messages are OnLoad and OnClosed. The messages types OnMove and OnMotionCommand are used to control the robot. UpdateFrame messages come from the web camera when new frames are ready for retrieval.

*Setting up these receivers ahead of time has no adverse effects. If no messages are ever sent to a particular port, then a receiver is wasted, but it does no harm to the service.*

The handlers for the game controller are set up separately once a subscription is successfully established, as discussed in the next section.

# Subscribing and Notifying

A common requirement in robotics is to "listen" for updates to sensor information. In MRDS you do this by *subscribing* to a service. The service then sends you *notification* messages, either at regular intervals or only when something changes.

The alternative to subscribing is to *poll* the sensors by periodically sending `Get` requests to the service that controls the robot (probably the "Brick" service). This is an inefficient approach because you might find that nothing has changed, especially if you are polling faster than new data arrives.

---

### To Poll or Not to Poll

It is important to understand that at some level all sensor updates are the result of polling because you do not have a direct connection to the devices attached to the robot (unless the PC is mounted on the robot, and even then it has to go through some sort of hardware interface). Even if the robot sends periodic sensor updates automatically, the onboard firmware in the robot is still polling based on a timer.

If you let another service get the data and send you notifications, then it must poll in the background without you knowing about it. The advantage is that you don't have to bother with a timer and continual `Get` requests that might not produce any new information. However, polling is still occurring behind the scenes.

Whether you should poll, and how often, should be based on the importance and usage of the sensor information. For example, reading switches on a robot to determine option settings is usually done during startup. After that, there is no need to poll them.

When you are using Bluetooth, a rough rule of thumb is not to poll any more frequently than once every 50 milliseconds. That's 20 times a second, and it should be fast enough for most applications. Of course, this assumes that your robot can keep up.

---

To summarize, whenever you want to get data, you have two options:

❑    Send a request to the robot and wait for the data to be sent back.

❑    Let another service get the data and tell you when it is available.

The remainder of this section discusses the second option — using subscriptions, i.e., letting somebody else do the polling for you.

# Subscribing to State Changes

Subscribing to a partner service is quite simple. Partners that offer a subscription service must implement the Subscribe operation, so you simply send a Subscribe message. This is one of the fundamental message types defined in the DSSP service model.

The steps to subscribe to a partner service are:

1. Set up a partnership, which can be done declaratively or dynamically in the code.
2. Send a Subscribe message with appropriate parameters.
3. Set up receivers and the corresponding handlers for notification messages.

Each of these steps is explained in more detail in the rest of this section using the Game Controller service as an example.

Consider the following code fragment from the TeleOperation service, which establishes a partnership with a game controller. As usual, you declare the partnership using an attribute and create a port to access the Game Controller service. (Remember that game is an alias declared in one of the using statements above).

```
[Partner("GameController", Contract = game.Contract.Identifier,
    CreationPolicy = PartnerCreationPolicy.UseExistingOrCreate)]
game.GameControllerOperations _gameControllerPort =
    new game.GameControllerOperations();
game.GameControllerOperations _gameControllerNotify =
    new game.GameControllerOperations();
```

This code is usually added at the top of your service class. It establishes a partnership that must always exist (because the policy is UseExistingOrCreate), although if there are no game controllers then the partner service is not very useful. Notice here that a second port is also created to receive notification messages — _gameControllerPort is for sending and _gameControllerNotify is for receiving. Later in the code you can subscribe to the Game Controller service by posting a Subscribe message. You also need to set up a receiver to handle the incoming notification messages.

The following two statements basically show what is required to subscribe to the Game Controller service. However, the actual code is a little more sophisticated. (You can see the full subscription process in the SubscribeToGameController method shown later.)

```
_gameControllerPortPort.Subscribe(_gameControllerNotify);
Activate(Arbiter.Receive<game.Replace>(true, _gameControllerNotify,
        GameReplaceHandler));
```

This is the absolute minimum amount of code required to set up a subscription, and it is not well written — you should check for a successful subscription by looking at the response message from the Subscribe; and you should not create an entirely separate receiver because it does not participate in the main interleave. Look at SubscribeToGameController below.

Referring back to the partner declaration in the code again, there is a third port:

```
Port<Shutdown> _gameControllerShutdown = new Port<Shutdown>();
```

This port is used for unsubscribing, but it is optional. (Unsubscribing is covered in the next section.) If you want to be able to unsubscribe, then you must specify a valid shutdown port in the NotificationShutdownPort property of the Subscribe message. To unsubscribe later, you send a Shutdown message to this shutdown port, hence the need for yet another port.

Lastly, the Game Controller service can send three types of notifications, not just one, and you need to set up receivers for each of them.

Taking all of this into account, the subscription code can now be rewritten in a more robust way:

```
private IEnumerator<ITask> SubscribeToGameController()
{
    bool success = false;

    LogInfo("Subscribing to Game Controller");

    // Create a subscription message to subscribe to the Game ⬅
Controller service
    game.Subscribe msg = new game.Subscribe();
    msg.NotificationPort = _gameControllerNotify;
    // Specify a Shutdown port so we can unsubscribe later
    msg.NotificationShutdownPort = _gameControllerShutdown;
    // Post the message
    _gameControllerPort.Post(msg);
    // Wait for a response
    yield return Arbiter.Choice(
        msg.ResponsePort,
        delegate(SubscribeResponseType response) { success = true; },
        delegate(Fault fault) { LogError(fault); success = false; }
    );

    if (!success)
        yield break;      // Subscription failed

    // Add receivers to the main interleave for each of the possible
    // notification messages from the game controller.
    // If there are no game controllers, then there will be no messages!
    MainPortInterleave.CombineWith(new Interleave(
        new ExclusiveReceiverGroup(),
        new ConcurrentReceiverGroup
        (
            Arbiter.ReceiveWithIterator<game.Replace>(true,
_gameControllerNotify, GameReplaceHandler),
            Arbiter.ReceiveWithIterator<game.UpdateAxes>(true,
_gameControllerNotify, GameUpdateAxesHandler),
            Arbiter.ReceiveWithIterator<game.UpdateButtons>(true,
_gameControllerNotify, GameUpdateButtonsHandler)
        )
    ));

    LogInfo("Game Controller subscription successful");

}
```

As a matter of interest, a subscription to the Game Controller service always succeeds regardless of whether you have a gamepad or joystick connected to your PC. However, if you try to enumerate the game controllers, you will not find any. TeleOperation doesn't enumerate the controllers because the Game Controller service automatically selects the first controller (if there is one). The TeleOperation service does not give you any way to select a game controller if you have more than one.

Three receivers are set up here and merged with the Concurrent receiver group of the main interleave. (You need to write three iterator handlers for each of the different types of notification messages).

Most services just send a `Replace` message as a notification, but it is also possible to define additional messages that send subsets of the service state. These are based on the `Update` message type. For example, the game controller can send changes to the button states separately from the axes (movement of the joystick). However, the settings of the buttons are also included in a `Replace` message (which contains the whole of the game controller state).

In addition to a `Subscribe` operation, some services offer a `ReliableSubscribe` operation. With a normal subscription, if the receiving service dies or is dropped, the sender just continues to send notifications. With a reliable subscription, however, the sender stops sending notifications if the receiver becomes unreachable. The subscription is added to a suspended list, and every so often an attempt is made to ping the receiver. If the receiver comes back to life (perhaps it was temporarily overloaded), then notification messages resume.

Because of this additional feature, there is a parameter called `suspensionInterval` that can be specified using `ReliableSubscribe`. You won't always need reliable subscriptions, so don't just use it because it is there. The following example shows how to use `ReliableSubscribe`:

```
// Subscribe to the drive
_driveShutdown = new Port<Shutdown>();
drive.ReliableSubscribe subscribe = new drive.ReliableSubscribe(
    new ReliableSubscribeRequestType(10)
);
subscribe.NotificationPort = _driveNotify;
subscribe.NotificationShutdownPort = _driveShutdown;

_drivePort.Post(subscribe);

yield return Arbiter.Choice(
    subscribe.ResponsePort,
    delegate(SubscribeResponseType response)
    {
        LogInfo("Subscribed to " + service);
    },
    delegate(Fault fault)
    {
        _driveShutdown = null;
        LogError(fault);
    }
);
```

# Unsubscribing from State Change Notifications

It is always a good idea to unsubscribe from your partners in your service's Drop handler. Otherwise, your service might not be able to shut down cleanly. In addition, if you are dynamically connecting to services and disconnecting again, then you need to be able to unsubscribe. (Otherwise, you might have the strange situation where you are no longer talking to a particular service but it is still talking to you! There is no actual connection between services, just messages traveling back and forth.)

The following code fragment unsubscribes from the Webcam service (if one is in use):

```
// Already connected?
if (_webCamPort != null)
{
    // Unsubscribe
    if (_webCamShutdown != null)
        yield return PerformShutdown(ref _webCamShutdown);
}
```

Because unsubscribing is done from several places in the code, a function is defined in the TeleOperation service to do this:

```
Choice PerformShutdown(ref Port<Shutdown> port)
{
    Shutdown shutdown = new Shutdown();
    port.Post(shutdown);
    port = null;

    return Arbiter.Choice(
        shutdown.ResultPort,
        delegate(SuccessResult success) { },
        delegate(Exception e)
        {
            LogError(e);
        }
    );
}
```

That's all there is to unsubscribing — just send a Shutdown message to the port you supplied when you subscribed. Of course, you should wait for the response to ensure that the unsubscribe has completed.

# Building in Support for Subscriptions and Notifications

The TeleOperation service has no need to handle subscriptions from other partners, so an example is required from elsewhere in the book code. The following code snippets are from BSBumper.cs, which is in Chapter 14. It implements the "bumpers" for the Boe-Bot, and consists of two infrared sensors (which only register on and off) and two "whiskers."

Microsoft provides a Subscription Manager Service as part of MRDS. This makes it easy to handle subscriptions because you don't have to keep track of all the services that have subscribed to your service or worry about how to send notification messages to all of them. The following discussion outlines the steps that a service must follow in order to accept subscriptions:

**1.** You need a `using` statement to simplify access to the Subscription Manager:

```
using submgr = Microsoft.Dss.Services.SubscriptionManager;
```

**2.** Add a Subscription Manager partner at the top of your service class:

```
[Partner("SubMgr", Contract=submgr.Contract.Identifier,
    CreationPolicy=PartnerCreationPolicy.CreateAlways, Optional=false)]
private submgr.SubscriptionManagerPort _subMgrPort =
    new submgr.SubscriptionManagerPort();
```

**3.** You need a handler for `Subscribe` messages:

```
/// <summary>
/// Subscribe Handler
/// </summary>
/// <param name="subscribe"></param>
/// <returns></returns>
[ServiceHandler(ServiceHandlerBehavior.Exclusive)]
public virtual void SubscribeHandler(bumper.Subscribe subscribe)
{
    base.SubscribeHelper(_subMgrPort, subscribe.Body,
            subscribe.ResponsePort);
```

The `SubscribeHelper` method takes care of the subscription process for you. Any number of other services can subscribe to your service, but you do not need to keep track of them.

**4.** It is a good idea at this stage to immediately send a notification message to the new subscriber. This initializes its state. Otherwise, it might have to wait a while before the first notification message.

```
foreach (bumper.ContactSensor bumper in _state.Sensors)
{
    SendNotification<bumper.Update>(_subMgrPort,
            subscribe.Body.Subscriber,
            new bumper.Update(bumper));
}
}
```

This overload of `SendNotification` specifies a particular subscriber, rather than sending to all subscribers. This example is a little complicated because the individual bumpers in the contact sensor array are sent one at a time. In general, you would send the entire state using a `Replace` message.

**5.** In order for the `SubscribeHandler` to be called, you must add the `Subscribe` type to your main operations port. The message type in this case is `bumper.Subscribe` because this service implements the generic Contact Sensor Array service. When you implement a generic service, you use the operations that are defined in the generic contract. Consequently, the `Subscribe` message type is already declared. If you look carefully at the top of the service class declaration, you will see that it implements an alternate contract:

```
[Contract(Contract.Identifier)]
[AlternateContract(bumper.Contract.Identifier)]
[DisplayName("Boe-Bot Generic Contact Sensor")]
[Description("Provides access to the Parallax BASIC Stamp 2 Boe-Bot ↩
infrared sensor used as a bumper.\n(Uses Generic Contact Sensors contract.)")]
    public class BumperService : DsspServiceBase
```

Generic contracts are discussed toward the end of this chapter in "Inheriting from Abstract Services."

**6.** You can optionally define a `ReliableSubscribe` handler as well:

```
/// <summary>
/// ReliableSubscribe Handler
/// </summary>
/// <param name="subscribe"></param>
/// <returns></returns>
[ServiceHandler(ServiceHandlerBehavior.Exclusive)]
public virtual void ReliableSubscribeHandler(
        bumper.ReliableSubscribe subscribe)
{
    base.SubscribeHelper(_subMgrPort, subscribe.Body,
            subscribe.ResponsePort);
    foreach (bumper.ContactSensor bumper in _state.Sensors)
    {
        SendNotification<bumper.Update>(_subMgrPort, ↩
subscribe.Body.Subscriber, new bumper.Update(bumper));
    }
}
```

Although this looks identical to the Subscribe handler, note that the message type is different and the `SubscribeHelper` acts accordingly. Because the code looks the same at a quick glance, it is easy to overlook the difference between the two operations.

The preceding steps cover the process for supporting subscriptions in the Bumper service. However, the updates to the sensor information have to come from somewhere. The Bumper service subscribes to the BASICStamp2 "Brick" service (the Boe-Bot's brain) for `SensorsChanged` messages. There is a partner declaration at the top of the code, and a method called `SubscribeToBasicStamp2` does the subscribing. The process is similar to the game controller described earlier, so it is not repeated here.

When a notification arrives from the Boe-Bot brick, the Bumper service checks whether any of the sensors have changed since the last update; if so, it issues a notification to all of its subscribers. The last step in the `SensorsChangedHander` is as follows:

```
if (changed)
    this.SendNotification<bumper.Update>(_subMgrPort,
        new bumper.Update(bumper));
```

That's it. The Subscription Manager sends a `bumper.Update` message to all the subscribers (or does nothing if no other services have subscribed).

Make sure that you don't flood your partners with notification messages. Always check incoming data to see if anything has changed. If there are no changes, don't send a notification! This is particularly important for the Game Controller service, for example. Imagine that you let go of the joystick and it springs back to the (0,0) position. If the Game Controller service kept sending updates with axis values of (0,0), then you would not be able to use the buttons on the TeleOperation Form because the game controller would continually override them and stop the robot. Therefore, the game controller only sends notifications when you move the joystick.

If you have implemented a `Replace` message handler in your service, then you must modify it to send new state information to all of the subscribers using `SendNotification`. `Replace` messages are discussed in Chapter 3. The code is not shown here, but you can look at it in Visual Studio.

If your service state is quite large, and especially if it can be broken into logical subgroupings, then consider having more than one type of notification message. This Boe-Bot example is trivial, but it is conceivable that there could be a message type for only the IR sensors and a different message type for just the whiskers. A subscriber might choose to listen only to the IR data, and ignore messages about the whiskers. In fact, the updated firmware from Parallax for use with a SpinStamp microcontroller stops the motors whenever a whisker is pressed. The MRDS service has no say in the matter because this happens aboard the robot.

# User Interfaces

Although a primary objective of the robotics field is to create autonomous robots, almost all robots have to interact with humans. Therefore, user interfaces are an essential element of the equation. You have two different approaches available to you for creating user interfaces for MRDS services:

❑ Windows Forms (WinForms)

❑ Web Forms

Which approach you take depends a lot on the amount of user interaction that is required. In general, more complex or frequently used interfaces are best written using Windows Forms. However, a Windows Form will only be visible on the local computer that is running the DSS node. If you want to allow users to make changes to service parameters remotely, then you have to use a Web Form. You could write yet another service that displays a WinForm and run it as a client on another computer to talk to the main service, but this is getting ridiculous — where do you draw the line? In any case, this would require users to have MRDS installed on their computer instead of just a web browser.

In the example for this chapter, both types of interface are used. However, as you will see when you use it, the TeleOperation service would probably be easier to use if the option settings were in a Windows Form, rather than a Web Form. For comparison, the Dashboard service, also included with this chapter, uses a Windows Form for option settings.

In terms of "best practice," it is a good idea to implement a web page to display the service state, i.e., a `HttpGet` operation using an XSLT transform. This makes the state information much easier to read, and it looks more professional. Whether you decide to implement a Web Form, i.e., support for the `HttpPost` operation, to allow users to update fields in the state is a different issue. Once you have worked through the next few sections you will be able to make an informed decision based on your users' needs and your own programming skills.

# Using Windows Forms

This section assumes that you are familiar with Windows Forms (or WinForms, for short) in the same way that it is assumed you are already a C# programmer.

The MRDS Robotics Tutorial 4 (Drive-By-Wire) uses a simple form with four buttons to control a robot. The TeleOperation service in this chapter is much more sophisticated than this. However, you should read Robotics Tutorial 4 in conjunction with this section of the book.

> *If you plan to use a Windows Form on a CF (Compact Framework) device, e.g. a PDA, then you should read Chapter 16. There are some considerations that are specific to the CF environment. The details are omitted here in order to keep the discussion as simple as possible at this stage. A slimmed-down version of TeleOperation, called Drive-By-Wire, is provided with the code for this chapter, which includes a CF version (not discussed here).*

To see how the WinForms work, follow these steps:

1. Start the TeleOperation service in the debugger. It takes a little while because the default manifest is set up to start the simulator.

2. Select `localhost` as the node name and `50001` as the port number. (These values are stored in the `config` file, so they should already appear in the window.)

3. Click the Connect button. You should see another window appear with the view from the robot's camera, as shown in Figure 4-4. You can move the two windows around independently. The service was deliberately designed to use a second WinForm for the camera view so that you can still use TeleOperation when the robot has no camera.

**Figure 4-4**

> *If you don't see the simulated camera view in the webcam window, make sure that you have installed the V1.5 Refresh.*

4. You can drive the robot around using the arrow buttons on the main form, the arrow keys on the keyboard, a gamepad, or a joystick. If you close the WebCam View window, you can reopen it by clicking the Connect button again. If you close the TeleOperation window, then the service should shut down and take the DSS node with it. However, the DSS node sometimes doesn't shut down for reasons that are not apparent.

## How WinForms Work under MRDS

A quick overview of how Windows Forms work with DSS services is appropriate at this point. In a normal Windows Forms application, all of the relevant code is often included directly in the form. However, for use with MRDS, "control" of the service is done in the main service implementation code and not the form.

Windows Forms operate off the main Win32 message queue for the application. They handle events such as `MouseMove`, `ButtonClick`, `KeyDown`, and so on. These events need to be sent back to the main DSS service (possibly after some pre-processing, or perhaps not at all if they only affect the internal state of the form).

Therefore, a port is created by the main service to allow the WinForm to send messages back to the main service. The message types have to be defined in the same way as they would for any service. In this sense, the WinForm acts something like an internal partner service, but it does not have a Proxy.

When it is necessary to execute some code in the context of the WinForm, the main service has to use an approach that is similar to `PlatformInvoke` for calling unmanaged code. This is done by sending a `FormInvoke` message to the `WinFormsServicePort` specifying a delegate to execute.

Windows Forms are in a sense "legacy" code. (Eventually, WinForms might disappear and be replaced by the new Windows Presentation Foundation, WPF. However, as of V1.5 of MRDS the WPF is not supported.) WinForms run in a Single-Threaded Apartment model and do not fit nicely into the multi-threaded model of the CCR. Because WinForms have thread affinity — i.e., they store state information into the thread's local store — they require special treatment. Therefore, Microsoft defined a `WinFormsServicePort` in the `DsspServiceBase` class that is used to control Windows Forms.

Following is a summary of the steps for adding a WinForm to your service (steps 2–4 are similar to setting up a new service):

1. Create a new Form in your service project.

2. Define the request messages that the Form can send and a `PortSet` containing these message types. These messages usually correspond to each of the event handlers in the Form. You can place these message classes in the Form source file if they are declared as public, or add them to the `ServiceTypes.cs` file for the service.

3. Define a port in the main service (using the Form's `PortSet`) to receive the messages from the Form.

4. Write handlers for each of the Form message types and add appropriate receivers to the main interleave.

5. Modify the Form's constructor to accept a port as a parameter and create a variable to store it in.

6. Edit the Form source code to add appropriate public properties and methods to enable information to be passed back to the Form from the main service.

7. Update the `Start` method of the service to post a `RunForm` message to the `WinFormsServicePort`. This creates a new instance of the Form. The Form creation code should pass the Form port to the Form constructor. Save the handle (pointer) to the new Form instance so that you can access it later.

8. To send messages from the Form to the service, post them to the port that was supplied to the constructor.

9. To send information from the service to the Form, either copy data into public variables inside the Form or post a `FormInvoke` message to the `WinFormsServicePort` so that a delegate can execute public methods inside the Form using the Form handle.

This might seem like a daunting task, but once you get the hang of it you will find that it is not that complicated. These steps are explained in more detail in the following sections.

## Creating a WinForm

Creating a Windows Form and then setting it up to interact with your main service is a fairly involved process. However, once you have done it a couple of times you should not have any trouble.

As you saw in the last section, you need to add a reference to `System.Windows.Forms` to your project, and a `using` statement at the top of your code.

You create a new Windows Form for your service in exactly the same way as you do for any Visual Studio project, i.e., click Project ⇨ Add Windows Form. Figure 4-5 shows the Add New Item dialog for adding the `DriveControl` Form to the TeleOperation project.

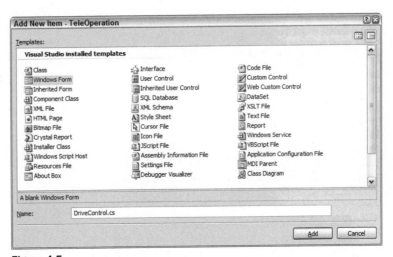

Figure 4-5

The TeleOperation service has two forms, each of which is associated with a source file where you place the event handlers:

❑ `DriveControl`: This enables you to control the robot. The `DriveControl` Form is discussed in this section. It is the window on the left in Figure 4-4.

❑ `WebCamForm`: This displays the live video feed. This is covered in the section on using web cameras. It is the window on the right in Figure 4-4.

For the `DriveControl` Form, then, at the top of the main service, you declare a variable to hold a handle (a reference to the class instance) for the Form, `_driveControl`, as well as a port for messages from the Form, `_eventsPort`:

```
// Handle to the main WinForm UI
DriveControl _driveControl;
// Port for the UI to send messages back to here (main service)
DriveControlEvents _eventsPort = new DriveControlEvents();
```

The `DriveControlEvents` class is discussed in the next section and is defined in `DriveControl.cs`. For now, you only need to know that it is a `PortSet` that enables the `DriveControl` Form to send information back to the main service.

There is nothing magical about these classes or variables. You can name them whatever you like, and you can change the messages that are accepted by the `DriveControlEvents PortSet`. This is just one example of how to implement communication between a service and a WinForm. Other samples in MRDS use different approaches. However, the authors believe this is a consistent and manageable approach.

In the `Start` method, you have to create a new instance of the `DriveControl` Form. (The `WebCamForm` is created only after a Webcam service has been found). This is done as follows:

```
// Create the WinForm UI
WinFormsServicePort.Post(new RunForm(CreateForm));
```

The `RunForm` message specifies a delegate to execute to create the new Form. This delegate should return a handle to the Form as follows:

```
System.Windows.Forms.Form CreateForm()
{
    // NOTE: Modify the constructor in the Form to pass these parameters
    return new DriveControl(_eventsPort, _state);
}
```

This routine is trivial and could be specified as an anonymous delegate in the `RunForm` message instead of as a separate method. If you are an experienced Windows Forms programmer, you might have noticed that the code does not save the handle to the new Form instance. It can be done here, but for illustrative purposes it is done later.

Note that when the `DriveControl` Form is created it is passed two parameters. Normally the constructor for a WinForm does not take any parameters. You must modify the automatically generated code for the Form to accept these parameters (in `DriveControl.cs`):

```
public partial class DriveControl : Form
{
    // This port is passed across to the constructor
    // It allows messages to be sent back to the main service
    DriveControlEvents _eventsPort;
    // This is part of the main service State
    GUIOptions options;
```

```
public DriveControl(DriveControlEvents EventsPort,
        TeleOperationState state)
{
    InitializeComponent();

    // Remember the port to use to send back messages
    _eventsPort = EventsPort;

    // Copy the option settings
    options = state.Options;
```

All the constructor needs to do is copy the parameters to variables inside the Form. Other code within the Form can then use these variables as necessary. Clearly, the EventsPort is required so that the Form can post back messages to the main service. The reason for passing across the service state is not so obvious — it enables the Form to access the option settings that are stored in the config file.

You have not entirely finished with Form creation because you still need the Form handle. This last step is covered in the next section.

## Passing Information Between a WinForm and a Service

Information needs to be passed in both directions between the main service and the WinForm. This section discusses how to pass information in each of these directions and the different approaches required:

❑ **Sending to the Form:** A WinForm is a separate module, not a service in its own right. Because it operates as a Single-Threaded Apartment model, it cannot wait on CCR ports to receive messages. However, the main service needs to update information on the Form in response to notification messages such as game controller updates. Sending information from the main service to the Form is done using FormInvoke.

❑ **Receiving from the Form:** The Form needs to pass back commands to the main service. When you interact with the Form, events fire inside the Form code. These WinForm events are not related to the CCR in any way, but the event handlers in the Form can send CCR messages back to the main service by posting to the _eventsPort, which was created especially for this purpose in the code in the previous section.

What FormInvoke does is execute code from the main service in the context of a WinForm. It does this by posting a message to the WinFormsServicePort, which means that the delegate executes asynchronously with respect to the caller. In other words, any code that follows a FormInvoke in the same routine cannot assume that the FormInvoke has completed execution.

A good example is the handler for axes updates from the game controller. This handler sends the new axes information to the DriveControl Form so that the onscreen trackball and the associated X, Y, and Z values can be updated:

```
IEnumerator<ITask> GameUpdateAxesHandler(game.UpdateAxes update)
{
    if (_driveControl != null)
    {
        WinFormsServicePort.FormInvoke(
```

*(continued)*

175

(continued)

```
                    delegate()
                    {
                        _driveControl.UpdateGameControllerAxes(update.Body);
                    }
                );
            }
            yield break;
    }
```

Note the following regarding the preceding code:

❑ The code checks to make sure that the `DriveControl` Form is active and then calls `FormInvoke` on the `WinFormsServicePort`. The delegate takes advantage of its access to local variables in the handler to pass `update.Body` to the `UpdateGameControllerAxes` method in `DriveControl.cs`, i.e., inside the Form. Notice the use of `_driveControl`, which is a pointer to the Form instance, to execute a public method inside the Form.

❑ `UpdateGameControllerAxes` is a public method in the Form code (`DriveControl.cs`). It is too involved to explain the details of how it works here. In short, it updates the current position of the trackball and then processes the new position in exactly the same way as when you use the mouse to move the trackball. This usually results in a new power setting for the drive motors, and `UpdateGameControllerAxes` posts an `OnMove` message back to the `_eventsPort`. The `OnMove` message is discussed later, but basically it causes the main service to send a `SetDrivePower` message to the differential drive.

If you only want to read data, you can also extract information from the Form by grabbing the values of public properties in the Form. (Technically, you can also write to public variables, but then you might have concurrency issues). Visual Studio creates all controls on a Form as private properties, so you cannot access the controls directly. However, you can declare your own public properties, and many properties at the Form level are public.

The following code snippet is from the handler that saves the state to a `config` file:

```
// Grab the current window location
if (_driveControl != null)
{
    _state.Options.WindowStartX = _driveControl.Location.X;
    _state.Options.WindowStartY = _driveControl.Location.Y;
}
```

The `Location.X` and `Location.Y` properties are the screen coordinates of the top-left corner of the window. This enables the window to be started at the same position the next time the service is run. Note that gathering data this way is an *on-demand approach*, rather than an *event-driven approach*.

The basic process for communication from the Form back to the main service is via the Form's event handlers, which generate messages that are posted to the `_eventsPort`. Some events in the Form might not result in a message being sent, but instead set internal variables inside the Form, so there is no one-to-one correspondence between events and messages sent to the `_eventsPort`.

Strictly speaking, the `_eventsPort` is not necessary. You could define the requests that the `DriveControl` Form requires as additional operations on the main operations port for the service.

However, it makes sense to separate WinForm interactions from the operations that the service provides because you don't want to expose these operations via the service Proxy. It also helps to make the WinForm handling code reusable.

Windows Forms often have dozens of controls. If you defined a message type for each possible event that could occur on the Form, it would quickly get out of hand. A better approach is to define a general class to handle WinForms interaction and then subclass this as necessary for message types that are specific to particular Forms. It is also advisable to try to group messages together to share message types.

Look in `TeleOperationTypes.cs` in the General Form Operations region at the bottom of the file:

```
#region General Form Operations
// This is the base class for all Form event messages
public class FormEvent
{
    private Form _theForm;

    public Form Form
    {
        get { return _theForm; }
        set { _theForm = value; }
    }

    public FormEvent(Form form)
    {
        _theForm = form;
    }
}

public class OnLoad : FormEvent
{
    public OnLoad(Form form)
        : base(form)
    {
    }
}

public class OnClosed : FormEvent
{
    public OnClosed(Form form)
        : base(form)
    {
    }
}

#endregion
```

The class called `FormEvent` is the base class for any event information that is passed back from a WinForm to the main service. This base class contains a property that identifies the WinForm by its handle. In the TeleOperation example, there are two WinForms.

Two other subclasses are also defined: `OnLoad` and `OnClosed`. These events are common to all WinForms so it is sensible to define them here. In the code for a WinForm, you add event handlers similar to the following. Notice that these handlers simply post a message of the appropriate type to the _eventsPort for the WinForm:

```
private void DriveControl_Load(object sender, EventArgs e)
{
    _eventsPort.Post(new OnLoad(this));
}

private void DriveControl_FormClosed(object sender, FormClosedEventArgs e)
{
    _eventsPort.Post(new OnClosed(this));
}
```

> Do not insert these routines by typing in the code because they won't work unless you edit the Designer-generated code and manually hook them to the events. You should allow Visual Studio to create the empty methods for you because then they will be hooked up to the events properly. Once you have the empty routines, you can insert code inside them.

If you are not familiar with adding event handlers, follow these steps:

1. Go to the Design View for the Form.

2. Click the background of the Form to select the entire Form.

3. In the Properties panel, shown in Figure 4-6, click the lightning bolt icon. This shows all the possible events that can be associated with the Form.

Figure 4-6

**4.** If you double-click one of the event handler names in the Properties panel, an empty method is added to your WinForm code.

*Figure 4-6 shows event handlers defined for* `FormClosed`, `KeyDown`, `KeyUp` *and* `Load`. *Note that* `KeyPress` *is not used. The key events are discussed in the next section.*

For buttons and other controls on the Form, you don't need to go to this much trouble because you can simply double-click the control in the Design View and the default event handler is added automatically. However, the keypress events cannot be added this way because there are no controls to click for keypress.

The `OnLoad` handler in the main service for `OnLoad` messages from the `DriveControl` Form is as follows (there is a different `OnLoad` handler for the `WebCamForm`):

```
/// <summary>
/// On Load Handler for completion of Form Load
/// </summary>
/// <param name="onLoad"></param>
/// <returns></returns>
IEnumerator<ITask> OnDriveControlLoadHandler(OnLoad onLoad)
{
    // Save a handle to the form
    _driveControl = (DriveControl)onLoad.Form;

    LogInfo("Drive Control Form Loaded");

    // Subscribe to the joystick
    yield return Arbiter.ExecuteToCompletion(Environment.TaskQueue,
        Arbiter.FromIteratorHandler(SubscribeToGameController));
}
```

This handler saves the Form handle at last. It is not necessary to do this here — it could have been saved when the Form was initially created, but it is done here as an example.

Then the handler tries to subscribe to a game controller. Other initialization can also be included at this point, such as updating labels or textbox controls on the Form with initial information. The main point to note is that this handler is not executed until the `OnLoad` message has been received, which is an indication that the Form has been created and is running.

The `OnClosed` event fires if the user clicks the Close button in the Form's title bar. There is no need to provide an Exit button on the Form, but one has been included as a convenience. The Exit button handler simply closes the Form, which fires the `OnClosed` event anyway, and this causes an `OnClosed` message to be sent to the main service.

The `OnClosed` handler in the main service attempts to shut down by posting a `Drop` message to itself. (The `Drop` handler is not discussed here).

```
/// <summary>
/// Form Closed Handler for when Drive Control Form has closed
/// </summary>
/// <param name="onClosed"></param>
void OnDriveControlClosedHandler(OnClosed onClosed)
{
    if (onClosed.Form == _driveControl)
    {
        LogInfo("Main Form Closed");

        // Send a Drop message to ourselves
        _mainPort.Post(new DsspDefaultDrop(DropRequestType.Instance));
    }
}
```

Notice here that the code checks the handle in the message against the handle in the Form to ensure that the message has been received from the correct Form. This isn't actually necessary, but it illustrates the point that one handler can potentially handle multiple Forms by using the Form handle to identify the Form.

## Keyboard Handling

If you refer back to Figure 4-6 you can see that there are event handlers for the KeyDown and KeyUp events. These handlers are used to enable you to drive the robot using the arrow keys on the keyboard. However, there are a couple of tricks involved in setting this up.

The keyboard handlers are in a region called Keyboard Handlers toward the bottom of the Form code, DriveControl.cs. Note that the event handlers were first created in the Properties panel by double-clicking the events, and then the code was added to each of the routines.

Start by looking at the KeyDown event handler:

```
#region Keyboard Handlers

// NOTE: The arrow keys will not normally appear in a KeyDown event.
// This is because they will be pre-processed. However, the operation
// under CF seems to be different and it does no harm to leave them
// in here anyway.

private void DriveControl_KeyDown(object sender, KeyEventArgs e)
{
    switch ((Keys)e.KeyValue)
    {
        case Keys.Up:
            Forward();
            e.Handled = true;
            break;

        case Keys.Down:
            Backward();
            e.Handled = true;
            break;
```

```
        case Keys.Left:
            TurnLeft();
            e.Handled = true;
            break;

        case Keys.Right:
            TurnRight();
            e.Handled = true;
            break;

        default:
            break;
    }

}
```

It turns out that this code is *not* responsible for the movements of the robot, even though it looks like it should be. It is called for keypresses in general, but not for the arrow keys. You could add other case statements to the switch to use alternative keys. For example, the A, S, D, and W keys are commonly used in first-person shooter games to move around.

Are you being misled? No. When you get to Chapter 16, you will find that Windows Mobile operates a little differently. In that case, the preceding code is called. The "arrow keys" on a PDA correspond to the "rocker switch" that is used to move the cursor around.

With desktop operating systems, you have to take a different approach to process the arrow keys. You first need to set the KeyPreview property on the Form to True. This indicates that you want to handle the "special" keys, rather than use the default handling. The arrow keys are normally used to move from one control to another, or to move side to side in a TextBox. By setting the KeyPreview flag, your code can take over.

Next, you must create a handler to override the default ProcessDialogKey method. Your handler can do whatever it likes with incoming keystrokes before they are passed on to the Form. If you wanted to be really nasty, you could throw away all "2" keystrokes and the poor users would think they had a broken key. Let's not do that.

ProcessDialogKey has to process the keys that it is interested in and return true to indicate that the keystrokes have been used up. Any keys that it is not interested in can just be passed to the original ProcessDialogKey handler, where they are processed as usual. In other words, you are intercepting keystrokes *before* they arrive at the Form:

```
// Overriding ProcessDialogKey allows us to trap the arrow keys
protected override bool ProcessDialogKey(Keys keyData)
{
    switch (keyData)
    {
        case Keys.Up:
            Forward();
            return true;

        case Keys.Down:
```

*(continued)*

*(continued)*

```
                Backward();
                return true;

            case Keys.Left:
                TurnLeft();
                return true;

            case Keys.Right:
                TurnRight();
                return true;

            default:
                break;
        }

        return base.ProcessDialogKey(keyData);
    }
```

Notice that the last step in the code is to pass the `keyData` back to the normal `ProcessDialogKey` handler and return whatever result it gives you. It is quite likely that the original `ProcessDialogKey` handler won't do anything with the keystroke (because it is not a special character), in which case it returns `false` and the `keyData` is sent to the Form for processing.

The functions that are called in response to each of the arrow keys are trivial, so only the `Forward` function is shown here:

```
    void Forward()
    {
        _eventsPort.Post(new OnMove(this,
            (int)options.MotionSpeed, (int)options.MotionSpeed));
    }
```

It is worth noting here that `MotionSpeed` is an option setting. This is part of the service state, so it is saved in the `config` file. The Form doesn't have direct access to the service state, but the state is passed to the Form when it is created.

`OnMove` messages are processed in the main service by the following handler. As you would expect, it issues a `SetDrivePower` request to the differential drive:

```
    /// <summary>
    /// Handle basic moves (motor power settings)
    /// </summary>
    /// <param name="onMove"></param>
    /// <returns></returns>
    IEnumerator<ITask> OnMoveHandler(OnMove onMove)
    {
        if (_drivePort != null)
        {
            // Create a drive request
            // There is a more concise syntax, but we need to access some
            // additional properties
```

```
drive.SetDrivePowerRequest request =
        new drive.SetDrivePowerRequest();
request.LeftWheelPower =
        (double)onMove.Left * MOTOR_POWER_SCALE_FACTOR;
request.RightWheelPower =
        (double)onMove.Right * MOTOR_POWER_SCALE_FACTOR;
drive.SetDrivePower sdp = new drive.SetDrivePower(request);
// Set a timeout so that this does not wait forever
sdp.TimeSpan = TimeSpan.FromMilliseconds(1000);
_drivePort.Post(sdp);

yield return Arbiter.Choice(
        sdp.ResponsePort,
        delegate(DefaultUpdateResponseType response)
        {
            //Console.WriteLine("Power updated");
        },
        delegate(Fault f)
        {
            // Log an error (most probably a timeout)
            LogError(f);
            if (f.Code.Subcode.Value ==
                        DsspFaultCodes.ResponseTimeout)
                Console.WriteLine("Timeout on Move");
            else
                Console.WriteLine(f.Detail);
        }
    );
}
}
```

Another point to note about the SetDrivePower request is that a timeout is set on it. All DSS requests have a TimeSpan property. If you use a shorthand method for sending a request, rather than explicitly creating the request object, then you do not have the opportunity to set the TimeSpan. The code is therefore a little more involved.

It is important to have a timeout here so that the TeleOperation service doesn't hang. Some robot services are not well-behaved when something goes wrong with the robot. For example, try turning your robot off and see whether your service stops responding to requests.

The last point about this code is that the motor power settings from the Form are scaled by the MOTOR_POWER_SCALE_FACTOR, which is 0.001. This is because the Game Controller service sends axis values in the range −1000 to +1000. However, the SetDrivePower request must use values in the range −1.0 to +1.0.

At this point, the code gets the robot moving, but you also need to stop it. The simplest way to do this is to stop the robot as soon as the user releases the key. This means that as long as the arrow key is held down, the robot continues to move in that direction, which is intuitively easy to use.

The `KeyUp` event handler therefore calls `Stop` and is trivial:

```
private void DriveControl_KeyUp(object sender, KeyEventArgs e)
{
    Stop();
}
```

A `KeyUp` event occurs when any key is released. The handler for `KeyUp` immediately sends a `Stop` command without even determining which key was released.

*If you "tap" a key, the robot might jerk a little, but it won't keep driving. You must hold a key down. Pounding on the keyboard will not achieve anything, and it certainly won't make the robot go faster.*

While you are holding a key down, the keyboard automatically repeats the keystroke, typically a few times per second. These additional "keystrokes" simply execute the same `KeyDown` code, but they are not necessary for the process to work — the first `KeyDown` event starts the robot moving, and the key release stops it.

As noted above, if you are working with a PDA running Windows Mobile, only two event handlers (`KeyDown` and `KeyUp`) are required provided that you change the `KeyPreview` property for the Form to true. The `ProcessDialogKey` method is not required for Windows Mobile.

## Button Handling

There are arrow buttons on the `DriveControl` Form for controlling the robot. The way that these work is actually quite simple: The event handlers for the buttons post messages to the main service, and the service posts appropriate messages to the differential drive.

Consider the event handler for the "Forward" button (the up arrow):

```
private void btnForward_Click(object sender, EventArgs e)
{
    if (chkFixedMoves.Checked)
        _eventsPort.Post(new OnMotionCommand(this, ⏎
MOTION_COMMANDS.Translate, options.DriveDistance / 1000, options.MotionSpeed));
    else
        Forward();
}
```

The code first checks whether the Fixed Moves checkbox is enabled or not. If it is, then `DriveDistance` and `RotateDegrees` are used to control the robot. These functions are handy because they only move the robot by a fixed amount and it can't run away from you. However, some robot services don't implement these functions, which would make the buttons useless.

Notice that there is another message type for the fixed motions called `OnMotionCommand`. The enum called `MOTION_COMMANDS` has three possible values: `Translate`, `Rotate`, and `Stop`. The second parameter to this request is a `double` that is either a distance or an angle (depending on the request type) and the last parameter is the drive power.

If the Fixed Moves checkbox is *not* clicked, the Forward method shown earlier is called. This just turns on the motors; you have to click the Stop button to stop the robot. In this mode, the buttons behave like they do in Robotics Tutorial 4, but they are much prettier buttons.

# Using Web Forms

This section explains how to set up your service to handle HttpGet and HttpPost requests. Both of these require you to write an XSLT (Extensible Stylesheet Language Transformation) file, which is used to format the data. Because XSLT might be new to some readers, also included here is a brief explanation of how it works and how to set up your development environment.

Following that are examples demonstrating how to display service state on a web page and how to update service state using a Web Form. The final section discusses using JavaScript to do client-side validation of input data before it is submitted to update the state. This is not essential, but it makes your Web Form much more user-friendly.

As you should know by now, DssHost implements a web server, and you can examine the state of services using a web browser. You can also start and stop services via a browser. If you don't know how this works, read Chapter 3.

The TeleOperation service uses several option settings that are stored in its state. These can be changed by editing the config file prior to running the service. However, if you are trying to adjust the motor settings for your particular robot, it is very annoying to have to edit the config, run the service, shut down, edit the config, run the service, and so on.

There are two solutions to this problem: use a Windows Form or a Web Form. You can add another Windows Form to your service to update the configuration. This is how the Dashboard works. Although there are some extra steps involved in passing the configuration data between the Form and the main service, you have already learned how to create a Windows Form and exchange information between the service and the Form.

The alternative is to use a Web Form. This section explains the process. You should also look at the MRDS Service Tutorial 6.

## Accessing Services from Web Pages

When you enter a URL that refers to a service into the address bar of a web browser, DssHost gets the state information from the service on your behalf and sends back a web page. You will see either some raw XML code or a nicely formatted web page.

If the service does not implement the HttpGet request type, then DssHost makes a Get request. If you have declared the HttpGet message type in your main operations port but you did not supply a handler, then the handler in the DsspServiceBase class is used instead.

If you add a `HttpGet` handler to your code but forget to add the operation to the operations port, you get a run-time error something like the following:

```
*** Dssp Operation handler has been marked with the
ServiceHandlerAttribute but
its operation type is not on the operations
port.Method:System.Collections.
Generic.IEnumerator`1[Microsoft.Ccr.Core.ITask]
HttpGetHandler(Microsoft.Dss.Core.Dss
pHttp.HttpGet) [01/16/2008 17:42:56][http://koala:50000/teleoperation]
```

To see the difference between a formatted page and raw XML, follow these steps:

1. Start the TeleOperation service in the debugger again, enter `localhost` and `50001`, and click Connect. Note that TeleOperation connects to the TCP port, not the HTTP port (which is 50000).

2. Open a web browser and enter the following URL: **http://localhost:50000/teleoperation**.

   This requests the current state of the TeleOperation service. Because the TeleOperation service has an XSLT (Extensible Stylesheet Language Transformation) file that defines how to display the data, you see a formatted web page, as shown in Figure 4-7.

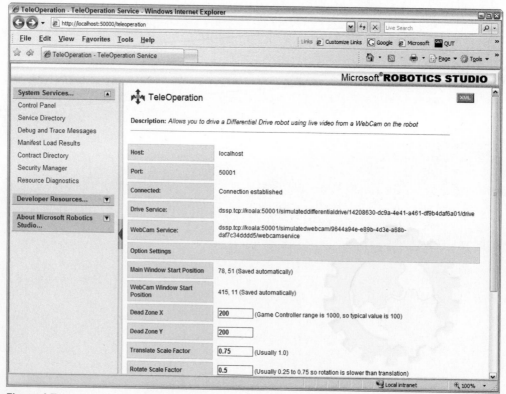

Figure 4-7

Figure 4-7 shows the values of all the state properties. It indicates that the service is connected and lists the full URIs of the Differential Drive and Webcam services that it is using. Although they are both simulated services, they implement the generic contracts so TeleOperation can connect to them.

If you have investigated MRDS in a web browser before, then you should recognize that this page is in the standard format for MRDS web pages. This is because it uses the Master Page (template) that is provided by MRDS for consistency.

**3.** Enter the following URL in the address bar of the web browser:
   **http://localhost:50000/teleoperation/raw**.

This displays the state as an XML file without using the XSLT file to format it. (It is the same as clicking the orange XML button in the top right-hand corner of the window in Figure 4-7). The XML code is as follows:

```
<?xml version="1.0" encoding="utf-8" ?>
<TeleOperationState
    xmlns:s="http://www.w3.org/2003/05/soap-envelope"
    xmlns:wsa="http://schemas.xmlsoap.org/ws/2004/08/addressing"
    xmlns:d="http://schemas.microsoft.com/xw/2004/10/dssp.html"
    xmlns="http://www.promrds.com/contracts/2008/01/teleoperation.html">
  <Host>localhost</Host>
  <Port>50001</Port>
  <Connected>true</Connected>
  <DriveService>dssp.tcp://koala:50001/simulateddifferentialdrive/↵
68d363c7-39a9-41b2-b90d-2e24a966015b/drive</DriveService>
  <WebCamService>dssp.tcp://koala:50001/simulatedwebcam/53f5c3a7-3f08-↵
4f64-800f-8edddeabe68a/webcamservice</WebCamService>
  <Options>
    <WindowStartX>78</WindowStartX>
    <WindowStartY>51</WindowStartY>
    <WebCamStartX>415</WebCamStartX>
    <WebCamStartY>11</WebCamStartY>
    <DeadZoneX>200</DeadZoneX>
    <DeadZoneY>200</DeadZoneY>
    <TranslateScaleFactor>0.75</TranslateScaleFactor>
    <RotateScaleFactor>0.5</RotateScaleFactor>
    <MotionSpeed>500</MotionSpeed>
    <DriveDistance>300</DriveDistance>
    <RotateAngle>45</RotateAngle>
  </Options>
</TeleOperationState>
```

Which of these two versions, a formatted web page or raw XML, would you prefer to see as a user of the TeleOperation service? Obviously, a nicely formatted screen gives a more professional look to your service and is much easier to use.

Now that you have seen both the raw state information and a formatted web page, here is a summary of the steps for using a Web Form:

**1.** Create an XSLT file that transforms the XML service state into a more readable HTML format.

**2.** Add the XSLT file to your assembly as an embedded resource, or place it into the appropriate directory under MRDS.

3. Add or modify the `HttpGet` handler in your service to use the transform. To simply display the state, you don't need to go any further.

4. If you want to use a Web Form to update the state, add appropriate `input` tags to the XSLT file, and a `Submit` button.

5. Add an `HttpPost` handler to your service and write code for it to process the input data and update the state (and maybe save the state too).

6. Optionally, add JavaScript code to your XSLT file to do client-side validation *before* the data is submitted to the service.

If you are an experienced web developer, you should find this process easy to follow. If not, please keep reading.

## XSLT Overview

If you don't have a background in web development, then you might find this section useful. If you are already an experienced web developer, skim through to the next section (and ignore the little "white lies").

As you know, data is transferred in messages between services using XML (Extensible Markup Language). This is a much stricter language than HTML (Hypertext Markup Language), which is used for writing web pages. In particular, XML is case sensitive. It is assumed that you have some basic familiarity with HTML.

An XSLT file defines how to take an XML file as input and reformat it to create a new output file. Note that the output does not have to be HTML, although this is usually the case. In theory, XSLT can be used to create a PDF file, a Word document, and so on. The definition of XSLT can be found on the World Wide Web Consortium website at www.w3.org/TR/xslt.

A lot of the code in an XSLT file is just HTML, which is destined to be displayed on the output page. Embedded in this are `<xsl>` elements that control what information from the XML data stream is displayed, such as the `<xsl:value-of>` element, which displays a value by name.

Visual Studio understands XSL syntax, so when you open an XSLT file you can use IntelliSense and the online help. For example, when you enter `<xsl:`, Visual Studio pops up a list of available XSL elements (as soon as you press the ":" key).

XSL includes logic and repetition elements, such as `<xsl:if>`, `<xsl:choose>`, and `<xsl:for-each>`. You can even create variables in XSL. Place the cursor over one of these commands in the editor in Visual Studio and press F1 to view the online help.

### Setting Up Your Development Environment for XSLT

Before you start working with XSLT, you should follow the instructions provided in this section. They will help you to set up your environment, and they explain a little bit about how XSLT is used by MRDS.

1. Start up `DssHost` with no manifest:

```
C:\Microsoft Robotics Studio (1.5)>dsshost /p:50000 /t:50001
```

2. Open a web browser and browse to the following:

```
http://localhost:50000/resources/dss/Microsoft.Dss.Runtime.Home.Template.xml
```

You should see a message that says that this page is a template. Click the link to `Template.xslt`. The full URL is as follows:

```
http://localhost:50000/resources/dss/Microsoft.Dss.Runtime.Home.Template.xslt
```

**3.**  Assuming that your copy of MRDS is installed on the C: drive, create a directory called `C:\Resources`. Under this, create a directory called `DSS`. (Substitute the appropriate drive letter if your MRDS installation is on a different drive).

**4.**  Save the XSLT file from the web browser into this new directory. The path `C:\Resources\DSS` corresponds to the path in the URL. The entire template is shown here:

```
<!--
This file is a template for xslt files that use the default
Microsoft Robotics Studio layout to represent a DSSP service state.
-->
<?xml version="1.0" encoding="utf-8" ?>
<xsl:stylesheet
    version="1.0"
    xmlns:xsl="http://www.w3.org/1999/XSL/Transform"
    xmlns:soap="http://www.w3.org/2003/05/soap-envelope"
    xmlns:dssp="http://schemas.microsoft.com/xw/2004/10/dssp.html"
    xmlns:svc="http://schemas.tempuri.org/app/2007/1/template.html"
    >

  <xsl:import href="/resources/dss/Microsoft.Dss.Runtime.Home.MasterPage.xslt" />

  <xsl:template match="/">
    <xsl:comment><!-- Service Header Info --></xsl:comment>
    <xsl:variable name="title">
      Service Page Title
    </xsl:variable>
    <xsl:variable name="serviceName">
      Service Name
    </xsl:variable>
    <xsl:variable name="description">
      Service Description
    </xsl:variable>

    <xsl:call-template name="MasterPage">
      <xsl:with-param name="serviceName" select="$serviceName" />
      <xsl:with-param name="description" select="$description" />
      <!-- If title is not provided, serviceName will be used instead. -->
      <xsl:with-param name="title">
        <xsl:value-of select="$serviceName" />
        <xsl:if test="$title != ''">
          <xsl:text> - </xsl:text>
          <xsl:value-of select="$title" />
        </xsl:if>
      </xsl:with-param>
      <!-- Possible values for navigation are: 'Open', 'Closed', and 'None'
           'Open' is the default value. -->
```

*(continued)*

*(continued)*

```
              <xsl:with-param name="navigation" select="'Open'" />
              <!-- The contents of head param will be placed just before the </head> tag in
        html. -->
              <xsl:with-param name="head">
                <style type="text/css">
                  /* Service-specific stylesheet goes here */
                </style>
                <script language="javascript" type="text/javascript">
                  <![CDATA[<!--

/* Service-specific script goes here */

dssRuntime.init = function()
{
  // Add page initialization code here.
  // This function is attached to the window.onload event.
  // Do not override window.onload.
}

//-->      ]]>
                </script>
              </xsl:with-param>
            </xsl:call-template>
          </xsl:template>

          <!-- Match service state's document element. -->
          <xsl:template match="/svc:Template">
            <xsl:comment><!-- Service State Contents --></xsl:comment>
            <form name="DssForm" method="post">
              <xsl:copy-of select="." />
            </form>
          </xsl:template>

        </xsl:stylesheet>
```

**5.** If you open this file in Visual Studio, you might see an error:

```
Unexpected XML declaration. The XML declaration must be the first node in the
document and no white space characters are allowed to appear before it.
```

If you see this error, delete the comment lines at the top of the file just before the `<?xml>` line and save the file.

Now when you want to create a new XSLT file for MRDS in Visual Studio, you can easily use this template. This is the approach that Microsoft recommends. Bear in mind that you might need to update this template whenever a new version of MRDS is released.

**6.** Looking at the XML code, note the `<xsl:import>` element that refers to the following:

```
http://localhost:50000/resources/dss/Microsoft.Dss.Runtime.Home.MasterPage.xslt
```

Enter this URL into your web browser. Again, you should see an XML file. Save this file to the `C:\Resources\DSS` folder along with the template.

**7.** This master page refers to yet another file:

```
http://localhost:50000/resources/dss/Microsoft.Dss.Runtime.Home.Navigation.xslt
```

Enter this URL and then save the XML file to `C:\Resource\DSS` too.

By saving these files to your hard drive, you avoid some other error messages that appear if you try to edit an XSLT file. This step is not essential, but it gets rid of some annoying errors.

If you are interested, there are other embedded resources for MRDS that you can examine using a web browser:

```
http://localhost:50000/resources/dss/Microsoft.Dss.Runtime.Home.Styles.Common.css
http://localhost:50000/resources/dss/Microsoft.Dss.Runtime.Home.JavaScript
.Common.js
```

There are also some images that are not listed here, but you don't need them. Unfortunately, you cannot simply browse `/resources/dss` to view the available embedded resources. In order to see an embedded resource, you must know its full URI. Otherwise, `DssHost` presents you with a blank page.

## Using XSLT for Enhanced Information Display

To use an XSLT file to display your service state you have two options:

❑ **Use an external file:** Under the MRDS root directory is a `store` folder containing a `transforms` folder. You can place your XSLT files here if you wish. However, this makes deployment a little more complicated and it's possible for the XSLT file to become separated from the service. Conversely, it is much easier to edit an external XSLT file because you don't need the service source code and you don't have to recompile.

❑ **Use an embedded resource:** Embedded resources are linked into the service assembly. When the service starts, these resources are "mounted" into the `/resources` pseudo-folder on the DSS node so that they are accessible via URIs. The advantages of this approach are that nobody can mess with your XSLT and it can never get lost because it is part of the service DLL. Conversely, it is more difficult to change the XSLT file and it requires recompilation of the service.

From here on, the examples use embedded resources because this is the authors' preferred approach. The format of the URI for embedded resources is as follows:

```
/resources/Assembly-Name/Default-Namespace.Path.Filename
```

You can open the project properties for your service and look on the Application tab to find the assembly name and the default namespace. If your XSLT file is in the same folder as the rest of your source files, then there is no `Path`. The `Filename` is just the name of the XSLT file. The URI is not case sensitive.

It is common practice to locate XSLT files in a `Resources` subfolder under your source directory. The TeleOperation service follows this convention, so the full URI for the XSLT file is as follows (ignoring the wrap-around):

```
/resources/TeleOperation.y2008.m01/ProMRDS.Robotics.TeleOperation.↵
Resources.TeleOperation.xslt
```

## Building an XSLT File from Scratch

To create a new XSLT file you have two options:

❑ **Create the entire XSLT file yourself:** Building your own XSLT file gives you complete flexibility in the design. You can develop your own "look and feel," but it won't be compatible with the existing MRDS pages.

❑ **Use the MRDS master page and template:** Using the MRDS template saves you some work and ensures that the layout and formatting of the page match the other MRDS pages.

Obviously, this section covers building the XSLT file from scratch; the next section discusses using the MRDS template. If you want to start with a blank file and embed it in your service assembly, proceed as follows:

**1.** Click Visual Studio ⇨ Project ⇨ Add New Item and then select XSLT file from the dialog. The new XSLT file should appear in the Solution Explorer.

**2.** Click on the new file in Solution Explorer and then look in the Properties panel. Change the Build Action from Content to Embedded Resource.

*This is an important step. If you don't set the Build Action to Embedded Resource, you won't be able to see your embedded XSLT file in a web browser.*

**3.** The XSLT file that you create can contain a complete web page, giving you total flexibility to format the page any way you want. The basic layout in this case is similar to the following example:

```
<?xml version="1.0" encoding="UTF-8" ?>
<xsl:stylesheet version="1.0"
        xmlns:xsl="http://www.w3.org/1999/XSL/Transform"
        xmlns:svc="http://schemas.tempuri.org/yyyy/mm/servicename.html">
  <xsl:output method="html"/>

  <xsl:template match="/svc:ServiceState">
    <html>
      <head>
        <title>Service Name</title>
        <link rel="stylesheet" type="text/css"
href="/resources/dss/Microsoft.Dss.Runtime.Home.Styles.Common.css" />
      </head>
      <body style="margin:10px">
        <h1>Service Name</h1>
        <table border="1">
          <tr class="odd">
            <th colspan="2">Service State</th>
          </tr>
          <tr class="even">
            <th>Property 1:</th>
            <td>
              <xsl:value-of select="svc:property1"/>
            </td>
          </tr>
          <tr class="odd">
            <th>Property 2:</th>
            <td>
```

```
            <xsl:value-of select="svc:property2"/>
          </td>
        </tr>
      </table>
    </body>
  </html>
  </xsl:template>
</xsl:stylesheet>
```

The service contract identifier is listed as one of the xmlns (XML namespace) parameters at the top of the file using the alias svc so that you can easily refer to it throughout the rest of the file. In the <xsl:template> element, the match attribute specifies the ServiceState. This must be the actual class name of your service state. Remember that the XML file output by your service is a serialized version of the service state.

*Be very careful with capitalization in the contract identifier and the names of the service state and the properties. XML is case sensitive.*

Inside the template section is a complete HTML page. Property values are displayed from the state using the <xsl:value-of> element. All properties must be referred to using a full URI. This is why they are shown as svc:property1 and svc:property2. If you use nested structures within your state, then you cannot just use the "dot notation" that you use in C#. For example, the state for the TeleOperation service contains a class called Options, which in turn contains a number of other properties. To access Options.MotionSpeed you must write the following:

```
<xsl:value-of select="svc:Options/svc:MotionSpeed"/>
```

Notice that a slash (/) replaces the dot (.) in the property name and that the contract identifier (svc:) is repeated. If you make a mistake in the syntax of the select attribute, the value quietly disappears. No errors are reported — the value is just missing from the web page.

In addition to formatting the layout of your state properties using XSLT, you can use a Cascading Style Sheet (CSS) file to control the formatting of the HTML tags, i.e., the appearance of the elements on the page.

*It is beyond the scope of this book to explain how to write CSS code, so it is assumed that you already know how, or you can learn from a tutorial website on the Internet. The full definition of CSS is also available on the World Wide Web Consortium website, but it is quite complex to read.*

A CSS file can be embedded in your service in exactly the same way as an XSLT file. (Add it to your project and then mark it as an embedded resource). Then you just need to know the appropriate URI. In the preceding example, the built-in MRDS Cascading Style Sheet is used for formatting. The full URI for this is in the href attribute of the <link> tag. This MRDS CSS file defines the odd and even classes that are used on the table rows to make them easier to read by changing the background highlighting.

## Creating an XSLT file Using the MRDS Template

If you want to save yourself some time and promote page uniformity, instead of building an XSLT file from scratch as shown in the last section, you can always create one using the MRDS template. To use the MRDS template, follow these steps:

**1.** Copy the template from the C:\Resources\DSS folder (where you saved it earlier in the section "Setting Up Your Development Environment for XSLT") into your project directory and give it an appropriate name. Click Project ⇨ Add Existing Item to add it to your project.

*Remember that once you have added the XSLT file to your project, you must change the properties so that the Build Action is set to Embedded Resource.*

2. Edit the XSLT file. You need to insert your service's contract identifier at the top of the file:

```
<?xml version="1.0" encoding="utf-8" ?>
<xsl:stylesheet
    version="1.0"
    xmlns:xsl="http://www.w3.org/1999/XSL/Transform"
    xmlns:soap="http://www.w3.org/2003/05/soap-envelope"
    xmlns:dssp="http://schemas.microsoft.com/xw/2004/10/dssp.html"
    xmlns:svc="http://www.promrds.com/contracts/2008/01/teleoperation.html"
    >
```

For convenience, the alias for your service is `svc`. You can change it if you like, but it is easier to leave it alone.

3. In the first `<xsl:template>` section, change the Service Title, Name, and Description as appropriate for your service:

```
<xsl:template match="/">
    <xsl:comment><!-- Service Header Info --></xsl:comment>
    <xsl:variable name="title">
        TeleOperation Service
    </xsl:variable>
    <xsl:variable name="serviceName">
        <img
src="/resources/teleoperation.y2008.m01/ProMRDS.Robotics.TeleOperation.Resources.
icon_32x32.gif" align="middle" /> TeleOperation
    </xsl:variable>
    <xsl:variable name="description">

        Allows you to drive a Differential Drive robot using live video from a WebCam
on the robot
    </xsl:variable>
```

4. Adding images to your page is also relatively easy. You can use embedded resources once again. The file format is not particularly important as long as it is one that the web browser understands. Note that BMP files are not understood by all web browsers. (This is a Windows format; and in any case it is not compressed, so the images can be quite large). Use the PNG, GIF, or JPG file format for your images.

Notice that the TeleOperation service displays an icon in the page header (in the preceding code) by adding an image to the `serviceName` variable. (Ignore the wrap-around in the URI).

In the MRDS template, you can either define CSS styles directly (between the `<style>` tags) or include a style sheet file. Look at the comments carefully. They indicate that the `head` parameter is placed in the HTML `<head>` section of the web page, and there is also a comment showing where to insert style definitions directly.

In the following example, an embedded style sheet file is used:

```
    <!-- The contents of head param will be placed just before the </head> tag in
html. -->
```

```
  <xsl:with-param name="head">

  <link rel="stylesheet" href="/resources/teleoperation.y2008.m01/ProMRDS.Robotics ⬋
.TeleOperation.Resources.TeleOperation.css" type="text/css" />
    <style type="text/css">
      /* Service-specific stylesheet goes here */
    </style>
```

Using your own style sheet means that your page might not have the same appearance as other MRDS pages. Notice here that the CSS file is also an embedded resource in the `Resources` folder of the project. The standard MRDS style sheet is automatically included by the master page.

At the bottom of the first template is another `<xsl:template>` section. You need to change the match criterion from `/svc:Template` to use the correct name for your service state class. This is shown in the following example (ignore the `<form>` tag for now, as it is discussed later):

```
  <xsl:template match="/svc:TeleOperationState">
    <form name="DssForm" method="post" onsubmit="return checkform(this);">
      <div class="Content">
    <table width="100%" border="0" cellpadding="5" cellspacing="5">
      <tr>
        <th>Host:</th>
        <td>
          <xsl:value-of select="svc:Host"/>
        </td>
      </tr>
      <tr>
        <th>Port:</th>
        <td>
          <xsl:value-of select="svc:Port"/>
        </td>
      </tr>
      <tr>
        <th>Connected:</th>
        <td>
          <xsl:choose>
            <xsl:when test="svc:Connected = 'true'">
              Connection established
            </xsl:when>
            <xsl:otherwise>
              Not connected
              (<b>Tip:</b>
              <span class="greyText">
                Enter a Host name and a Port and click on Connect
              </span>)
            </xsl:otherwise>
          </xsl:choose>
        </td>
      </tr>
  ...
      <tr>
        <th colspan="2">Option Settings</th>
```

*(continued)*

*(continued)*

```
            </tr>
            <tr>
              <th>Main Window Start Position</th>
              <td><xsl:value-of select="svc:Options/svc:WindowStartX"/>,
    <xsl:value-of select="svc:Options/svc:WindowStartY"/> (Saved automatically)
            </td>
            </tr>
            <tr>
    ...
        </table>
        </div>
        </form>
      </xsl:template>

    </xsl:stylesheet>
```

For this code, note the following:

❑   Usually the data is displayed in a table with two columns — the left column consists of Table Headings (th) and the right column is Table Data (td).

❑   Values are inserted into the page using the <xsl:value-of> element. Remember to use the correct syntax for nested structures in your state, such as for the WindowStartX property.

❑   Notice the use of <xsl:choose> to display different text depending on the state of the Connected property. You can have several <xsl:when> elements as well as <xsl:otherwise>, so it is conceptually similar to a switch statement. Alternatively, you could also use <xsl:if>. However, there is no else part, so you would have to use two if statements.

The best way to learn how to create XSLT code is probably to look at existing files. Search the MRDS samples folder for XSLT files.

## Displaying the State Using XSLT

You are not finished yet. To display your state using XSLT, you must create a handler for HttpGet requests and add this message type to your main operations port (if it is not there already). The handler specifies the XSLT file when it returns the state.

To refer to your embedded XSLT file in the HttpGet handler, you define a string variable at the top of the source file. It is usually called _transform, but it doesn't have to be. You can use the [EmbeddedResource] attribute to construct the full URI for the XSLT file as shown here for the TeleOperation service:

```
    /// <summary>
    /// Embedded XSLT file for formatting State on a web page
    /// </summary>
    [EmbeddedResource("ProMRDS.Robotics.TeleOperation.Resources.TeleOperation.xslt")]
        string _transform = null;
```

The appropriate prefix is added to the URI and the resulting string is assigned to _transform when you compile the code.

The code for the `HttpGet` handler is quite simple:

```
/// <summary>
/// HttpGet Handler
/// </summary>
/// <param name="get"></param>
/// <returns></returns>
[ServiceHandler(ServiceHandlerBehavior.Concurrent)]
public virtual IEnumerator<ITask> HttpGetHandler(HttpGet httpGet)
{
    // Format the response using a transform
    httpGet.ResponsePort.Post(new HttpResponseType(
        HttpStatusCode.OK,
        _state,
        _transform)
    );
    yield break;
}
```

The response posted back is created using `_state` and `_transform`.

That's all there is to formatting your state information. However, if you make a mistake in the URI of your XSLT file in the [`EmbeddedResource`] attribute, you might see an error like the one shown in Figure 4-8.

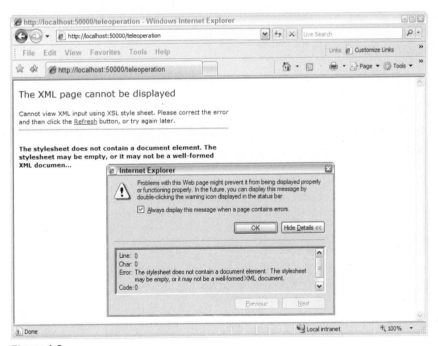

Figure 4-8

You can check whether your embedded XSLT file is present in your assembly by entering the full URI into the address bar of the web browser. If it is not found, a blank page is displayed. In that case, you need to double-check your spelling, the path, and the placement of dots. The URI is not case sensitive, so that cannot be the problem.

## Creating a Web Form for Data Input

Creating a Web Form is not much different from what you have already seen except that you wrap your HTML code in a `<form>` tag and put the property values into `<input>` tags. You also have to add a `Submit` button; otherwise, there is no way to post the Form! If you are familiar with web development, then you should not have any trouble:

1.  Looking at the `TeleOperation.xslt` file again, there is a `<form>` tag at the top of the template:

    ```
    <xsl:template match="/svc:TeleOperationState">
      <form name="DssForm" method="post" onsubmit="return checkform(this);">
    ```

    Notice that the method is `post`. You cannot (easily) use the `get` method because the `HttpGet` and `HttpPost` handlers are separate. The Form `name` is not important, and the `onsubmit` attribute is discussed in the next section.

2.  In the main part of the HTML table, you can use `<input>` tags to get values from the user. The following example displays the current value of the `DeadZoneX` property in a textbox:

    ```
    <tr>
      <th>Dead Zone X</th>
      <td>
        <input type="text" name="DeadZoneX" class="TextBox">
          <xsl:attribute name="value">
            <xsl:value-of select="svc:Options/svc:DeadZoneX"/>
          </xsl:attribute>
        </input>
        (Game Controller range is 1000, so typical value is 100)
      </td>
    </tr>
    ```

Note that the XSL code sets the `value` attribute of the `<input>` tag to show the current value.

*The* **dead** *zone is a region at the center of the joystick's movement where the drive speed is set to zero. Most joysticks do not return exactly to the zero position when you release them. The small amount by which they are off-center causes very small set power commands to be sent to the robot's wheels, but it is unlikely to move due to inertia. In some cases, this causes the motors to squeal, which can be quite annoying!*

3.  The last row in the table is a Submit button. The Form data is sent to the service (as a `HttpPost` request) when you click the button, which is labeled `"Save"`:

    ```
    <tr>
      <td>
        <input type="submit" name="Save" value="Save" />
      </td>
    </tr>
    ```

The Submit (Save) button is visible at the very bottom of Figure 4-9.

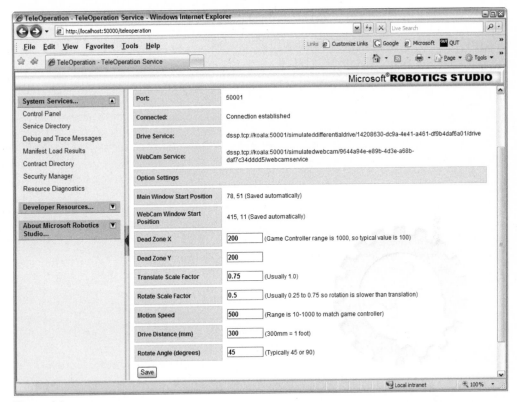

Figure 4-9

Notice in Figure 4-9 that the values above the Option Settings heading cannot be edited; they are for informational purposes only. The window positions cannot be edited either. They are obtained directly from the windows, so you just need to position the windows where you want them before you save the state.

All of the remaining properties have textboxes for entering new values. It is possible to make them right-aligned, but this has not been done. It is simply a matter of changing the TextBox style in the CSS file.

You can have more than one <form> tag on a web page, and you can have more than one Submit button in a Form (with different names). In that case, the code in your HttpPost handler needs to check the Form name and/or the Submit button name.

After you click Save, the data is sent in an HttpPost request to your handler. Before looking at the handler though, you need to add some support utilities. At the top of the service, add the following statement:

```
DsspHttpUtilitiesPort _httpUtilities;
```

This port can be used to request support services for processing HTTP forms. Do *not* be tempted to create a new port like this:

```
DsspHttpUtilitiesPort _httpUtilities = new DsspHttpUtilitiesPort();
```

Using new does not work. The new port must be created in the Start method:

```
// Needed for HttpPost
_httpUtilities = DsspHttpUtilitiesService.Create(Environment);
```

*If you forget to create the HTTP utilities port and simply use* new, *you will fall into a common trap of MRDS — sending messages when there is nobody home! You can send as many messages as you like to a port, but if no service is listening on the other end then your requests will go unanswered. You will wait a very long time for the HTTP utilities to do their job!*

The TeleOperation HttpPost handler begins like this:

```
/// <summary>
/// Http Post Handler for Web Form inputs
/// </summary>
[ServiceHandler(ServiceHandlerBehavior.Concurrent)]
public virtual IEnumerator<ITask> HttpPostHandler(HttpPost httpPost)
{
    string ErrorMessage = String.Empty;
    Fault fault = null;
    NameValueCollection parameters = new NameValueCollection();

    // Use helper to read form data
    ReadFormData readForm = new ReadFormData(httpPost.Body.Context);
    _httpUtilities.Post(readForm);

    // Wait for result
    yield return Arbiter.Choice(
        readForm.ResultPort,
        delegate(NameValueCollection col)
        {
            parameters = col;
        },
        delegate(Exception e)
        {
            fault = Fault.FromException(e);
            LogError(null, "Error processing form data", fault);
            ErrorMessage += e.Message;
        }
    );

    if (fault != null)
    {
        httpPost.ResponsePort.Post(fault);
        yield break;
    }
```

The first thing the handler does is send a ReadFormData request to the HTTP utilities. Because the utilities operate like a service, you have to wait for a response and be prepared to handle a Fault.

There is an alternative to using these utilities: You can write your own code to extract the form parameters. The following code is not seriously proposed as a replacement, but it shows how the parameters are obtained. The code performs no error handling, and it fails if the total amount of data exceeds 10K:

```
NameValueCollection parameters = new NameValueCollection();

// Extract the parameters from the Request data stream
int length = 0;
byte[] data = new byte[10240];
length = httpPost.Body.Context.Request.InputStream.Read(↵
data, 0, 10240);
StringBuilder sb = new StringBuilder();
for (int i = 0; i < length; i++)
    sb.Append((char)data[i]);

string request = sb.ToString();
char[] ampersand = { '&' };
char[] equals = { '=' };
string[] pairs = request.Split(ampersand);
for (int i = 0; i < pairs.Length; i++)
{
    string[] namevalue = pairs[i].Split(equals);
    parameters.Add(namevalue[0], namevalue[1]);
}
```

Each of the parameters is checked in turn to see if it is valid. You must do this even if you use client-side validation in JavaScript because users can turn off JavaScript in their web browsers. For example, the following code confirms that the user entered a valid number for MotionSpeed:

```
double MotionSpeed = _state.Options.MotionSpeed;
bool validValue = false;

if (!string.IsNullOrEmpty(parameters["MotionSpeed"]))
{
    try
    {
        MotionSpeed = double.Parse(parameters["MotionSpeed"]);
        validValue = true;
    }
    catch (Exception e)
    {
        string msg = "Could not parse Motion Speed: " + e.Message;
        LogError(msg);
        ErrorMessage += msg;
    }
}

if (validValue && MotionSpeed >= 10 && MotionSpeed <= 1000)
{
    _state.Options.MotionSpeed = MotionSpeed;
}
```

The code also checks whether `MotionSpeed` is within a reasonable range, and finally assigns the new value to the property in the state. (Because it modifies the state, the `HttpPost` handler should be Exclusive. However, most implementations mark it as a Concurrent handler. Make sure that you mark it as Exclusive if it changes the state.)

Similar checks are performed for the rest of the properties. Notice that any error messages are appended to the `ErrorMessage` variable. Once all of the updates have completed, it is a simple matter to determine whether the `ErrorMessage` is empty and call the appropriate routine to post success or failure:

```
// Finally, process the result
if (ErrorMessage == string.Empty)
{
    HttpPostSuccess(httpPost);
}
else
{
    HttpPostFailure(httpPost, ErrorMessage);
}

yield break;
}
```

The `HttpPostSuccess` routine does not require much comment. Just note that it saves the option settings back into public variables in the WinForms if they are active. In effect, it passes information from the main service to the WinForms. Then it saves the state to a `config` file:

```
/// <summary>
/// Send Http Post Success Response
/// </summary>
private void HttpPostSuccess(HttpPost httpPost)
{
    // Grab the current window location
    if (_driveControl != null)
    {
        _state.Options.WindowStartX = _driveControl.Location.X;
        _state.Options.WindowStartY = _driveControl.Location.Y;
    }
    if (_cameraForm != null)
    {
        _state.Options.WebCamStartX = _cameraForm.Location.X;
        _state.Options.WebCamStartY = _cameraForm.Location.Y;
    }

    // Update the Forms with the new option settings
    // This is another implicit method of communication with WinForms
    if (_driveControl != null)
        _driveControl.options = _state.Options;
    if (_cameraForm != null)
        _cameraForm.options = _state.Options;

    // Post a response
    HttpResponseType rsp =
        new HttpResponseType(HttpStatusCode.OK, _state, _transform);
```

```
        httpPost.ResponsePort.Post(rsp);

        // Save the state now because it might have changed
        SaveState(_state);
    }
```

Conversely, the `HttpPostFailure` routine does require some explanation. If it simply posts back a `Fault` (as in the code that is commented out), then XML code is displayed on the screen. Instead, it posts a `Fault` response but specifies an XSLT file (`_faultTransform`):

```
/// <summary>
/// Send Http Post Failure Response
/// </summary>
private void HttpPostFailure(HttpPost httpPost, string failureReason)
{
    // Create a new Fault based on the error message
    Fault fault = Fault.FromCodeSubcodeReason(FaultCodes.Receiver, ↵
DsspFaultCodes.OperationFailed, failureReason);

    //HttpResponseType rsp = new HttpResponseType(fault);
    //httpPost.ResponsePort.Post(rsp);

    // Post it back but use the Fault Transform
    httpPost.ResponsePort.Post(new HttpResponseType(
        HttpStatusCode.OK,
        fault,
        _faultTransform
        )
    );
}
```

The `Fault.xslt` file isn't covered here. It formats a standard SOAP fault (which is an XML message) and provides a couple of buttons so that users are not left at a dead end with nowhere to go. You can examine this file yourself.

## Using JavaScript for Client-Side Scripting

A Web Forms approach whereby all the data on the Form is sent to the web server for validation does not result in a good user experience. It is much better to have JavaScript code in the Form to do validation prior to submitting the Form.

If you do not already know JavaScript, it might be a good investment in your job skills because it is a key component of web development. Numerous websites on the Internet offer tutorials on JavaScript, such as www.w3schools.com/js/default.asp. Because it is a C-like language, you should not have trouble learning enough to do basic validation. There are also websites that provide JavaScript code for a variety of tasks, such as www.dynamicdrive.com.

The position in the XSLT file where you place your JavaScript code is indicated quite clearly in the comments in the template. You can either enter the code directly into the XSLT file or create another embedded resource and add your own `<script>` tag to reference it as follows:

```
<script language="javascript" type="text/javascript" src="URI-for-JS-file" />
```

In `TeleOperation.xslt` you can see the validation code, which is split across two routines. The first routine, simply called `validate`, is used to confirm that a value is numeric and falls within a specified range:

```
// Check that a value is numeric and within a specified range
// Returns a string which is either empty or an error message
function validate(val, fieldname, minval, maxval)
{
  var msg = "";
  var num;
  if (val.length == 0)
  {
     msg += "Please enter a value for " + fieldname + "\n";
  }
  else
  {
    num = parseFloat(val);
    if (isNaN(num))
      msg += fieldname + " must be a number\n";
    else if (num < minval || num >= maxval)
      msg += fieldname + " is out of range (" + minval + " to " + maxval +")\n";
  }
  return msg;
}
```

The `validate` function returns an error message if there is one, or an empty string. It is called by the `checkform` routine:

```
// Check the form before submission
// NOTE: Return true if OK to proceed, or false if there are errors
function checkform(f)
{
  var msg = "";

  // Accumulate all of the error messages (if there are any)
  msg += validate(f.DeadZoneX.value, "Dead Zone X", 0, 900);
  msg += validate(f.DeadZoneY.value, "Dead Zone Y", 0, 900);
  msg += validate(f.TranslateScaleFactor.value, "Translate Scale Factor", ↵
0.01, 100);
  msg += validate(f.RotateScaleFactor.value, "Rotate Scale Factor", 0.01, 100);
  msg += validate(f.MotionSpeed.value, "Motion Speed", 10, 1000);
  msg += validate(f.DriveDistance.value, "Drive Distance", 10, 2000);
  msg += validate(f.RotateAngle.value, "Rotate Angle", 5, 360);

  // If there were any errors at all, display them
  if (msg != "")
  {
    // Pop up a dialog to tell the user what is wrong
    alert(msg);
    return false;
  }

  // If we got here, then the data is OK
  return true;
}
```

In addition, you need to modify the `<form>` tag so that the validation routine is called prior to the Form being submitted. To do this, add the `onsubmit` attribute to the tag:

```
<form name="DssForm" method="post" onsubmit="return checkform(this);">
```

Note the syntax used in the `onsubmit` attribute — you must return the Boolean value from `checkform` so that the Form submission can either proceed or be suppressed. In addition, note that JavaScript is case sensitive, so if you write `"return CheckForm(this);"`, you will receive an error when you try to submit the Form.

Validating Form data this way cannot catch every possible type of error. There might still be errors that occur only after the Form has been submitted and the service has had a chance to process it. In any case, the user can turn off JavaScript in the web browser, so you always have to check on the server side. This is just basic Web Development 101.

# Using a Camera

This section discusses how to set up a webcam and display live video in a WinForm. You have already seen in Figures 4-3 and 4-4 that TeleOperation displays a separate window for the video feed.

Before you start working with a webcam, it is a good idea to test it. Some webcams don't work with MRDS because they must support DirectX. In particular, older camera drivers might not be suitable if they support only the obsolete Video for Windows (VFW) standard.

To test the camera, plug it in and make sure that the drivers are loaded. (Read the manufacturer's instructions for this. There might even be a test program included with the software so that you can see if the camera works.)

In Chapter 3, you learned how to start `DssHost` without a manifest and manually start up a Webcam service. If you need detailed instructions, refer back to Chapter 3. In summary:

1. Start up `DssHost` from a MRDS Command Prompt window without a manifest.

2. Open a web browser and browse to the Control Panel.

3. Locate the Webcam service in the list of services.

4. Click the Create button beside the Webcam service.

5. Browse to the Service Directory.

6. Click the Webcam service in the list of running services.

This displays a page that shows an image from your webcam. You can click the Start button to get a continuous video feed. If you worked through Chapter 3, then you might have already tried out your webcam in the section "Service Directory."

If your camera requires a particular service to be running — for example, the Surveyor SRV-1 with the modified service supplied with this book — then you have to start `DssHost` with the appropriate manifest. However, the Swann Microcam2 plugs into a USB Video Capture Device (VCD) and is automatically detected by the standard Webcam service.

If you change webcams, you might encounter a problem trying to get the new one to work. In this case, try renaming (or deleting) the `config` file, which is `samples\Config\webcam.Config.xml`. This file contains the last selected webcam, so if you switch cameras, the Webcam service sometimes gets confused. This seems to be the case with the Belkin USB VCD.

> *You can change the settings on your webcam via the Webcam state page even while the TeleOperation service is running. TeleOperation does not provide facilities to do this. If the default resolution is not appropriate, then you have to change it manually. Once you have made the change, it is recorded in the* config *file and you should not have to worry about it again.*

## Adding a Camera to a Service

The following is a quick summary of the steps that are required to add a webcam to a new service. You should first test the camera as outlined above and make sure that it works with MRDS.

1. Set up the webcam as a partner, or write code to locate it in the service directory and connect to it dynamically.

2. If you want to use an existing WinForm to display the video images, skip to step 4.

3. Create a WinForm (see the section "Creating a WinForm"). Set up the necessary communication with the main service as explained earlier in the chapter. This Form does not need to send messages back to the main service, so the standard `OnLoad` and `OnClosed` messages (discussed above) should be sufficient.

4. Add a `PictureBox` to the WinForm with an appropriate size. Cameras usually have an aspect ratio of 4:3, e.g., 160 × 120, 320 × 240, etc. Set the `SizeMode` property to `Zoom` so that the image aspect ratio is preserved (unless you like stretched images).

5. Add a public property or method to the WinForm to set the bitmap in the `PictureBox`.

6. Add ports to your main service to communicate with the webcam and receive notifications. You should add a shutdown port as well.

7. Add a receiver to the main interleave to handle notification messages from the webcam. This receiver listens for `UpdateFrame` messages from the webcam.

8. Write a handler for `UpdateFrame` messages. This handler sends `QueryFrame` messages to the webcam to get the actual bitmap data, and then puts it into the `PictureBox` on the Form using `FormInvoke`.

9. Add code to your main service to subscribe to the webcam either when you connect or in the `Start` method if the partnership is established in the manifest.

10. Optionally, add code to your `Drop` handler to unsubscribe from the webcam before shutting down. (This is best practice).

Most of the code you can just copy and paste from another service such as TeleOperation or the Dashboard.

# Setting Up a WinForm for the Video Feed

If you require a particular web camera for your service to operate, you can declare the generic Webcam service as a partner at the top of your main service source file. Then add a partner in the manifest, and when your service starts it should find the appropriate camera.

However, for the TeleOperation service there are two problems with this approach:

❑ This would lead to a static definition and you would have to modify the manifest or the source code to use a different DSS node or a different camera.

❑ The service might not start if the webcam partner cannot be found.

The TeleOperation service does not connect to a webcam until you click the Connect button. The button click event handler in DriveControl.cs posts a message to the main service requesting a connection to the host and port specified on the WinForm. The OnConnect message contains the URI of the appropriate directory service in the Service property.

The code to handle the connection request in TeleOperation.cs is in two parts. The OnConnectHandler is called first in response to the message from the DriveControl Form; then the ConnectDrive and/or ConnectWebCam routines are called to make the actual connections.

## Handling Connection Requests

OnConnectHandler searches the directory on the specified remote host (it does not have to be localhost) for generic Differential Drive and Webcam services. It is possible that there are no matching services, so this must be taken into account:

```
/// <summary>
/// Connect Handler
/// </summary>
/// <param name="onConnect"></param>
/// <returns></returns>
IEnumerator<ITask> OnConnectHandler(OnConnect onConnect)
{
    if (onConnect.Form == _driveControl)
    {
        string ErrorMessage = null;

        // The service here is the Directory on the specified host:port
        UriBuilder builder = new UriBuilder(onConnect.Service);
        builder.Scheme = new Uri(ServiceInfo.Service).Scheme;

        ds.DirectoryPort port = ↩
ServiceForwarder<ds.DirectoryPort>(builder.Uri);
        ds.Get get = new ds.Get();

        port.Post(get);
        ServiceInfoType[] list = null;

        yield return Arbiter.Choice(get.ResponsePort,
            delegate(ds.GetResponseType response)
```

*(continued)*

*(continued)*

```
                {
                        list = response.RecordList;
                },
                delegate(Fault fault)
                {
                        list = new ServiceInfoType[0];
                        LogError(fault);
                }
        );
```

If you did not have to allow for the possibility of remote hosts, then it would not be necessary to create a `ServiceForwarder` for the directory service. Instead, you could just use the `DirectoryPort`, which is defined in the `DsspServiceBase` class.

A simple `Get` request to the directory service returns an array of `ServiceInfoType` records. The code extracts the host name and port number from the first record, although this is not really necessary. It also handles the situation where no services are found:

```
ServiceInfoType driveInfo = null;
ServiceInfoType webcamInfo = null;

try
{
    if (list.Length > 0)
    {
        UriBuilder node = new UriBuilder(list[0].Service);
        node.Path = null;
        string nodestring = node.Host + ":" + node.Port;
        LogInfo(nodestring);
    }
    else
    {
        LogError("No services found!");
        ErrorMessage = "No services found!\n";
        _state.Connected = false;
    }
```

Next, the code loops through all of the services in the list, comparing the Contract ID with the IDs for the Differential Drive service and the Webcam service. If either of these is found, the service info is remembered:

```
string driveUriPath = null;
string webcamUriPath = null;
foreach (ServiceInfoType info in list)
{
        if (driveInfo == null && info.Contract == ↵
drive.Contract.Identifier)
        {
                driveInfo = info;
        }

        if (webcamInfo == null && info.Contract == ↵
webcam.Contract.Identifier)
```

```
                {
                    webcamInfo = info;
                }

                if (driveInfo != null && webcamInfo != null)
                    break;
            }
        }
        catch (Exception ex)
        {
            string msg = "Service search error: " + ex.Message;
            LogError(msg);
            ErrorMessage += msg + "\n";
            _state.Connected = false;
        }
```

An error message is constructed based on the outcome of the search. If either or both of the services were *not* found, a MessageBox displays. The MessageBox is invoked via a public method in the DriveControl Form. The ShowErrorMessage method is trivial — it contains a single statement, which displays a MessageBox using the string parameter it was given. This illustrates one way to display message boxes from within a service:

```
        if (driveInfo == null)
            ErrorMessage += "No Drive service found\n";
        if (webcamInfo == null)
            ErrorMessage += "No WebCam service found\n";

        if (ErrorMessage != null)
        {
            if (_driveControl != null)
            {
                WinFormsServicePort.FormInvoke(
                    delegate()
                    {
                        _driveControl.ShowErrorMessage(ErrorMessage);
                    }
                );
            }
        }
```

Lastly, the appropriate connect routine is called for each of the services (if found):

```
        if (driveInfo != null)
            SpawnIterator<string>(driveInfo.Service, ConnectDrive);

        if (webcamInfo != null)
            SpawnIterator<string>(webcamInfo.Service, ConnectWebCam);

        // We are "connected" if either service was found
        if (driveInfo != null || webcamInfo != null)
            _state.Connected = true;
    }
}
```

## Connecting a Webcam

Before looking at the ConnectWebCam routine, there are several global variables declared at the top of TeleOperation.cs that are used in this routine:

```
// Ports for the Web Camera
webcam.WebCamOperations _webCamPort;
webcam.WebCamOperations _webCamNotify = new webcam.WebCamOperations();
Port<Shutdown> _webCamShutdown = null;
// Form to display the video in
WebCamForm _cameraForm;
// Port for the WebCam Form to communicate on
WebCamFormEvents _webCamEventsPort = new WebCamFormEvents();
// Flag to indicate that the form is loaded and ready
bool _webCamFormLoaded = false;
```

For this code, note the following:

❑ _webCamPort is used to send requests to the camera service.

❑ The purpose of _webCamNotify should be obvious: It receives UpdateFrame messages from the camera. These messages do not contain the actual bitmap data — you must make a request to the camera to get the data in response to a notification. This avoids sending a lot of data around if there is a backlog.

❑ Likewise, the purpose of _webCamShutdown is obvious, as is the _cameraForm handle for the WinForm instance.

❑ The _webCamEventsPort is used for communication from the Webcam Form. In order to ensure that you can drive the robot using the keyboard regardless of which window has the input focus, the keyboard handling code is duplicated in the Webcam Form. It is a different port from the DriveControl Form, but it uses the same handler.

❑ _webCamFormLoaded is an important flag. It indicates whether the Webcam Form is active or not. Due to timing issues, it is possible for frames to arrive from the camera before the Form is properly initialized, or after the user has closed down the Form. The point of this flag is to prevent access violations caused by trying to update the PictureBox when it does not exist.

The rest of the code is in the Camera region at the bottom of TeleOperation.cs.

ConnectWebCam takes the name of the service as a string. If a connection is already open to a camera, then it unsubscribes:

```
// Handler for connecting to WebCam
IEnumerator<ITask> ConnectWebCam(string camera)
{
    //ServiceInfoType info = null;
    Fault fault = null;
    SubscribeResponseType s;
    //String camera = Opt.Service;

    // Already connected?
    if (_webCamPort != null)
    {
```

```
        // Unsubscribe
        if (_webCamShutdown != null)
            yield return PerformShutdown(ref _webCamShutdown);
}
```

Next, it creates a new operations port and subscribes to the webcam:

```
// Create a new port
_webCamPort = ServiceForwarder<webcam.WebCamOperations>(camera);

// Subscribe to the webcam
webcam.Subscribe subscribe = new webcam.Subscribe();
subscribe.NotificationPort = _webCamNotify;
subscribe.NotificationShutdownPort = _webCamShutdown;

_webCamPort.Post(subscribe);

yield return Arbiter.Choice(
    //_webCamPort.Subscribe(_webCamNotify),
    subscribe.ResponsePort,
    delegate(SubscribeResponseType success)
    { s = success; },
    delegate(Fault f)
    {
        fault = f;
    }
);

if (fault != null)
{
    LogError(null, "Failed to subscribe to webcam", fault);
    yield break;
}
```

If the subscription is successful, then the state is updated with the full URI of the Webcam service and a new Webcam View Form is created:

```
// Put the service URI into the state for visibility
_state.WebCamService = camera;
LogInfo("Connected WebCam to " + camera);

// Now that we have found the service and subscribed,
// create a form to display the video
RunForm runForm = new RunForm(CreateWebCamForm);

WinFormsServicePort.Post(runForm);

yield return Arbiter.Choice(
    runForm.pResult,
    delegate(SuccessResult success) { },
    delegate(Exception e)
    {
        fault = Fault.FromException(e);
    }
```

*(continued)*

*(continued)*

```
        );

        if (fault != null)
        {
            LogError(null, "Failed to Create WebCam window", fault);
            yield break;
        }

        yield break;

    }
```

At this stage, the TeleOperation service is waiting for messages to arrive from the webcam and a new WinForm should be visible on the screen, as shown in Figures 4-3 and 4-4.

## Processing Video Frames

When an `UpdateFrame` message arrives from the webcam, the handler must issue a `QueryFrame` request to get the image data. Notice that a timeout is set on the request so that the handler does not get bogged down if the Webcam service dies or is too slow responding:

```
// Handler for new frames from the camera
IEnumerator<ITask> WebCamUpdateFrameHandler(webcam.UpdateFrame update)
{
    webcam.QueryFrameResponse frame = null;
    Fault fault = null;

    // Don't do anything if the form has not loaded or has been closed!
    // Race conditions can arise when the form is first created, or if
    // the user closes the form. These result in access violations unless
    // we are careful not to execute the rest of the code.
    if (!_webCamFormLoaded)
        yield break;

    // Throw away the backlog
    // This does no harm because we are throwing away notifications,
    // not webcam images
    Port<webcam.UpdateFrame> p = ↩
(Port<webcam.UpdateFrame>)_webCamNotify[typeof(webcam.UpdateFrame)];
    if (p.ItemCount > 2)
    {
        Console.WriteLine("Webcam backlog: " + p.ItemCount);
        p.Clear();
    }

    webcam.QueryFrame query = new webcam.QueryFrame();
    // Set a timeout so that this cannot wait forever
    query.TimeSpan = TimeSpan.FromMilliseconds(1000);
    _webCamPort.Post(query);

    // Wait for response
    yield return Arbiter.Choice(
        query.ResponsePort,
```

```
                    delegate(webcam.QueryFrameResponse success)
                    {
                        frame = success;
                    },
                    delegate(Fault f)
                    {
                        fault = f;
                    }
                );

                if (fault != null)
                {
                    LogError(null, "Failed to get frame from camera", fault);
                    yield break;
                }
```

Once a response is received successfully, the data (which is a raw array of bytes) can be turned into a `Bitmap` and inserted into the `PictureBox` on the Form:

```
                // Create a bitmap from the webcam response and display it
                Bitmap bmp = MakeBitmap(frame.Size.Width, frame.Size.Height, ↵
        frame.Frame);
                // Display the image in the WinForm
                SpawnIterator<Bitmap>(bmp, DisplayImage);

                yield break;
            }
```

The code for creating a `Bitmap` from a byte array is general and can be used anywhere that you need to do this type of conversion. Note that the image dimensions must be supplied separately because a byte array does not contain this information:

```
        Bitmap MakeBitmap(int width, int height, byte[] imageData)
        {
            // NOTE: This code implicitly assumes that the width is a multiple
            // of four bytes because Bitmaps have to be longword aligned.
            // We really should look at bmp.Stride to see if there is any padding.
            // However, the width and height come from the webcam and most cameras
            // have resolutions that are multiples of four.

            Bitmap bmp = new Bitmap(width, height, PixelFormat.Format24bppRgb);

            BitmapData data = bmp.LockBits(
                new Rectangle(0, 0, bmp.Width, bmp.Height),
                ImageLockMode.WriteOnly,
                PixelFormat.Format24bppRgb
            );

            Marshal.Copy(imageData, 0, data.Scan0, imageData.Length);

            bmp.UnlockBits(data);

            return bmp;
        }
```

Finally, the new `Bitmap` image is copied into the `PictureBox` using a `FormInvoke`. This code explicitly creates a `FormInvoke` message, and then posts it:

```
// Display an image in the WebCam Form
IEnumerator<ITask> DisplayImage(Bitmap bmp)
{
    Fault fault = null;

    // Insurance in case the form was closed
    if (!_webCamFormLoaded)
        yield break;

    FormInvoke setImage = new FormInvoke(
        delegate()
        {
            if (_webCamFormLoaded)
                _cameraForm.CameraImage = bmp;
        }
    );

    WinFormsServicePort.Post(setImage);

    yield return Arbiter.Choice(
        setImage.ResultPort,
        delegate(EmptyValue success) { },
        delegate(Exception e)
        {
            fault = Fault.FromException(e);
        }
    );

    if (fault != null)
    {
        LogError(null, "Unable to set camera image on form", fault);
    }
    else
    {
        // LogInfo("New camera frame");
    }
    yield break;
}
```

There are only a couple of lines of code in the `FormInvoke` delegate. First, a test is done to make sure that the Webcam Form is still valid. Then the `Bitmap` is assigned to a public property in the Form.

The code inside the Webcam Form that handles the `Bitmap` is as follows:

```
private Bitmap _cameraImage;

public Bitmap CameraImage
{
    get { return _cameraImage; }
    set
    {
```

```
                    _cameraImage = value;

                    Image old = picCamera.Image;
                    picCamera.Image = value;

                    // Dispose of the old bitmap to save memory
                    // (It will be garbage collected eventually, but this is faster)
                    if (old != null)
                    {
                        old.Dispose();
                    }
                }
            }
        }
```

As well as keeping a private copy of the Bitmap, the code assigns it to the PictureBox and then disposes of the previous Bitmap to save memory.

That completes the processing of a camera frame. The result is that you see images updating continuously on the screen, unless the Webcam service has a problem.

# Inheriting from Abstract Services

There has been a fair amount of discussion on the MRDS Forum about object-oriented concepts and inheritance. The design of DSS does not really allow for traditional inheritance, but DSS has a form of inheritance through *generic contracts*.

Generic contracts have been mentioned several times previously. The concept is quite simple: You define a service state, a set of message types, and an operations port. (You can also include enum data types as part of the data contract). Note that there is no executable code included in a generic contract.

Developers can implement generic contracts for different brands and models of robots so that they all have a common interface. This enables applications like the Dashboard and TeleOperation to work on a variety of robots. In fact, the applications do not even know what type of robot they are talking to because all robots look the same from an API point of view.

When a developer implements a generic contract, the new service is like a device driver in an operating system — it hides the details about how to control a physical device and presents a "virtual" device to the operating system that accepts a standard set of commands. Generic contracts are covered in the MRDS Service Tutorials 8 and 9. You should read these tutorials for further information.

You are more likely to implement a generic contract than you are to define one, so the next section covers implementing contracts; building generic contracts is covered in the following section.

## Implementing a Generic Service

In the ProMRDS\Chapter16 folder is a service called StingerPWMDrive. This example implements the Generic Differential Drive service for the Stinger robot. The original services from RoboticsConnection for the Stinger did not support the generic interface, which meant that it would not work with the TeleOperation service.

StingerPWMDrive is a "wrapper" that translates generic drive operations into requests to the Serializer Services. (The Serializer is the on-board brains of a Stinger robot). By using StingerPWMDrive instead of the drive service supplied by RoboticsConnection, the TeleOperation service can treat a Stinger like other types of robots because StingerPWMDrive has a generic interface. In fact, TeleOperation is not even aware that it is talking to a Stinger. Without this service, you cannot control the Stinger using TeleOperation.

*The StingerPWMDrive service uses the Pulse Width Modulation (PWM) interface on the Stinger, not the Proportional, Integral, and Derivative (PID) interface. Therefore, it does not make use of the wheel encoders. The* DriveDistance *and* RotateDegrees *operations just use a timer, which is not very accurate. However, it is still a useful example.*

To create a new service based on a generic contract, you use DssNewService with the /alt parameter to specify the alternate contract. You must also include the /i parameter to specify which assembly to look at for the alternate service. (The bold text in the following code indicates what you type.)

```
C:\Microsoft Robotics Studio (1.5)\ProMRDS\Chapter16>dssnewservice
/service:"StingerPWMDrive" /namespace:"ProMRDS.Robotics.Stinger.PWMDrive" ↵
/year:"2008" /month:"01" ↵
/alt:"http://schemas.microsoft.com/robotics/2006/05/drive.html" ↵
/i:"..\..\bin\RoboticsCommon.dll"
```

The new service has the [AlternateContract] attribute with the contract identifier of the generic service:

```
    /// <summary>
    /// Provides access to a differential drive (that coordinates two motors that ↵
function together).
    /// </summary>
    [DisplayName("Stinger Generic Differential Drive")]
    [Description("Provides access to the Stinger Drive\n(Uses the Generic ↵
Differential Drive contract)\n(Partners with Stinger 'brick')")]
    [Contract(Contract.Identifier)]
    [AlternateContract("http://schemas.microsoft.com/robotics/2006/05/drive.html")]
    public class StingerPWMDriveService : DsspServiceBase
    {
```

Because it is implementing an existing service, the main port and the state both use data types from the generic service:

```
        /// <summary>
        /// Main Port
        /// </summary>
        /// <remarks>Note: The main port is an instance of the Generic Differential ↵
Drive Operations Port</remarks>
        [ServicePort("/stingerpwmdrive", AllowMultipleInstances=false)]
        private drive.DriveOperations _mainPort = new drive.DriveOperations();

        /// <summary>
        /// Stinger PWMDrive Service State
```

```
        /// </summary>
        /// <remarks>Note: The State is an instance of the Generic Differential ↵
Drive State</remarks>
        [InitialStatePartner(Optional=true, ↵
ServiceUri="Stinger.PWMDrive.Config.xml")]
        private drive.DriveDifferentialTwoWheelState _state =
new drive.DriveDifferentialTwoWheelState();
```

Stubs are created for all of the operations in the generic contract, but they throw exceptions saying that they are not implemented, as shown in this example:

```
        /// <summary>
        /// HttpPost Handler
        /// </summary>
        /// <param name="submit"></param>
        /// <returns></returns>
        [ServiceHandler(ServiceHandlerBehavior.Concurrent)]
        public virtual IEnumerator<ITask> HttpPostHandler(dsphttp.HttpPost submit)
        {
            // TODO: Implement Submit operations here.
            throw new NotImplementedException("TODO: Implement Submit operations ↵
here.");
        }
```

The `StingerPWMDriveTypes.cs` file only contains the contract identifier. There is no need to define a main operations port or request types because these are all in the generic contract.

However, if you want to extend the state or add more operations, then you can subclass your state or `PortSet` off the generic ones. It is therefore possible to add more fields to the state, or more operations to the `PortSet`. The Microsoft Tutorials refer to this as "extending" a generic contract.

What if you find yourself writing very similar services, e.g., one for simulation and one for real hardware, and you want to avoid duplicating code? The simplest approach is to use a Helper DLL or service.

If you abstract the common routines and place them into a separate DLL, then you avoid the maintenance nightmare of keeping two copies of the code in sync. Whether you choose to use a DLL with a conventional library interface or a service is up to you, although your decision should be guided by the need to implement queuing, which ports are very good at.

## Building Virtual Services

Generic contracts are intended to be used across a range of different hardware. The most common example is the Generic Differential Drive contract. If you are building services for a single robot and have no plans to support other robots, you probably don't need to create any generic contracts. However, you should try to use existing generic contracts in this case.

The source files for most of the generic contracts in MRDS are located under the MRDS root directory in the folder `samples\Common`. These are good examples of how to create generic services. You should become familiar with these generic services and try to use them whenever possible, rather than define your own.

Creating a generic contract is very easy, assuming that you have already planned what data you need in the state and the types of operations you want to perform. Note that Chapter 17 walks through the steps for creating a new Generic Brick contract, but they are outlined here for completeness.

To create a generic contract, follow these steps:

1. Create a new service, such as `MyGenericContract`. There are no special requirements for creating the new service. Note that there is no need to make any changes to the `Contract` class that is automatically generated when you create a new service.

2. Open `AssemblyInfo.cs` and modify the `ServiceDeclaration` attribute. The `ServiceDeclaration` should initially contain the following:

```
[assembly: ServiceDeclaration(DssServiceDeclaration.ServiceBehavior)]
```

This indicates an implementation of a service. However, this new service is a generic service so it has no implementation. Change this declaration to the following:

```
[assembly: ServiceDeclaration(DssServiceDeclaration.DataContract)]
```

This revised declaration says that the service contains only a `DataContract`, i.e., it is generic.

If you want to reduce the number of DLLs in your applications, you might choose to have generic services and service implementations in the same solution. If so, you can modify the `ServiceDeclaration` so that both types of service can coexist in the same DLL:

```
[assembly: ServiceDeclaration(DssServiceDeclaration.DataContract |
DssServiceDeclaration.ServiceBehavior)]
```

*If you combine both types of services in a single DLL, then you must use different namespaces for each of the services.*

3. Remove the service implementation source file, `MyGenericContract.cs`, from the solution and delete the file. A generic contract, as its name implies, is simply a data contract; therefore, it does not contain any executable code, i.e., there is no actual implementation of the service.

4. Update the state in `MyGenericContractTypes.cs`. Defining the generic service state is no different from a normal service. You use the `[DataContract]`, `[DataMember]`, and `[DataMemberConstructor]` attributes in the same way as usual.

   If you want, you can split this file into two parts and create `MyGenericContractState.cs`. This is not required but it might be easier for users of your contract to understand. Many of the standard MRDS generic contracts are organized this way.

5. Define all the necessary data types for your generic service operations in the file `MyGenericContractTypes.cs`. This process is identical to how you define operations for a normal service, as explained in Chapter 3.

6. Add a main operations port that lists all of the operations. The main operations port is similar to a normal service.

*If you plan to use the generic service to create services to run under the .NET Compact Framework (CF), then you have to be careful about how you declare the operations port. In particular, if there are more than eight data types in the* PortSet *declaration, then you must use* typeof. *Refer to Chapter 16 for more information on CF services.*

Generic services do not explicitly include a version number. (It is possible to specify a version number in the Assembly Information, but this is not used for locating services.) You can change the year and month in the contract identifier if you make substantial changes to a generic contract. However, existing services cannot use the new generic contract without being modified and recompiled. Therefore, try to design your generic services carefully to avoid possible changes in the future.

The MRDS Service Tutorial 9 discusses how to extend existing generic contracts. The procedure is straightforward, so it is not discussed here.

# More on Debugging

Debugging is discussed briefly in Chapter 3. This section provides some more tips on how to debug services under MRDS.

## Read the Documentation First

It sounds obvious, but read the documentation. Also read all of the messages that are displayed in the Console window, and the Debug and Trace messages in a web browser. Often the answer is right under your nose. Even if the problem is not clear, you can use part of an error message to search the MRDS Discussion Forum or even Google it.

The MRDS Discussion Forum contains a wealth of information. You might not get an immediate answer if you post a question there, but in general you will find the information you need, often from an expert or one of the people on the MRDS Development Team.

As with all forums, make sure you can clearly define your problem, and if possible narrow it down to a small code snippet. (You can mark code when you post it to the Discussion Forum so that it doesn't end up with smiley faces all through it). You are more likely to get a response if your posting demonstrates that you have made an effort to solve the problem yourself and you list the things you have already tried. If you post a question like "Why doesn't this code work?" followed by 100 lines of code, or you ask a question that has been discussed several times before, then you might not get an answer.

## Use the Visual Studio Debugger

The obvious way to debug a service is using the Visual Studio Debugger! What might not be so obvious is that you can actually set breakpoints inside multiple services so you can see the effect of messages bouncing back and forth. All you have to do is open the relevant source code files from the other projects and set breakpoints. (This assumes that you have the source code and that the current assemblies in the bin folder were compiled from those sources).

The debugger is not always the solution to finding bugs. Some bugs occur due to subtle timing issues, and as soon as you stop in the debugger you change the timing. In addition, because the environment is multi-threaded, stopping at a breakpoint might leave other code running, with the result that messages pile up in a port.

Finally, remember that there is a Threads window in the Debugger. This might help you to identify what is happening.

## Examine the State of a Node and Services

You have already seen how to examine the state of a service, even if it is displayed in XML. This works across the network, so you can examine services on remote hosts.

Don't be afraid to add debugging information to your state. The values of key variables, especially counters such as the number of packets sent and the number of packet errors, can be invaluable for verifying that the service is operating correctly or locating the source of errors. Instrumenting your service in this way is a good idea. You can always use conditional compilation to remove these variables later.

## Traditional Debugging Techniques

Of course, if you really want to, you can resort to the classic debugging technique called "debug print statements" whereby you place `Console.WriteLine` statements at strategic places in your code. However, this is the lazy approach.

You should be aware that `Console.WriteLine` is quite slow, and it can actually hold up your services, resulting in behavior that might not be typical. You should not use it in time-critical code, inside loops, or in handlers that are called frequently. However, it is useful during initialization to indicate progress and for catastrophic errors that force the service to close down.

## Using Trace Level and Logging

You might be familiar with the .NET `Debug` and `Trace` classes in `System.Diagnostics` that can be used to conditionally output messages. The `DsspServiceBase` class has a set of built-in methods that you can use to log information. In addition to the general-purpose `Log` method, specific methods are associated with each of the trace levels. These methods also output to the Debug and Trace Messages page shown in Chapter 3.

The methods and their associated trace levels are shown in the following table:

| Method | Trace Level |
| --- | --- |
| LogError | 1 |
| LogWarning | 2 |
| LogInfo | 3 |
| LogVerbose | 4 |

Tracing is a feature of .NET, so the trace levels are set in the .NET application configuration file for `DssHost`, which is `bin\dsshost.exe.config`. Enabling tracing at one level also enables all trace messages at a lower level. The default trace levels should be appropriate. If you are interested in changing the tracing behavior, read the Visual Studio help on the subject.

Each of these methods has several overloads. You can use the Object Browser to look at their definitions. Alternatively, type **LogInfo** into a Visual Studio source window, place the cursor over it, and press F1. This invokes the online help for Visual Studio, but since the V1.5 Refresh, the MRDS Class Reference is also available via this method, so you will see a description of `LogInfo`. However, most of the Class Reference is automatically generated and it might not give you much more information than the Object Browser.

The primary advantage of using tracing is that it can be enabled or disabled without recompiling the code.

# Where to Go from Here

The TeleOperation service covers a wide variety of tasks that a service can perform, but it is by no means complete. You can use it as a starting point for building your own applications, or continue to improve it.

Another service called Drive-By-Wire is included with the code for the book. It is an abbreviated version of TeleOperation designed to run on a PDA. For example, it works with the Boe-Bot if you have a PDA with built-in Bluetooth. However, it won't work with a LEGO NXT on a Dell Axim 50v because the NXT requires a baud rate of 115200, and Bluetooth on the Axim will not run that fast.

Here are a few more enhancements for the TeleOperation service that you might consider:

❑ Add controls to one of the WinForms (and the necessary code) to change the camera resolution, or select a different camera. Remember to add this information to the state so it is written into the `config` file. On startup, set the resolution based on the `config` file.

❑ Add a drop-down list to select the game controller. (This is already implemented in the Dashboard service, so you can cheat and look in there).

❑ Allow the Webcam View window to be resized, and adjust the size of the `PictureBox` so that the aspect ratio is maintained. You could send a message to the main service to change the camera resolution to try to match the window size.

❑ Add some color or blob tracking code so the robot can follow an object. A sample Blob Tracker service is included with MRDS — look in `samples\Misc\BlobTracker`.

❑ Incorporate some vision processing using the MRDS services from RoboRealm (`www.roborealm.com`).

The possibilities are endless! Let your imagination run free, and remember that you can use the simulator if you don't have real hardware.

# Summary

Many services do not require a user interface. Those that do have two options: Windows Forms or Web Forms (using XSLT to format the state information). Both of these were covered in detail in this chapter. This chapter has also shown you how to use a web camera as part of the TeleOperation service for remotely driving a robot.

This is the end of the introductory part of the book. It has covered all the basics of MRDS, and by now you should be comfortable writing your own services.

The next part of the book discusses using the MRDS simulator. It offers a great environment for testing services without a real robot. This is a significant benefit if you don't have a lot of money. It also means that if you make a mistake and crash your robot, it doesn't matter.

# Part II
# Simulations

# The MRDS Visual Simulation Environment

The Visual Simulation Environment, a key part of Microsoft Robotics Developer Studio, uses 3D graphics to render a virtual world and a physics engine to approximate interactions between objects within that world. At first glance, it looks something like a game engine. However, after a little bit of use, it quickly becomes apparent that it is much different; whereas many actions and events are scripted in a game environment, the things that happen in the simulator rely completely on the interactions and motions of objects in the environment.

This chapter explains how to use the Visual Simulation Environment along with all of the simulation objects and services provided in the SDK. Subsequent chapters explain in detail how you can add your own objects and environments to the simulator to explore your own robotic designs.

In this chapter, you'll learn the following:

- ❏ How the simulator can help you prototype new algorithms and robots
- ❏ What hardware and software are required to run the simulator
- ❏ How to use the basic simulator functions
- ❏ How to use the Simulation Editor to place robots and other entities in a new environment
- ❏ What built-in entities are provided with the SDK and how to use them

This chapter focuses on using the simulator functionality with the environments and entities provided as part of the SDK. The next chapters go into more detail about how you define your own environment and entities.

# The Advantages of Simulation

The simulator is a great place to prototype new robot designs, offering many advantages:

❑ Prototyping new robots in the real world is a task full of soldering irons and nuts and bolts. Moving from one iteration of a design to the next can take weeks or months. In the simulator, you can easily make several changes and refinements to a design in just a day or two.

❑ The simulator enables you to easily design and debug software. It is often difficult to have a debugger connected to a robot under motion, making it difficult to determine just what went wrong when the robot suddenly decides to ram the nearest wall. When your robot is running in the simulation environment, it is easy to set breakpoints on the services that control it and to find the bugs in your code. Once you have your algorithms debugged and tuned in the simulator, it is often possible to run the exact same code on your actual robot.

❑ The simulator can be particularly useful in a classroom scenario in which you have several students who need to use a robot and only limited hardware available. Students can spend most of their time writing software and debugging the behavior of a simulated robot before testing the final version on the actual hardware. This is also a very useful feature when a new robot is being developed for which only one or two prototypes are available but several people need to write software for the new robot.

❑ Another advantage to simulation is personal safety. An intern once demonstrated how he had programmed a real-world Pioneer 3DX robot to determine the direction of his voice using a microphone array. The robot would respond to spoken commands such as "Come here!" or "Stop!" Unfortunately, the robot was a little slow to respond to the stop command and it nearly pinned the intern against the wall before he could hit the reset switch. The simulator provides a way to debug your robot without risking bodily injury. Moreover, if your robot suddenly achieves self-awareness and starts running amok while yelling "Destroy all humans!" you won't need to call out the National Guard to subdue him if he is confined to the simulator.

# The Difficulties with Simulation

It would be nice to able to tell you that the simulation environment will bring world peace and solve embarrassing hygiene problems, but you should be aware that it does have some limitations:

❑ The physics engine is remarkably good at solving complicated object dynamics and collisions. The real world, however, is a very complicated place, and it is certainly possible to define a simulation scene that is difficult for the physics engine to process in its allotted time. The physics model for a scene is, by necessity, much more simplified than its real-world counterpart. A complicated object composed of many pieces may be modeled as a simple cube or sphere in the simulation world. You need to make careful decisions about where to add detail to your simulated world to increase the fidelity of the simulation while still allowing it to run at a reasonable rate.

❑ The real world is a very noisy place. Real sensors often return noisy data. Real gears and wheels sometimes slip, and real video cameras often return images with bad pixels. Good robotics algorithms need to account for these issues, but they are often not present in the simulator. A simulated IR sensor is only as noisy as you have programmed it to be. A simulated camera will return more consistent and noise-free images than a real camera. In addition, the simulated world is often more visually simple than the real world, so image-processing algorithms have an easier time interpreting images from the simulated world.

*It is theoretically possible to program the same noise and other real-world issues into the simulated sensors and motors. It is more common to initially write algorithms that work in simulation and then tweak them to handle the additional complexity and noise of the real world.*

❑ You should consider the difference in computing power between the simulation environment and the actual robot. The simulator may run on a powerful desktop computer with a very fast CPU, while the actual robot might have a much slower CPU with slower data busses. A simulated camera may deliver 60 frames per second of video while a real camera might be limited to 1–2 frames per second. When designing algorithms, it is important to consider the performance and limitations of the final hardware.

# Prerequisites

Your experience with the simulator will be much better if you have the proper software and hardware to run it well. If the simulator cannot run, it displays a red window with some text indicating what problem it encountered. The problem may be due to insufficient hardware capability or a software configuration problem. This section details the hardware and software requirements you'll need to run the simulator.

When the simulator displays the red screen, it will be unable to draw the graphical environment on the screen but will continue to run the physics part of the simulation. As you will see, it is quite easy to query the state of the simulation to find out what is happening. In other words, the simulator may still be useful even if it is unable to display the graphics.

## Hardware Requirements

At a minimum, you must have a graphics card that is capable of running pixel and vertex shader programs of version 2.0 or higher. Video cards are often rated according to the version of DirectX that they support. If your card supports only DirectX 8, then it will not run the simulator. If your card supports DirectX 9, then you need to check which versions of pixel and vertex shaders are supported. Some older cards only support version 1.0 or 1.1 and will not run the simulator. If your card supports version 2.0 or 3.0 shaders or it supports DirectX 10, it should run the simulator well. Most mid- or upper-range graphics cards sold within the past couple of years have sufficient power.

*One caveat here is that some newer laptops and even desktop machines have mobile or other specialized chipsets that may not run the simulator even though they are fairly new. This is because these machines are primarily intended for word processing or web browsing and they don't have high-performance 3D graphics capabilities.*

The best way to determine the capabilities of your graphics card is to do the following:

1. Right-click the Windows desktop to bring up the Display Properties.

2. Click the Settings tab to see the name of your video card. Alternatively, you can bring up the Hardware Device Manager from the Windows Control Panel.

**3.** Expand the Display Adapters item to see the name of your video card.

**4.** Go to the website of the card manufacturer and search on the video card name. The manufacturer often provides detailed specifications for fairly recent cards.

If you are unable to find more information directly from the manufacturer, several websites list the capabilities of various video cards. You can go to the following website to find information on nVidia, ATI, and Matrox video cards: www.techpowerup.com/gpudb.

If your video card appears to meet the requirements just described and the simulator still fails to run, you may have a software configuration problem.

## Software Requirements

All of the software required to run the simulator is installed as part of the Microsoft Robotics Developer Studio SDK. Figure 5-1 shows the final screen of the installer. When you install the SDK, be sure to check the box to install the AGEIA PhysX engine, as well as Microsoft XNA and Microsoft DirectX. It is recommended that you also check the boxes to install the other components.

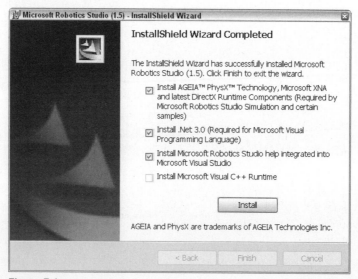

Figure 5-1

The XNA and DirectX installers require an active Internet connection. If your machine did not have an Internet connection during installation, it may be necessary to go back and manually install these components. Alternatively, you can uninstall and reinstall the SDK when you do have an Internet connection.

Version 1.5 of the MRDS Visual Simulation Environment requires the following additional software:

❑    AGEIA PhysX engine

❑    Microsoft XNA

❑    Microsoft DirectX

The following sections discuss this additional software.

## The AGEIA PhysX Engine

The simulator uses the AGEIA PhysX engine to simulate physical interactions within the virtual environment. It requires version 2.7.0 of the PhysX engine. You can see which version you have installed by selecting the AGEIA item from the Start menu and then selecting ageia PhysX Properties. Make sure that V2.7.0 is displayed as one of the installed versions on the Info tab. You can also run the simple demo on the Demo tab to verify that the engine is working properly.

One of the great features of the AGEIA PhysX engine is that it can be accelerated with specialized hardware. You can find out more about this option on the AGEIA website at www.ageia.com. You can also download the latest AGEIA run-time drivers from www.ageia.com/drivers/drivers.html.

You may find it useful to download the entire AGEIA development SDK to learn more about how the physics engine works. While the PhysX APIs are not directly exposed by the simulator, an understanding of the engine and the objects it supports is very helpful. It is necessary to register before downloading the PhysX SDK from the following website: www.ageia.com/developers/downloads.html.

## Microsoft XNA

The Microsoft XNA library is a managed code implementation of the DirectX APIs. The simulator uses the Microsoft XNA Framework, which is a set of APIs that provide access to the underlying DirectX subsystem from managed code.

MRDS version 1.5 requires version 1.0 of the XNA Framework or the version 1.0 refresh, which was made available on April 24, 2007. You can find a link to the download page for the XNA Framework on the XNA Game Studio Express page at http://msdn2.microsoft.com/en-us/xna/aa937795.aspx. Alternatively, you can go to the XNA home page (www.microsoft.com/xna) and search for the framework download.

You may also find it helpful to download the XNA Game Studio Express SDK, which provides documentation on XNA classes and APIs. Some of these are used in the simulation entity sample code provided in the MRDS SDK. A link to the download page for XNA Game Studio Express can be found on the web page referenced above.

## Microsoft DirectX

DirectX is a set of system APIs that provide access to graphics, sound, and input devices. The simulator in MRDS version 1.5 requires a DirectX runtime at least as recent as April, 2007. You can find the download link for this by loading the following page and searching for "DirectX Redist (April 2007)" on www.microsoft.com/downloads.

You may also find it helpful to download the entire DirectX SDK. This SDK includes a variety of tools that aid in making texture maps and other content for simulator scenes. You can find a download link for this by loading the aforementioned page and searching for "DirectX SDK – (April 2007)."

# Starting the Simulator

The easiest way to get the simulator up and running is to run it from the Start menu. Select the Visual Simulation Environment item from the Microsoft Robotics Developer Studio menu to see the available simulation scenarios. Some of these scenarios, described in the following list, are tutorials provided by Microsoft with the SDK:

- ❑ **Basic Simulation Environment:** Contains sky and ground along with an earth-textured sphere and a white cube. This is the scene associated with Simulation Tutorial 1.

- ❑ **iRobot Create Environment:** Contains a simulated iRobot Create and a few other shapes

- ❑ **KUKA LBR3 Arm:** Contains a simulated KUKA LBR3 robotic arm along with some dominos to tip over. This scene is discussed in more detail in Chapter 7, where you learn about articulated entities. This is the scene associated with Simulation Tutorial 4.

- ❑ **LEGO NXT Tribot Simulation:** Contains a simulated LEGO NXT Tribot and a few other shapes

- ❑ **Multiple Simulated Robots:** Contains a table with a simulated LEGO NXT Tribot and a simulated Pioneer 3DX robot with a laser range finder, bumpers, and a camera. This is the scene associated with Simulation Tutorial 2.

- ❑ **Pioneer 3DX Simulation:** Contains a simulated Pioneer 3DX robot and a few other shapes

- ❑ **Simulation Environment with Terrain:** Contains a simulated Pioneer 3DX robot, a terrain entity, and a number of other types of entities. This is the scene associated with Simulation Tutorial 5.

It is also possible to start the simulator by running a manifest that starts a service that partners with the SimulationEngine service. The following command can be executed in a Microsoft Robotics Developer Studio command window to start the Simulation Tutorial 1 service, which also starts the SimulationEngine service:

```
dsshost -p:50000 -t:50001 -m:"samples\config\SimulationTutorial1.manifest.xml"
```

Yet another way to start the simulator is to start a node and then use the ControlPanel service to start a manifest. You can start a node without running a manifest with the following command:

```
bin\dsshost -p:50000 -t:50001
```

Now open a browser window and open the following URL:

```
http://localhost:50000/controlpanel
```

This opens a page that shows all available services that can run on the node. To limit the services shown to only those containing the word simulation, type **simulation** in the Search box. Figure 5-2 shows the ControlPanel service page, filtered to show only simulation services.

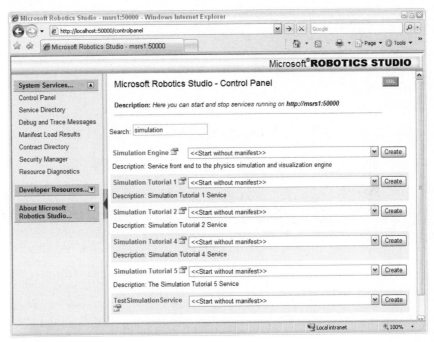

Figure 5-2

Each service is displayed with a drop-down box listing all of the manifests installed on the system that reference that service. You can run one of these manifests or you can start the service on its own. If you click the Create button next to Simulation Tutorial 1, the Control Panel will start that service and in turn start the SimulationEngine service.

# Using the Simulator

The controls for the simulator are fairly intuitive. The following sections explain the basics and provide some background information about how the simulator works. This section uses the ProMRDS.Marbles service to illustrate some of the simulator features. If you have installed the ProMRDS Package from the CD, you can start the service by typing the following in the MRDS command window: **bin\Marbles .cmd**. This file starts a DSS node and runs `Marbles.manifest.xml`. It should bring up the scene of the Marbles sample shown in Figure 5-3.

Figure 5-3

The Marbles sample doesn't actually do a lot of useful robotics-type things, but it helps to illustrate some of the simulator's features. In this sample, marbles are dropped from above and bounce through the grid of pegs until they roll down the ramp and hit the blades of the turnstile, causing the turnstile to turn. After 200 marbles have been dropped, instead of adding more marbles to the scene, the Marbles service picks them up from the ground and drops them.

## Navigating the Simulation Environment

When the simulator starts, it displays the simulated world as seen from the Main Camera. There is always a Main Camera object in the scene. Its position and orientation are often referred to as the *eyepoint*. Cameras are discussed in the section "The Camera Menu" later in this chapter.

Looking around in the simulator is very simple. Just press the left mouse button in the simulator window and drag the pointer around to change the viewpoint. This changes the direction in which the Main Camera is pointing.

You move yourself around in the simulator by pressing keys. It is easiest to navigate in the simulator with one hand on the keyboard and the other hand on the mouse. The following table shows the keys that move the eyepoint in the environment:

| Key | Action |
| --- | --- |
| W | Move forward |
| S | Move backward |
| A | Move to the left |
| D | Move to the right |
| Q | Move up |
| E | Move down |

Notice that there are six keys and they work in pairs. W and S move you forward and backward. A and D move you left and right. Q and E move you up and down.

Holding down the Shift key while pressing one of the motion keys multiplies that motion by 10 times. This is very useful when you are in a hurry to get from one place to another. You can adjust the speed of mouse and keyboard movements in the Graphics Settings dialog, which is described shortly in the section "Graphics Settings."

*The movements are always relative to the current eyepoint orientation. This enables you to combine key presses with mouse drags for more complicated motions. For example, you can press A to move to the left while simultaneously dragging the mouse pointer to the right to move in a circle around an object.*

You can use the following function keys as shortcuts to commonly used functions:

| Key | Action |
| --- | --- |
| F2 | Change the render mode |
| F3 | Toggle the physics engine on and off |
| F5 | Toggle between edit mode and run mode |
| F8 | Change the active camera |

Practice moving around in the environment until it becomes natural to you. Move yourself below the ground (that's right, there's nothing down there). Go as high as you can and see how the level of detail on the ground gradually fades into the fog. Practice zooming up to objects and moving right through them like a ghost. In the section "Physics Menu," you'll learn how to change the eyepoint so that it won't pass through objects.

## *The Status Bar*

Don't worry about getting lost in the simulator. If you ever wonder where you are, you can display the status bar (see Figure 5-4). Click View ⇨ Status Bar to make it visible.

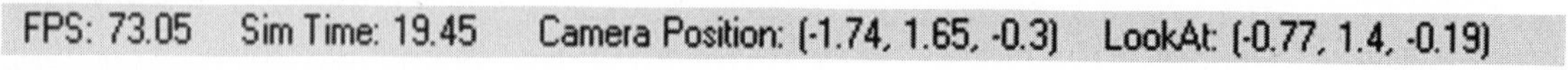

FPS: 73.05    Sim Time: 19.45    Camera Position: (-1.74, 1.65, -0.3)    LookAt: (-0.77, 1.4, -0.19)

**Figure 5-4**

The status bar provides the position of the current camera, as well as the point that it is looking at. Every point in the simulation environment is represented by three coordinates: X, Y, and Z. The Y coordinate represents altitude, while the X and Z coordinates represent directions parallel to the ground plane. The simulation environment uses a right-handed coordinate system. This means that when the +X axis is pointing to the right and the +Y axis is pointing upward, the +Z axis is pointing out of the screen and the eyepoint is looking in the direction of the -Z axis (into the screen), as shown in Figure 5-5.

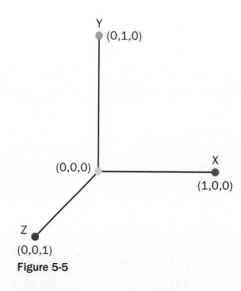

**Figure 5-5**

### Left-handed vs. Right-Handed Coordinate Systems

Left-handed versus right-handed coordinate systems have been a source of confusion in computer graphics systems for years. Direct3D typically uses a left-handed coordinate system, whereas OpenGL and several major modeling systems use a right-handed system. The MRDS Visual Simulation Environment uses a right-handed system so that it is easier to export models from modeling programs. You won't have to worry about this issue unless you need to use models built for a left-handed system, in which case you will need to convert them.

Also of interest on the status bar is the number of frames being processed per second and the number of seconds that have passed in simulation time since the simulator started.

You can click View ⇨ Look along to become more familiar with the directions of the axes in the environment. Each item in the submenu orients the camera so that it is looking along a particular axis without changing its location.

## The Help Menu

As you might expect, the Help menu gives you a couple of options to find out more about the simulator. The Help Contents option in the menu brings up the documentation file for the simulator. The second option, About Microsoft Visual Simulation Environment, can be used to display the version number of the simulator and specific information about your video card (see Figure 5-6):

❏   The manufacturer and model

❏   The name and version of the display driver

❏   The size and format of the desktop

❏   The highest supported versions of the vertex and pixel shaders

If your card supports version 3.0 of the shaders (VS_3_0 and PS_3_0) or higher, then you can display more realistic scenes, which include specular highlights and shadows.

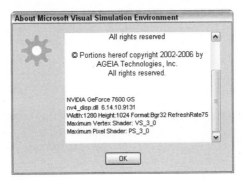

Figure 5-6

## The Camera Menu

Finally, here is your chance to be in two places at the same time. You can set up as many cameras in the simulation environment as you like and then instantly and easily switch between them, using the Camera menu. All cameras in the scene appear on the menu regardless of type. You can select the desired camera from the menu or you can press F8 to switch from one camera to the next. There are two types of behavior for cameras: real-time and non-real-time:

❏   **Real-time cameras:** These render the environment from their viewpoint every frame. As you can imagine, multiple real-time cameras can increase the graphics load on a machine substantially. A real-time camera is appropriate when you have a need to retrieve images from that camera on a regular basis. Real-time cameras have a horizontal and vertical resolution associated with them. When the scene is rendered for that camera, it is rendered at that resolution.

❑ **Non-real-time cameras:** These are simply placeholders in the 3D environment. Only the selected camera is rendered each frame. As a result, these cameras don't reduce graphics performance. They are useful when you want to define a viewpoint in the scene so that you can switch to it quickly. The Main Camera is a special instance of a non-real-time camera. There is always a Main Camera in the scene, and its resolution always matches the resolution of the display window. When the window changes, the Main Camera adjusts accordingly. Other non-real-time cameras have a set horizontal and vertical resolution. If the resolution of the display window doesn't match the resolution of the camera, the rendered image is stretched to fit.

The Marbles environment has several non-real-time cameras set up. Use the F8 key or the Camera menu to switch between them.

# The Rendering Menu

You've already seen how you can fly through the air faster than a speeding bullet in the simulator. You've seen how you can leap tall buildings in a single bound. (Okay, so we haven't had any buildings yet. Stay with us.) Now we'll give you X-ray vision.

The Rendering menu controls how the simulated world is displayed. Different rendering modes give you different information about how the world is put together. The following sections discuss the various modes.

## Visual Rendering Mode

Shown in Figure 5-7, this is the default rendering mode, and you are already well acquainted with it. In this mode, the world is rendered to look its best — with fully rendered 3D meshes and lighting effects. This is the mode that you will probably use most often.

Figure 5-7

## Wireframe Rendering Mode

In this mode, shown in Figure 5-8, the world is still rendered with 3D meshes and lighting effects, but only the outline of each triangle is drawn. This quickly gives you a good idea whether the 3D meshes you are displaying have the proper resolution. Some meshes are so dense with polygons that they look about the same in wireframe mode as in the visual mode, which means they contain too many polygons and may be loading your graphics card more than necessary. Many modeling programs provide a way to reduce the number of polygons in a model.

Another benefit of wireframe rendering mode is that you can see through the objects in the scene. This is particularly helpful when you have just added a great new robot to your scene but it fails to appear. Switching over to wireframe mode shows that you accidentally embedded it in the couch.

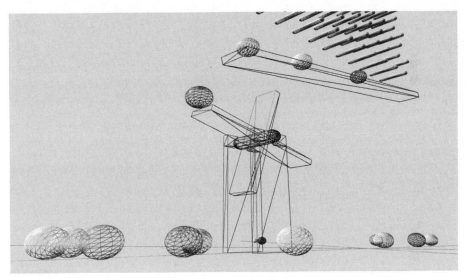

Figure 5-8

## Physics Rendering Mode

The physics rendering mode (see Figure 5-9) shows you the physics shapes that make up the scene. This is the view that shows the physics shapes that the physics engine works with. Complex objects can sometimes be represented by very simple physics shapes such as boxes or spheres. The physics rendering mode enables you to see the scene the same way the physics engine sees it. If you notice that objects aren't colliding or moving the way you might expect, then switch to the physics rendering mode to see their underlying physics shapes.

It might be surprising but this mode can tax your graphics card more than the visual rendering mode even though less geometry is being drawn. This is because the shapes are composed of individual lines, which are drawn less efficiently than 3D meshes.

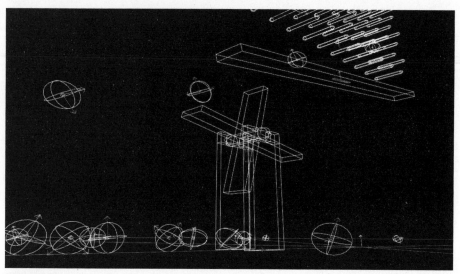

**Figure 5-9**

In Figure 5-9, you can see the three most common physics shapes: the sphere, the box, and the capsule. *Spheres* are defined by their radius. *Boxes* are defined by their dimensions along the X, Y, and Z axes. *Capsules* are cylinders with rounded ends, and they are defined by the radius of the cylinder and their height along the Y axis. Each of these shapes can also have a position offset and a rotation around each axis.

Notice in Figure 5-9 that each shape has coordinate axes drawn at its origin. You can deduce from the display that the origin of a sphere is its center. Although you can't see the color in the book, on the screen the X axis is drawn in red, the Y axis is green, and the Z axis is blue.

The simulation environment is composed of objects called *entities*, which typically represent a single rigid body such as a robot chassis that may contain multiple shapes. Each entity contains an optional 3D mesh and zero or more physics shapes. The visual representation of the object and the physics representation exist side-by-side. Entities are always positioned and rotated as a single object.

> *Entities are the basic building blocks of simulator scenes. The properties of the entities provided with the simulator are described in this chapter. Additional details about the implementation of entities are provided in Chapter 6.*

In Figure 5-9, each entity has a cube that represents its center of mass. On the screen, the cube is red if the entity is controlled manually (kinematic), and white if the entity is controlled by the physics simulator. Each of the marbles in the scene is controlled by the physics engine; but the ramp and the mill supports have a red center-of-mass box, indicating that they are controlled by the program and not by the physics engine. All the pegs that the balls fall through are contained by a single entity, and you can see that this entity is kinematic by its red center-of-mass cube.

## Dynamic, Kinematic, and Static Entities

The physics rendering mode has three types of entities: dynamic, kinematic, and static. An entity is dynamic if it is under the control of the physics engine. The marbles in the marbles scene are a good example of a dynamic entity. Once they are created, the marbles service doesn't do anything to control them. Their position and motion are determined only by the effect of gravity and the forces from collisions with other objects.

A kinematic entity, conversely, interacts with the physics environment but is positioned externally. The ramp is a good example of a kinematic entity. If it were a dynamic entity, it would fall to the ground because it has no supports. Because it is set to be kinematic, it stays where it is put. Other objects can collide with it but it won't move unless its position is manually changed.

The physics engine also supports static entities, which are composed of shapes that have zero mass. These entities are intended to remain in the same position during the entire simulation. Static entities can collide with other entities, but they should not be moved because the physics engine assumes that they will never move. For example, if a stack of dynamic entities is on top of a static cube entity and you delete the static cube entity, the dynamic entities will continue to hover in mid-air until some other force affects them because the physics engine assumes that static objects will never move or disappear.

## Combined Rendering Mode

This mode (see Figure 5-10) combines the visual mode with the physics mode. In some cases, it is easier to see which outlines go with which shapes in this mode. The outlines of the physics shapes are drawn as if the visual rendering were transparent so that you can still see physics shapes that are obscured by other entities. This rendering mode will typically tax your graphics card the most, and you may notice the frame rate drop.

Figure 5-10

### No Rendering Mode

This is what your graphics scene looks like with the lens cover on the camera. At night. This mode is for those hardy souls who want to use the simulator completely for physics simulation and don't want to waste any time rendering. Turn the status bar on and select this rendering mode to see how many frames are processed per second. You'll learn how to use this mode a little later.

## Graphics Settings

The last option on the Rendering menu takes you to the Graphics Setting dialog. These settings control how the graphics hardware is used to render the scene:

❏ **Exposure:** This is much like the exposure on a camera. Increasing the exposure brightens the scene.

❏ **Anti-aliasing:** This setting exposes the anti-aliasing modes supported by your graphics hardware. Selecting no anti-aliasing can make some cards run faster, but more anti-aliasing makes the edges of objects look much better. You have to experiment to find the best setting for your hardware.

---

**Anti-aliasing**

Aliasing in a graphics context means that the edge of an object must follow pixel outlines. This can give the edges of some objects a jagged appearance (known in graphics-speak as *jaggies*). Anti-aliasing algorithms are now supported on newer graphics hardware, and they enable edges to be rendered on subpixel boundaries, resulting in objects that look cleaner and more realistic.

---

❏ **Rotation and Translation Movement Scale:** These numbers scale the movement for mouse and keyboard navigation. If you are one of those people who likes to turn your mouse movement up so high that you can zip across the screen with a millimeter of mouse movement, feel free to crank these values up.

❏ **Quality Level:** Modern graphics hardware runs XNA shader programs to render objects. The simulator is equipped with version 3.0, 2.0, and 1.0 shaders. If you have newer and more powerful hardware, you can run version 3.0 shader programs and you will be rewarded with a nicer-looking scene and shiny specular highlights. If you have older hardware that only supports version 2.0 shaders, you will still get a nice-looking scene but it will lack specular highlights and some fog and lighting effects. The Quality Level option enables you to select which version of the shaders is used. The version that your card supports is marked as "Recommended" but you can switch between versions without any problem. A 2.0 shader card on one machine was able to run the 3.0 shaders, albeit a little slowly. (If you try this and your graphics card melts into a pool of glowing slag, forget that it was ever suggested.) If you find that your frame rate is lower than 10–20 frames per second, then try lowering the quality level to see whether it helps.

> ### Persistent Settings
>
> Entities in the scene are considered simulator state, and they are persisted to an XML file when the scene is saved. Other settings, such as those on the Graphics Setting dialog, are persisted to a separate configuration file and restored each time the simulator is started. This means that if you bump up the exposure by a notch or two, that same exposure will be used the next time you run the simulator.
>
> Settings on the Graphics Settings dialog are automatically saved to `config\ SimulationEditor.config.xml`. The only other settings that are persisted in this file are the simulation time settings on the Physics menu.

## The Physics Menu

Just as there is a Graphics menu to control the way things look, there is also a Physics menu to control how things behave. The Physics Settings dialog is shown in Figure 5-11. You display it by clicking Physics ⇨ Settings.

Figure 5-11

This dialog contains the following options:

❑ **Enable rigid body for default camera:** This checkbox enables you to associate a physics sphere shape with the Main Camera. It is fine to observe what is going on in the simulation environment but sometimes you just have to poke or push something. If this option is checked, the camera can do just that.

*You can try this out in the Marble environment by switching the render mode to physics mode. Enable the rigid body in the Physics Settings dialog and press OK. You should see a sphere appear just in front of the camera. You can verify that it works by maneuvering the camera down toward the turnstile, which is moved by the falling marbles. Position the camera on the opposite side of the turnstile from where the marbles are falling. Eventually you should be able to jam the mill and keep it from turning. Switch to the Mill View Camera and you should see a physics sphere shape blocking the mill from turning. This is the sphere shape associated with the camera.*

❑ **Gravity:** If you're planning to simulate the Mars Rover or the robot arm on the space shuttle, you'll want to use this setting to adjust gravity appropriately for your environment. The value is specified as a force along the Y axis, and it is negative because the Y axis is defined to go in the opposite direction as gravity. The gravity value is specified as meters per second, which works out to about –9.81 on Earth and –3.72 on Mars. If you're just aching to do it, go ahead and set it to a positive value. The gravity value is persisted as a part of the scene. When you load a previously saved scene, it will have the same gravity as when it was saved.

❑ **Time Base Run-time Settings:** These settings control how simulator time is related to real time. They are the only settings on this dialog that are persisted from session to session independent of the scene that is currently loaded. To simulate the movement or collision of a robot with improved accuracy, you can slow down simulator time relative to real time. This causes the physics engine to process more steps per second, which improves accuracy. You can select a real-time scale of 0.1 to slow down the action in the simulator by 10 times. In this mode, simulator time is still related to the actual time that passes. You can select a fixed time interval (in seconds) to completely separate simulator time from real time. This is useful when you want to run a detailed physics simulation over a period of time and you aren't concerned with the graphics rendering during that time. You can choose a small fixed-time interval like 50 microseconds (0.00005) per frame and then set the rendering mode to No Rendering to increase the frame rate of the simulator to achieve a very accurate simulation. You can enable rendering periodically to see how the simulation is progressing.

---

### What Happens in a Frame

The simulator divides time into discrete chunks called *frames*. At the start of each frame, the simulator calculates the simulator time for that frame, which may or may not be related to the real amount of time that has passed since the last frame. First, the simulator retrieves the previous frame results from the physics engine. Then the Update method is called for every entity in the scene. After all of the Update methods have completed, the scene is rendered once for each real-time camera in the scene. This is done by calling the Draw method on every entity. Finally, the scene is rendered from the eyepoint of the Main Camera, and the physics engine begins processing the next frame.

---

You can turn off the physics engine by selecting the Enabled option on the Physics menu. If the physics engine is enabled, a check appears next to this menu option. When the physics engine is disabled, the simulator does not call the physics engine asking for scene updates. All dynamic entities stop moving but you can still move the camera around the scene. As far as the physics engine is concerned, no time passes, but the simulation time is still incremented on the status bar.

The physics rendering mode and the combined rendering mode are not supported while the physics engine is turned off. If either of these modes is selected when the physics engine is disabled, visual mode will be selected.

You can quickly toggle the physics engine between enabled and disabled by pressing F3. When the physics engine is disabled, the Physics Menu item on the Main menu bar is red.

You can disable the physics engine to temporarily freeze the scene. It is also useful to disable the physics engine while moving objects around in the scene with the Simulation Editor, as described in the "Simulation Editor" section coming up shortly.

## Saving and Loading Scenes

A simulation scene can be saved or serialized to an XML file and reloaded later. When the scene is reloaded, the entities will have the same position and velocity as they had when the scene was saved. Click File ⇨ Save Scene As to save the current scene to a file, and File ⇨ Open Scene to load a scene into the simulator.

When a scene is saved, two files are actually written. The first contains the scene state and the .xml is appended to the specified filename. The second contains a manifest for the scene and it has .manifest.xml appended to the specified filename. This file is described in the next section.

Loading a scene is called *deserialization*. In order for it to be successful, all of the types referenced in the XML scene file must be currently defined in the CLR environment. If a type is not defined, the simulator will not be able to load that entity into the scene and it will display a warning dialog containing a list of all the entities that could not be instantiated.

> *If you run the SimulationTutorial2 manifest (or select Visual Simulation Environment ⇨ Multiple Simulated Robots scene from the Start menu), a table and two robots will be displayed. The table is defined in the SimulationTutorial2 service. If you save this scene to a file, run Simulation Tutorial 1 (Basic Simulation Environment on the Start menu), and then attempt to load the previously saved scene, the simulator will complain that it is unable to instantiate the table because the service that defines the table is not loaded into the CLR. You can avoid this problem by running the SimulationTutorial2 service along with the SimulationTutorial1 service, adding code to SimulationTutorial1 to manually load the SimulationTutorial2 assembly, or by defining the table entity in a third DLL that is linked with both SimulationTutorial2 and SimulationTutorial1.*

## Saving and Loading Manifests

In a typical simulation scene, one or more services are associated with entities in the scene. Loading or saving a scene does nothing to start or stop services. For example, if you have a service that is driving your robot around at the time the scene is saved and then you reload the scene later, after the service has been terminated, that service will not be restarted.

The solution to this problem is the manifest. When a scene is saved, a manifest for the scene is also written to a file. This manifest contains service records for services that should be started. The manifest contains an entry for the SimulationEngine service, along with a state partner with the same filename as the scene file that was saved. In addition, it contains service records for other services that were associated with entities in the scene. Click File ⇨ Open Manifest to load a manifest that in turn will load the associated scene file and start the associated services.

## Other File Menu Items

You can save a snapshot of the current screen by clicking File ⇨ Capture Image As. The screen image can be saved to a file using either the JPEG, BMP, PNG, or TIFF image format.

The Exit Simulator item on the File menu actually does more than just shut down the SimulationEngine service. It also posts a Drop message to the `ControlPanel` port, which causes the entire node to shut down.

# The Simulation Editor

The simulator has a built-in editor mode that makes it possible to modify or build a scene. Click Mode ⇨ Edit to enable the simulation editor, or you can press F5 to toggle between Run mode and Edit mode.

When the editor starts, the physics engine is automatically disabled. The physics engine does not have to be disabled while the editor is running, but it is often much easier to work with entities in the scene if it is turned off.

While the editor is running, the display window is divided into three panes, as shown in Figure 5-12:

❑ The main window renders the scene. When this window has input focus, it has a blue border around it and keystrokes affect the movement of the camera, just as in Run mode.

❑ The upper-left pane, or the Entities pane, shows an alphabetical list of all the entities in the scene. Each entity has a checkbox next to it that selects it for various operations. If an entity has one or more child entities, a small box with a plus sign appears next to it. Clicking this box will show the child entities in the list.

❑ The lower-left pane, or the Properties pane, shows all of the properties associated with the entity currently selected in the Entities pane.

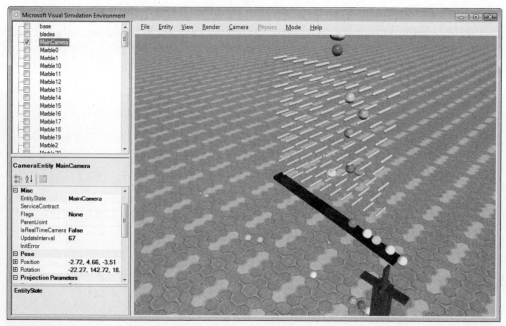

Figure 5-12

## Selecting Entities

Clicking the name of an entity highlights it and causes its properties to appear in the lower-left pane, the Properties pane. The Properties pane shows the name and type of the entity currently selected in the Entities pane. Below that, all of the properties of the selected entity are shown. Many of these properties can be modified.

When an entity is selected, it can be highlighted in the Display pane by pressing the Ctrl key. A highlighted circle is drawn around the entity that is not obscured by other entities (see Figure 5-13).

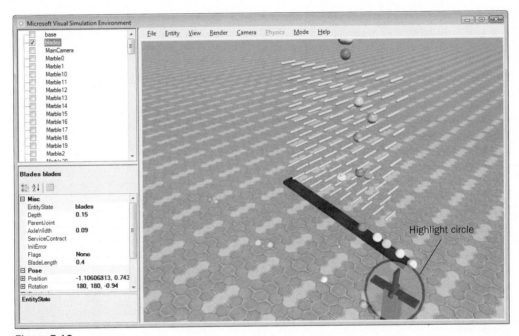

**Figure 5-13**

Another way to select an entity is to click with the right mouse button over the entity in the Display pane. The simulator determines which entities are under the position of the mouse click and selects the closest one. The ground and sky entities cannot be selected in this way.

It is very easy to identify each entity in the scene using this method. Hold down the Ctrl key so that the selected entity is highlighted and then right-click on various entities in the scene. The name of the entity and all of its properties appear in the Properties pane.

*The Entities pane maintains two different entity selections. The entity that is highlighted in the Entities pane is the one that is shown in the Properties pane, and it is also highlighted in the Display pane. Entities can also be selected by clicking the checkbox by the name. It is possible to select multiple entities using the checkboxes, although only the highlighted entity is shown in the Properties pane. Operations that affect multiple entities will operate on all entities that are checked.*

## *Manipulating Selected Entities*

When an entity is selected, a number of additional operations are enabled on the entity. For example, it is a common problem to lose track of an entity and to be unable to see it in the scene. Sometimes the entity can be quickly located by selecting it in the Entities pane and then pressing the Ctrl key to highlight it. If the highlight cannot be seen, the eyepoint can be moved to the entity by holding down the Ctrl key and pressing one of the arrow keys, as described in the following table:

| Key | Eyepoint Position |
| --- | --- |
| Up Arrow | Above the entity (+Y) |
| Shift + Up Arrow | Below the entity (−Y) |
| Left Arrow | To the right of the entity (+X) |
| Shift + Left Arrow | To the left of the entity (−X) |
| Right Arrow | In front of the entity (+Z) |
| Shift + Right Arrow | Behind the entity (−Z) |

You can move an entity around in the 3D environment by selecting its position property in the Properties pane. You can enter three values separated by commas to represent the new X, Y, and Z coordinates of the entity, or you can expand the position property and modify each coordinate separately.

Sometimes it is easier to position an entity graphically. To do this, select the position property and then hold down the Ctrl key to highlight the selected entity. Now press the left mouse button in the Display pane and drag the mouse pointer. The entity will move to follow the pointer. If you only want to move the entity along the X axis, simply select the X coordinate of the expanded position property, and when you drag the mouse pointer, the entity will only move along the X axis. The behavior is similar for the Y and Z coordinates.

*The screen is only two-dimensional, so it is not really possible to move an entity in all three dimensions using the mouse pointer. The axes of movement are perpendicular to the vector formed by subtracting the camera Location from the camera LookAt point. You can move an entity parallel to the ground plane by positioning the camera directly above the entity looking down by pressing the Ctrl key and the up arrow and then dragging the mouse pointer with the left button held down.*

You can rotate an entity in a similar way by selecting the rotation property of that entity and then dragging the mouse pointer while holding the Ctrl key. Selecting a single coordinate of the rotation property will constrain the rotation to be only around that axis.

Moving and rotating entities while the physics engine is enabled can be difficult. Gravity and other constraints in the environment can prevent or limit the movement of entities. You can disable the physics engine by pressing F3 or by toggling the Enabled item on the Physics menu, as described in the previous section.

# The Entities Menu

You may have noticed that when you put the simulator in Edit mode, the Entities menu appears on the main menu. The items on this menu operate on one or more entities and are only available in Edit mode. The following sections briefly discuss each item on the menu.

## Saving and Loading Entities

You've already seen how to save and load an entire scene. It is also possible to save and load a subset of a scene. Entities can be saved to a file by clicking one or more checkboxes in the Entities pane and then clicking Entities ⇨ Save Entities. A dialog appears, prompting for a filename, and the entities are serialized to that file. These entities can be loaded later by clicking Entities ⇨ Load Entities.

Sometimes it is more convenient to cut or copy and then paste selected entities. This is also done using the Entities menu. You can select one or more entities using the checkboxes in the Entities pane and then select either Cut or Copy from the Entities menu. At this point, the entities are serialized to a file called `SimEditor.Paste` in the `Store` directory. If you select Cut, the entities are deleted from the scene. When you click Entities ⇨ Paste, the entities in the `SimEditor.Paste` file are deserialized and inserted into the scene. The usual keyboard shortcuts (Ctrl+X for cut, Ctrl+C for copy, and Ctrl+V for paste) can also be used.

---

### Overlapping Objects

If it is true that nature abhors a vacuum, then nature must really hate it when more than one physical object is occupying the same space at the same time. The physics engine tries to correct this situation by applying a force to overlapping objects that will quickly separate them. This can be a source of fun in the simulated environment. Disable the physics engine on the Physics menu and then select a dynamic entity in a scene and press Ctrl+ to copy it to the paste buffer. If you are using the Marbles scene, select one of the marbles. Now press Ctrl+ multiple times to paste that entity back into the same position repeatedly. Next, enable the physics engine by pressing F3 and watch the fireworks. The more entities that overlap, the more energetic is the explosion that separates them. Keep this behavior in mind if you see an entity behaving erratically or moving quickly just after it is created. This is usually a good indication that one or more of the physics shapes in the entity are overlapping and the physics engine is trying to separate them.

---

## Creating New Entities

You can click Entities ⇨ New to create new entities in your scene using the New Entity dialog. You must first select the assembly that defines the entity you want to create. The <Executing assembly> selection will show all entities currently defined by all assemblies currently loaded. The second drop-down list shows all the entities that are defined in the selected assembly. In order for an entity to be displayed, it must inherit from `VisualEntity`. Once an entity has been selected, it must be given a unique name in the Name textbox. If the entity is to be created as a child of another entity, then the parent entity must be selected in the last drop-down list.

Press OK when all of the selections are correct. The simulator will then display a dialog with properties for each parameter in the constructor of the new entity. When there are several constructors for an entity, the constructor defined immediately after the default constructor is selected. Press OK when all of the constructor parameters have been defined, and the entity will be created.

Occasionally, you must set some of the entity properties before you can successfully create the entity. For example, a `SingleShapeEntity` requires a shape to be defined but it cannot be specified in the constructor that is presented. The entity will appear in the Entities pane but it will have a red exclamation point next to it, indicating that there was an error in its creation. Look at the `InitError` property of the entity to see whether it provides some information about why the entity could not be initialized. In the case of the `SingleShapeEntity`, clicking one of the shape properties and defining a valid shape will clear the error, enabling the entity to be initialized.

### Child Entities

The simulator supports the idea of child entities, which are entities that are defined relative to a parent entity. If you add a `SingleShapeEntity` as a child entity of another `SingleShapeEntity`, the child will be attached to the parent with a fixed joint that effectively glues the two pieces together. The physics engine treats the pair as if they were one entity.

When the simulator is in Edit mode, you can move and rotate the parent entity and the child entity moves with it. When you move or rotate the child entity, it changes the attachment of the child to the parent to accommodate the new position or rotation. This is how you can change the way in which the child is attached to the parent.

You can make one entity a child of another using the following steps:

1. Select the entity you want to make a child and press Ctrl+ to cut it from the scene.

2. Select the entity you want to be the parent.

3. Click Entity ⇨ Paste as Child. The cut entity will be added to the scene as a child of the selected entity.

You can also make an entity a child when you create it using the New Entity dialog described earlier by selecting the parent entity in the appropriate drop-down list.

## Modifying Entity State

Every entity in the simulation environment shares a common entity state. The entity state contains properties that affect the visual appearance of the entity as well as its physical behavior. Select an entity in the Entities pane. Its state is the property called `EntityState`, which has a value equal to the name of the entity. Click the value and an ellipsis (...) appears. When you click that ellipsis, a dialog box appears, enabling you to change the values of the entity state.

If you modify one or more entity state values and then select OK, the entity is removed from the scene and then reinitialized and inserted into the scene with the new values. This operation is similar to cutting the entity and then pasting it back into the scene.

The following sections discuss each part of the entity state.

## *Graphics Assets*

Graphics assets determine how the entity looks. You can specify a new `DefaultTexture`, `Effect`, or `Mesh` to change the appearance of an entity:

❑ **DefaultTexture:** If no mesh is specified for the entity, the mesh is generated from the physics shapes. In this case, the `DefaultTexture` specifies which texture map to apply to the generated meshes. The texture map should be a 2D texture and it can be any of the supported texture formats (.dds, .bmp, .png, .jpg, .tif).

❑ **Effect:** This is the shader program that should be used to render the entity. In most cases, you will just want to specify the default effect: `SimpleVisualizer.fx`. Advanced users who want to create a special look for an entity can write their own shader program and specify it here.

❑ **Mesh:** If a mesh file is specified here, it is rendered for the entire entity. Only a single mesh file is supported per entity. As mentioned above, if no mesh is specified, then a default mesh is constructed from the physics shapes that make up the entity. Mesh files must be in .obj or .bos format.

*The simulator reads .obj meshes. The .obj format (also called the* Alias Wavefront *format) is a popular 3D format that many modeling programs are able to export. These files usually have one or more material files with an extension of .mtl that specify the rendering characteristics of the objects. The format of .obj and .mtl files is ASCII, so they are easy to modify but less efficient to read, and large meshes can take some time to load. When an .obj file is read, the simulator automatically generates an optimized form of that file and writes it to disk with the same filename but with a .bos extension. The next time the .obj file is loaded, the simulator loads the .bos file instead, which is faster. If the obj file is modified, the simulator will not read the obsolete .bos file but will instead generate a new one. The MRDS SDK includes a command-line tool called Obj2Bos.exe, which can convert .obj files to .bos files without having to run the simulator. Microsoft has not made the .bos format public, so no tools outside of the simulator can read this format.*

## *Misc*

The following entity state properties appear under the Misc heading:

❑ **Flags:** These bitwise flags define some physics behavior characteristics for the entity. If no flags are defined, then the default value for this field is `Dynamic`, meaning the entity behaves as a normal dynamic entity, which is completely controlled by the physics engine. The other flags are as follows:

   ❑ **Kinematic:** If this flag is set, the entity is kinematic, meaning its position and orientation are controlled by a program external to the physics engine. The entity is not affected by gravity and it is not moved due to collisions with other objects, but it may influence the movement of other objects due to collisions.

   ❑ **IgnoreGravity:** This one is self-explanatory. Gravity will have no effect on the entity if this flag is set.

   ❑ **DisableRotationX, DisableRotationY, DisableRotationZ:** These flags prevent the entity from rotating around each of the specified axes. They are only relevant if the entity is not kinematic.

❑ **Name:** This is the name of the entity. Each entity must have a unique name. Simulation services use the name of the entity to identify it, so it is important that no two entities have the same name. You can modify the name of an entity by changing it here, but any simulation services that previously connected to the entity will no longer work properly.

❑ **Velocity:** This vector holds the magnitude and direction of the velocity of the entity. You can specify a new velocity in the state, and the entity will be updated with this new velocity. This state property is updated each frame, much like the `Position` and `Rotation` properties.

## Physical Properties

These properties in the entity state define physical characteristics of the entity:

❑ **AngularDamping/LinearDamping:** These damping coefficients are numbers that range from zero to infinity. They provide a way to model real-world effects such as air friction, which tend to slow down the linear and angular motion of a body. The higher the number, the more the linear or angular motion slows. For example, entities with the default value of zero (0) for damping will continue to roll along the ground until some other force stops them.

❑ **Mass/Density:** These properties are really two ways of specifying the same thing. If the mass is nonzero for one or more of the physics shapes in an entity, the physics engine will calculate the total mass and density of the entity from the sum of the mass of the shapes. In this case, these two values are ignored when the entity is initialized. If the sum of the masses of the shapes is zero, then the physics engine uses one of these values to determine the mass of the entity. The `Mass` value provides a shortcut for specifying the total mass of an entity. The physics engine calculates the center of mass assuming that the density of the shapes in the entity is uniform. The `Density` value provides another shortcut to specify the overall mass. If `Density` is specified, the physics engine calculates the total mass based on the volume of the shapes in the entity. Again, the engine assumes that the density is uniform throughout the entity.

❑ **InertiaTensor:** This property is a way to specify the mass distribution of the entity. The inertia tensor expresses how hard it is to rotate the shape in various directions. Shapes such as cylinders are naturally easier to rotate around their longitudinal axis but more difficult to rotate around the other two axes due to their mass distribution. If this vector is not specified, the physics engine will calculate it based on the position and relative size of the shapes in the entity.

## Other Common Entity Properties

The following properties are common to all entities. Just like the other entity state, they can be edited in the Properties pane of the Simulation Editor.

❑ **ServiceContract:** This property is an optional URI for a service that controls or interacts with this entity. When the scene is saved and a manifest is written, the simulator will add this service to the list of services that run and specify the name of this entity as a partner.

❑ **InitError:** If an entity fails to initialize, it is displayed in the Entities pane with a red exclamation point. When this happens, the error displays in the `InitError` property.

❑ **Flags:** These are miscellaneous bitwise flags that apply to the entire entity:

  ❑ **UsesAlphaBlending:** Set this flag if your entity is partly transparent. The simulator renders transparent entities last.

❑    **DisableRendering:** Set this flag to skip rendering of this entity. This is very useful for entities that are large and cover up other entities. Even though the entity is not drawn, it still is active in the scene.

❑    **InitializedWithState:** This flag is supposed to indicate that the entity was deserialized from a file, rather than built with a nondefault constructor. The simulator does not appear to set or pay attention to this flag currently.

❑    **DoCompletePhysicsShapeUpdate:** This flag specifies that the physics engine should update the pose information for each physics shape in the entity each frame. By default, the shape state is not updated each frame.

❑    **Ground:** This flag indicates that the entity is the ground. This flag is set by some entities but the simulator does not currently do anything with it.

❑   **ParentJoint:** This is the joint that connects a child entity to its parent. It is discussed in more detail in Chapter 8 when we get to articulated entities.

❑   **Position and Rotation:** These properties were previously discussed in the section "Misc" earlier in the chapter. Together they specify the position and orientation of the entity in the environment.

❑   **MeshScale/MeshTranslation/MeshRotation:** These properties make it easier to fit a 3D mesh to an entity. Because different modeling tools have different defaults for axis orientation and scale, it is sometimes difficult to export a model that exactly matches an entity in the simulation environment. Perhaps the mesh is rotated by 90 degrees or perhaps the modeling package uses centimeters rather than meters. These properties make it easy to scale, move, and rotate the mesh so that it fits the physics shapes that define the entity.

❑   **Meshes:** This property shows information about the meshes that are loaded for the entity. These meshes may have been constructed from the physics shapes that make up the entity, or they may have been loaded from a .obj file. There is not much you can change about the meshes, but you can change the materials that go with the meshes. This is a great way to tune the appearance of entities within the environment. You can make them darker or lighter, or you can change how shiny they look. The color of an object in the simulator is a combination of ambient, diffuse, and specular light. The simulator uses a high–dynamic range lighting model, so it is fine if these values sum to a color greater than (1,1,1). At the end of all the lighting calculations, the color of the object is clamped to the range that the graphics hardware can display.

It is easy to tune the way objects look in the environment by modifying the following material properties. The material properties are part of the entity state, but they are usually set when a mesh is loaded because they describe the appearance of the mesh. For example, a .obj mesh file usually has an associated .mtl material file that defines the material properties for each mesh. If you change the material properties using the Simulation Editor and you want those changes to persist between sessions, then you must tell the simulator to save the material changes back to the source files. To do so, click File ⇨ Save Material Changes.

❑   **Ambient:** This color value determines how much light is reflected from the object due to ambient light, which is light that is scattered throughout the environment from other objects (see Figure 5-14a). Typical values for the color components of the ambient color range from 0 to 0.2.

❑   **Diffuse:** This color value determines how much light is reflected from an object due to being lit by a diffuse light (see Figure 5-14b). This value represents the maximum amount of light that will be reflected for the part of an object that faces directly toward the light. Parts of the object that are more edge-on to the light are lit less by that light, and parts of the object that are perpendicular to the light are not lit at all. Diffuse lighting is what gives a 3D look to shapes in the environment because the amount of light that is reflected depends on the orientation of the shape with respect to the light. Values for diffuse color typically range between 0.4 and 0.8.

❑ **Specular:** This color value determines how much light is reflected from an object in a specular highlight (see Figures 5-14c and d), which is basically a hack to make an object look shiny, as if it were reflecting the actual shape and color of the light. The specular highlight color is typically the same color as the light, whereas the diffuse color is typically the color of the object.

❑ **Power:** This value determines the "tightness" of the specular highlight. The higher the power, the smaller the highlight. Figure 5-14c shows a specular highlight power of 2, whereas Figure 5-14d shows a specular highlight power of 128. Typical values range from 32 to 256.

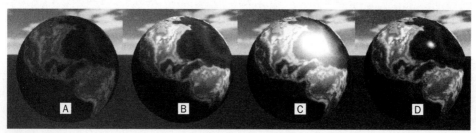

**Figure 5-14**

# Built-in Simulation Entities

The simulator exists to provide a way for you to create your own robots and environments. The Simulation Editor discussed in the previous section enables you to compose and modify environments using the built-in entities that are provided with the simulator. The following sections go into more detail about these entities. The next chapter describes how you can define your own custom entities.

In the following sections, the built-in entities have been divided into four categories:

❑ Sky and ground entities

❑ Light entities

❑ General-purpose entities

❑ Robot and sensor entities

Only wheeled robot entities are discussed in this chapter. Jointed entities, such as arms and walkers, are discussed in Chapter 8.

All of the 3D meshes and textures used by these entities can be found in Store ➪ Media. When you add your own media files, you should place them there as well.

# Sky and Ground Entities

The simulator provides two built-in entities to define the sky, as well as a number of entities to define various types of terrain. You can build a scene without these entities if you prefer, but they make the environment look more realistic, particularly for outdoor scenes.

## SkyEntity and SkyDomeEntity

You don't really pay a lot of attention to the sky until there isn't one. The sky provides diffuse scattered light that illuminates objects in the scene. It is also sometimes useful to simulate bright and overcast outdoor conditions. The difference between the Sky entity (see Figure 5-15a) and the SkyDome entity (see Figure 5-15b) is the way they are drawn:

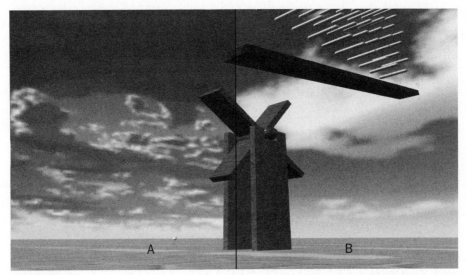

**Figure 5-15**

❏ The Sky entity surrounds the entire scene with a 3D sphere mesh. A cube map, specified in the VisualTexture property of the Sky entity, provides the color of the mesh. This method of drawing the sky is somewhat inefficient because the ground obscures half the sky mesh that is drawn. Half the texture stored in memory is never seen. The texture used by the simulation tutorials for the sky is simply called Sky.dds.

❏ The SkyDome entity is more efficient because it uses a half dome to surround only the part of the environment above the ground. It is colored by a 2D texture map such as Skydome.dds. The one disadvantage to the Skydome entity is that if you rise too high in altitude, you may see a gap between the sky and the ground. The Sky entity is preferable for high-altitude simulations.

Both the Sky entity and the SkyDome entity can add to the diffuse lighting of the scene. The color and intensity of light that they cast on the entities in the scene is defined by a special texture map called a *cube map*. To visualize how a cube map works, imagine yourself on the inside of a giant cube. Each of the six faces of the cube has an image on it so no matter which direction you look, you see a pixel in one of the images. The lighting cube map surrounds the entire scene and defines the color and intensity of the light that shines on the entities from each direction. It is common to define a diffuse light map that lights entities brightly from above and less from the sides. The illumination coming from below is usually a little more than from the sides to simulate light scattering from the ground.

*You can see how this cube map affects the lighting of the scene by clicking Start ⇨ Visual Simulation Environment ⇨ Multiple Simulated Robots. Start the Simulation Editor by pressing F5 and then select the Sun entity in the Entities pane and delete it by pressing the Delete key. Although there is no light in the scene, the robots, table, and ground are lit by the light coming from the sky. You can remove this light from the scene by clearing the texture specified by the LightingTexture property on the Sky entity. The sky will still be lit, but the objects in the scene will be very dark.*

The Sky and SkyDome entities also provide a way to set the fog in the scene. The FogColor property defines the color of the fog. For a night scene, you want to use black fog, and for a day scene, grayish or white fog. Using the FogStart property, you can also specify the distance at which fog begins to appear. The FogEnd property specifies the distance at which entities in the scene are completely fogged. The fog increases linearly between the FogStart distance and the FogEnd distance.

One diffuse sky texture is provided with the MRDS kit. It is called, oddly enough, sky_diff.dds. The .dds format is associated with DirectX, and the extension was originally an abbreviation for Direct Draw Surface. These days, .dds files are most often used as texture maps for Direct3D. You can edit and create .dds texture maps using the DirectX Texture Tool, which is part of the DirectX SDK described previously in this chapter under the section "Software Requirements."

The simulator changes the Y offset of the Sky and Skydome entities to follow the eyepoint of the Main Camera. This prevents the camera from ever going above the sky. Do you remember that episode of *Star Trek* in which a whole colony of people were living inside a hollowed-out asteroid that was a spaceship, except they didn't know it was a spaceship until an old guy climbed a mountain and touched the sky? You don't have to worry about that in the simulation environment.

*The SkyDome entity is used in Simulation Tutorial 1. You can run this environment by clicking Start ⇨ Visual Simulation Environment ⇨ Basic Simulation Environment. Notice how the skydome texture is less blocky than the sky texture. It also consumes less video card memory. You can purchase additional textures that will work with the SkyDome entity from www.hyperfocaldesign.com.*

If you find yourself occasionally becoming disoriented in the simulation environment, you can literally paint a compass in the sky using the Sky entity. Start the Simulation Editor by pressing F5 and then select the Sky entity. In the VisualTexture property, select Directions.dds. This texture will paint the direction you are facing on the sky (see Figure 5-16). For example, if you are looking along the +Z axis, then you should see +Z ahead of you.

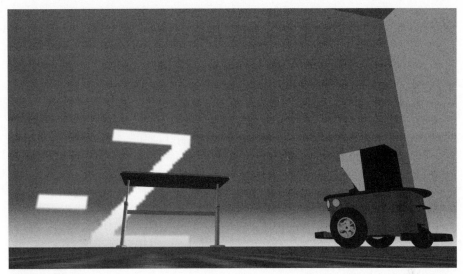

Figure 5-16

## HeightFieldEntity

The HeightFieldEntity is used as a ground plane in many of the simulation tutorials provided with the SDK. It is one of the more important components in the scene because without it, everything falls into an abyss.

A HeightFieldEntity is flat, with an elevation of 0 meters. It may seem infinitely large but it is really just 49,000 square meters, ranging from (−4000, 0, −4000) to (3000, 0, 3000). Columbus would not have fared well in the simulation environment.

A HeightField shape contains a grid of height values. The HeightFieldEntity uses a grid of 8 rows by 8 columns, and it sets the altitude of each height value to 0.

The most remarkable feature of the HeightFieldEntity is the texture that is applied to it. The default texture used in most of the SDK samples is 03RamieSc.dds, which looks a little like industrial carpet. It is a little depressing to look at 49,000 square meters of carpet, so you are encouraged to find another texture for your scenes. The only requirement is that the texture should tile properly without any seams, because it is repeated across the ground many times. The default scaling for the texture on the ground plane is 1 meter by 1 meter. This scale can be changed programmatically but it can't be changed using the Simulation Editor.

*You can change the texture associated with the ground by selecting the ground and then opening its* EntityState *property. Enter a new 2D texture filename in the* DefaultTexture *property.*

The friction and restitution (or bounciness) of the ground can be changed by editing the HeightFieldShape property. This brings up a dialog that enables you to edit its shape properties, among which are StaticFriction, DynamicFriction, and Restitution. These shape properties are described more fully in the upcoming section, which describes SingleShapeEntities.

## *TerrainEntity*

The `TerrainEntity` is a much more interesting way to describe the ground than the `HeightFieldEntity`. The `HeightFieldEntity` is fine for simulating indoor scenes, but the challenge that many robots face is navigating uneven ground in an outdoor environment. The `TerrainEntity` provides a way to construct such an environment.

You can take a look at a `TerrainEntity` by clicking Start ⇨ Visual Simulation Environment ⇨ Simulation Environment with Terrain. The `TerrainEntity` uses the `HeightField` physics shape just like the `HeightFieldEntity`, but it sets the `HeightField` grid values to varying heights based on the pixel values of a grayscale image file. The pixel values range from 0 to 256 and the terrain entity converts these values to –12.8 meters to 12.7 meters, with height samples spread one meter apart by default. Figure 5-17 shows a `TerrainEntity` built from the `terrain.bmp` image file with the terrain_tex.jpg texture map to color it.

**Figure 5-17**

There aren't many `TerrainEntity` properties that you can modify in the Simulation Editor, but you can change the image that generates the `HeightField` and the texture that you apply to it. Start the Simulation Editor by pressing F5 and select the entity named Terrain. Change `TerrainFileName` to another image filename, such as `particle.bmp`. Nothing immediately happens because this particular entity does not automatically refresh when a property is changed. With the `TerrainEntity` still selected, press Ctrl+X and then Ctrl+V to refresh the terrain. You'll notice that the ground level has changed noticeably, and when you leave the Simulation Editor by pressing F5, the physics engine will pop the buried objects out of the ground.

You can change the texture that is applied to the `TerrainEntity` by changing the `DefaultTexture` in the `EntityState` property.

### TerrainEntityLOD

The `TerrainLODEntity` is identical to the `TerrainEntity` except that it does some extra tricks to try to reduce the load on the graphics system. It calculates a reduced resolution version of each section of the height field and then uses that version when the section is far away from the eyepoint. This reduces the number of triangles that must be drawn, as shown in Figure 5-18. The left side of the figure (a) shows the `TerrainEntity` and the right side (b) shows the `TerrainEntityLOD`. It does not reduce the number of `HeightField` samples used by the physics engine, so it does not affect physics behavior.

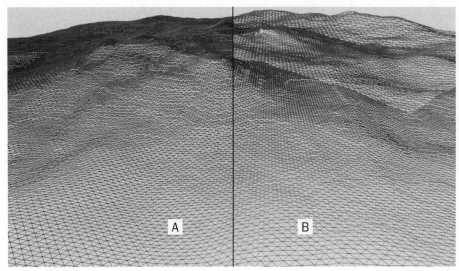

**Figure 5-18**

Unfortunately, the `TerrainEntityLOD` may not be suitable for many applications because it changes the height values for portions of the terrain in the distance, which may cause some entities to be incorrectly visible or obscured in the distance. In addition, it makes no attempt to seam together the edges of each section, so you may see unattractive cracks between sections. Nonetheless, if your graphics card is struggling to render a scene with terrain, the `TerrainEntityLOD` is an option for you.

## Lights and Shadows

The simulator provides three different types of lights to illuminate the scene. They are implemented as a single `LightEntity`, which has three different modes of operation, described further in this section. You can define up to three lights in the scene, but be aware that each additional light creates extra work for the graphics chip and may reduce your frame rate. You can continue to use the simulation environment with Terrain to see how lights work in the simulator.

### Directional Lights

You use a directional light to model a light that is very far away, such as the sun. The position of the light doesn't matter, only its direction. Most of the tutorial and sample scenes provided with the SDK include a directional light named `Sun`. Put the simulator into Edit mode by pressing F5 and select the `Sun` entity.

Now select its `Rotation` property and drag the left mouse button while holding the Ctrl key to see how changing the direction of the light affects the lighting in the scene.

When the simulator is in Edit mode, `LightEntities` are represented by simple meshes, as shown in Figure 5-19 (from left to right, directional lights, omni lights, and spot lights).

Figure 5-19

The `Color` property defines the color of the light for all light types. The `SpotUmbra`, `FalloffEnd`, and `FalloffStart` properties do not affect directional lights.

## Omni Lights

Omni lights are so named because they send light out in every direction. Consider light bulbs. These lights are usually very close to the objects that they illuminate, so their position is very important; but because they cast light in every direction, their rotation is unimportant.

You can change the `Sun` to be an omni light by changing its `Type` property. Try changing its rotation just as you did for the directional light. No effect. Now select its `Position` property and move the light around. You can see that even small changes in the position of the light change the way entities are lit in the scene.

If you ever need to position a light in the scene but you can't find it, you can use the same trick that works for every other entity. Select it in the Entities pane and then hold the Crtl key and press the up arrow. The eyepoint will be moved to be just above the selected light.

The `FalloffStart` and `FalloffEnd` properties affect both omni lights and spot lights. They define the distance at which the light starts to diminish and the distance after which the light has no effect, respectively. Unlike the sun, light bulbs tend to have limited range here on earth.

## Spot Lights

Finally, a light for which position and rotation are both important. In the real world, many light bulbs have reflectors and shades to direct the light in a particular direction. That is exactly what a spot light does. It projects a cone of light with the apex at its position. The angle swept out by the cone is specified by the `SpotUmbra` property. These types of lights are good for headlights or emphasis lights.

## Shadows

Light shadows provide cues to our brain regarding the position of objects in the world. They can be a help to robot vision processing algorithms and they can also be an added challenge. The simulator supports casting shadows from one light. Each `LightEntity` has a `CastsShadows` property, which, when set to `True`, causes that light to cast shadows. Switch the `Sun` back to an omni light and turn on shadows. The first time shadows are turned on, the simulator must generate special shadow meshes for each mesh in the scene. If a mesh was read from a file, the shadow mesh is saved to a file in the `store\ media` directory with the same name as the mesh file but with a `.shadowMesh.bin` extension. It may take several seconds to generate shadow meshes for very complex or large models. For example, the LEGO Tribot can take 20–30 seconds on a fast machine. The next time the shadow meshes are needed, they are read from the disk, rather than generated. If the original mesh is changed, its shadow mesh is updated.

Move the omni light around the scene and note how the shadows are updated (Figure 5-20 shows shadows cast from a single omni light). The entities in the scene are self-shadowing, which means that one part of an entity can cast a shadow on another part.

If more than one light has the `CastsShadows` property set to `True`, the simulator will cast shadows from the first light encountered as it is drawing the scene. Be aware that drawing shadows is expensive and can slow down the rendering of the scene substantially.

Figure 5-20

# General-Purpose Entities

Some entities in the environment don't justify extensive modeling. If you need to test how well your robot avoids obstacles, a few entities made of a single box shape will probably suffice. You can build furniture or stairs or ramps out of multiple box shapes.

The simulator provides the `SingleShapeEntity`, the `MultiShapeEntity`, the `SimplifiedConvexMeshEnvironmentEntity`, and the `TriangleMeshEnvironmentEntity` to make it easier for you to add objects to your scene. Each of these entities is used in the scene built by Simulation Tutorial 5 and is discussed in detail in this section. You can run this scene by clicking Visual Simulation Environment ⇨ Simulation Environment with Terrain.

## SingleShapeEntity

As the name implies, this entity is wrapped around a single physics shape — either a sphere, a box, or a capsule. This is the easiest way to add a single object to the environment: Decide which shape will best represent the object, specify its dimensions, and add it to the scene. You can even specify a mesh to make it look more like the object you are modeling.

The `bluesphere`, the `goldenscapsule`, and the `greybox` are good examples of `SingleShape` entities that use the various physics shapes. In the Simulation Editor, you'll notice that `SingleShape` entities have three properties under the Shape category representing each of the three possible shapes. You can edit the properties for the shape that is defined, or you can click the value of one of the other shape properties to change the entity to use that shape. This is probably a good time to talk about general shape properties because they are used in nearly every entity.

Expand the shape properties associated with the sphere in the `bluesphere` entity. The properties are divided into the following categories: MassDensity, Material, Misc, and Pose. The Material category is where the friction and restitution properties of the shape are specified. Several shapes may share the same material, and changing the material for one shape will affect the other shapes as well.

| **MassDensity** | |
| --- | --- |
| `AngularDamping/LinearDamping` | These properties mean the same thing as the corresponding properties in the `EntityState` but they only apply to a single shape. These damping coefficients are numbers that range from 0 to infinity. They provide a way to model real-world effects, such as air friction, that tend to slow down the linear and angular motion of a body. The higher the number, the more the linear or angular motion is slowed. |
| `Density/Mass` | You specify one or the other. If `Mass` is specified, that is the mass of the shape. If `Density` is specified, the mass of the shape is calculated by multiplying by the volume of the shape. |
| `CenterOfMass` | This is the position of the center of mass of the shape relative to the origin of the shape. The center of mass is the point about which the shape naturally rotates. If this vector is (0,0,0), then the physics engine computes the center of mass assuming a constant density. Specifying a different center of mass is a way to indicate that one part of the shape is heavier than another part of the shape. |
| `InertiaTensor` | This can be thought of as the angular equivalent of the center of mass. It is similar to the same property, which is part of the `EntityState`, but it applies only to this shape. `InertiaTensor` expresses how hard it is to rotate the shape in various directions. If this vector is (0,0,0), then the physics engine calculates it assuming a constant density. |

| **Material** | |
|---|---|
| DynamicFriction | This indicates the friction of the shape when it is moving against another shape. It ranges from 0 to infinity, with 0 representing no friction. It must be less than the static friction specified for the shape. |
| StaticFriction | This indicates the friction of the shape when it is not moving against another object. You may have noticed when dragging an object that it is initially harder to get it moving than it is to move it once it is moving. This is the effect of static friction. It ranges from 0 to infinity, with 0 representing no friction. It must be greater than or equal to the dynamic friction of the shape. |
| Restitution | This is the "bounciness" of the shape. It ranges from 0 to 1, with 0 representing no bounciness. The AGEIA documentation notes that values close to or above 1 may cause stability problems and/or increasing energy. |
| MaterialIndex/MaterialName | These properties identify the material. As mentioned earlier, several shapes can share the same material. |
| **Misc** | |
| DiffuseColor | Each shape has a default color that is used to create a visual mesh if no other mesh is specified in the EntityState. This is how the goldencapture got to be so golden. |
| Dimensions/Radius | A box shape uses all three dimensions to specify its size along each axis. The sphere shape uses only the Radius. The capsule uses the Radius and the Y coordinate of the Dimensions for its height. |
| EnableContactNotifications | Shapes can be set to provide a notification when they contact another shape. This option is not very useful in the Simulation Editor but it can be very useful when writing a service. For more on services, see Chapters 2, 3, and 4. |
| Name | It is not necessary to set the name but it can be useful when the shape is passed to a notification function and you need to identify which entity it is part of. |
| ShapeId | This identifies the type of the shape. |
| TextureFileName | Like DiffuseColor, this defines how the mesh built from the shape will be rendered if no other 3D mesh is specified. This is how the texbox, texcapsule, and texsphere entities in the scene were textured. |
| **Pose** | |
| Position/Rotation | This specifies the offset and orientation of the shape compared to the origin of the entity. This becomes more important in the MultiShapeEntity when several shapes are used to define a single entity. |

Several of the entities in the terrain scene do not specify a 3D mesh, so their visual mesh is built from the shape itself. The StreetCone entity does specify a 3D mesh. Switch to the combined view and look at this entity, and you will see that it is just a single cube for physics purposes but looks like a traffic cone visually.

## MultiShapeEntity

The MultiShapeEntity is much like the SingleShapeEntity except that it contains — yes, you guessed it — multiple shapes. It only supports boxes and spheres but you can have as many of them as you like. The table_mesh entity in the scene is built from multiple box shapes, but you would never know it until you select the physics view because it also uses a 3D mesh to give it a nicer visual appearance.

The state for a MultiShapeEntity is very similar to the state for a SingleShapeEntity except that the BoxShapes property is a collection of boxes and the SphereShapes property is a collection of spheres.

## SimplifiedConvexMeshEnvironmentEntity

This entity enjoys the distinction of having the most syllables of any other built-in entity so it must be important. In some cases, it is not sufficient to have a single box represent the physical shape of an object because it just doesn't collide with other objects the right way. The simulator can construct a physics object called a *simplified convex mesh* from any 3D mesh. A convex mesh is built from the points of a convex hull, which is the smallest polyhedron containing all of the points of the original mesh. You can think of generating a simplified convex mesh as similar to taking an arbitrary mesh and wrapping it with shrink wrap. When you heat the shrink wrap and it pulls tightly around the object, the resulting shape is very close to the convex mesh representation. (Of course, the physics SDK does this using math, not shrink wrap.)

SimplifiedConvexMeshEnvironmentEntities interact with the physics environment just like spheres, boxes, and capsules except that they are more expensive to calculate. The table_phys_m is an example of this type of entity. It is shown in Figure 5-21 (right) along with a TriangleMeshEnvironmentEntity (left).

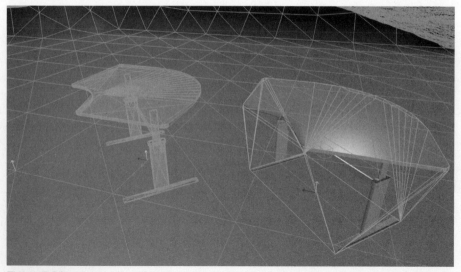

Figure 5-21

The state associated with a `SimplifiedConvexMeshEnvironmentEntity` is very similar to the other entities except that it must have a 3D mesh specified in the `EntityState`. This is the mesh that will be used to construct the convex mesh. The maximum number of polygons in the simplified convex mesh is 256.

## TriangleMeshEnvironmentEntity

This entity is very similar to the `SimplifiedConvexMeshEnvironmentEntity` except that the entire 3D mesh is used for the physics shape instead of constructing a convex mesh. Because the physics shape is identical to the visual shape, the entity will collide and move just like the visual shape.

You may be tempted to use a `TriangleMeshEnvironmentEntity` for every object in the scene, but you should be aware that this type of entity has some limitations. Collision detection is not as robust as it is for other shapes. If a shape moves fast enough that its center is inside the triangle mesh shape, then a collision may not be registered. In addition, `SimplifiedConvexMeshEnvironmentEntities` must be static entities, meaning they cannot be moved around in the environment, and they act as if they have infinite mass.

# Robot Entities

Let's see, what was this book about again? Oh yeah, robots! We should talk about some robots. This chapter covers simulating differential drive robots. In Chapter 8, you'll look at robots with joints. A *differential drive* consists of two wheels with independent motors. A differential drive robot can drive forward or backward by driving both wheels in the same direction, and it turns left or right by driving the wheels in opposite directions. Usually, one or two other wheels or castors provide stability.

In this section, you will learn how to drive the iRobot Create, the LEGO NXT, and the Pioneer 3DX robots in the simulation environment. Each of these robots is based on the `DifferentialDriveEntity`. You'll also learn how to use sensors such as bumpers, laser range finders, and cameras.

## Out for a Drive with the iRobot Create

The best way to become familiar with differential drive robots is to drive one around. Click Start ⇨ Visual Simulation Environment ⇨ iRobot Create Simulation to start up an environment that contains a number of environmental entities and an iRobot Create robot. This manifest starts several services, listed as follows, including the SimpleDashboard service (shown in Figure 5-22):

- ❑ **SimulationEngine:** This is the simulator and it has a state partner that causes the simulator to load its initial scene from the file `iRobot.Create.Simulation.xml`.

- ❑ **SimulatedDifferentialDrive:** This service connects with the iRobot Create entity in the simulation environment and then drives each of its motors according to the commands that it receives.

- ❑ **SimulatedBumper:** This service sends a notification to other services when the bumpers on the iRobot Create make contact with another object.

- ❑ **SimpleDashboard:** This service provides a Windows Forms user interface that can be used to send commands to the DifferentialDrive service to drive the robot.

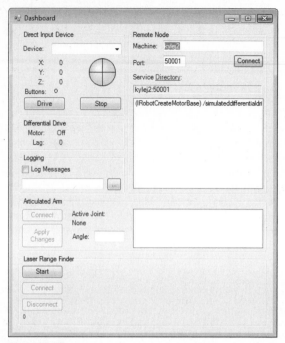

Figure 5-22

Follow these steps to connect the dashboard to the differential drive service and drive the iRobot Create in the environment:

1. Type your machine name in the Machine textbox. Alternatively, you can just type **localhost** to select the current machine.

2. Click the Connect button. The (IRobotCreateMotorBase)/simulateddifferentialdrive service should appear in the list labeled Service Directory.

3. Double-click the (IRobotCreateMotorBase) /simulateddifferentialdrive/ entry. The Motor label should change to On in the Differential Drive group box.

4. Click the Drive button in the Direct Input Device group box to enable the drive control. *Don't forget this step!*

5. Drag the trackball graphic forward, backward, to the left, and to the right, to drive the robot in the environment. Go wild. Crash into entities to your heart's content and take satisfaction in knowing that no one will present you with a bill for a damaged blue sphere.

*You can also drive simulated robots and actual robots using a wired Xbox 360 controller. Plug the controller into an open USB port. You will be able to drive the robot using the left thumbstick. If you are using the simulator, you can also move the camera around the environment using the right thumbstick. The only downside is that it's hard to convince your colleagues that you are doing serious robotics work while holding an Xbox 360 controller.*

## The LEGO NXT Tribot

The LEGO NXT Tribot is the first robot you build with the LEGO NXT kit. It has two drive wheels and a castor in the back. It also has a number of sensors that you can mount to the robot, but the one in the simulation model is the touch sensor, or bumper. You can run an environment that contains a LEGO NXT Tribot, a Pioneer 3DX robot, and a lovely table by selecting Visual Simulation Environment ➪ Multiple Simulated Robots.

When you connect the Dashboard, you'll see three services. You can drive the Tribot by double-clicking the (LEGONXTMotorBase)/simulateddifferentialdrive/ service. This is the service that is used with all three differential drive robots in the simulator. It provides a connection between the `DifferentialDriveEntity` in the simulator and services outside the simulator. When you examine the `ServiceContract` property of the LEGO NXT Tribot, you will notice that it references the contract ID for the SimulatedDifferentialDrive service.

Put the simulator into Edit mode by pressing F5. You'll notice that the Physics menu turns red, indicating that the physics engine has been disabled. You'll also notice that you can no longer drive the robot. When you can't drive a robot in the simulation, ensure that the physics engine is enabled.

## BumperArrayEntity

The `LEGONXTMotorBase` entity in the Entities pane has a single child entity, `LEGONXTBumpers`, which is a `BumperArrayEntity`. This entity has an array of box shapes that define where the bumpers are located. The LEGO NXT Tribot uses only a single box shape. The `ServiceContract` property of the `LEGONXTBumpers` entity is set to the contract identifier for the SimulatedBumper service. This service connects with the `BumperArrayEntity` and provides a way for other services to be notified when the bumpers are in contact with another shape.

## DifferentialDriveEntity

The iRobot Create, the LEGO NXT Tribot, and the Pioneer 3DX entities all inherit from the `DifferentialDriveEntity`. This is what enables the SimulatedDifferentialDrive service to work with all three of them. Most of the properties of the `DifferentialDriveEntity` can only be set programmatically, but a few are useful in the Simulation Editor:

❑ **CurrentHeading:** This is a read-only property that indicates the current heading of the robot in radians. The heading is 0 when the robot faces along the –Z axis and it increases as the robot turns left.

❑ **MotorTorqueScaling:** This factor multiplies the torque applied to the wheels when `SetMotorTorque` is called.

❑ **IsEnabled:** This is `True` if the drive is enabled.

Not all robots in the simulation environment need to inherit from the `DifferentialDriveEntity`. It simply serves as a convenient way to provide common services and properties for two-wheeled differential drive robots. You'll learn how to implement more complicated robots in the next chapter.

## Pioneer3DX

The Pioneer 3DX Robot is manufactured by Mobile Robots, Inc. It is interesting because it has an onboard laser range finder and an onboard computer, so it is capable of autonomous movement. It also has bumpers, and the version in the simulator has a mounted webcam.

You've already learned about cameras in the scene but this camera is mounted to the robot as a child entity. When an entity is added as a child of another entity, its position and orientation are relative to the position and orientation of the parent. When the parent moves, the child moves with it. Adding a `CameraEntity` as a child entity of the Pioneer3DX entity causes the camera and its image to follow the robot.

*Try this out by double-clicking the (P3DMotorBase)/simulateddifferentialdrive/ service in the* `Service` *directory list of the Dashboard. Click the Drive button to enable the drive and verify that you can drive the Pioneer by dragging the trackball icon. Now switch the simulator view to the camera mounted to the Pioneer by selecting robocam from the Camera menu. Drive the robot around and you will see what the robot sees.*

The `ServiceContract` property of the robocam entity shows that it is associated with the SimulatedWebcam service. This service reads the images generated by the real-time camera and makes them available to other services or to users via a browser window. You can watch updated images from any real-time camera in the scene in a browser window by following these steps:

1. Make sure the Visual Simulation Environment ⇨ Multiple Simulated Robots scene is running.

2. Open a browser window and point it to `http://localhost:50000`. The image from the SimulatedWebcam service appears in a browser window (see Figure 5-23).

3. Select Service Directory from the left pane.

4. Click either service that begins with /simulatedwebcam/.

5. Select a refresh interval and click Start. The web page should periodically update with a new image from the camera.

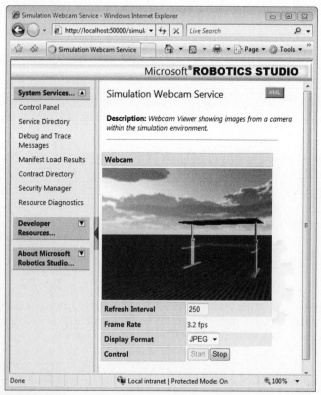

Figure 5-23

### LaserRangeFinderEntity

A laser range finder is a sophisticated and expensive device for measuring distance. It uses a laser to sweep a horizontal line in front of the robot. The laser measures the distance to objects in front of the robot by shining a low-power laser and then recording the light that is bounced off of objects. As the laser sweeps in a horizontal arc, the number of samples that it takes defines its angular resolution.

The `LaserRangeFinderEntity` simulates a laser range finder by casting rays into the simulation environment that correspond to laser samples. These rays are intersected with the physics shapes in the environment, and an array of distances is returned. Small, flashing, red circles are drawn in the environment at each intersection point. You may notice that these hit points are sometimes drawn in mid-air where no object is visible. This is usually because the physics shape that represents an entity is different from the visual mesh that represents the entity. This is the case with the table in the scene. Switching to combined rendering mode shows that the box shapes defining the sides of the table are quite different from the visual mesh of the table. The laser always operates on physics shapes.

The Dashboard user interface provides a way to visualize the laser range finder data. Follow these steps to enable this feature:

1. Click Start ⇨ Visual Simulation Environment ⇨ Multiple Simulated Robots.

2. Connect the Dashboard by entering the computer name or **localhost** in the Machine textbox.

3. Double-click the (P3DXLaserRangeFinder)/simulatedlrf service in the Service Directory list.

4. An image of two box shapes should appear in the Laser Range Finder window at the bottom of the dialog. Darker pixel colors represent objects that are nearer to the robot.

# Summary

This chapter has been a user guide to the simulator. The functionality provided by the simulator has been explained and the built-in entities have been described. Several example scenes have been demonstrated. You might want to remember the following key points as you head into other chapters:

❑ The Microsoft Robotics Developer Studio Visual Simulation Environment is a 3D virtual environment with a powerful physics engine. It can be used to prototype algorithms and new robot designs.

❑ The simulator provides a user interface to move through the environment, as well as to change the rendering mode and save and restore scenes.

❑ The simulator has a built-in editor that enables entire scenes to be defined. Robots and other entities can be created and placed in the environment.

❑ The simulator provides several built-in entities, including robots and sensors.

❑ Custom entities can be defined and used in the environment.

The environments and robots provided with the SDK are interesting, but the real utility and fun comes from building something new. That is what is covered in the next four chapters, where you will learn about adding new entities and simulation services and creating new environments.

# Extending the MRDS Visual Simulation Environment

The previous chapter showed how to use the MRDS Visual Simulation Environment, including making simple edits to the environment using the Simulation Editor. The robot entities and environments provided with the MRDS SDK are great, but they only tap a small part of the simulator's potential.

This chapter demonstrates how to add your own custom entities and services to the simulation environment. You will define a new four-wheel-drive robot with a camera and IR distance sensors, along with the services needed to drive the motors and read the sensor values. In the next chapter, you'll use this robot in a simulation of the SRS Robo-Magellan contest.

By the time you complete these two chapters, you will know how to build an entire simulation scenario, complete with special environmental entities, a custom robot, services to interface with the entities, and a high-level orchestration service to control the behavior of the robot. Figure 6-1 shows the Corobot entity defined in this chapter.

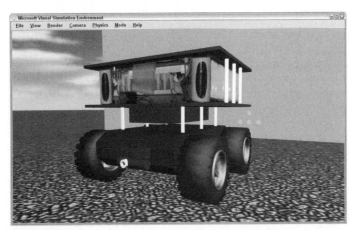

Figure 6-1

# Simulation DLLs and Types

Before you set out on any great adventure, it pays to know what resources are available to you. When you're writing services that interact with the simulation engine, you need to create classes and use types that are defined in the following DLLs.

## *RoboticsCommon.DLL*

This DLL defines a number of common types and generic contracts that both hardware and simulation services can use. Most simulation services will need to reference this DLL so that they can use at least some of the types defined in the `Microsoft.Robotics.PhysicalModel` namespace. Some of the types you'll use in this chapter are as follows:

❑   `Vector2`, `Vector3`, `Vector4`: A structure that contains the indicated number of floating-point values. Vectors are typically used to represent 2D, 3D, or homogeneous coordinate vectors. `Vector3` and `Vector4` are also sometimes used to represent colors.

❑   `Quaternion`: 3D rotations are represented by the physics engine as quaternions. It is beyond the scope of this chapter to completely explain the math behind quaternions, but you can reference the following link to learn more: `http://en.wikipedia.org/wiki/Quaternion`. The `Quaternion` type contains four floating-point values: `X`, `Y`, `Z`, and `W`.

❑   `Pose`: A `Pose` defines the position and orientation of an entity within the simulation environment. It consists of a `Vector3` position and a `Quaternion` orientation.

❑   `Matrix`: As you might expect, this is a 4 × 4 array of floating-point values that represents the transform for a point in the simulation environment. There is also a matrix type provided by the XNA library. This matrix type is used more commonly than the `RoboticsCommon` matrix type because it supports more built-in operations.

❑   `ColorValue`: This contains four floating-point values (`Alpha`, `Red`, `Green`, and `Blue`) that range from 0 to 1 to define a color.

In addition to these basic types, `RoboticsCommon.dll` also defines a number of generic contracts, which define particular operations but don't necessarily associate any behavior with the operations. It is often useful to write a hardware service and a simulation service that implement a generic contract. The orchestration service that drives the robot can then work properly in simulation and in the real world by making a simple manifest change. Some of the generic contracts defined in `RoboticsCommon` are as follows:

❑   `AnalogSensor`: A generic analog sensor that returns a floating-point value representing the current state of a continuously varying hardware sensor

❑   `AnalogSensorArray`: Multiple analog sensors

❑   `Battery`: A generic battery contract that enables the system to report on the state of the battery and provide notifications when the battery level falls below a critical threshold.

❑   `ContactSensor`: A generic sensor that has a pressed and unpressed state. This is suitable for bumpers, pushbuttons, and other similar sensors.

❑   `Drive`: A generic two-wheel differential drive that provides operations such as `DriveDistance`, `RotateDegrees`, `AllStop`, and so on. It provides control for two wheels that are driven independently.

❑ Encoder: A generic wheel encoder sensor that provides information about the current encoder state

❑ Motor: Provides a way to control a generic single motor

❑ Sonar: Exposes a generic sonar device, including information about the current distance measurement and angular range and resolution

❑ Webcam: Provides a way to retrieve images from a generic camera such as a webcam

You'll be using the AnalogSensor, Drive, and Webcam generic contracts as we develop the Robo-Magellan simulation.

## SimulationCommon.DLL

This DLL defines types that are used only in the simulation environment. The types specific to the simulation engine are contained in the Microsoft.Robotics.Simulation namespace. Some of these types include the following:

❑ Entity: The base type for all entities in the simulation environment. This type contains all of the information common to both the simulation engine and the physics engine.

❑ EntityState: This type contains information about the entity such as its Pose, Velocity, Angular Velocity, and Name. In addition, it contains a list of all the physics primitives associated with the entity, as well as physical properties such as the mass and density of the object and visual properties such as the default texture, mesh, and rendering effect. Miscellaneous flags are provided to control various aspects of the rendering or physics behavior of the entity.

❑ LightEntity: This is a deprecated type, only present in MRDS 1.5 to provide backward compatibility. Lights are now represented by LightEntities.

❑ SimulationState: This is what is returned from a Get operation on the simulator. It contains information about the Main Camera, and some other information such as the current render mode and whether the physics engine is paused. The most important thing it contains is a list of all the entities in the simulation environment.

The types that the physics engine uses are defined under the namespace Microsoft.Robotics.Simulation.Physics. These types are too numerous to completely list here but the most commonly used types are as follows:

❑ BoxShapeProperties, CapsuleShapeProperties, SphereShapeProperties, ConvexMeshShapeProperties, TriangleMeshShapeProperties, HeightFieldShapeProperties, WheelShapeProperties: These types hold state information about each of the shape objects supported by the physics engine. Some of the information is specific to a particular shape, such as the ProcessedMeshResource in the ConvexMeshShapeProperties. Much of the information is common to all or some shapes, such as Dimensions, Radius, LocalPose, etc. These shape properties are covered in more detail in the previous chapter.

❑ BoxShape, CapsuleShape, SphereShape, ConvexMeshShape, TriangleMeshShape, HeightFieldShape, WheelShape: These types are the shapes created from their associated shape properties. They contain a reference to the actual physical shape representation in the AGEIA physics engine.

❑   UIMath: This type contains static methods that you can use to convert between an Euler angle rotation representation and a quaternion rotation representation. A method is also provided to round a double value to the nearest hundredth.

---

### Using Visual Studio Help to find out more about MRDS types

When the Microsoft Robotics Developer Studio SDK is installed, it integrates its API help with the Visual Studio help file. You can bring up Visual Studio help by selecting Help ➪ Index. When the help index is displayed, select (unfiltered) from the Filtered by drop-down menu. Type the name of an MRDS type or method in the Look for: textbox and information about that type will be displayed.

---

# SimulationEngine.DLL

This DLL contains most of the simulator functionality. From a programmer's perspective, the most important types that it contains are in the Microsoft.Robotics.Simulation.Engine namespace. These are the built-in entity types provided with the simulator, such as CameraEntity, SkyDomeEntity, SingleShapeEntity, and so on. Many of these entities are described in the previous chapter. The full source code for all of these entities is in samples\simulation\entities\entities.cs. You can use this code as an example for your own custom entities.

Note that all of these entities inherit from the VisualEntity class. Some of the members and properties of the VisualEntity class are important to understand when creating new entities and services for the simulation environment.

## VisualEntity Methods and Members

Initialize, Update, and Render are three virtual methods on VirtualEntity that can be overridden in a subclass to define new behavior for the entity. This section describes these methods, as well as other important methods and member variables on the VisualEntity class.

❑   Initialize: This method is called after the entity has been inserted into the simulation environment. In this method, the state values of the entity are used to instantiate run-time objects, such as shapes and meshes, which enable the entity to function in the simulator. It is important to keep all of the code that creates run-time objects in the Initialize function and not in the constructor. When an entity is deserialized from an XML file, such as when it is pasted using the Simulation Editor, the deserializer calls the default constructor for the entity. This is the constructor with no parameters. It then sets all of the state variables according to the XML data and calls Initialize on the entity to instantiate run-time objects. If there are run-time objects that are initialized in a nondefault constructor, the entity will fail to initialize properly when it is deserialized. It is a good idea to enclose most of the code in the Initialize method within a Try/Catch block and to set the value of the InitError field with the text of any errors that are encountered.

❑   Update: This method is called once each frame while the physics engine is not processing the frame. This is important because some physics engine functions cannot be called while the physics engine is actively processing the frame. In the Update method, the entity calculates its transformation matrix, which is used in rendering. Custom behavior can be implemented here as well, such as setting the pose of an entity or the axle speed of a wheel, for example.

❑ Render: This method is called once each frame after the Update method for all the other entities has completed. This is where the mesh is rendered so that the entity appears on the screen. Several entities override the default behavior for this method to implement specialized rendering effects.

❑ State: This class is actually defined as part of the Entity class from which VisualEntity is subclassed, but it is important enough to mention it here. The EntityState contains important information about the entity, such as its name, its pose, its velocity, its physics shapes, and its rendering assets. More information is provided about the EntityState in Chapter 5.

❑ InsertEntity, InsertEntityGlobal, RemoveEntity: These methods are used to add children entities. InsertEntity assumes that the pose of the child entity is relative to the parent entity, whereas InsertEntityGlobal assumes that the pose of the child is given in global coordinates. RemoveEntity removes a child entity. An example of a child entity is a camera that is mounted to a parent robot. The camera doesn't have a physics shape associated with it so its position is updated in its Update method, enabling it to move with its parent entity. Other entities such as the BumperArrayEntity do have a physics shape, so they are moved by the physics engine each frame. To keep them attached to their parent entity, a joint is created between the parent entity and the child entity. This joint is stored in the ParentJoint field of the child entity. Joints are covered in more detail in Chapter 7.

❑ PhysicsEntity: This class represents the link between this entity and the physics engine. If the entity has no physics shapes associated with it, then this member will be null. The more common case is that the entity will have one or more physics shapes associated with it and the PhysicsEntity member is initialized with a call to CreateAndInsertPhysicsEntity after all of the physics shapes have been created and added to the State.PhysicsPrimitives list. The PhysicsEntity member contains several methods that affect the entity within the simulation environment, such as SetPose, SetLinearVelocity, SetAngularVelocity, ApplyForce, and ApplyTorque.

❑ DeferredTaskQueue: This is a list of tasks that need to be executed during the Update method. The Update method is the only time when the physics engine is guaranteed to not be busy processing a frame. Most of the methods on the PhysicsEntity object cannot be called when the physics engine is busy. When one of these methods needs to be called outside of the Update method, it is added as a task to the DeferredTaskQueue and its execution is deferred until Update runs again.

❑ LoadResources: This method is used to load the mesh, texture, and effect resources associated with the entity. It is typically called from the Initialize method and it stores references to the loaded meshes in the Meshes field.

## The SimulationEngine Class

The SimulationEngine class defines a static member called GlobalInstance, which holds a pointer to the single instance of the SimulationEngine. The DssHost environment will not allow multiple instances of the SimulationEngine class to be instantiated. Public properties such as a reference to the graphics device and a reference to the ServiceInfo class associated with the SimulationEngine service can be accessed from GlobalInstance. Public methods such as IntersectRay and IsEntityNameInUse can also be accessed from this reference.

The `SimulationEngine` class defines another static member called `GlobalInstancePort`, which contains a reference to the simulation engine port that can be used to insert, update, and remove entities. These two global variables are useful to services that must interact closely with the simulation engine.

# SimulationEngine.Proxy.DLL

This DLL is necessary to access the service elements of the simulation engine. Items such as contracts and port definitions should always be accessed from the Proxy DLL.

# PhysicsEngine.DLL

This DLL contains the definitions for the objects used in the simulation environment that represent objects in the physics engine. The `PhysicsEntity` object described in the previous section is defined in this DLL, so your service must add a reference to `PhysicsEngine.DLL` if you want to call any of the methods on the `PhysicsEntity` object.

This DLL also defines the `PhysicsEngine` object. This object represents the physics engine scene that contains all of the entities in the simulation environment. The methods on this object are not typically called by a service directly, but the `PhysicsEngine` object must be passed into several of the physics engine APIs.

# Microsoft.Xna.Framework.DLL

This DLL contains all of the rendering code. It provides a managed DirectX interface and eventually calls the underlying native DirectX DLLs. You must have both DirectX and XNA installed for the simulator to work properly. Some simulation services must use the XNA types, so they must have a reference to the XNA Framework DLL.

In some cases, it is necessary to convert between XNA types and their counterparts defined in `RoboticsCommon.DLL`. There are several functions defined in the `Microsoft.Robotics.Simulation.Engine` namespace in the `TypeConversion` class that make this easy. This class provides a `FromXNA` method for `Vector3`, `Vector4`, `Quaternion`, and `Matrix` that converts a type from XNA to its `RoboticsCommon` counterpart. Similar `ToXNA` methods are provided to convert in the opposite direction.

There are two namespaces that contain types used in the simulator: `Microsoft.Xna.Framework` and `Microsoft.Xna.Framework.Graphics`. Both are discussed in the following sections.

## The Microsoft.Xna.Framework Namespace

You will see several of the basic types defined in this namespace used in various simulation services:

❑  `Vector2`, `Vector3`, `Vector4`: These are functionally equivalent to the corresponding types found in `RoboticsCommon.DLL` but these types often have more utility methods associated with them. In addition, some APIs work with the XNA typed vectors and other APIs work with the `RoboticsCommon` vectors. It is sometimes necessary to convert between XNA types and their `RoboticsCommon` counterparts.

❑  `Matrix`: This is a standard 4 × 4 matrix, but the XNA version has more utility functions associated with it than does the `RoboticsCommon` version.

❑  `Quaternion`: The XNA `Quaternion` equivalent.

## *The Microsoft.Xna.Framework.Graphics Namespace*

This is where most of the graphics-related types are defined. Because there are so many types defined in this namespace, only a few are covered in this section:

❏   GraphicsDevice: This object represents the hardware graphics device. It is initialized by the simulator, so it is rare for a service to interact directly with this object. However, if you override the Initialize method, you have to include a reference to the XNA DLL because the GraphicsDevice is passed as a parameter along with a reference to the PhysicsEngine object.

❏   Effect, IndexBuffer, VertexBuffer, RenderState, Texture2D, TextureCube: These types are used to represent data that is directly accessed by the graphics hardware. Effects are used to tell the graphics hardware how to draw geometry. IndexBuffers and VertexBuffers hold the geometry to be drawn. RenderStates further specify how the graphics hardware should draw the objects. Texture2D and TextureCube objects hold texture map data.

It is beyond the scope of this book to completely cover the XNA graphics APIs. Numerous books and online articles cover XNA thoroughly. You can find the Microsoft XNA forums at http://forums.xna.com, and the forum that deals specifically with the framework APIs is at http://forums.xna.com/56/ShowForum.aspx.

---

### Adding a Reference to Microsoft.Xna.Framework.DLL

With the exception of Microsoft.Xna.Framework.DLL, all the DLLs mentioned in this section can be found in c:\Microsoft Robotics Studio (1.5)\bin. The Microsoft.Xna.Framework.DLL is installed in the .NET Global Assembly Cache (GAC). You must find it in a subdirectory of the Windows directory in order to add it as a reference. Look under \WINDOWS\assembly\GAC_32\Microsoft.Xna.Framework. You may find multiple copies installed in subdirectories under this directory. Choose the most recent version. Windows Explorer does not allow you to browse to this directory, so you will find it easier to use a command window to access it. You can then copy the DLL to your own directory and then add it as a reference. To make this easier, install the entire XNA SDK. Then it will contain a copy of the Xna.Framework.DLL that you can use as a reference. By default, MRDS only installs the run-time version of the XNA DLLs, so the steps outlined above are necessary.

---

# *Using Statements and DLL References*

In summary, add the following DLLs as references to your service to enable it to work with the simulator:

```
Microsoft.Xna.Framework
PhysicsEngine
RoboticsCommon
SimulationCommon
SimulationEngine
SimulationEngine.proxy
```

Add the following `using` statements to properly resolve references to objects in these DLLs:

```
#region Simulation namespaces
using Microsoft.Robotics.Simulation;
using Microsoft.Robotics.Simulation.Engine;
using engineproxy = Microsoft.Robotics.Simulation.Engine.Proxy;
using Microsoft.Robotics.Simulation.Physics;
using Microsoft.Robotics.PhysicalModel;
using xna = Microsoft.Xna.Framework;
using xnagrfx = Microsoft.Xna.Framework.Graphics;
#endregion
```

# Building Your Own SRS-Robo-Magellan Simulation

In the following sections, you'll begin building an SRS Robo-Magellan simulation from the ground up. If you are the type of person who likes to skip to the end of the book to see how it all turns out, feel free to go to the `Chapter6` directory of the sample software and build and run the solution to see the simulated Corobot in action. Chapter 7 includes a referee and orchestration service to complete the Robo-Magellan simulation.

This chapter describes how to build a custom robot, the CoroWare Corobot, along with its sensors and associated simulation services. You'll also define an environment to test the Corobot. The next chapter describes how to use the robot you've defined in a custom simulation scenario, the simulated SRS Robo-Magellan scenario.

If you prefer to follow the step-by-step implementation of these services, it's recommended that you make a new directory parallel to the `Chapter6` directory. You can call it something like `MyChapter6`.

## Simulation Services

A simulation service is a service that creates, manipulates, or reads data from entities in the simulation environment. It typically specifies the SimulationEngine service as a partner, and it references the DLLs listed in the previous sections. A simulation service follows the same rules as other services and it only communicates with the SimulationEngine service through a `SimulationEnginePort`. It can insert, replace, or delete entities by sending messages to this port. It can also subscribe to a particular entity by name so that it receives a notification when that entity is inserted, replaced, or deleted.

Simulation services that are running on the same node as the SimulationEngine service can take advantage of a shortcut that dramatically improves the performance of setting and retrieving entity data. When a service is running on the same node as the simulator and it receives an insert notification for a particular entity, the body of the insert notification message contains a reference to the actual entity in the simulation environment. This object is "live," meaning its fields are updated as each frame is processed in the simulator. In this case, it is only necessary for a simulation service to receive a single notification when an entity is inserted in the environment, after which it can use the reference to that entity for all subsequent operations. Because simulation services typically need to interact with their associated entities frequently, this can speed things up substantially.

A typical simulation scenario has one or more of these services running on the same node as the simulator and one or more higher-level orchestration services that may or may not be running on the same node. Only the simulation services interact directly with entities in the simulator. The simulation and orchestration services for the Robo-Magellan simulation are shown in Figure 6-2.

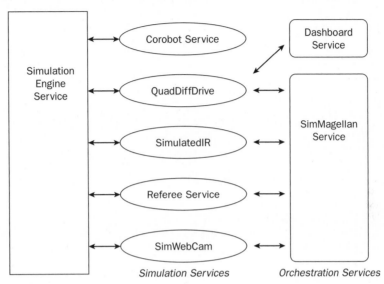

Figure 6-2

The SimulationEngine service, the Dashboard service, and the SimulatedWebCam service are provided as part of the SDK. The rest of the services will be developed in this chapter and in Chapter 7.

Notice that neither the Dashboard service nor the SimMagellan service interacts directly with the simulator. These services rely on information passed to them from the simulation services. The orchestration services can run on a different node or even a different machine from the simulator, but simulation services must run on the same node.

## Creating a Simulation Service

The first simulation service you'll create is the Corobot service. This service will add sky, ground, and a few other simulation entities to the simulation environment. Eventually, it will contain the definition for a new robot entity modeled after the CoroWare Corobot.

One way to create a new simulation service is to use the DssNewService utility. Make sure that you have installed the ProMRDS samples from the CD. Open an MRDS command prompt by clicking Start ⇨ Microsoft Robotics Studio (1.5) ⇨ Command Prompt. Create a new directory and run DssNewService as follows (bold text indicates text that you type):

```
C:\Microsoft Robotics Studio (1.5)>
C:\Microsoft Robotics Studio (1.5)>cd ProMRDS
C:\Microsoft Robotics Studio (1.5)\ProMRDS>mkdir MyChapter6
C:\Microsoft Robotics Studio (1.5)\ProMRDS>cd MyChapter6
C:\Microsoft Robotics Studio (1.5)\ProMRDS\MyChapter6>dssnewservice
/Service:"Corobot" /Namespace:"ProMRDS.Simulation.Corobot" /year:"2007" /month:"07"
```

The /Service parameter specifies the name of the service. The /Namespace parameter specifies the namespace that will be used in the service. The /year and /month options are typically not necessary because they default to the current year and month. They are used to make the contract identifier for the service. In this case, you want the contract identifier to match the Corobot service in the Chapter6 directory so you specify the year and month in which this contract was created.

After executing the command, a directory called Corobot is created with all of the files necessary to build a complete service. The contract identifier associated with the service is as follows:

```
"http://schemas.tempuri.org/2007/07/corobot.html"
```

Note that several files have been generated for you:

❑ **AssemblyInfo.cs:** This file identifies the assembly as a Dss service.

❑ **Corobot.cs:** This file contains the service implementation.

❑ **Corobot.manifest.xml:** This is a manifest that can be used to start the Corobot service. Visual Studio is set up to run this manifest when you press F5 to start debugging your service. This is set up in the file Corobot.csproj.user.

❑ **CorobotTypes.cs:** This file contains the contract identifier and the definition of the state associated with the service, and the operations defined on the main port.

Go ahead and build and run the service. At this point, it doesn't do much, but you can start a browser window and navigate to http://localhost:50000 and select Service Directory from the left side of the window and verify that the /corobot service is indeed running. You can even click it to retrieve its state, although you haven't put anything interesting in the service yet.

Now you'll modify the service so that it can interact with the simulator:

1. Begin by adding the following using statements to the top of Corobot.cs as described in the preceding section:

```
#region Simulation namespaces
using Microsoft.Robotics.Simulation;
using Microsoft.Robotics.Simulation.Engine;
using engineproxy = Microsoft.Robotics.Simulation.Engine.Proxy;
using Microsoft.Robotics.Simulation.Physics;
using Microsoft.Robotics.PhysicalModel;
using xna = Microsoft.Xna.Framework;
using xnagrfx = Microsoft.Xna.Framework.Graphics;
#endregion
```

2. Now add references to the following DLLs. Remember that the XNA DLL has to be handled specially according to the instructions provided in the section "Microsoft Xna.Framework.DLL," earlier in this chapter.

```
Microsoft.Xna.Framework
PhysicsEngine
RoboticsCommon
SimulationCommon
SimulationEngine
SimulationEngine.proxy
```

3.  After you have added a reference to a new external DLL, select the DLL so that its properties are displayed. Set the `CopyLocal` and `Specific Version` properties to `False`. Build the service again to ensure that no mistakes were made. Change the `Description` attribute, which precedes the `CorobotService` class, to something more descriptive if you like.

4.  At the top of the `CorobotService` class, add the following partner specification:

```
[Partner("Engine",
Contract = engineproxy.Contract.Identifier,
CreationPolicy = PartnerCreationPolicy.UseExistingOrCreate)]
private engineproxy.SimulationEnginePort _engineStub =
    new engineproxy.SimulationEnginePort();
```

This identifies the SimulationEngine service as a partner to the Corobot service and instructs DSS to start this service if it isn't already running. At this point, when you run the service, it will at least start up the simulation engine even though the simulation environment is empty.

5.  Now you'll add code to the `Start` method to set up the simulation environment. The first thing to do is define the Main Camera initial view. Insert the following code after the call to `base.Start`. This code defines a `CameraView` message, sets it `EyePosition` and `LookAtPoint` and then sends it to the SimulationEngine service. This is all pretty straightforward until the last line where you post the `CameraView` message to the `SimulationEngine` `.GlobalInstancePort`. The SimulationEngine service holds a static global pointer to its own main operation port. It is often convenient for other simulation services to interact with the SimulationEngine through this port.

```
// MainCamera initial view
CameraView view = new CameraView();
view.EyePosition = new Vector3(-1.65f, 1.63f, -0.29f);
view.LookAtPoint = new Vector3(0, 0, 0);
SimulationEngine.GlobalInstancePort.Update(view);
```

6.  Adding a sky to the simulation environment is as easy as adding the following two lines of code:

```
// Add a SkyDome.
SkyDomeEntity sky = new SkyDomeEntity("skydome.dds", "sky_diff.dds");
SimulationEngine.GlobalInstancePort.Insert(sky);
```

Here, you create a new `SkyDomeEntity` and specify a visual texture of `"skydome.dds"` and a lighting texture of `"sky_diff.dds"`. When the entity has been created, you insert it into the simulation environment.

7.  Now you'll add a directional light to simulate the sun. Without this light source, entities in the simulation environment are only lit by ambient light and by the light from the `SkyDome` entity.

```
LightSourceEntity sun = new LightSourceEntity();
sun.State.Name = "Sun";
sun.Type = LightSourceEntityType.Directional;
sun.Color = new Vector4(0.8f, 0.8f, 0.8f, 1);
sun.Direction = new Vector3(0.5f, -.75f, 0.5f);
SimulationEngine.GlobalInstancePort.Insert(sun);
```

**8.** The next step is to add a ground entity. If you forget to add this entity, your other simulation entities will fall into nothingness as soon as they are inserted in the environment. The ground is a good thing.

```
// create a large horizontal plane, at zero elevation.
HeightFieldEntity ground = new HeightFieldEntity(
    "Ground", // name
    "Gravel.dds", // texture image
    new MaterialProperties("ground",
        0.2f, // restitution
        0.5f, // dynamic friction
        0.5f) // static friction
    );
SimulationEngine.GlobalInstancePort.Insert(ground);
```

Here, you create a `HeightFieldEntity` with zero-elevation height specified for each height field sample, which gives you a nice flat ground plane. Notice that you're using a gravel texture for the ground plane instead of the infinite carpet texture that most of the MRDS Simulation Tutorials use. There is something vaguely unsettling about a world of infinite carpet. You also specify a material for the ground, consisting of restitution and dynamic and static friction values, as described in Chapter 5.

**9.** Finally, add a giant box to the environment. Giant boxes can be handy. This particular giant box is positioned so that its center is exactly 29 inches + 1 meter in the +Z direction from the origin of the simulation world. The origin of all `SingleShapeEntities` is at their center, so you must specify a Y offset equal to half the height of the box to keep the box from being partially buried in the ground. You might surmise that the reason the box is offset 29 inches + 1 meter in the +Z direction is that you really want its edge to be about 29 inches from the origin. This will be useful later.

```
Vector3 dimensions = new Vector3(2f, 2f, 2f); // meters
SingleShapeEntity box = new SingleShapeEntity(
    new BoxShape(
        new BoxShapeProperties(
        100, // mass in kilograms.
        new Pose(), // relative pose
        dimensions)), // dimensions
        new Vector3(0, 1f, 29f * 2.54f / 100f + 1f));

// Name the entity. All entities must have unique names
box.State.Name = "box";

// Insert entity in simulation.
SimulationEngine.GlobalInstancePort.Insert(box);
```

Let's see—you've got a sky, a ground, and a giant box. Build and run your service and verify that it looks something like what is shown in Figure 6-3.

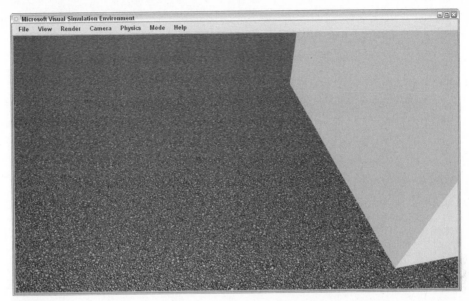

Figure 6-3

If the scene does not look like you expect, you can start the Simulation Editor by pressing F5 and then examine the list of entities in the upper-right pane. If any of the entities have a red exclamation point next to them, they did not initialize properly for some reason. This is often due to a missing texture or mesh file. Select the entity with a problem and look at the InitError property to find more information about the error.

If your entity fails to show up in the list at all, the simulation engine could not create it. This is often because you have forgotten to specify an entity name or you have duplicated a name already in use. Examine the DssHost window to see whether it contains any error messages that can shed some light on the problem.

Congratulations! You have just created your first simulation environment. Now let's build a robot. If you ran into any trouble building your service or if your results are different, compare your Corobot.cs file with the corresponding file in the Chapter6 directory.

## Defining a Custom Robot Entity

Granted, a world of infinite gravel and a giant box is not very exciting, so let's define a new robot entity to move some of that gravel around (Figure 6-4 shows the CoroWare Corobot in the real world). In this section, you'll model a new robot entity similar to the Corobot manufactured by CoroWare. You can find more information about this robot at www.corobot.net.

Figure 6-4

The features of this robot that are most interesting in the simulator are its four-wheel differential drive, its front and rear infrared distance sensors, and its front-mounted camera.

The first step in creating this robot in the simulator is to define a new class that inherits from VisualEntity as follows:

```
/// <summary>
/// An entity which represents a Corobot. This entity is created facing
/// in the +Z direction meaning that its front has positive Z
/// coordinates. Front and Rear IR sensors are included along with
/// a front-facing camera.
/// </summary>
[DataContract]
[DataMemberConstructor]
public class CorobotEntity : VisualEntity
{
}
```

The [DataContract] attribute indicates that this entity has state that needs to be serialized when it is sent across nodes. The [DataMemberConstructor] attribute indicates that the nondefault constructor defined in the class should also appear in the generated proxy class. All of the other two-wheel differential drive entities in the simulator inherit from DifferentialDriveEntity. This entity will be different enough that it makes sense to incorporate the functionality of DifferentialDriveEntity without subclassing it.

The next thing you need to do is define constructors for this entity. Every simulation entity must have a default constructor that has no parameters. This constructor is called when the entity is deserialized, which occurs when the entity is inserted into the simulator from a remote node or when an entity is pasted into the environment using the Simulation Editor. It also occurs when an entire scene is loaded into the simulator. Here are the Corobot constructors:

```
/// <summary>
/// Default constructor used when this entity is deserialized
/// </summary>
public CorobotEntity()
{
}

/// <summary>
/// Initialization constructor used when this entity is built
/// programmatically.
/// </summary>
/// <param name="initialPos"></param>
public CorobotEntity(string name, Vector3 initialPos)
{
    base.State.Name = name;
    base.State.Pose.Position = initialPos;
}
```

The nondefault constructor enables the entity name and initial position to be specified. The only action the constructor takes is to modify its state with the passed parameters. The parameters do not need to be initialized when the default constructor is called during deserialization because the entity state will be restored to the value it had when the entity was serialized.

Next, you need to define some dimensions. Add the following code to the top of the CorobotEntity class:

```
private static float InchesToMeters(float inches)
{
    return (float)(inches * 2.54 / 100.0);
}

static float mass = 3.63f; // kg
static float chassisClearance = InchesToMeters(1.5f);
static float wheelGap = InchesToMeters(3f / 8f);
static float wheelWidth = InchesToMeters(2.2f);
static float wheelDiameter = InchesToMeters(4.75f);
static float wheelMass = 0.1f; // kg
static float platformClearance = InchesToMeters(5.75f);

static Vector3 platformDimensions = new Vector3(
    InchesToMeters(11.0f),
    InchesToMeters(3.0f),
    InchesToMeters(8.5f));
static Vector3 chassisDimensions = new Vector3(
    platformDimensions.X - 2 * wheelWidth - 2 * wheelGap,
    InchesToMeters(2.5f),
    platformDimensions.Z);
static Vector3 wheelFRPosition = new Vector3(
    chassisDimensions.X / 2.0f + wheelGap + wheelWidth / 2.0f,
    wheelDiameter / 2.0f,
    -InchesToMeters(5.75f - 2.125f));
```

*(continued)*

*(continued)*

```
static Vector3 wheelFLPosition = new Vector3(
    -wheelFRPosition.X,
    wheelFRPosition.Y,
    wheelFRPosition.Z);
static Vector3 wheelRRPosition = new Vector3(
    wheelFRPosition.X,
    wheelFRPosition.Y,
    -wheelFRPosition.Z);
static Vector3 wheelRLPosition = new Vector3(
    -wheelFRPosition.X,
    wheelFRPosition.Y,
    -wheelFRPosition.Z);
```

The motor box is called the *chassis* and the upper box containing the processor is called the *platform*. The height of the chassis above the ground is defined by `chassisClearance`, and the height of the platform is defined by `platformClearance`. The width of the chassis is calculated to be just wide enough to allow the outer edge of the wheels to be even with the sides of the platform. The position of the front right wheel is calculated and the positions of the other three wheels are derived from it. This gives us enough information to specify the basic physics shapes that make up the physics model of the Corobot.

The wheels are more than just shapes—they are full entities with their own meshes and physics shapes. You need a place to store all four wheel entities when you create them. Add this code just before the constructors:

```
// instance variables
WheelEntity _wheelFR;
WheelEntity _wheelFL;
WheelEntity _wheelRR;
WheelEntity _wheelRL;

[Category("Wheels")]
[DataMember]
public WheelEntity FrontRightWheel
{
    get { return _wheelFR; }
    set { _wheelFR = value; }
}

[Category("Wheels")]
[DataMember]
public WheelEntity FrontLeftWheel
{
    get { return _wheelFL; }
    set { _wheelFL = value; }
}

[Category("Wheels")]
[DataMember]
public WheelEntity RearRightWheel
{
    get { return _wheelRR; }
    set { _wheelRR = value; }
}
```

```
[Category("Wheels")]
[DataMember]
public WheelEntity RearLeftWheel
{
    get { return _wheelRL; }
    set { _wheelRL = value; }
}
```

The `Category` attribute just groups these properties together so that they are better organized in the Simulation Editor view. Each `WheelEntity` property has the `[DataMember]` attribute. This attribute tells the proxy generator that this is a property that must be serialized when the entity is sent to a remote node or when it is saved to disk. Only public properties can be marked with this attribute. You'll test the entity later to ensure that it can be properly serialized and deserialized.

These wheel entities and chassis and platform shapes are created in the `Initialize` method. Add the following code to override the base class `Initialize` method:

```
public override void Initialize(
    xnagrfx.GraphicsDevice device,
    PhysicsEngine physicsEngine)
{
    try
    {
        // chassis
        BoxShapeProperties chassisDesc = new BoxShapeProperties(
            "chassis",
            mass / 2.0f,
            new Pose(new Vector3(
                0,
                chassisClearance + chassisDimensions.Y / 2.0f,
                0)),
            chassisDimensions);

        chassisDesc.Material =
            new MaterialProperties("chassisMaterial", 0.0f, 0.5f, 0.5f);

        BoxShape chassis = new BoxShape(chassisDesc);
        chassis.State.Name = "ChassisShape";
        base.State.PhysicsPrimitives.Add(chassis);

        // platform
        BoxShapeProperties platformDesc = new BoxShapeProperties(
            "platform",
            mass / 2.0f,
            new Pose(new Vector3(
                0,
                platformClearance + platformDimensions.Y / 2.0f,
                0)),
            platformDimensions);
```

*(continued)*

*(continued)*

```
          platformDesc.Material = chassisDesc.Material;
          BoxShape platform = new BoxShape(platformDesc);
          platform.State.Name = "PlatformShape";
          base.State.PhysicsPrimitives.Add(platform);
```

The first thing the `Initialize` method does is create two box shapes to represent the chassis and the platform. The mass of both objects is assumed to be the same, so the total mass is split between them. They are given the dimensions and position specified. These positions are relative to the origin of the entity. Both box shapes are given the same material definition, which specifies a restitution of 0 and mid-range static and dynamic friction. After each shape is created, it is added to the `PhysicsPrimitives` list:

```
          base.CreateAndInsertPhysicsEntity(physicsEngine);
          base.PhysicsEntity.SolverIterationCount = 128;
```

When the physics entity is created, it passes all of the objects in the `PhysicsPrimitives` list to the physics engine, which creates its own representation for them based on the attributes specified in the `BoxShape` objects. The `SolverIterationCount` determines how many iterations the physics engine will use to resolve constraints on the entity, such as contact points or joints. The higher the count, the more accurate the results become. A value of 128 is probably overkill because the AGEIA documentation states that the default value is 4 and AGEIA developers have never needed to use a value higher than 30. This code follows the example of the `DifferentialDriveEntity` in the SDK in setting this value to 128:

```
          // Wheels
          WheelShapeProperties wheelFRprop = new WheelShapeProperties(
              "FrontRightWheel", wheelMass, wheelDiameter / 2.0f);
          WheelShapeProperties wheelFLprop = new WheelShapeProperties(
              "FrontLeftWheel", wheelMass, wheelDiameter / 2.0f);
          WheelShapeProperties wheelRRprop = new WheelShapeProperties(
              "RearRightWheel", wheelMass, wheelDiameter / 2.0f);
          WheelShapeProperties wheelRLprop = new WheelShapeProperties(
              "RearLeftWheel", wheelMass, wheelDiameter / 2.0f);

          wheelFRprop.Flags |= WheelShapeBehavior.OverrideAxleSpeed;
          wheelFLprop.Flags |= WheelShapeBehavior.OverrideAxleSpeed;
          wheelRRprop.Flags |= WheelShapeBehavior.OverrideAxleSpeed;
          wheelRLprop.Flags |= WheelShapeBehavior.OverrideAxleSpeed;

          wheelFRprop.InnerRadius = 0.7f * wheelDiameter / 2.0f;
          wheelFLprop.InnerRadius = 0.7f * wheelDiameter / 2.0f;
          wheelRRprop.InnerRadius = 0.7f * wheelDiameter / 2.0f;
          wheelRLprop.InnerRadius = 0.7f * wheelDiameter / 2.0f;

          wheelFRprop.LocalPose = new Pose(wheelFRPosition);
          wheelFLprop.LocalPose = new Pose(wheelFLPosition);
          wheelRRprop.LocalPose = new Pose(wheelRRPosition);
          wheelRLprop.LocalPose = new Pose(wheelRLPosition);
```

Next, the `WheelShapeProperties` for each wheel are specified. The `OverrideAxleSpeed` flag tells the physics engine not to calculate the axle speed based on motor torque and friction. Instead, you specify the axle speed each frame. This turns out to be a better way to simulate the types of motors that typically drive wheeled robots. The `LocalPose` of each shape is useful when an entity contains multiple shapes. The `LocalPose` specifies the position and orientation of each shape relative to the origin of the entity. In the preceding code, each pose is initialized with a position vector and the orientation part of the pose defaults to 0.

```
_wheelFR = new WheelEntity(wheelFRprop);
_wheelFR.State.Name = base.State.Name + " FrontRightWheel";
_wheelFR.Parent = this;
_wheelFR.Initialize(device, physicsEngine);

_wheelFL = new WheelEntity(wheelFLprop);
_wheelFL.State.Name = base.State.Name + " FrontLeftWheel";
_wheelFL.Parent = this;
_wheelFL.Initialize(device, physicsEngine);

_wheelRR = new WheelEntity(wheelRRprop);
_wheelRR.State.Name = base.State.Name + " RearRightWheel";
_wheelRR.Parent = this;
_wheelRR.Initialize(device, physicsEngine);

_wheelRL = new WheelEntity(wheelRLprop);
_wheelRL.State.Name = base.State.Name + " RearLeftWheel";
_wheelRL.Parent = this;
_wheelRL.Initialize(device, physicsEngine);
```

After the `WheelShapeProperties` have been initialized, you can create the `WheelEntities`. Each `WheelEntity` is given a name based on the parent entity name.

The parent reference of each `WheelEntity` is set to the `CorobotEntity`. The `WheelEntity` uses this reference in an unusual way. You can see this code in the `Initialize` method of the `WheelEntity` in `samples\entities\entities.cs`. Instead of calling `CreateAndInsertPhysicsEntity`, the `WheelEntity` calls `InsertShape` on its parent's `PhysicsEntity`. This adds the `WheelEntity` shape to the set of shapes that makes up the parent. As far as the physics engine is concerned, the wheel shapes are just part of the parent entity. This reduces the amount of computation required by the physics engine because it doesn't have to calculate the interactions between the wheels and the chassis and platform shapes. It assumes that they are rigidly joined.

You must explicitly call the `Initialize` method for each `WheelEntity` because these entities have not been inserted into the parent entity as children using the `InsertEntity` method.

---

### Child Entity or Not?

When you make an entity a child of another entity, and both entities have physics shapes in them, the child is joined to the parent with a joint. The properties of that joint are contained in the `ParentJoint` property of the child entity. (Joints are discussed in Chapter 7). The `Initialize` method of each child entity is called from the base class `Initialize` method. Likewise, the `Update` and `Render` methods are automatically called from the corresponding methods in the parent. In some cases, it is not desirable to have the child entity joined to the parent with a joint, such as in the case of the `WheelEntities` above. In this case, the `WheelEntities` are not added as children to the parent and `Initialize`, `Update`, and `Render` are explicitly called from the corresponding methods in the parent. Both methods of dealing with dependent entities are acceptable.

```
        base.Initialize(device, physicsEngine);
    }
    catch (Exception ex)
    {
        // clean up
        if (PhysicsEntity != null)
            PhysicsEngine.DeleteEntity(PhysicsEntity);
        HasBeenInitialized = false;
        InitError = ex.ToString();
    }
}
```

---

Finally, you call the `base.Inialize` method, which, among other things, loads any meshes or textures associated with the entity. If no mesh was specified in `State.Assets.Mesh`, simple meshes are constructed from the shapes in the entity.

Now that you have the `Initialize` method completed, only one thing remains to add the Corobot entity into the simulation environment. Add the following lines of code to the `Start` method just after the code inserting the giant box into the simulation environment:

```
// create a Corobot
SimulationEngine.GlobalInstancePort.Insert(
    new CorobotEntity("Corobot", new Vector3(0, 0, 0)));
```

This creates a new `CorobotEntity` with the name "Corobot" at the simulation origin. Compile and run the service. You should see something similar to what is shown in Figure 6-5.

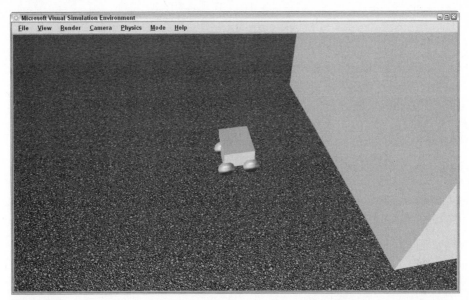

**Figure 6-5**

You can test your new entity by going into the Simulation Editor and selecting it by holding down the Ctrl key while pressing the right mouse button with the mouse pointer on the entity. You can examine the fields in the `EntityState` and each of the `WheelEntities` to verify they are correct. You can even press Ctrl+X and then Ctrl+V to cut and paste the entity back into the simulator to verify that serialization and deserialization of the entity work properly.

You still need to add some properties to the entity to make it compatible with the `DifferentialDriveEntity`. The `MotorTorqueScaling` property scales the speed of the motors to model the gear ratio on the physical robot. The `IsEnabled` property allows other services to enable or disable the drive. Add the following properties to the `CorobotEntity` class:

```
bool _isEnabled;
/// <summary>
/// True if drive mechanism is enabled
/// </summary>
[DataMember]
[Description("True if the drive mechanism is enabled.")]
public bool IsEnabled
{
    get { return _isEnabled; }
    set { _isEnabled = value; }
}

float _motorTorqueScaling;
/// <summary>
/// Scaling factor to apply to motor torque requests
/// </summary>
[DataMember]
```

*(continued)*

*(continued)*

```
[Description("Scaling factor to apply to motor torque requests.")]
public float MotorTorqueScaling
{
    get { return _motorTorqueScaling; }
    set { _motorTorqueScaling = value; }
}
```

Next, you initialize the `MotorTorqueScaling` in the nondefault constructor. The value set here is an arbitrary value. You'll reexamine this value later in the section "Tuning MotorTorqueScaling."

```
_motorTorqueScaling = 20f;
```

The `CurrentHeading` property simulates a compass sensor. It returns the heading of the entity based on `State.Pose`:

```
public float CurrentHeading
{
    get
    {
        // return the axis angle of the quaternion
        xna.Vector3 euler = UIMath.QuaternionToEuler(State.Pose.Orientation);
        // heading is the rotation about the Y axis.
        return xna.MathHelper.ToRadians(euler.Y);
    }
}
```

The next few methods enable explicit control of the speed of each motor by setting the left and right target velocities. If one of these methods is called, then any current `DriveDistance` or `RotateDegrees` commands are terminated with an error:

```
float _leftTargetVelocity;
float _rightTargetVelocity;
public void SetMotorTorque(float leftWheel, float rightWheel)
{
    ResetRotationAndDistance();
    SetAxleVelocity(
        leftWheel * _motorTorqueScaling,
        rightWheel * _motorTorqueScaling);
}

public void SetVelocity(float value)
{
    ResetRotationAndDistance();
    SetVelocity(value, value);
}

/// <summary>
/// Sets angular velocity on the wheels
/// </summary>
/// <param name="left"></param>
/// <param name="right"></param>
public void SetVelocity(float left, float right)
{
```

```
    ResetRotationAndDistance();
    if (_wheelFR == null || _wheelFL == null)
        return;

    left = ValidateWheelVelocity(left);
    right = ValidateWheelVelocity(right);

    // v is in m/sec - convert to an axle speed
    //  2Pi(V/2PiR) = V/R
    SetAxleVelocity(
        left / _wheelFR.Wheel.State.Radius,
        right / _wheelFL.Wheel.State.Radius);
}

private void SetAxleVelocity(float left, float right)
{
    _leftTargetVelocity = left;
    _rightTargetVelocity = right;
}

const float MAX_VELOCITY = 20.0f;
const float MIN_VELOCITY = -MAX_VELOCITY;

float ValidateWheelVelocity(float value)
{
    if (value > MAX_VELOCITY)
        return MAX_VELOCITY;
    if (value < MIN_VELOCITY)
        return MIN_VELOCITY;

    return value;
}
```

All that remains are the DriveDistance and RotateDegrees methods and their associated helper functions and variables:

```
Pose _startPoseForDriveDistance;
double _distanceToTravel;
SuccessFailurePort _driveDistancePort = null;

public void DriveDistance(
    float distance,
    float power,
    SuccessFailurePort responsePort)
{
    // reset drivedistance or rotatedegrees commands not yet completed
    ResetRotationAndDistance();

    // keep track of the response port for when we complete the request
    _driveDistancePort = responsePort;

    // handle negative distances
    if (distance < 0)
```

*(continued)*

*(continued)*

```
        {
            distance = -distance;
            power = -power;
        }
        _startPoseForDriveDistance = State.Pose;
        _distanceToTravel = distance;
        SetAxleVelocity(
            power * _motorTorqueScaling,
            power * _motorTorqueScaling);
    }

    // DriveDistance and RotateDegrees variables
    Queue<double> progressPoints = new Queue<double>();
    const int averageKernel = 6;
    const int decelerateThreshold = 6;
    const float twoPI = (float)(2 * Math.PI);

    // RotateDegrees variables
    double _targetRotation = double.MaxValue;
    double _currentRotation = 0;
    double _previousHeading = 0;
    const float acceptableRotationError = 0.005f;
    SuccessFailurePort _rotateDegreesPort = null;

    public void RotateDegrees(
        float degrees,
        float power,
        SuccessFailurePort responsePort)
    {
        // reset drivedistance or rotatedegrees commands not yet completed
        ResetRotationAndDistance();

        // keep track of the response port for when we complete the request
        _rotateDegreesPort = responsePort;

        _targetRotation = xna.MathHelper.ToRadians(degrees);
        _currentRotation = 0;
        _previousHeading = CurrentHeading;

        if (degrees < 0)
            SetAxleVelocity(
                power * _motorTorqueScaling,
                -power * _motorTorqueScaling);
        else
            SetAxleVelocity(
                -power * _motorTorqueScaling,
                power * _motorTorqueScaling);
    }

    void ResetRotationAndDistance()
    {
        progressPoints.Clear();
        _distanceToTravel = 0;
```

```
    _targetRotation = double.MaxValue;
    if (_driveDistancePort != null)
    {
        _driveDistancePort.Post(
            new Exception("Request superceded prior to completion."));
        _driveDistancePort = null;
    }
    if (_rotateDegreesPort != null)
    {
        _rotateDegreesPort.Post(
            new Exception("Request superceded prior to completion."));
        _rotateDegreesPort = null;
    }
}
```

## The Drive Methods

The DifferentialDriveEntity provides several methods that drive its two wheels. Your entity needs to implement these same methods with four wheels in mind. The DifferentialDriveEntity keeps track of a target velocity for each wheel. Each frame, the axle speed of each wheel is adjusted according to this target velocity. The Update method also adjusts the target speed of each wheel if a DriveDistance or RotateDegrees command is currently being executed. It attempts to slow the wheels as the target distance or heading approaches so that the target is not overshot.

Add the following Update method override to the CorobotEntity class:

```
const float SPEED_DELTA = 0.5f;
public override void Update(FrameUpdate update)
{
    // update state from the physics engine
    PhysicsEntity.UpdateState(true);
```

This call updates the entity state Pose and Velocity from the physics engine:

```
    if (_distanceToTravel > 0)
    {
        // DriveDistance update
        double currentDistance =
            Vector3.Length(State.Pose.Position -
                        _startPoseForDriveDistance.Position);
        if (currentDistance >= _distanceToTravel)
        {
            _wheelFR.Wheel.AxleSpeed = 0;
            _wheelFL.Wheel.AxleSpeed = 0;
            _wheelRR.Wheel.AxleSpeed = 0;
            _wheelRL.Wheel.AxleSpeed = 0;
            _leftTargetVelocity = 0;
            _rightTargetVelocity = 0;
            _distanceToTravel = 0;
            // now that we're finished, post a response
            if (_driveDistancePort != null)
```

*(continued)*

(continued)

```
            {
                SuccessFailurePort tmp = _driveDistancePort;
                _driveDistancePort = null;
                tmp.Post(new SuccessResult());
            }
        }
        else
        {
            // need to drive further, check if we should slow down
            if (progressPoints.Count >= averageKernel)
            {
                double distanceRemaining =
                    _distanceToTravel - currentDistance;
                double framesToCompletion =
                    distanceRemaining * averageKernel /
                    (currentDistance - progressPoints.Dequeue());
                if (framesToCompletion < decelerateThreshold)
                {
                    _leftTargetVelocity *= 0.5f;
                    _rightTargetVelocity *= 0.5f;
                    progressPoints.Clear();
                }
            }
            progressPoints.Enqueue(currentDistance);
        }
    }
```

The preceding code handles the behavior of the entity while a `DriveDistance` command is being executed. Suffice it to say that the code attempts to slow the entity as it approaches the distance goal so that it doesn't overshoot it. This is a better implementation than the one found in the `DifferentialDriveEntity` but there is still room for improvement:

```
else if (_targetRotation != double.MaxValue)
{
    // RotateDegrees update
    float currentHeading = CurrentHeading;
    double angleDelta = currentHeading - _previousHeading;
    while (angleDelta > Math.PI)
        angleDelta -= twoPI;
    while (angleDelta <= -Math.PI)
        angleDelta += twoPI;
    _currentRotation += angleDelta;
    _previousHeading = currentHeading;   // for next frame

    float angleError;
    if (_targetRotation < 0)
        angleError = (float)(_currentRotation - _targetRotation);
    else
        angleError = (float)(_targetRotation - _currentRotation);

    if (angleError < acceptableRotationError)
```

```
        {
            // current heading is within acceptableError or has overshot
            // end the rotation
            _targetRotation = double.MaxValue;
            _wheelFR.Wheel.AxleSpeed = 0;
            _wheelFL.Wheel.AxleSpeed = 0;
            _wheelRR.Wheel.AxleSpeed = 0;
            _wheelRL.Wheel.AxleSpeed = 0;
            _leftTargetVelocity = 0;
            _rightTargetVelocity = 0;
            // now that we're finished, post a response
            if (_rotateDegreesPort != null)
            {
                SuccessFailurePort tmp = _rotateDegreesPort;
                _rotateDegreesPort = null;
                tmp.Post(new SuccessResult());
            }
        }
        else
        {
            if (angleDelta != 0)
            {
                // need to turn more, check if we should slow down
                if (progressPoints.Count >= averageKernel)
                {
                    double framesToCompletion =
                        Math.Abs(angleError * averageKernel /
                                (_currentRotation - progressPoints.Dequeue()));
                    if (framesToCompletion < decelerateThreshold)
                    {
                        _leftTargetVelocity *= 0.5f;
                        _rightTargetVelocity *= 0.5f;
                        progressPoints.Clear();
                    }
                }
                progressPoints.Enqueue(_currentRotation);
            }
        }
    }
```

The following code handles the RotateDegrees command. Just like DriveDistance, it attempts to slow the rotation of the entity as the target heading approaches:

```
float left = _wheelFL.Wheel.AxleSpeed + _leftTargetVelocity;
float right = _wheelFR.Wheel.AxleSpeed + _rightTargetVelocity;

if (Math.Abs(left) > 0.1)
{
    if (left > 0)
        _wheelFL.Wheel.AxleSpeed -= SPEED_DELTA;
    else
        _wheelFL.Wheel.AxleSpeed += SPEED_DELTA;
}
```

(continued)

*(continued)*

```
if (Math.Abs(right) > 0.1)
{
    if (right > 0)
        _wheelFR.Wheel.AxleSpeed -= SPEED_DELTA;
    else
        _wheelFR.Wheel.AxleSpeed += SPEED_DELTA;
}
```

The `AxleSpeed` is the negative of the target velocity. When the two are nearly equal, left and right will be close to zero. If they are not nearly equal, then the axle speed is adjusted by `SPEED_DELTA`.

Here, the four-wheel-drive is implemented. The rear wheels are given axle speeds comparable to the front wheels:

```
// match the rear wheels with the front wheels
_wheelRL.Wheel.AxleSpeed = _wheelFL.Wheel.AxleSpeed;
_wheelRR.Wheel.AxleSpeed = _wheelFR.Wheel.AxleSpeed;
```

Finally, the `Update` method for the wheel entities is called, along with the base `Update` method:

```
// update entities in fields
_wheelFL.Update(update);
_wheelFR.Update(update);
_wheelRL.Update(update);
_wheelRR.Update(update);

// sim engine will update children
base.Update(update);
}
```

# The SimulatedQuadDifferentialDrive Service

Now you know how to create a custom simulation entity, but let's face it: it's pretty boring. It just sits there doing nothing. Wouldn't it be great to be able to drive it around in the simulation environment? That is the topic of this next section.

A SimulatedDifferentialDrive service is provided with the MRDS SDK. It can be controlled with the SimpleDashboard service and it drives two-wheeled robots that subclass the `DifferentialDriveEntity` class. However, you have a four-wheeled robot that doesn't use the `DifferentialDriveEntity` class, so you are going to build a custom service that supports the same generic drive contract, enabling you to use the SimpleDashboard to control it.

Just as you previously used `DssNewService.exe` to create the Corobot service, you now use it to create your SimulatedQuadDifferentialDrive service. However, there is a twist this time. You're going to tell `DssNewService` that you want the new service to support the generic drive contract as an alternate contract. This means that your service will support two different ports, each identified with a different contract. The alternate port will look just like the port that the SimulatedDifferentialDrive service supports.

Go back to the `MyChapter6` directory and use the following command line (shown in bold below) to create the SimulatedQuadDifferentialDrive service. Each command-line option to `dssnewservice` is shown on a separate line due to word wrap, but you should type them all on one line. The bold code indicates text that you should type:

```
C:\Microsoft Robotics Studio (1.5)\ProMRDS\MyChapter6>dssnewservice
/s:SimulatedQuadDifferentialDrive
/i:"\Microsoft Robotics Studio (1.5)\bin\RoboticsCommon.dll"
/alt:"http://schemas.microsoft.com/robotics/2006/05/drive.html"
/Namespace:"ProMRDS.Simulation.QuadDifferentialDrive"
/year:"2007" /month:"07"
```

The `/s` parameter names the new service. The `/alt` parameter specifies the contract that you want to implement as an alternate contract, and the `/i` parameter specifies where that contract is implemented. You use the `/year` and `/month` parameters to ensure that the contract for this service will be the same as the SimulatedQuadDifferentialDrive service in the `Chapter6` directory.

Take a moment to look at the service code that has been generated. DssNewService has generated a service call SimulatedQuadDifferentialDrive service that supports an alternate contract identified by `"http://schemas.microsoft.com/robotics/2006/05/drive.html"`. This is the generic drive contract you specify on the command line. It has defined a `_mainPort` of type `pxdrive.DriveOperations` and a service state of type `pxdrive.DriveDifferentialTwoWheelState`. You're going to shuffle things a bit because you want to support two ports: an alternate port that supports the generic drive contract, and the main port that supports the new SimulatedQuadDifferentialDrive contract.

Begin modifying `SimulatedQuadDifferentialDrive.cs` by replacing the `using` statements at the top with the following to prepare you to access the simulator and the `Corobot` entity you just defined:

```
using Microsoft.Ccr.Core;
using Microsoft.Dss.Core;
using Microsoft.Dss.Core.Attributes;
using Microsoft.Dss.ServiceModel.Dssp;
using Microsoft.Dss.ServiceModel.DsspServiceBase;
using Microsoft.Dss.Services.SubscriptionManager;
using System;
using W3C.Soap;
using System.Collections.Generic;
using dssphttp = Microsoft.Dss.Core.DsspHttp;
using pxdrive = Microsoft.Robotics.Services.Drive.Proxy;
using xml = System.Xml;
using xna = Microsoft.Xna.Framework;
using submgr = Microsoft.Dss.Services.SubscriptionManager;
using simtypes = Microsoft.Robotics.Simulation;
using simengine = Microsoft.Robotics.Simulation.Engine;
using physics = Microsoft.Robotics.Simulation.Physics;
using corobot = ProMRDS.Simulation.Corobot;
using Microsoft.Robotics.PhysicalModel;
```

The new project will already have a reference to `RoboticsCommon.Proxy.dll` because it is referencing the generic drive contract from that DLL. You'll also need to add references to the following DLLs.

Don't forget to edit the properties of each DLL you add to set `Copy Local` and `Specific Version` to False:

```
Corobot.Y2007.M07.dll
Microsoft.Xna.Framework.dll
PhysicsEngine.dll
RoboticsCommon.dll
SimulationCommon.dll
SimulationEngine.dll
```

You may be wondering why you need a reference to `RoboticsCommon.dll` when the project already has a reference to `RoboticsCommon.proxy.dll`. The definition of the generic drive contract is drawn from the proxy DLL, and the common types such as `Vector3` are drawn from `RoboticsCommon.dll`.

Now you add a port to receive the generic drive commands:

```
// Port for receiving generic differential drive commands
[AlternateServicePort(
    AllowMultipleInstances = true,
    AlternateContract = pxdrive.Contract.Identifier)]
private pxdrive.DriveOperations _diffDrivePort = new
    Microsoft.Robotics.Services.Drive.Proxy.DriveOperations();
```

Change the definition of `_mainPort` to support the extended quad drive operations:

```
/// <summary>
/// Main service port for quad drive commands
/// </summary>
[ServicePort("/simulatedquaddifferentialdrive",
    AllowMultipleInstances = true)]
private QuadDriveOperations _mainPort = new QuadDriveOperations();
```

You also need to extend the state for this service because you need to keep track of four wheels now, rather than two. Replace the `_state` declaration with the following:

```
[InitialStatePartner(
    Optional = true,
    ServiceUri = "SimulatedQuadDifferentialDriveService.Config.xml")]
private DriveDifferentialFourWheelState _state =
    new DriveDifferentialFourWheelState();
```

You'll also add the following definition of this class in `SimulatedQuadDifferentialDriveTypes.cs` just after the `Contract` class definition:

```
public class DriveDifferentialFourWheelState : pxdrive.
DriveDifferentialTwoWheelState
{
    private pxmotor.WheeledMotorState _rearLeftWheel;
    private pxmotor.WheeledMotorState _rearRightWheel;
    private Vector3 _position;

    [Description("The rear left wheel's state.")]
    [DataMember]
```

```
            public pxmotor.WheeledMotorState RearLeftWheel
            {
                get { return _rearLeftWheel; }
                set { _rearLeftWheel = value; }
            }
            [DataMember]
            [Description("The rear right wheel's state.")]
            public pxmotor.WheeledMotorState RearRightWheel
            {
                get { return _rearRightWheel; }
                set { _rearRightWheel = value; }
            }
            [DataMember]
            [Description("The current position of the entity.")]
            public Vector3 Position
            {
                get { return _position; }
                set { _position = value; }
            }
        }
```

DriveDifferentialFourWheelState inherits from pxdrive.DriveDifferentialTwoWheelState, so all you need to do is add a definition for the rear wheels. You'll also add the current position of the entity, which you'll use as a crude way to simulate a GPS system later.

As long as you're modifying SimulatedQuadDifferentialDriveTypes.cs, you may as well add the definition for QuadDriveOperations, which are the operations supported by the main port. It will support DsspDefaultLookup, DsspDefaultDrop, HttpGet, Get, and a new operation: SetPose. Add this code for the operations after the state definition:

```
/// <summary>
/// QuadDrive Operations Port
/// </summary>
[ServicePort]
public class QuadDriveOperations :
    PortSet<DsspDefaultLookup,DsspDefaultDrop,HttpGet,Get,SetPose>
{
}

/// <summary>
/// Operation Retrieve Drive State
/// </summary>
[Description("Gets the drive's current state.")]
public class Get :
    Get<GetRequestType, PortSet<DriveDifferentialFourWheelState, Fault>>
{
}
/// <summary>
/// Operation Set Entity Pose
/// </summary>
[Description("Sets the pose of the quadDifferentialDrive entity.")]
public class SetPose :
    Update<SetPoseRequestType, PortSet<DefaultUpdateResponseType, Fault>>
```

*(continued)*

*(continued)*

```
    {
    }

    /// <summary>
    /// Set entity pose request
    /// </summary>
    [DataMemberConstructor]
    [DataContract]
    public class SetPoseRequestType
    {
        Pose _entityPose;

        [DataMember]
        public Pose EntityPose
        {
            get { return _entityPose; }
            set { _entityPose = value; }
        }

        public SetPoseRequestType()
        {
        }
    }
```

Notice that you've defined a new operation, SetPose, which is an Update operation. It has a SetPoseRequestType class as its body, which in turn contains an EntityPose. Before you leave the file, replace the using statements at the top with the following:

```
using Microsoft.Ccr.Core;
using Microsoft.Dss.Core.Attributes;
using Microsoft.Dss.ServiceModel.Dssp;
using Microsoft.Dss.Core.DsspHttp;
using System;
using System.Collections.Generic;
using System.ComponentModel;
using W3C.Soap;
using pxdrive = Microsoft.Robotics.Services.Drive.Proxy;
using pxmotor = Microsoft.Robotics.Services.Motor.Proxy;
using Microsoft.Robotics.PhysicalModel;
```

To summarize, you've changed the _mainPort to be a QuadDriveOperations port and added another port called _diffDrivePort, which supports the generic drive operations. You also changed the service state to DriveDifferentialFourWheelState, which includes all of the generic drive state as well as two additional wheels and a Pose.

## Simulation Entity Notifications

All simulation services have one thing in common: They need to manipulate or read data from entities in the simulation environment. This is easy if the service created and inserted the entity, as was the case with the Corobot service. The SimulatedQuadDifferentialDrive service needs to interact with the Corobot entity in the simulation environment, so it needs to request a notification from the

SimulationEngine service when the entity it needs is inserted into the environment. It does this by sending a subscribe message to the SimulationEngine service, which includes the entity name and a port to receive the notification. To support this, add the following variables at the top of the `SimulatedQuadDifferentialDriveService` class definition in `SimulatedQuadDifferentialDrive.cs`:

```
#region Simulation Variables
corobot.CorobotEntity _entity;
simengine.SimulationEnginePort _notificationTarget;
#endregion
```

`_entity` will eventually hold a reference to the `Corobot` entity in the simulation environment, and `_notificationTarget` is the port that will receive the notification.

The subscribe message is sent to the simulator in the `Start` method even before the service calls `base.Start` to insert itself into the service directory:

```
protected override void Start()
{
    if (_state == null)
        CreateDefaultState();

    _notificationTarget = new simengine.SimulationEnginePort();

    // PartnerType.Service is the entity instance name.
    simengine.SimulationEngine.GlobalInstancePort.Subscribe(
        ServiceInfo.PartnerList, _notificationTarget);

    // don't start listening to DSSP operations, other than drop,
    // until notification of entity
    Activate(new Interleave(
        new TeardownReceiverGroup
        (
            Arbiter.Receive<simengine.InsertSimulationEntity>(
                false,
                _notificationTarget,
                InsertEntityNotificationHandlerFirstTime),
            Arbiter.Receive<DsspDefaultDrop>(
                false,
                _mainPort,
                DefaultDropHandler),
            Arbiter.Receive<DsspDefaultDrop>(
                false,
                _diffDrivePort,
                DefaultDropHandler)
        ),
        new ExclusiveReceiverGroup(),
        new ConcurrentReceiverGroup()
    ));
}
```

As a convenience, the Subscribe method on the simulation engine port takes a PartnerList as a parameter. The name of the entity is contained in the partner list because it is specified as a partner to the SimulatedQuadDifferentialDrive service in the manifest. Update the manifest line at the top of the Corobot manifest to create a simcommon namespace to save typing:

```
<Manifest
    xmlns="http://schemas.microsoft.com/xw/2004/10/manifest.html"
    xmlns:dssp="http://schemas.microsoft.com/xw/2004/10/dssp.html"
    xmlns:simcommon="http://schemas.microsoft.com/robotics/2006/04/simulation.html"
    >
```

Add the following lines to the Corobot manifest to start the drive service and to specify the entity named "Corobot" as a partner. This ServiceRecordType is inserted immediately after the </ServiceRecordType> line that ends the Corobot service record. ProMRDS\config\Corobot .manifest.xml provides an example.

```
<ServiceRecordType>

<dssp:Contract>http://schemas.tempuri.org/2007/07/simulatedquaddifferentialdrive
.html</dssp:Contract>
    <dssp:PartnerList>
        <dssp:Partner>
            <!--The partner name must match the entity name-->
            <dssp:Service>http://localhost/Corobot</dssp:Service>
            <dssp:Name>simcommon:Entity</dssp:Name>
        </dssp:Partner>
    </dssp:PartnerList>
</ServiceRecordType>
```

The entity name is in the form of a URI, so an entity name of "Corobot" becomes "http://localhost/ Corobot".

When a partner is defined this way in the manifest, it shows up in the PartnerList and the simulation engine will parse the passed PartnerList until it finds an entity name. When it receives the subscribe request, it scans through all of the entities in the environment. If that entity already exists in the environment, then the simulation engine immediately sends a notification to the _notificationTarget port with a reference to the named entity. If the entity does not exist in the environment, then the simulation engine waits until it is inserted before sending the notification. If the entity is never inserted, then no notification is ever sent.

In the Start method, a new Interleave is activated, which activates handlers for the DsspDefaultDrop message on each port and a handler for an InsertSimulationEntity handler on the _notificationTarget port. No other messages are processed and the service won't even show up in the service directory until it receives a notification from the simulation engine.

The next step is to add the InsertEntityNotificationHandlerFirstTime method, which handles the InsertSimulationEntity notification:

```
void InsertEntityNotificationHandlerFirstTime(
    simengine.InsertSimulationEntity ins)
```

```
{
    // insert ourselves into the directory
    base.Start();

    InsertEntityNotificationHandler(ins);
}
```

When this handler is called, the service finally lets the rest of the world know that it exists by calling base.Start and then calling InsertEntityNotificationhandler. This is the method that will handle subsequent insert notifications because base.Start only needs to be called the first time a notification arrives.

Add the following code for the InsertEntityNotificationHandler method:

```
void InsertEntityNotificationHandler(
    simengine.InsertSimulationEntity ins)
{
    _entity = (corobot.CorobotEntity)ins.Body;
    _entity.ServiceContract = Contract.Identifier;
```

You store a reference to the entity in _entity so that it can be used by the other handlers. You also set the ServiceContract property on the entity to be the contract of your service to indicate that it is being controlled by this service.

```
    // create default state based on the physics entity
    xna.Vector3 separation =
        _entity.FrontLeftWheel.Position - _entity.FrontRightWheel.Position;
    _state.DistanceBetweenWheels = separation.Length();

    _state.LeftWheel.MotorState.PowerScalingFactor = _entity.MotorTorqueScaling;
    _state.RightWheel.MotorState.PowerScalingFactor = _entity.MotorTorqueScaling;
```

Your service state is updated based on the properties of the entity. These values don't change over time so they are copied to the state only at this initialization time.

```
    // enable other handlers now that we are connected
    Activate(new Interleave(
        new TeardownReceiverGroup
        (
            Arbiter.Receive<DsspDefaultDrop>(false, _mainPort, DefaultDropHandler),
            Arbiter.Receive<DsspDefaultDrop>(
                false, _diffDrivePort, DefaultDropHandler)
        ),
        new ExclusiveReceiverGroup
        (
            Arbiter.Receive<SetPose>(true, _mainPort, SetPoseHandler),
            Arbiter.ReceiveWithIterator<pxdrive.DriveDistance>(
                true, _diffDrivePort, DriveDistanceHandler),
            Arbiter.ReceiveWithIterator<pxdrive.RotateDegrees>(
                true, _diffDrivePort, RotateHandler),
            Arbiter.ReceiveWithIterator<pxdrive.SetDrivePower>(
                true, _diffDrivePort, SetPowerHandler),
```

*(continued)*

*(continued)*

```
                    Arbiter.ReceiveWithIterator<pxdrive.SetDriveSpeed>(
                        true, _diffDrivePort, SetSpeedHandler),
                    Arbiter.ReceiveWithIterator<pxdrive.AllStop>(
                        true, _diffDrivePort, AllStopHandler),
                    Arbiter.Receive<simengine.InsertSimulationEntity>(
                        true, _notificationTarget, InsertEntityNotificationHandler),
                    Arbiter.Receive<simengine.DeleteSimulationEntity>(
                        true, _notificationTarget, DeleteEntityNotificationHandler)
                ),
            new ConcurrentReceiverGroup
                (
                    Arbiter.ReceiveWithIterator<dssphttp.HttpGet>(
                        true, _mainPort, MainPortHttpGetHandler),
                    Arbiter.ReceiveWithIterator<Get>(true, _mainPort, MainPortGetHandler),
                    Arbiter.ReceiveWithIterator<dssphttp.HttpGet>(
                        true, _diffDrivePort, HttpGetHandler),
                    Arbiter.ReceiveWithIterator<pxdrive.Get>(
                        true, _diffDrivePort, GetHandler),
                    Arbiter.ReceiveWithIterator<pxdrive.Subscribe>(
                        true, _diffDrivePort, SubscribeHandler),
                    Arbiter.ReceiveWithIterator<pxdrive.ReliableSubscribe>(
                        true, _diffDrivePort, ReliableSubscribeHandler),
                    Arbiter.ReceiveWithIterator<pxdrive.EnableDrive>(
                        true, _diffDrivePort, EnableHandler)
                )
        ));
    }
```

Finally, all of the handlers for _mainPort and _diffDrivePort are activated. Any messages queued up on these ports can now be processed because you have connected to the entity in the simulator. The handlers are also activated for subsequent DeleteSimulationEntity and InsertSimulationEntity notifications. This enables the service to handle the case where entities are deleted from the environment and added again.

DeleteSimulationEntity sets the global _entity reference to null and then disables all handlers except Drop and InsertEntityNotification:

```
    void DeleteEntityNotificationHandler(simengine.DeleteSimulationEntity del)
    {
        _entity = null;

        // disable other handlers now that we are no longer connected to the entity
        Activate(new Interleave(
            new TeardownReceiverGroup
                (
                    Arbiter.Receive<simengine.InsertSimulationEntity>(
                        false, _notificationTarget,
                        InsertEntityNotificationHandlerFirstTime),
                    Arbiter.Receive<DsspDefaultDrop>(false, _mainPort, DefaultDropHandler),
                    Arbiter.Receive<DsspDefaultDrop>(
                        false, _diffDrivePort, DefaultDropHandler)
```

```
            ),
            new ExclusiveReceiverGroup(),
            new ConcurrentReceiverGroup()
        ));
    }
```

Before you add the handler code, add the following methods to initialize and update the state. CreateDefaultState is called from the Start method if no configuration file is present. UpdateStateFromSimulation is called from handlers to update the state before it is broadcast:

```
    void CreateDefaultState()
    {
        _state = new DriveDifferentialFourWheelState();
        _state.LeftWheel =
            new Microsoft.Robotics.Services.Motor.Proxy.WheeledMotorState();
        _state.RightWheel =
            new Microsoft.Robotics.Services.Motor.Proxy.WheeledMotorState();
        _state.LeftWheel.MotorState =
            new Microsoft.Robotics.Services.Motor.Proxy.MotorState();
        _state.RightWheel.MotorState =
            new Microsoft.Robotics.Services.Motor.Proxy.MotorState();
        _state.LeftWheel.EncoderState =
            new Microsoft.Robotics.Services.Encoder.Proxy.EncoderState();
        _state.RightWheel.EncoderState =
            new Microsoft.Robotics.Services.Encoder.Proxy.EncoderState();
        _state.RearLeftWheel =
            new Microsoft.Robotics.Services.Motor.Proxy.WheeledMotorState();
        _state.RearRightWheel =
            new Microsoft.Robotics.Services.Motor.Proxy.WheeledMotorState();
        _state.RearLeftWheel.MotorState =
            new Microsoft.Robotics.Services.Motor.Proxy.MotorState();
        _state.RearRightWheel.MotorState =
            new Microsoft.Robotics.Services.Motor.Proxy.MotorState();
    }

    void UpdateStateFromSimulation()
    {
        if (_entity != null)
        {
            _state.TimeStamp = DateTime.Now;
            _state.LeftWheel.MotorState.CurrentPower =
                _entity.FrontLeftWheel.Wheel.MotorTorque;
            _state.RightWheel.MotorState.CurrentPower =
                _entity.FrontRightWheel.Wheel.MotorTorque;
            _state.RearLeftWheel.MotorState.CurrentPower =
                _entity.RearLeftWheel.Wheel.MotorTorque;
            _state.RearRightWheel.MotorState.CurrentPower =
                _entity.RearRightWheel.Wheel.MotorTorque;
            _state.Position = _entity.State.Pose.Position;
        }
    }
```

All that is left now is to implement the message handlers for each port. The following handlers are methods on the SimulatedQuadDifferentialDrive service:

```
public IEnumerator<ITask> SubscribeHandler(
    pxdrive.Subscribe subscribe)
{
    Activate(Arbiter.Choice(
        SubscribeHelper(
            _subMgrPort,
            subscribe.Body,
            subscribe.ResponsePort),
        delegate(SuccessResult success)
        {
            _subMgrPort.Post(new submgr.Submit(
                subscribe.Body.Subscriber,
                DsspActions.UpdateRequest, _state, null));
        },
        delegate(Exception ex) { LogError(ex); }
    ));

    yield break;
}

public IEnumerator<ITask> ReliableSubscribeHandler(
    pxdrive.ReliableSubscribe subscribe)
{
    Activate(Arbiter.Choice(
        SubscribeHelper(
            _subMgrPort,
            subscribe.Body,
            subscribe.ResponsePort),
        delegate(SuccessResult success)
        {
            _subMgrPort.Post(new submgr.Submit(
                subscribe.Body.Subscriber,
                DsspActions.UpdateRequest, _state, null));
        },
        delegate(Exception ex) { LogError(ex); }
    ));
    yield break;
}
```

These two handlers use the SubscriptionManager to handle subscribe requests from other services. This is described in more detail in the MRDS documentation in Service Tutorial 4.

```
public IEnumerator<ITask> MainPortHttpGetHandler(dssphttp.HttpGet get)
{
    UpdateStateFromSimulation();
    get.ResponsePort.Post(new dssphttp.HttpResponseType(_state));
    yield break;
}

public IEnumerator<ITask> MainPortGetHandler(Get get)
{
    UpdateStateFromSimulation();
```

```
        get.ResponsePort.Post(_state);
        yield break;
    }
```

These two handlers support `Get` and `HttpGet` requests on the main port. The state is updated with the most current information from the `Corobot` entity and then it is posted as a response to the request.

The following handler supports the `SetPost` update request:

```
public void SetPoseHandler(SetPose setPose)
{
    if (_entity == null)
        throw new InvalidOperationException(
            "Simulation entity not registered with service");

    Task<corobot.CorobotEntity, Pose> task = new
        Task<corobot.CorobotEntity,Pose>(
            _entity, setPose.Body.EntityPose, SetPoseDeferred);

    _entity.DeferredTaskQueue.Post(task);
}

void SetPoseDeferred(corobot.CorobotEntity entity, Pose pose)
{
    entity.PhysicsEntity.SetPose(pose);
}
```

If you are connected to an entity, it creates a task that sets the pose of that entity to the pose specified in the request. The task is added to the `DeferredTaskQueue` on the entity so that it executes during the `Update` method when the physics engine is not running.

The next two handlers respond to `Get` and `HttpGet` requests on the `_diffDrive` port:

```
public IEnumerator<ITask> HttpGetHandler(dssphttp.HttpGet get)
{
    UpdateStateFromSimulation();
    pxdrive.DriveDifferentialTwoWheelState _twoWheelState =
        (pxdrive.DriveDifferentialTwoWheelState)
        ((pxdrive.DriveDifferentialTwoWheelState)_state).Clone();
    get.ResponsePort.Post(new dssphttp.HttpResponseType(
        _twoWheelState));
    yield break;
}

public IEnumerator<ITask> GetHandler(pxdrive.Get get)
{
    UpdateStateFromSimulation();

  pxdrive.DriveDifferentialTwoWheelState _twoWheelState =
        (pxdrive.DriveDifferentialTwoWheelState)
        ((pxdrive.DriveDifferentialTwoWheelState)_state).Clone();
    get.ResponsePort.Post(_twoWheelState);
    yield break;
}
```

It is not valid to return the service state in response to a Get request on the _diffDrive port. It must first be converted to a pxdrive.DriveDifferentialTwoWheelState class by casting it to that type and then calling the Clone method to make a copy. This ensures that services which send a Get message to this port will receive the state they are expecting.

```
public IEnumerator<ITask> DriveDistanceHandler(pxdrive.DriveDistance driveDistance)
{
    SuccessFailurePort entityResponse = new SuccessFailurePort();
    _entity.DriveDistance(
        (float)driveDistance.Body.Distance,
        (float)driveDistance.Body.Power,
        entityResponse);

    yield return Arbiter.Choice(entityResponse,
        delegate(SuccessResult s)
        {
            driveDistance.ResponsePort.Post(DefaultUpdateResponseType.Instance);
        },
        delegate(Exception e)
        {
            driveDistance.ResponsePort.Post(new W3C.Soap.Fault());
        });

    yield break;
}

public IEnumerator<ITask> RotateHandler(pxdrive.RotateDegrees rotate)
{
    SuccessFailurePort entityResponse = new SuccessFailurePort();
    _entity.RotateDegrees(
        (float)rotate.Body.Degrees,
        (float)rotate.Body.Power,
        entityResponse);

    yield return Arbiter.Choice(entityResponse,
        delegate(SuccessResult s)
        {
            rotate.ResponsePort.Post(DefaultUpdateResponseType.Instance);
        },
        delegate(Exception e)
        {
            rotate.ResponsePort.Post(new W3C.Soap.Fault());
        });

    yield break;
}
```

Both of these handlers forward the request to the entity by calling either DriveDistance or RotateDegrees. They wait for the asynchronous response from the entity before posting their own response.

```
public IEnumerator<ITask> SetPowerHandler(pxdrive.SetDrivePower setPower)
{
    if (_entity == null)
        throw new InvalidOperationException(
            "Simulation entity not registered with service");

    // Call simulation entity method for setting wheel torque
    _entity.SetMotorTorque(
        (float)(setPower.Body.LeftWheelPower),
        (float)(setPower.Body.RightWheelPower));

    UpdateStateFromSimulation();
    setPower.ResponsePort.Post(DefaultUpdateResponseType.Instance);

    // send update notification for entire state
    _subMgrPort.Post(new submgr.Submit(_state, DsspActions.UpdateRequest));
    yield break;
}

public IEnumerator<ITask> SetSpeedHandler(pxdrive.SetDriveSpeed setSpeed)
{
    if (_entity == null)
        throw new InvalidOperationException(
            "Simulation entity not registered with service");

    _entity.SetVelocity(
        (float)setSpeed.Body.LeftWheelSpeed,
        (float)setSpeed.Body.RightWheelSpeed);

    UpdateStateFromSimulation();
    setSpeed.ResponsePort.Post(DefaultUpdateResponseType.Instance);

    // send update notification for entire state
    _subMgrPort.Post(new submgr.Submit(_state, DsspActions.UpdateRequest));
    yield break;
}
```

These two handlers change the axle speed of the wheels, and send a notification to any services that have subscribed that the service state has changed.

The following handler sets the IsEnabled property on the entity according to the request and sends a notification to subscribers:

```
public IEnumerator<ITask> EnableHandler(pxdrive.EnableDrive enable)
{
    if (_entity == null)
        throw new InvalidOperationException(
            "Simulation entity not registered with service");

    _state.IsEnabled = enable.Body.Enable;
    _entity.IsEnabled = _state.IsEnabled;

    UpdateStateFromSimulation();
```

*(continued)*

*(continued)*

```
        enable.ResponsePort.Post(DefaultUpdateResponseType.Instance);

        // send update for entire state
        _subMgrPort.Post(new submgr.Submit(_state, DsspActions.UpdateRequest));
        yield break;
    }
```

The `AllStopHandler` sets the axle speed of each wheel to 0 and sends a notification to subscribers:

```
    public IEnumerator<ITask> AllStopHandler(pxdrive.AllStop estop)
    {
        if (_entity == null)
            throw new InvalidOperationException(
                "Simulation entity not registered with service");

        _entity.SetMotorTorque(0, 0);
        _entity.SetVelocity(0);

        UpdateStateFromSimulation();
        estop.ResponsePort.Post(DefaultUpdateResponseType.Instance);

        // send update for entire state
        _subMgrPort.Post(new submgr.Submit(_state, DsspActions.UpdateRequest));
        yield break;
    }
```

## Testing the SimulatedQuadDifferentialDrive Service

The service is now at a point where you can use it to drive around the simulation environment. Modify the `Corobot.manifest.xml` file to start the SimpleDashboard service by adding the following `ServiceRecord`:

```
    <!-- Start the Dashboard service -->
    <ServiceRecordType>
        <dssp:Contract>
            http://schemas.microsoft.com/robotics/2006/01/simpledashboard.html
        </dssp:Contract>
    </ServiceRecordType>
```

When you run this manifest, it should start three separate services: the Corobot service, the SimulatedQuadDifferentialDrive service, and the SimpleDashboard service. Once the simulator starts up and the Corobot service inserts a `Corobot` entity in the environment, the SimulatedQuadDifferentialDrive service should make itself visible in the service directory.

Type **localhost** into the Machine textbox and press Enter. A service should be displayed in the service list, as shown in Figure 6-6.

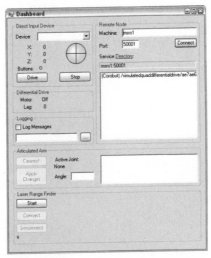

**Figure 6-6**

The SimpleDashboard service recognizes the SimulatedQuadDifferentialDrive service because it implements the generic `Drive` contract as an alternate port. Double-click the service and click the Drive button. You should now be able to drive the `Corobot` entity around the environment by dragging the trackball icon forward and backward. (The trackball icon is the circle with the crossbar in it.) If you have trouble compiling either of the services or driving the `Corobot` entity, compare the behavior of your service with the corresponding service in the `Chapter6` directory to find the problem.

You should also verify that the proper state is returned in response to an `HttpGet` request to either port on the service. Run a web browser and navigate to `http://localhost:50000`. Select Service Directory from the left column. You should see something similar to Figure 6-7, which shows the Corobot and SimulatedQuadDifferentialDrive services.

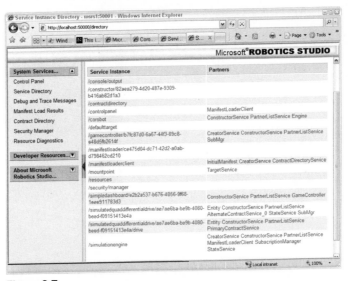

**Figure 6-7**

311

Click the first /simulatedquaddifferentialdrive service listed and verify that DriveDifferentialFourWheelState is displayed. Click the second /simulatedquaddifferentialdrive service listed and verify that DriveDifferentialTwoWheelState is displayed.

## Testing DriveDistance and RotateDegrees

The two most complex methods on the Corobot entity are DriveDistance and RotateDegrees, so it is appropriate to expend some effort in testing them. A simple way to do this is to add a short test routine to the SimulatedQuadDifferentialDrive service, which runs after the notification from the simulation engine. This test routine is normally commented out.

Add this line as the last line of InsertEntityNotificationHandlerFirstTime:

```
SpawnIterator(TestDriveDistanceAndRotateDegrees);
```

Now add the following method to the SimulatedQuadDifferentialDriveService class:

```
// Test the DriveDistance and RotateDegrees messages
public IEnumerator<ITask> TestDriveDistanceAndRotateDegrees()
{
    Random rnd = new Random();
    bool success = true;

    // drive in circles
    while (success)
    {
        double distance = rnd.NextDouble() * 1 + 0.5;
        double angle = rnd.NextDouble() * 90 - 45;

        // first leg
        yield return Arbiter.Choice(
            _diffDrivePort.RotateDegrees(angle, 0.2),
            delegate(DefaultUpdateResponseType response) { },
            delegate(Fault f) { success = false; }
        );

        yield return Arbiter.Choice(
            _diffDrivePort.DriveDistance(distance, 0.2),
            delegate(DefaultUpdateResponseType response) { },
            delegate(Fault f) { success = false; }
        );

        // return
        yield return Arbiter.Choice(
            _diffDrivePort.RotateDegrees(180, 0.2),
            delegate(DefaultUpdateResponseType response) { },
            delegate(Fault f) { success = false; }
        );

        yield return Arbiter.Choice(
            _diffDrivePort.DriveDistance(distance, 0.2),
            delegate(DefaultUpdateResponseType response) { },
            delegate(Fault f) { success = false; }
        );
```

```
            // reset position
            yield return Arbiter.Choice(
                _diffDrivePort.RotateDegrees(180-angle, 0.2),
                delegate(DefaultUpdateResponseType response) { },
                delegate(Fault f) { success = false; }
            );
        }
    }
```

This test method sends `DriveDistance` and `RotateDegrees` messages to the `_diffDrivePort` to drive the Corobot a random distance in a random direction. It then turns around and returns to its starting point. Over time, the errors in rotation and driving distance accumulate and the robot will fail to return to its exact starting point. This code is a good example of how the Corobot can be driven from another service. You'll see more examples of this later.

## Tuning the Corobot Entity

Once you have a basic robot entity defined, it is a good idea to tune its properties so that it more closely resembles its real-world counterpart. In the following sections, you'll adjust the top speed of the `Corobot` entity, the grip of its tires, and its appearance. You'll also adjust the model so that its wheels actually turn.

### Tuning MotorTorqueScaling

The `MotorTorqueScaling` property of the entity is a multiplier on the axle speed. When the entity is moved by calling `SetMotorTorque`, the torque that is passed as a parameter is scaled by `MotorTorqueScaling` to set the axle speed. A torque value of 1.0 represents the greatest torque that the motor can apply to the axle. Therefore, `MotorTorqueScaling` determines the speed of the entity when its torque is at the maximum. The CoroWare engineers indicate that the top speed of the Corobot is somewhere around two feet per second. In the initial implementation of the entity, a value of 20 was arbitrarily chosen to initialize `MotorTorqueScaling`. Now it is time to determine what the actual value should be.

You need to cause the entity to move forward at its maximum speed and somehow measure that speed to determine whether it is close to the rated top speed of two feet per second. The first step is to add the following line as the last line of the `Corobot` entity's `Initialize` method:

```
SetMotorTorque(1, 1);
```

This will cause the Corobot to move forward at its maximum speed. Change the first few lines of the `Update` method as follows to measure the speed of the entity:

```
const float SPEED_DELTA = 0.5f;
Pose startPose;
double totalTime = -1;
public override void Update(FrameUpdate update)
{
    if (totalTime < 0)
    {
        startPose = State.Pose;
```

*(continued)*

**313**

(continued)

```
            totalTime = update.ElapsedTime;
    }
    else
    {
        double distance = Vector3.Length(
            State.Pose.Position - startPose.Position);
        double speed = distance / totalTime;
        totalTime += update.ElapsedTime;
        Console.WriteLine(speed.ToString());
    }
```

The first time `Update` is called, the pose of the entity is stored. On each subsequent call to `Update`, the total distance traveled is calculated, along with the accumulated time since the first call. The speed is calculated and displayed. The value displayed on the console will eventually converge to the actual speed of the entity.

An initial run with this code yielded a speed of approximately 1.2 meters per second. Two feet per second is equal to 0.61 meters per second, so the entity's top speed with a `MotorTorqueScaling` value of 20 is about twice as fast as it should be. The new `MotorTorqueScaling` value can be calculated by multiplying the initial value (20) by the ratio of the desired top speed to the current top speed:

```
NewMotorTorqueScaling = InitialValue * TopSpeedDesired / TopSpeedMeasured
```

The proper `MotorTorqueScaling` factor should be about 10.16. When that value is updated in the `CorobotEntity` constructor and the test is executed again, the top speed comes very close to 0.61 meters per second. After you have determined the appropriate value for `MotorTorqueScaling`, you can comment out this code so that it doesn't interfere with normal operation of the entity.

## Tuning the Tire Friction

After driving your Corobot around in the simulation environment, you may notice that the tires grip the ground very well. If you drive up to the giant box, the tires grip the ground and the box so well that the Corobot flips on its back. Furthermore, when the Corobot is turning, it shakes and jitters because the tires grip the ground very tightly and slip very little. This is different behavior than what is observed on the actual Corobot, so you need to adjust the tire friction.

As discussed in Chapter 5, the simulation environment allows each shape to have a material definition describing its bounciness (restitution) and its static and dynamic friction. The wheel shapes have a more advanced friction model that enables the friction along the longitudinal direction (in the direction of rotation) to be specified separately from the lateral direction (perpendicular to the direction of rotation). The friction model utilizes a spline function to determine the amount of wheel slippage as a function of the amount of force acting on the wheel.

Four parameters specify the spline function: `ExtremumSlip`, `ExtremumValue`, `AsymptoteSlip`, and `AsymptoteValue`. A fifth factor, `StiffnessFactor`, acts as a multiplier on the tire forces. Higher values cause the wheel to grip the ground more strongly.

The AGEIA SDK documentation describes the effect of modifying the four spline function parameters. Changing the `StiffnessFactor` from its default value of 1000000.0 in the lateral direction will have the desired effect for the simulation. The values for the spline function are the same as the AGEIA defaults.

Add the following code just before the first `WheelEntity` is created in the Corobot `Initialize` method:

```
TireForceFunctionDescription LongitudalFunction =
    new TireForceFunctionDescription();
LongitudalFunction.ExtremumSlip = 1.0f;
LongitudalFunction.ExtremumValue = 0.02f;
LongitudalFunction.AsymptoteSlip = 2.0f;
LongitudalFunction.AsymptoteValue = 0.01f;
LongitudalFunction.StiffnessFactor = 1000000.0f;

TireForceFunctionDescription LateralFunction = new TireForceFunctionDescription();
LateralFunction.ExtremumSlip = 1.0f;
LateralFunction.ExtremumValue = 0.02f;
LateralFunction.AsymptoteSlip = 2.0f;
LateralFunction.AsymptoteValue = 0.01f;
LateralFunction.StiffnessFactor = 100000.0f;

wheelFRprop.TireLongitudalForceFunction = LongitudalFunction;
wheelFLprop.TireLongitudalForceFunction = LongitudalFunction;
wheelRRprop.TireLongitudalForceFunction = LongitudalFunction;
wheelRLprop.TireLongitudalForceFunction = LongitudalFunction;

wheelFRprop.TireLateralForceFunction = LateralFunction;
wheelFLprop.TireLateralForceFunction = LateralFunction;
wheelRRprop.TireLateralForceFunction = LateralFunction;
wheelRLprop.TireLateralForceFunction = LateralFunction;
```

The `LateralFunction` reduces the slip by a factor of 10. This has the desired effect of making the turns smoother due to tire slippage.

It is difficult to measure the exact tire friction of the actual robot, so these parameters must typically be tuned by hand until the simulated behavior closely matches the observed real-world behavior.

## Making the Wheels Turn

One of the first things you might have noticed about your `Corobot` entity is that its tires don't actually turn as the robot moves. This doesn't affect the simulation behavior but it does reduce the visual realism of the scene quite a bit. In addition, you likely noticed that the wheels never turn on the robots in the MRDS simulation tutorials, either. Let's fix that problem now.

You need to define a new `WheelEntity` that can keep track of its current rotation and adjust the rendering of its mesh accordingly. The following code shows how to create a new entity called `RotatingWheelEntity`, which inherits from `WheelEntity` but adds this additional functionality:

```
[DataContract]
public class RotatingWheelEntity : WheelEntity
{
    const float rotationScale = (float)(-1.0 / (2.0 * Math.PI));
    public float Rotations = 0;

    public RotatingWheelEntity()
    {
    }
```

*(continued)*

*(continued)*

```
public RotatingWheelEntity(WheelShapeProperties wheelShape)
    : base(wheelShape)
{
}

public override void Initialize(
    Microsoft.Xna.Framework.Graphics.GraphicsDevice device,
    PhysicsEngine physicsEngine)
{
    base.Initialize(device, physicsEngine);
}

public override void Update(FrameUpdate update)
{
    base.Update(update);

    // set the wheel to the current position
    Wheel.State.LocalPose.Orientation =
        TypeConversion.FromXNA(
            xna.Quaternion.CreateFromAxisAngle(
                new xna.Vector3(-1, 0, 0),
                (float)(Rotations * 2 * Math.PI)));

    // update the rotations for the next frame
    Rotations += (float)(Wheel.AxleSpeed *
        update.ElapsedTime * rotationScale);
}
}
```

Add the code for this entity outside of the Corobot entity definition but within the same namespace. This entity has a public data member called `Rotations`. For each frame, the `LocalPose` of the wheel shape is set according to the current rotation of the wheel. The rotation of the wheel for the next frame is calculated by converting the wheel axle speed to radians per second and multiplying by the elapsed time. The `Rotation` variable serves as an encoder for the current wheel rotation.

The `Render` function for the `WheelEntity` is already set up to take the `LocalPose` of the wheel shape into account when rendering the mesh; all you have left to do is ensure that the `Render` function for each wheel is called. Add the following override to the `Render` method in the `Corobot` class to accomplish this:

```
public override void Render(
    RenderMode renderMode,
    MatrixTransforms transforms,
    CameraEntity currentCamera)
{
    base.Render(renderMode, transforms, currentCamera);
    _wheelFL.Render(renderMode, transforms, currentCamera);
    _wheelFR.Render(renderMode, transforms, currentCamera);
    _wheelRL.Render(renderMode, transforms, currentCamera);
    _wheelRR.Render(renderMode, transforms, currentCamera);
}
```

This method first takes care of rendering the `Corobot` entity by calling the `base.Render` method. It then it takes care of rendering each `WheelEntity` by explicitly calling the `Render` method for each wheel. Remember that you must do this explicitly because the `WheelEntities` are not actually children of the `Corobot` entity.

Additionally, at this point you need to change the definition of `_wheelFR` (and the others) to use `RotatingWheelEntity` instead of `WheelEntity`. The accessor functions for the wheels must be changed to return a `RotatingWheelEntity` instead of a `WheelEntity`. In addition, the initialization of the private wheel members (`_wheelFR`, etc.) must be changed to create a `RotatingWheelEntity`. The code changes for the `FrontRightWheel` are shown here:

```
RotatingWheelEntity _wheelFR;

[Category("Wheels")]
[DataMember]
public RotatingWheelEntity FrontRightWheel
{
    get { return _wheelFR; }
    set { _wheelFR = value; }
}

_wheelFR = new RotatingWheelEntity(wheelFRprop);
```

Recompile your Corobot project and run the manifest again. You should now notice the wheels move as the robot moves around. It may be a little difficult to tell if they are moving because each wheel is just a uniform gray disc. You'll fix that soon.

## Adding Encoders

Now that each wheel has a concept of its current rotation, it is possible for you to add simulated wheel encoders. The Corobot has encoders on its front wheels. Each encoder has a resolution of 600 ticks per revolution of the wheel. Add the following two properties to access the simulated encoder values on these wheels:

```
[Category("Wheels")]
[DataMember]
public int FrontRightEncoder
{
    get { return (int)(_wheelFR.Rotations * 600f); }
    set { _wheelFR.Rotations = (float)value / 600f; }
}

[Category("Wheels")]
[DataMember]
public int FrontLeftEncoder
{
    get { return (int)(_wheelFL.Rotations * 600f); }
    set { _wheelFL.Rotations = (float)value / 600f; }
}
```

These properties will show up in the Simulation Editor when you display the properties of the `Corobot` entity so you can easily verify that the values are correct.

SimulatedQuadDifferentialDrive provides a way to expose the encoder values in its state. Add the following two lines of code to the `UpdateStateFromSimulation` method in the `SimulatedQuadDifferentialDriveService` class:

```
_state.LeftWheel.EncoderState.CurrentReading = _entity.FrontLeftEncoder;
_state.RightWheel.EncoderState.CurrentReading = _entity.FrontRightEncoder;
```

## Making It Look Real

One might argue that the visual appearance of the entity has little bearing on the usefulness of the simulation. However, one gets tired of looking at a lot of gray boxes driving around. You can build a mesh using a graphics modeling tool such as 3D Studio Max, Maya, or Blender that provides a better visualization of the entity. This mesh displays in the simulator in place of the entity but it has no effect on the entity's physics behavior.

It doesn't matter which modeling tool you use to create the mesh as long as it can export to the Alias .obj format. This is a fairly universal, if somewhat basic, 3D geometry format. Each .obj file is typically accompanied by a .mtl file with the same name that specifies the material characteristics for the geometry.

It is beyond the scope of this book to discuss geometry modeling in any great detail, but here are a few lessons that were learned from building meshes for the Corobot model:

❑ **Modeling tools:** The Maya modeling tool was used to generate the geometry for this model. Other modeling and CAD packages such as 3D Studio Max, SolidWorks, and Blender have been successfully used to build models for the simulation environment. Blender is a reasonable option if money is a concern. You can find more information about Blender at www.blender .org.

❑ **Realism versus speed:** Because you want the wheels to move independently from the robot body, you must make both a wheel mesh and a separate body mesh. There is usually a trade-off between the realism and visual interest of the model and the number of polygons it takes to define the model. Too many polygons makes the simulator run more slowly, especially if you have many robots or objects in the scene and you are running on a less powerful graphics card. Always endeavor to keep your simulation running at 30 frames per second or faster. Certainly, if it is running at 20 fps or slower, it becomes much less usable. Also keep in mind that the AGEIA physics engine becomes less stable if it is asked to continuously simulate steps much larger than 16.6 ms. If your simulator is running at less than 30 frames per second, then each frame represents over 33 ms and you may begin to notice physics behavior problems.

❑ **Wheel geometry:** Because this model has four wheels, the wheel geometry affects the overall look of the model a great deal. For that reason, more polygons were budgeted for the wheels than for any other part of the robot model. The wheels were modeled by defining a wheel hub composed of a cylindrical tube with a disc in the center with five triangular holes. Maya's smoothing feature was used to round the edges of the hub. Finally, a cylindrical tire was modeled to fit over the hub. The resulting tire model is shown in Figure 6-8. The right part of the figure illustrates the number of polygons that were used. The wheel doesn't look exactly like the one on the robot but it is close enough.

Figure 6-8

❏ **Modeling the robot chassis and platform:** The dimensions of the physics model are a good place to start in modeling the visual geometry. The chassis is modeled as a simple box combined with hexagonal cylinders on the front and back to give it the beveled look of the actual robot. The platform is modeled as a simple box sandwiched between two larger thin boxes. The middle box represents the batteries and circuit boards carried by the robot. A couple of cylinders and a box represent the camera in front, and a very simple representation of the IR sensor was also added to the front and back of the chassis.

❏ **Circuit boards and battery:** Because a real Corobot was available, pictures were taken of the circuit boards and battery on the platform and added as a texture map to the platform box. The texture map is shown in Figure 6-9, and Figure 6-10 shows the model with the texture map applied. The simulation environment supports texture maps in a number of formats, including .bmp and .jpg. You can also use the DirectX Texture Tool, available in the DirectX SDK, to make .dds files that support *mip-maps*. A mip-map is a texture that contains multiple resolutions. The hardware automatically chooses the most appropriate resolution based on the mapping of the texture to the screen. You might have noticed that a .dds texture with mip levels is typically used on the ground plane. If you substitute a regular .bmp or .jpg texture map with only a single resolution, you will notice shimmering pixels in the distance as the camera moves around due to aliasing of the texture map.

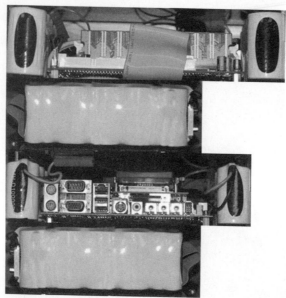

Figure 6-9

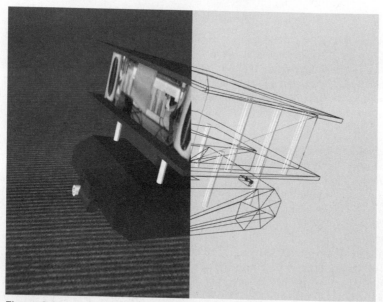

Figure 6-10

❑ **Tweaking the simulation environment:** Because different modeling packages handle materials and geometry in different ways, you will likely need to iterate several times to make the model look good. For example, some modeling packages define +Z as the upward axis, in which case you would need to rotate your model to make it look right in the simulation environment. Other packages use sophisticated lighting models that don't translate well to the .obj format, so you may have to go back and forth between the modeling tool and the simulation environment, tweaking the lighting parameters until the model looks as expected.

## Adding the Custom Mesh to the Corobot Entity

If no mesh is specified for an entity in `State.Assets.Mesh`, the simulator generates meshes from the physics shapes in the entity. If a mesh is specified, it is used. You need to change your `Corobot` entity to add these custom meshes. Add the following line for each of the `RotatingWheelEntities` just prior to the call to `Initialize`:

```
_wheelFR = new RotatingWheelEntity(wheelFRprop);
_wheelFR.State.Name = base.State.Name + " FrontRightWheel";
_wheelFR.Parent = this;
_wheelFR.MeshRotation = new Vector3(0, 180, 0);    // flip the wheel mesh
_wheelFR.State.Assets.Mesh = "CorobotWheel.obj";
_wheelFR.Initialize(device, physicsEngine);
```

The simulator looks in the `store\media` directory by default but you can also specify any other directory inside the SDK directory structure.

After adding the wheel mesh to each of the wheels, add the Corobot mesh to the `Corobot` entity just before `base.Initialize` is called in its `Initialize` method:

```
State.Assets.Mesh = "Corobot.obj";
base.Initialize(device, physicsEngine);
```

Run the Corobot service. You should now see the custom mesh in place of the old physics model. To display the physics representation or combine the two, select a different rendering mode by pressing F2. Figure 6-11 shows the full Corobot model in the simulation environment.

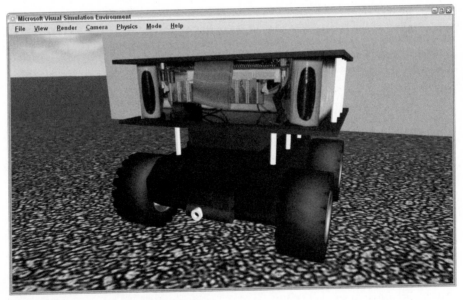

Figure 6-11

# Adding a Camera

The Corobot has a camera mounted on the front center of its chassis. Fortunately, a camera entity and its associated service are already provided in the SDK so it is fairly simple to add it to your entity.

## Adding the Camera Entity

Add the following lines of code to the `CorobotEntity` nondefault constructor:

```
CameraEntity frontCam = new CameraEntity(
    320,
    240,
    (float)(30.0 * Math.PI / 180.0));
frontCam.State.Name = name + "_cam";
frontCam.IsRealTimeCamera = true;     // update each frame
frontCam.State.Pose.Position = new Vector3(
    0,
    chassisDimensions.Y / 2.0f + chassisClearance - 0.01f,
    -chassisDimensions.Z / 2.0f);
InsertEntityGlobal(frontCam);
```

For this code, you first create a `CameraEntity` with a resolution of 320 by 240 pixels and a field of view of 30 degrees (converted to radians). You give it a name based on its parent entity and set it to be a real-time camera. Because the view from a real-time camera is rendered every frame, be careful not to add too many real-time cameras in a scene because they can greatly increase rendering time.

Finally, you give the camera a pose that positions it just a little below the center of the front of the chassis shape facing forward. Because the camera is a child of the `Corobot` entity, it follows its parent's orientation and position.

Compile and build with these changes and run the Corobot manifest. When the scene is displayed, select the Camera menu item. Both the main camera and the Corobot_cam should appear in the menu. When you select the Corobot_cam, you see the simulation environment from the point of view of that camera. The 320 × 240 image is stretched to fill the display window so it may look a little blocky, and objects may be distorted if the display window has a different aspect ratio. Drive the robot around with the dashboard and verify that the camera view changes.

## Adding the SimulatedWebcam Service

Just as the Corobot motors needed a SimulatedQuadDifferentialDrive service to drive them, the camera entity you just added needs a SimulatedWebCam service to retrieve frames from the camera for other services to use.

Like other simulation services, the SimulatedWebCam service is started with an entity name for a partner. It attempts to connect with the entity by making a subscription request to the simulation engine.

Add the following lines to the `Corobot.manifest.xml` file:

```
<!-- Start WebCam service -->
<ServiceRecordType>
<dssp:Contract>
```

```
                http://schemas.tempuri.org/2006/09/simulatedwebcam.html
        </dssp:Contract>
        <dssp:Service>http://localhost/Corobot/Cam</dssp:Service>
        <dssp:PartnerList>
          <dssp:Partner>
            <dssp:Service>http://localhost/Corobot_cam</dssp:Service>
            <dssp:Name>simcommon:Entity</dssp:Name>
          </dssp:Partner>
        </dssp:PartnerList>
        </ServiceRecordType>
```

The service is started by specifying its contract, and the entity name `Corobot_cam` is added as a partner (in the form of a URI). The service is given a run-time identifier of /Corobot/Cam. This is the identifier that will appear in the service directory.

Verify that the SimulatedWebCam service is working properly by running the manifest. Open a browser window and navigate to `http://localhost:50000`. Select Service Directory in the left column and then click the entry called /Corobot/Cam. A web page will be displayed showing the current view from the camera. If you select a refresh interval and press the Start button, the web page will be updated with new images at the specified rate.

# Adding Infrared Distance Sensors

The Corobot has infrared (IR) distance sensors mounted on the front and rear of it chassis. These sensors serve as virtual bumpers, telling the robot when it is about to collide with an obstacle. In this section, you modify the `LaserRangeFinderEntity` included with the MRDS SDK to become an IR sensor. You also add a service to read the value of an IR sensor and send notifications to other services when it changes.

## The CorobotIREntity

It is difficult to fully simulate all of the properties of an infrared sensor. The main purpose of the sensor is to return an approximate distance value based on the amount of infrared light reflected back to the sensor. The sensor can also be used as a reflectivity sensor because the value it returns changes with the reflectivity of different materials even if they are at the same distance. This property of the IR sensor is used to advantage in the MRDS Sumo Competition Package. The iRobot Create robots used as sumo players utilize the IR sensors mounted on their underside to detect changes in reflectivity in the outer region of the sumo ring so that they don't exceed that.

The simulation environment does not currently support modeling reflectivity, so this aspect of the sensor is hard to model. The `CorobotIREntity` that you'll create for the Corobot assumes that every material in the world has the same reflectivity, so it will simply return distance.

The `LaserRangeFinderEntity` provided in the SDK is a very accurate distance sensor that sweeps a laser horizontally across the environment and measures the light that is reflected back. Simulation Tutorial 2 (run by clicking Start ➪ Visual Simulation Environment ➪ Multiple Simulated Robots) contains a Pioneer3DX robot with a laser range finder mounted on top. The laser impact points are illuminated to help visualize the geometry that the laser is scanning.

You can easily modify this entity to become your `CorobotIREntity`. Copy the source code for the `LaserRangeFinderEntity` from samples\simulation\entities\entities.cs into `Corobot.cs`. Change the name of the entity to `CorobotIREntity` and include the following member variables:

```
public class CorobotIREntity : VisualEntity
{
    [DataMember]
    public float DispersionConeAngle = 8f; // in degrees
    [DataMember]
    public float Samples = 3f;  // the number of rays in each direction
    [DataMember]
    public float MaximumRange =
        (30f * 2.54f / 100.0f); // 30 inches converted to meters

    float _elapsedSinceLastScan;
    Port<RaycastResult> _raycastResultsPort;
    RaycastResult _lastResults;
    Port<RaycastResult> _raycastResults = new Port<RaycastResult>();
    RaycastProperties _raycastProperties;
    CachedEffectParameter _timeAttenuationHandle;
    float _appTime;
    Shape _particlePlane;

    /// <summary>
    /// Raycast configuration
    /// </summary>
    public RaycastProperties RaycastProperties
    {
        get { return _raycastProperties; }
        set { _raycastProperties = value; }
    }


    float _distance;
    [DataMember]
    public float Distance
    {
        get { return _distance; }
        set { _distance = value; }
    }
```

`DispersionConeAngle` is a new variable that sets the angle across which the infrared rays spread out from the emitter. The `Samples` variable specifies the number of distance samples to take horizontally and vertically. The `MaximumRange` variable specifies the farthest distance from which the sensor returns any meaningful data. If objects are farther than this distance, the sensor reports `MaximumRange` as the distance. You can use the `Distance` property to retrieve the last reading from the sensor. The rest of the variables are copied directly from the `LaserRangeFinderEntity`.

The constructors are renamed to match the new class name. `State.Assets.Effect` is set to `"LaserRangeFinder.fx"`. This effect is used when the laser impact points are rendered.

```
/// <summary>
/// Default constructor used when this entity is deserialized
/// </summary>
```

```
public CorobotIREntity()
{
}

/// <summary>
/// Initialization constructor used when this entity is built programmatically
/// </summary>
/// <param name="initialPos"></param>
public CorobotIREntity(string name, Pose initialPose)
{
    base.State.Name = name;
    base.State.Pose = initialPose;

    // used for rendering impact points
    base.State.Assets.Effect = "LaserRangeFinder.fx";
}
```

The Initialize method is not substantially different from the LaserRangeFinderEntity:

```
public override void Initialize(xnagrfx.GraphicsDevice device, PhysicsEngine
physicsEngine)
{
    try
    {
        if (Parent == null)
            throw new Exception(
                "This entity must be a child of another entity.");

        // make sure that we take at least 2 samples in each direction
        if (Samples < 2f)
            Samples = 2f;

        _raycastProperties = new RaycastProperties();
        _raycastProperties.StartAngle = -DispersionConeAngle / 2.0f;
        _raycastProperties.EndAngle = DispersionConeAngle / 2.0f;
        _raycastProperties.AngleIncrement =
            DispersionConeAngle / (Samples - 1f);
        _raycastProperties.Range = MaximumRange;
        _raycastProperties.OriginPose = new Pose();
```

The sensor calculates a distance value by casting rays out into the environment to see where they intersect with physics objects. The _raycastProperties structure specifies the number of rays and the angle over which they are cast.

```
        // set flag so rendering engine renders us last
        Flags |= VisualEntityProperties.UsesAlphaBlending;

        base.Initialize(device, physicsEngine);
```

The LaserRangeFinder.fx effect is created in base.Initialize. The following code creates the mesh that is used to render the laser impact point. You keep this functionality in your sensor because it is often useful in debugging services to see exactly where the IR sensor is pointing. The HeightFieldShape is created solely for the purpose of constructing a planar mesh that is two centimeters on a side. The mesh is created and added to the Meshes collection. It is given a texture map called `particle.bmp` that uses transparency to give the rendered plane a circular appearance:

```
            // set up for rendering impact points
            HeightFieldShapeProperties hf = new HeightFieldShapeProperties(
                "height field", 2, 0.02f, 2, 0.02f, 0, 0, 1, 1);
            hf.HeightSamples =
                new HeightFieldSample[hf.RowCount * hf.ColumnCount];
            for (int i = 0; i < hf.HeightSamples.Length; i++)
                hf.HeightSamples[i] = new HeightFieldSample();

            _particlePlane = new Shape(hf);
            _particlePlane.State.Name = "laser impact plane";

            // The mesh is used to render the ray impact points
            // rather than the sensor geometry.
            int index = Meshes.Count;
            Meshes.Add(SimulationEngine.ResourceCache.CreateMesh(
                device, _particlePlane.State));
            Meshes[0].Textures[0] =
                SimulationEngine.ResourceCache.CreateTextureFromFile(
                    device, "particle.bmp");

            if (Effect != null)
                _timeAttenuationHandle = Effect.GetParameter(
                    "timeAttenuation");

        }
        catch (Exception ex)
        {
            // clean up
            if (PhysicsEntity != null)
                PhysicsEngine.DeleteEntity(PhysicsEntity);

            HasBeenInitialized = false;
            InitError = ex.ToString();
        }
    }
```

This entity is unusual in that its single mesh is used to render impact points, rather than the geometry of the entity. This won't be a problem for the Corobot because the IR sensors are very small compared to the body of the robot and they are fixed to the body, so it isn't necessary to render them as a separate mesh.

Most of the work that the entity does is in the `Update` method:

```
public override void Update(FrameUpdate update)
{
    base.Update(update);
    _elapsedSinceLastScan += (float)update.ElapsedTime;
    _appTime = (float)update.ApplicationTime;
```

```
// only retrieve raycast results every SCAN_INTERVAL.
if ((_elapsedSinceLastScan > SCAN_INTERVAL) &&
    (_raycastProperties != null))
{
```

It is a fairly expensive operation to cast rays into the physics environment to determine which physics object they intersect, so this operation is not done every frame. The SCAN_INTERVAL constant determines the frequency with which the distance value is updated.

The position and orientation of the raycast pattern is set according to the Pose of this entity and the Pose of its parent:

```
_elapsedSinceLastScan = 0;

_raycastProperties.OriginPose.Orientation =
    TypeConversion.FromXNA(
        TypeConversion.ToXNA(Parent.State.Pose.Orientation) *
        TypeConversion.ToXNA(State.Pose.Orientation));

_raycastProperties.OriginPose.Position =
    TypeConversion.FromXNA(
        xna.Vector3.Transform(
            TypeConversion.ToXNA(State.Pose.Position),
            Parent.World));
```

You use the PhysicsEngine Raycast2D API to find the intersections of the rays with physics shapes in the environment. The first set of rays that you cast are in the horizontal plane:

```
// cast rays on a horizontal plane and again on a vertical plane
_raycastResultsPort =
    PhysicsEngine.Raycast2D(_raycastProperties);
```

If the first raycast was successful, then you cast a second set of rays in the vertical plane to form a cross pattern:

```
_raycastResultsPort.Test(out _lastResults);
if (_lastResults != null)
{
```

You combine the impact points, if any, from both sets of rays and then find the distance to the closest intersection:

```
RaycastResult verticalResults;

// rotate the plane by 90 degrees
_raycastProperties.OriginPose.Orientation =
    TypeConversion.FromXNA(
        TypeConversion.ToXNA(
            _raycastProperties.OriginPose.Orientation) *
        xna.Quaternion.CreateFromAxisAngle(
            new xna.Vector3(0, 0, 1),
            (float)Math.PI / 2f));
```

*(continued)*

*(continued)*

```
        _raycastResultsPort =
            PhysicsEngine.Raycast2D(_raycastProperties);
        _raycastResultsPort.Test(out verticalResults);

        // combine the results of the second raycast with the first
        if (verticalResults != null)
        {
            foreach (RaycastImpactPoint impact in
            verticalResults.ImpactPoints)
                _lastResults.ImpactPoints.Add(impact);
        }
```

That is the distance you return. If there is no intersection, then `MaximumRange` is returned.

```
        // find the shortest distance to an impact point
        float minRange = MaximumRange * MaximumRange;
        xna.Vector4 origin = new xna.Vector4(
            TypeConversion.ToXNA(
                _raycastProperties.OriginPose.Position), 1);

        foreach (RaycastImpactPoint impact in
        _lastResults.ImpactPoints)
        {
            xna.Vector3 impactVector = new xna.Vector3(
                impact.Position.X - origin.X,
                impact.Position.Y - origin.Y,
                impact.Position.Z - origin.Z);

            float impactDistanceSquared =
                impactVector.LengthSquared();
            if (impactDistanceSquared < minRange)
                minRange = impactDistanceSquared;
        }
        _distance = (float)Math.Sqrt(minRange);
    }
  }
}
```

The final two entity methods render the impact points of the rays:

```
public override void Render(
    RenderMode renderMode,
    MatrixTransforms transforms,
    CameraEntity currentCamera)
{
    if ((int)(Flags & VisualEntityProperties.DisableRendering) > 0)
        return;
```

Rendering of the impact points is disabled if the `DisableRendering` flag is set:

```
    if (_lastResults != null)
        RenderResults(renderMode, transforms, currentCamera);
}

void RenderResults(
    RenderMode renderMode,
    MatrixTransforms transforms,
    CameraEntity currentCamera)
{
    _timeAttenuationHandle.SetValue(
        new xna.Vector4(100 * (float)Math.Cos(
            _appTime * (1.0f / SCAN_INTERVAL)), 0, 0, 1));
```

This sets a value in the effect that causes the impact points to flash on and off. A local transform matrix is built that rotates the impact point mesh so that it faces the camera:

```
    // render impact points as a quad
    xna.Matrix inverseViewRotation = currentCamera.ViewMatrix;
    inverseViewRotation.M41 =
        inverseViewRotation.M42 =
            inverseViewRotation.M43 = 0;
    xna.Matrix.Invert(ref inverseViewRotation, out inverseViewRotation);
    xna.Matrix localTransform = xna.Matrix.CreateFromAxisAngle(
        new xna.Vector3(1, 0, 0),
        (float)-Math.PI / 2) * inverseViewRotation;
    SimulationEngine.GlobalInstance.Device.RenderState.
        DepthBufferWriteEnable = false;
```

The `DepthBuffer` is disabled because these impact points should not occlude other objects. The impact point mesh is adjusted to be a little closer to the ray emitter than the exact impact point:

```
    for (int i = 0; i < _lastResults.ImpactPoints.Count; i++)
    {
        xna.Vector3 pos = new
            xna.Vector3(_lastResults.ImpactPoints[i].Position.X,
                        _lastResults.ImpactPoints[i].Position.Y,
                        _lastResults.ImpactPoints[i].Position.Z);

        xna.Vector3 resultDir = pos - Parent.Position;
        resultDir.Normalize();
        localTransform.Translation = pos - .02f * resultDir;
        transforms.World = localTransform;
```

This helps the impact points to show up clearly instead of being rendered in the same plane as the shape they intersected.

```
        base.Render(renderMode, transforms, Meshes[0]);
    }
    SimulationEngine.GlobalInstance.Device.RenderState.
        DepthBufferWriteEnable = true;
}
```

Now that you've completely defined the `CorobotIR` entity, you want to add two of them to your Corobot—one in the front and one in the rear. Add the following code to the `CorobotEntity` nondefault constructor:

```
InsertEntityGlobal(
    new CorobotIREntity(
        name + "_rearIR",
        new Pose(new Vector3(
            0,
            chassisDimensions.Y / 2.0f + chassisClearance,
            chassisDimensions.Z / 2.0f))));
```

The default orientation for the IR entity is facing toward the +Z direction. That faces toward the rear of the `CorobotEntity`, so the rear IR sensor is inserted with a default orientation. The coordinates of the position vector place the sensor in the middle of the rear face of the chassis. The position is specified using world coordinates instead of coordinates relative to the parent entity because the `InsertEntityGlobal` method is used to add the child entity.

The call to insert the front IR entity is essentially the same except that the position coordinates place it in the center of the front face of the chassis shape, just above the camera:

```
InsertEntityGlobal(
    new CorobotIREntity(
        name + "_frontIR",
        new Pose(new Vector3(
            0,
            chassisDimensions.Y / 2.0f + chassisClearance,
            -chassisDimensions.Z / 2.0f),
        TypeConversion.FromXNA(
            xna.Quaternion.CreateFromAxisAngle(
                new xna.Vector3(0, 1, 0), (float)Math.PI)))));
```

In addition, the entity is created with a `Pose` that rotates it 180 degrees around the +Y axis so that it is facing toward the front of the entity.

With all of the rotations and transformations going on, it is important to test these new sensors to ensure that they are oriented and mounted correctly:

1. Run the Corobot manifest. You should see the rendered impact marks from the rear IR sensor flashing on and off on the side of the giant box.

2. Move the Corobot slightly forward. The impact points should disappear. This indicates that the giant box is farther from the sensor than 30 inches.

3. Move the entity closer to the giant box until the impact points are again visible.

4. Start the Simulation Editor by pressing F5, expand the Corobot entity in the Entities pane to see its children, and select the `Corobot_readIR` entity.

5. Set the `DisableRendering` flag in the entity flags. The impact points should disappear. This verifies that the rear IR sensor is actually generating those impacts.

6. Check the `Distance` property of the `Corobot_rearIR` sensor. It should change as the Corobot moves closer to the giant box.

Repeat these tests with the front IR sensor to verify that it is also working correctly. Figure 6-12 shows how the laser impact points should appear in the scene when the IR sensor is working properly. (Because this book is black and white, the impact points were enhanced; on the interface they appear read).

Figure 6-12

## Adding a SimulatedIR Service

Now that you have an IR entity, you also need a SimulatedIR service to go along with it. This one will be much simpler than the SimulatedQuadDifferentialDrive. You're going to use the existing `AnalogSensor` contract defined in `RoboticsCommon.dll` to reduce the amount of code you need to write. Unlike the SimulatedQuadDifferentialDrive service, you won't need to implement a `_mainPort` that supports different operations than the alternate contract. This `_mainPort` will implement only the `AnalogSensor` operations. This is analogous to subclassing an existing class and overriding methods to change behavior but adding no additional methods or public variables.

Start an MRDS command prompt and change to the `ProMRDS\MyChapter6` directory. Use the following command to generate the SimulatedIR service (bold code indicates something you should type):

```
C:\Microsoft Robotics Studio (1.5)>
C:\Microsoft Robotics Studio (1.5)>cd ProMRDS
C:\Microsoft Robotics Studio (1.5)\ProMRDS>cd MyChapter6
C:\Microsoft Robotics Studio (1.5)\ProMRDS\MyChapter6> dssnewservice /
Service:"SimulatedIR" /Namespace:"ProMRDS.Simulation.SimulatedIR"
/alt:"http://schemas.microsoft.com/robotics/2006/06/analogsensor.html" /i:"\
Microsoft Robotics Studio (1.5)\bin\RoboticsCommon.dll" /year:"07" /month:"08"
```

As you would by now expect, this generates a service called SimulatedIR, which supports the `AnalogSensor` contract. Open your newly generated `simulatedIR.csproj` from the command line so that Visual Studio inherits the environment from the MRDS command-line environment. Use the following steps to transform this generic service into a SimulatedIR service. Refer to the completed service in the Chapter6 directory as necessary.

1. Add the using statements and DLL references required for a simulation service just as you did in the section "The SimulatedQuadDifferentialDrive Service" earlier in this chapter. Don't forget to set the CopyLocal and SpecificVersion properties to false for each reference added.

2. Add the following additional using statement and a reference to the Corobot service:

```
using corobot = ProMRDS.Simulation.Corobot;
```

3. Change the DisplayName and Description attributes to describe the service.

4. Add two private class members to handle subscribing to the simulation engine:

```
corobot.CorobotIREntity _entity;
simengine.SimulationEnginePort _notificationTarget;
```

5. Change the AllowMultipleInstances parameter of the ServicePort attribute on the _ mainPort from false to true. You want multiple instances of this service running because you have multiple IR sensors to support.

```
[ServicePort("/simulatedir", AllowMultipleInstances=true)]
```

6. Add a SubscriptionManagerPort to handle interactions with the SubscriptionManager service:

```
[Partner("SubMgr",
    Contract = submgr.Contract.Identifier,
    CreationPolicy = PartnerCreationPolicy.CreateAlways)]
private submgr.SubscriptionManagerPort _submgrPort =
    new submgr.SubscriptionManagerPort();
```

7. Add the following code to the Start method to subscribe for a partner entity from the SimulationEngine service and to set up a handler for the notification just as you did in the SimulatedQuadDifferentialDrive service:

```
protected override void Start()
{
    _notificationTarget = new simengine.SimulationEnginePort();

    // PartnerType.Service is the entity instance name.
    simengine.SimulationEngine.GlobalInstancePort.Subscribe(
        ServiceInfo.PartnerList, _notificationTarget);

    // don't start listening to DSSP operations, other than drop,
    // until notification of entity
    Activate(new Interleave(
        new TeardownReceiverGroup
        (
            Arbiter.Receive<simengine.InsertSimulationEntity>(
                false,
                _notificationTarget,
                InsertEntityNotificationHandlerFirstTime),
            Arbiter.Receive<dssp.DsspDefaultDrop>(
```

```
                false,
                _mainPort,
                DefaultDropHandler)
        ),u
        new ExclusiveReceiverGroup(),
        new ConcurrentReceiverGroup()
    ));

    // start notification method
    SpawnIterator<DateTime>(DateTime.Now, CheckForStateChange);
}
```

8. The `SpawnIterator` call at the end of the `Start` method is used to start up a method that periodically checks for a change in the reading from the IR sensor and sends a notification to subscribers if necessary. Add this code for that method. When this method runs, it checks whether the `Distance` property on the entity has changed. If it has, the service updates its state from the entity and then sends a notification to all subscribed services. The method then sets itself to wake up again after 200 ms have elapsed.

```
float _previousDistance = 0;

private IEnumerator<ITask> CheckForStateChange(DateTime timeout)
{
    while (true)
    {
        if (_entity != null)
        {
            if (_entity.Distance != _previousDistance)
            {
                // send notification of state change
                UpdateState();
                base.SendNotification<Replace>(_submgrPort, _state);
            }
            _previousDistance = _entity.Distance;
        }
        yield return Arbiter.Receive(false, TimeoutPort(200), delegate { });
    }
}
```

9. Add the following code to create a default state for the service and to update the service state from the `CorobotIR` entity. The state is defined as part of the `AnalogSensor` contract. It includes a raw measurement, which is equal to the `Distance` property on the entity; a `RawMeasurementRange`, which reflects the maximum value the sensor can have; and a `NormalizedMeasurement` value, which is the raw value normalized against the maximum value:

```
private void CreateDefaultState()
{
    _state.HardwareIdentifier = 0;
    _state.NormalizedMeasurement = 0;
    _state.Pose = new Microsoft.Robotics.PhysicalModel.Proxy.Pose();
    _state.RawMeasurement = 0;
    _state.RawMeasurementRange = _entity.MaximumRange;
}
```

*(continued)*

*(continued)*

```
void UpdateState()
{
    // update our state from the entity
    _state.RawMeasurement = _entity.Distance;
    _state.NormalizedMeasurement =
        _state.RawMeasurement / _state.RawMeasurementRange;
    _state.TimeStamp = DateTime.Now;
}
```

**10.** Add the following code to receive notifications from the SimulationEngine service. These methods are nearly identical to the corresponding methods in the SimulatedQuadDifferentialDrive service:

```
void InsertEntityNotificationHandlerFirstTime(simengine.InsertSimulationEntity ins)
{
    InsertEntityNotificationHandler(ins);

    base.Start();

                // Add service specific initialization here.
    MainPortInterleave.CombineWith(
        new Interleave(
            new TeardownReceiverGroup(),
            new ExclusiveReceiverGroup(
                Arbiter.Receive<simengine.InsertSimulationEntity>(
                    true,
                    _notificationTarget,
                    InsertEntityNotificationHandler),
                Arbiter.Receive<simengine.DeleteSimulationEntity>(
                    true,
                    _notificationTarget,
                    DeleteEntityNotificationHandler)
            ),
            new ConcurrentReceiverGroup()
        )
    );
}

void InsertEntityNotificationHandler(simengine.InsertSimulationEntity ins)
{
    _entity = (corobot.CorobotIREntity)ins.Body;
    _entity.ServiceContract = Contract.Identifier;

    CreateDefaultState();
}
void DeleteEntityNotificationHandler(simengine.DeleteSimulationEntity del)

{
    _entity = null;
}
```

**11.** Add a call to UpdateState before the state is posted to the response port in the GetHandler method:

```
[ServiceHandler(ServiceHandlerBehavior.Concurrent)]
public virtual IEnumerator<ITask> GetHandler(pxanalogsensor.Get get)
{
  UpdateState();
    get.ResponsePort.Post(_state);
    yield break;
}
```

**12.** Add a call to SubscribeHelper in the SubscribeHandler to manage subscription requests. Handling subscriptions is described in Service Tutorial 4 in the SDK documentation.

```
public virtual IEnumerator<ITask> SubscribeHandler(pxanalogsensor.Subscribe
subscribe)
{
    SubscribeHelper(
        _submgrPort,
        subscribe.Body,
        subscribe.ResponsePort);
    yield break;
}
```

**13.** Add the following entries to the Corobot.manifest.xml file to start two copies of this new service. In the Corobot service, you gave the name "Corobot_frontIR" to the front IR sensor and "Corobot_rearIR" to the rear sensor. The entity partners associated with each of these services have those same names. You also used the <dssp:Service> attribute to specify a name for the service so that you can distinguish the front IR service from the rear one.

```
<!-- Start Front IR service -->
<ServiceRecordType>
  <dssp:Contract>http://schemas.tempuri.org/2007/08/simulatedir.html</dssp:Contract>
  <dssp:Service>http://localhost/Corobot/FrontIR</dssp:Service>
  <dssp:PartnerList>
    <dssp:Partner>
      <!--The partner name must match the entity name-->
      <dssp:Service>http://localhost/Corobot_frontIR</dssp:Service>
      <dssp:Name>simcommon:Entity</dssp:Name>
    </dssp:Partner>
  </dssp:PartnerList>
</ServiceRecordType>

<!-- Start Rear IR service -->
<ServiceRecordType>
  <dssp:Contract>http://schemas.tempuri.org/2007/08/simulatedir.html</dssp:Contract>
  <dssp:Service>http://localhost/Corobot/RearIR</dssp:Service>
  <dssp:PartnerList>
    <dssp:Partner>
      <!--The partner name must match the entity name-->
```

*(continued)*

*(continued)*

```
            <dssp:Service>http://localhost/Corobot_rearIR</dssp:Service>
            <dssp:Name>simcommon:Entity</dssp:Name>
        </dssp:Partner>
    </dssp:PartnerList>
</ServiceRecordType>
```

That's it. You should now have a working SimulatedIR service. Verify it by running the Corobot manifest. Start a browser window and navigate to `http://localhost:50000` and select Service Directory from the left column. You should see something similar to what is shown in Figure 6-13. Each service is listed twice because DssHost adds two entries to the service directory for services that support an alternate contract: one for their own contract and one for the alternate contract they support. In this case, both entries refer to the same port. Click each of the SimulatedIR services to see their current state. As you drive the Corobot around in the environment, refresh the service state to verify that the RawMeasurement field correctly reflects the distance from that IR sensor to the giant box.

| | |
|---|---|
| /corobot/frontir | Entity ConstructorService PartnerListService AlternateContractService_0 SubMgr |
| /corobot/frontir/analogsensor | Entity ConstructorService PartnerListService PrimaryContractService |
| /corobot/rearir | Entity ConstructorService PartnerListService AlternateContractService_0 SubMgr |
| /corobot/rearir/analogsensor | Entity ConstructorService PartnerListService PrimaryContractService |

**Figure 6-13**

# Summary

That completes the basic functionality for the Corobot entity and its associated services. This entity can now be used in a variety of simulation scenarios, such as the Robo-Magellan scenario covered in the next chapter.

This chapter began with an overview of the methods and types provided by the various simulation DLLs, including the characteristics of the VisualEntity type, which is fundamental to creating new simulation entities. The characteristics of a simulation service were also described, including their relationship to orchestration services such as the SimMagellan service.

You created a new simulation service called Corobot that added entities to the simulation environment. This eventually included a model of the Corobot robot. You then defined a SimulatedQuadDifferentialDrive service and used it to drive the Corobot around the environment using the SimpleDashboard service. You tuned the top speed of the entity and the tire friction and then you made the wheels turn as the robot moves. Finally, you added a detailed 3D mesh to the Corobot model to make it look more realistic, and then you added a camera and defined a Simulated IR Distance sensor entity and service so that you could add IR sensors to the front and rear of the Corobot.

The entities and services defined in this chapter provide examples for you to use as you create your own custom entities and their associated services.

# 7

# Using Orchestration Services to Build a Simulation Scenario

A simulation scenario is an environment and one or more robot models that is used to prototype a control algorithm. The extent to which the environment reflects real-world conditions and limitations depends on the desired results from the simulation. If it is important that the exact code you use in the simulator can run on a real robot, then the simulated world and services must carefully model every important aspect of the real world. If the simulator is to be used only to prototype or demonstrate an algorithm, then real-world fidelity is less important.

In this chapter, you'll use the new Corobot robot entity from the last chapter in a simulation scenario that mimics the Seattle Robotics Society's Robo-Magellan competition. A description of the competition and the rules can be found at `www.robothon.org/robothon/robo-magellan .php`. Before the competition begins, a referee places cones on the course and then provides GPS coordinates to the contestants for their location. Robots start at a predetermined place and navigate to each cone, avoiding obstacles along the way. The challenges that the robots face are navigating over outdoor terrain, avoiding obstacles, and dealing with GPS signal problems.

In this simulation, you'll mainly focus on navigation and obstacle avoidance. You haven't built a GPS service yet so you'll use the robot position reported by the `SimulatedQuadDifferentialDrive` to mimic this capability.

In the last chapter, you used the Corobot manifest to test the Corobot entity and its associated services. For this scenario, you'll define a new manifest that runs the Corobot services along with a referee service and the Robo-Magellan orchestration service. Figure 7-1 shows the completed Robo-Magellan scenario.

Figure 7-1

# The Robo-Magellan Referee Service

The first thing you'll need is a referee to set up the course and place the cones. In a scenario like this one, it is a good idea to separate the player functionality from the referee functionality. That way, the interface between the two is well defined and it is eventually easier to move the player service to an actual robot.

The referee service will be responsible for setting up all of the entities in the simulation environment, including the sky, sun, and ground entities as well as the obstacles in the world.

Instead of using DssNewService to create the referee service, you're going to copy Simulation Tutorial 1 from the SDK and modify it. Simulation Tutorial 1 is already set up to interact with the simulation environment and create environment entities, so it is already very close to what you want your referee to be. You can create a referee service using the following commands from the MRDS command prompt. The text that you type is shown in bold.

```
C:\Microsoft Robotics Studio (1.5)\ProMRDS>mkdir MyChapter7
C:\Microsoft Robotics Studio (1.5)\ProMRDS>cd MyChapter7

C:\Microsoft Robotics Studio (1.5)\ProMRDS\MyChapter7>mkdir Referee

C:\Microsoft Robotics Studio (1.5)\ProMRDS\MyChapter7>
copy ..\..\samples\simulationtutorials\tutorial1\*.* Referee

..\..\samples\simulationtutorials\tutorial1\AssemblyInfo.cs
..\..\samples\simulationtutorials\tutorial1\SimulationTutorial1.cs
..\..\samples\simulationtutorials\tutorial1\SimulationTutorial1.csproj
..\..\samples\simulationtutorials\tutorial1\SimulationTutorial1.csproj.user
..\..\samples\simulationtutorials\tutorial1\SimulationTutorial1.sln
        5 file(s) copied.
```

## Customizing the SimulationTutorial1 Service

Now you'll rename the files and change the namespace and contract identifier to make this service your own. Start by renaming the solution file and opening it from the command line as follows. Again, text that you type is shown in bold:

```
C:\Microsoft Robotics Studio (1.5)\ProMRDS\MyChapter7\Referee>ren
SimulationTutorial1.sln Referee.sln

C:\Microsoft Robotics Studio (1.5)\ProMRDS\MyChapter7\Referee> Referee.sln
```

Use the following steps to make the Referee project:

1. Rename the SimulationTutorial1 project to Referee and rename `SimulationTutorial1.cs` to `Referee.cs` inside Visual Studio.

2. Open `Referee.cs`. Change the namespace from `Robotics.SimulationTutorial1` to `ProMRDS.Simulation.MagellanReferee`.

3. Rename the `SimulationTutorial1` class to `MagellanReferee` and update the `DisplayName` and `Description` attributes.

4. Change the contract identifier at the bottom of the file to `http://schemas.tempuri.org/2007/08/MagellanReferee.htm`.

5. Do a global replace of SimulationTutorial1 with MagellanReferee.

6. Open the solution properties and change the assembly name to `SimMagellanReferee.Y2007.M08`, and change the default namespace to `ProMRDS.Simulation.MagellanReferee`.

7. The solution should now compile with no errors.

## Starting a Service from the Browser

You don't have a manifest to run this service yet, so you'll use a different method to run it. Start a Dss Node at port 50000 without specifying a manifest:

```
C:\Microsoft Robotics Studio (1.5)\ProMRDS\MyChapter7\Referee>cd ..\..\..
C:\Microsoft Robotics Studio (1.5)>dsshost -p:50000 -t:50001
*   Service uri:  [09/10/2007 08:12:33][http://msrs1:50000/directory]
*   Service uri:  [09/10/2007 08:12:33][http://msrs1:50000/constructor/ef2ee88f-
cf29-4052-9168-a6714fb53ae1]
*   No initial manifest supplied. [09/10/2007 08:12:34][http://msrs1:50000/
manifestloaderclient]
```

Open a browser window and navigate to `http://localhost:50000`. Select Control Panel from the left column and type **referee** in the Search box. You should see the new Magellan Referee service displayed with the `DisplayName` and `Description` you specified in the service. Click the Create button on the right and you should soon see the simulation window appear with the scene from Simulation Tutorial 1.

# Building the World and Adding Cameras

Now you'll modify the referee service to build a world for our scenario. Replace the contents of the PopulateWorld method with the following:

```
// Set up initial view
CameraView view = new CameraView();
view.EyePosition = new Vector3(-0.91f, 0.67f, -1f);
view.LookAtPoint = new Vector3(1.02f, 0.09f, 0.19f);
SimulationEngine.GlobalInstancePort.Update(view);
```

This sets the initial camera viewpoint to something convenient:

```
// Add another camera to view the scene from above
CameraEntity fromAbove = new CameraEntity(640, 480);
fromAbove.State.Name = "FromAbove";
fromAbove.Location = new xna.Vector3(4.3f, 20.59f, 0.86f);
fromAbove.LookAt = new xna.Vector3(4.29f, 18.26f, 0.68f);
fromAbove.IsRealTimeCamera = false;
SimulationEngine.GlobalInstancePort.Insert(fromAbove);
```

You also need to add a using statement, as shown in the following snippet, and a reference to Microsoft.Xna.Framework.DLL, as described in the previous chapter:

```
using xna = Microsoft.Xna.Framework;
```

Now add a second camera that looks at the whole scene from above. It is quick to switch to this camera using F8:

```
// Add a SkyDome.
SkyDomeEntity sky = new SkyDomeEntity("skydome.dds", "sky_diff.dds");
SimulationEngine.GlobalInstancePort.Insert(sky);

// Add a directional light to simulate the sun.
LightSourceEntity sun = new LightSourceEntity();
sun.State.Name = "Sun";
sun.Type = LightSourceEntityType.Directional;
sun.Color = new Vector4(1, 1, 1, 1);
sun.Direction = new Vector3(-0.47f, -0.8f, -0.36f);
SimulationEngine.GlobalInstancePort.Insert(sun);
```

# Adding a Sky, Sun, and Ground

You use a typical sky and sun:

```
HeightFieldShapeProperties hf = new HeightFieldShapeProperties("height field",
    64, // number of rows
    10, // distance in meters, between rows
    64, // number of columns
    10, // distance in meters, between columns
```

```
        1, // scale factor to multiple height values
        -1000); // vertical extent of the height field.

// create array with height samples
hf.HeightSamples = new HeightFieldSample[hf.RowCount * hf.ColumnCount];
for (int i = 0; i < hf.RowCount * hf.ColumnCount; i++)
{
    hf.HeightSamples[i] = new HeightFieldSample();
    hf.HeightSamples[i].Height = (short)(Math.Sin(i * 0.01));
}

// create a material for the entire field.
hf.Material = new MaterialProperties("ground", 0.8f, 0.5f, 0.8f);

// insert ground entity in simulation and specify a texture
SimulationEngine.GlobalInstancePort.Insert(
    new HeightFieldEntity(hf, "FieldGrass.dds"));
```

You use a ground plane substantially smaller than the standard flat ground used in the samples. This one is only 640 meters by 640 meters. Add the FieldGrass.dds texture to give the appearance of an open grassy field.

## Adding Barriers

Now that you have the basics in place, you need to add a few barriers to make life interesting for your Corobot. The following code adds several walls and a cement tower. By grouping the barrier parameters in an array, it is easier to add additional barriers or define several different scene configurations. If you are trying to create a general-purpose navigation algorithm, you probably want to test it against several different barrier configurations.

Add the following code to create and insert the barriers just after the ground definition in PopulateWorld:

```
// create barriers
foreach (Barrier bar in Barriers)
{
    SingleShapeEntity wall =
        new SingleShapeEntity(
            new BoxShape(
                new BoxShapeProperties(
                    0,  // no mass makes a static shape
                    new Pose(),
                    bar.Dimensions)), // dimensions
                bar.Position);
    wall.State.Pose.Orientation = bar.Orientation;
    wall.State.Name = bar.Name;
    wall.State.Assets.DefaultTexture = bar.Texture;
    SimulationEngine.GlobalInstancePort.Insert(wall);
}
```

Add this class definition as a peer to the `MagellanReferee` class:

```
public struct Barrier
{
    public string Name;
    public Vector3 Position;
    public Vector3 Dimensions;
    public string Texture;
    public Quaternion Orientation;
    public Barrier(string name, Vector3 position, Vector3 dimensions,
        string texture, Quaternion orientation)
    {
        Name = name;
        Position = position;
        Dimensions = dimensions;
        Texture = texture;
        Orientation = orientation;
    }
}
```

Finally, add the following barrier definitions as a member variable at the top of the `MagellanReferee` class:

```
public Barrier[] Barriers = new Barrier[]
{
    // Name, Position, Dimensions, Texture, Orientation
    new Barrier("Wall0", new Vector3(0, 0, -4), new Vector3(4f, 0.8f, 0.1f),
        "BrickWall.dds", new Quaternion(0,0,0,1)),
    new Barrier("Wall1", new Vector3(-2.05f, 0, -3.05f),
        new Vector3(2f, 0.8f, 0.1f),
        "BrickWall.dds", Quaternion.FromAxisAngle(0, 1, 0, (float)(Math.PI / 2))),
    new Barrier("Wall2", new Vector3(1.41f, 0, 2.46f), new Vector3(6f, 0.8f, 0.1f),
        "BrickWall.dds", Quaternion.FromAxisAngle(0, 1, 0, (float)(Math.PI / 2))),
    new Barrier("Tower", new Vector3(5.58f, 2f, -0.59f), new Vector3(2f, 4f, 2f),
        "MayangConcrP.dds", new Quaternion(0,0,0,1)),
};
```

## Running the Service

It's a good time to run the service again to ensure that everything looks right. Run the service using the instructions in the previous section and switch to the FromAbove camera using F8. You should see a layout something like what is shown in Figure 7-2, which shows the Sim-Magellan world with a few walls and a tower.

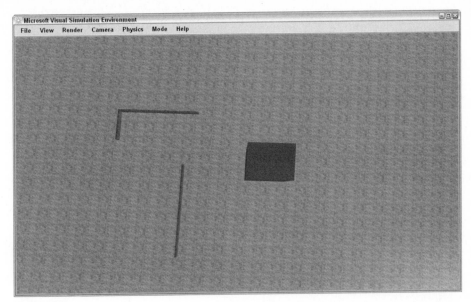

Figure 7-2

The next step is to add some goals for your little robot to achieve. In the next section, you're going to define some special entities that know when they've been touched.

## Building a Better Traffic Cone

The SRS Robo-Magellan rules state that waypoints are marked by cones. The robot must approach and touch the cone without knocking it over. To help the referee service detect when a cone has been touched, you define a special kind of entity that uses a trigger shape. A *trigger shape* is a special physics shape that does not collide with other shapes. However, when another shape is inside its volume, it sends a notification. This type of shape is used to detect goals in the soccer simulation and to detect when the Sumo robots leave the sumo ring in the sumo simulation. Here, you'll use them to detect when the Corobot gets close enough to the traffic cone to touch it. The following method builds traffic cones with embedded trigger shapes:

```
private VisualEntity CreateTriggerCone(
    string name,
    Vector3 position)
{
    Vector3 dimensions = new Vector3(0.35f, 0.7f, 0.35f);
    position.Y += dimensions.Y / 2f;
    SingleShapeEntity triggerShape = new SingleShapeEntity(
        new BoxShape(
            new BoxShapeProperties(
                0,
                new Pose(),
                dimensions)),
        position);

    // used to receive a single trigger notification from this cone
```

(continued)

*(continued)*

```
        Port<Shape> conePort = new Port<Shape>();


        triggerShape.State.Name = name;
        triggerShape.State.Assets.Mesh = "street_cone.obj";
        triggerShape.BoxShape.State.Name = name;
        triggerShape.BoxShape.State.Advanced =
            new ShapeAdvancedProperties();
        // make the box a trigger shape by adding a notification port
        triggerShape.BoxShape.State.Advanced.IsTrigger = true;
        triggerShape.BoxShape.State.Advanced.TriggerNotificationPort =
            conePort;
        triggerShape.State.MassDensity.Mass = 0;
        triggerShape.State.MassDensity.Density = 0;

        Activate(Arbiter.Receive<Shape>(false, conePort,
            delegate(Shape sh)
            {
                Console.WriteLine("Cone " + name + " was triggered.");
            }
        ));

        return triggerShape;
    }
```

You use a `SingleShapeEntity` as your trigger entity. The dimensions are specified to match the size of the traffic cone model included with the SDK. The Y position is adjusted so that the cone isn't halfway embedded in the ground. A port that receives `Shape` objects is also defined to receive notifications from the trigger shape. The entity is given a name and the `street_cone.obj` mesh is specified for it to use. Now it gets interesting. You allocate a `ShapeAdvancedProperties` class and then set the `IsTrigger` flag to `true`. You specify the port you just created as the `TriggerNotificationPort`. You also give the entity a mass of 0, which makes it a static entity. The physics engine will not move the entity and it expects the entity to have the same position and orientation for the duration of the simulation.

Finally, you activate a one-time receiver to listen on the `conePort` for any shape that intersects the trigger shape. It is a one-time receiver because the trigger shape sends a notification to the port once each frame for as long as a shape intersects. You're really only interested in being notified the first time this happens. When the port receives a `Shape` message, a message prints to the console indicating that the referee has ruled that the cone has been successfully touched. The entity is returned so that it can be inserted in the simulator.

Now you have a way to define goals for the robot. You also need a way to transfer the coordinates of the cones to the robot. You'll take advantage of the service architecture and the ease with which state can be passed between services by defining the state of the referee service to hold the list of cones in the environment. Add the following state class definition as a peer to the MagellanReferee service:

```
#region State
/// <summary>
/// Magellan Referee State
/// </summary>
[DataContract()]
```

```
public class MagellanRefereeState
{
    [DataMember]
    public Vector3[] Cones;
}
#endregion
```

Just to keep things simple, the state will consist of an array of positions that identify the location of each cone. Add a member variable to hold the service state just after the _engineStub port definition in the MagellanReferee service:

```
[InitialStatePartner(Optional = true, ServiceUri = "Magellan.Referee.config.xml")]
private MagellanRefereeState _state = new MagellanRefereeState();
```

By defining an InitialStatePartner, it is possible to change the configuration of the cones simply by changing the optional config file Magellan.Referee.config.xml. If no configuration file is specified, the state needs to be initialized with default values. Add a call to ValidateState at the beginning of the Start method to initialize the state if necessary. The ValidateState method generates a default state as follows:

```
private void ValidateState()
{
    if (_state == null)
        _state = new MagellanRefereeState();

    if ((_state.Cones == null) || (_state.Cones.Length == 0))
    {
        // default state
        _state.Cones = new Vector3[]
        {
            new Vector3(0, 0, -5),
            new Vector3(10, 0, 4),
            new Vector3(-3, 0, 2)
        };
    }
}
```

Here, three traffic cone positions are specified. You may change this to add more as you improve your navigation algorithm.

Now you can add code to the PopulateWorld method to add these cones to your environment. It is important to add these objects and the barrier objects after the ground has been added to the simulation environment. Otherwise, it is possible that the timing will be just right for the cones to be inserted and start falling before the ground is inserted:

```
for (int coneCount = 0; coneCount < _state.Cones.Length; coneCount++)
    SimulationEngine.GlobalInstancePort.Insert(
        CreateTriggerCone(
            "Cone" + coneCount.ToString(),
            _state.Cones[coneCount]));
```

Give each cone a unique name and create one cone for each position in the state. Now that the service has state, you have a reason to implement the GetHandler as a method on the MagellanReferee class:

```
[ServiceHandler(ServiceHandlerBehavior.Concurrent)]
public virtual IEnumerator<ITask> GetHandler(Get get)
{
    get.ResponsePort.Post(_state);
    yield break;
}
```

Change the MagellanRefereeOperations port definition to add the Get operation:

```
[ServicePort]
public class MagellanRefereeOperations : PortSet<DsspDefaultLookup,
DsspDefaultDrop, Get>
{
}

public class Get : Get<GetRequestType, PortSet<MagellanRefereeState, Fault>>
{
}
```

Other services can now query for the location of the cones simply by sending a Get request to the referee service. You'll need to add an additional using statement at the top of the SimMagellan.cs file to resolve the Fault type:

```
using W3C.Soap;
```

The only remaining task for the referee is to create the star of the show, the CorobotEntity. Add the following code at the end of the PopulateWorld method:

```
// create a Corobot
SimulationEngine.GlobalInstancePort.Insert(
    new corobot.CorobotEntity("Corobot", new Vector3(0, 0, 0)));
```

To tell the service what a CorobotEntity is, add the following using statement at the top of the file, and add a reference to the Corobot service. Remember to set Copy Local and Specific Version to false in the properties of the Corobot service reference:

```
using corobot = ProMRDS.Simulation.Corobot;
```

Compile and run the referee service again to verify that you have sky, grass, walls, tower, cones, and a nice-looking Corobot. Your opening scene should look something like Figure 7-1 at the beginning of this chapter. If there are some differences in the colors, it may be because your graphics card is only able to use version 1.1 shaders. If so, try changing the quality setting in the Graphics Settings dialog box (from the Rendering menu) to see if your card will support higher-quality rendering.

Switch to the combined physics and visual view by pressing F2 several times or by selecting it from the Render menu. You will notice that trigger shapes are drawn in violet on the screen. In Figure 7-3, they are the gray lines surrounding the cone.

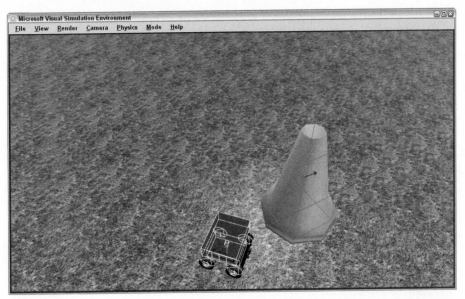

Figure 7-3

# The Robo-Magellan Orchestration Service

At last, you get to the service that implements the robot behavior. In this section, you'll generate a service that does not communicate directly with the simulator but instead gets its information about the simulated world from other services. You'll use the Manifest Editor to generate a manifest to run all of the services, and add a small Windows Forms user interface to enable user control and debugging of the service. You'll also use a very simple image processing method to enable the robot to see the cones in the environment.

## Generating the SimMagellan Service

Use the following steps to generate the SimMagellan service and to specify its references and ports:

1.  Use `DssNewService` to generate the service. The command line is simpler because you are not implementing an alternate contract. The following `dssnewservice` command is a single line but is shown on two lines due to formatting limitations. The text you type is shown in bold:

```
C:\Microsoft Robotics Studio (1.5)\ProMRDS>cd myChapter7

C:\Microsoft Robotics Studio (1.5)\ProMRDS\MyChapter7>dssnewservice
/s:"SimMagellan" /n:"ProMRDS.Simulation.Magellan" /year:"07" /month:"08"
C:\Microsoft Robotics Studio (1.5)\ProMRDS\MyChapter7>cd SimMagellan
C:\Microsoft Robotics Studio (1.5)\ProMRDS\MyChapter7>SimMagellan.csproj
```

2.  Edit the `DisplayName` and `Description` attributes in `SimMagellan.cs` to better describe the SimMagellan service.

**3.** Add the following using statements at the top of the SimMagellan.cs file:

```
using Microsoft.Robotics.PhysicalModel.Proxy;
using drive = Microsoft.Robotics.Services.Drive.Proxy;
using quadDrive = ProMRDS.Simulation.QuadDifferentialDrive.Proxy;
using webcam = Microsoft.Robotics.Services.WebCam.Proxy;
using referee = ProMRDS.Simulation.MagellanReferee.Proxy;
using irsensor = Microsoft.Robotics.Services.AnalogSensor.Proxy;
```

**4.** Add references to the following DLLs to the project. Notice that we are using the Proxy versions of each service DLL. RoboticsCommon and its Proxy are referenced because you make use of proxy and non-proxy types in the Microsoft.Robotics.PhysicalModel namespace. You reference proxy versions of the referee, the IR service, and the drive service. SimulationCommon.dll is referenced to pick up the types in Microsoft.Robotics.Simulation. Finally, you reference System.Drawing.dll because you will be using some bitmap types. Don't forget to set the Copy Local and Specific Version properties to false on each DLL reference.

```
RoboticsCommon
RoboticsCommon.proxy
SimMagellanReferee.Y2007.M08.Proxy.dll
SimulatedIR.Y2007.M08.Proxy.dll
SimulatedQuadDifferentialDrive.Y2007.M08.Proxy.dll
SimulationCommon.dll
System.Drawing.dll
```

**5.** Now add the following ports to the SimMagellan class. The service uses these ports to communicate with the other services:

```
// used to communicate with the QuadDifferentialDrive
[Partner(
    "QuadDrive",
    Contract = quadDrive.Contract.Identifier,
    CreationPolicy = PartnerCreationPolicy.UseExisting,
    Optional = false)]
quadDrive.QuadDriveOperations _quadDrivePort = new
    quadDrive.QuadDriveOperations();

// used to communicate with the DifferentialDrive alternate
[Partner(
    "Drive",
    Contract = drive.Contract.Identifier,
    CreationPolicy = PartnerCreationPolicy.UseExisting,
    Optional = false)]
drive.DriveOperations _drivePort = new drive.DriveOperations();

// used to communicate with the webcam on the Corobot
[Partner(
    "robotcam",
    Contract = webcam.Contract.Identifier,
    CreationPolicy = PartnerCreationPolicy.UseExisting,
    Optional = false)]
webcam.WebCamOperations _cameraPort = new webcam.WebCamOperations();
```

```
// used to communicate with the referee service
[Partner(
    "referee",
    Contract = referee.Contract.Identifier,
    CreationPolicy = PartnerCreationPolicy.UseExisting,
    Optional = false)]
referee.MagellanRefereeOperations _refereePort = new
    referee.MagellanRefereeOperations();

// used to communicate with the IR sensor service
[Partner(
    "irsensor",
    Contract = irsensor.Contract.Identifier,
    CreationPolicy = PartnerCreationPolicy.UseExisting,
    Optional = false)]
irsensor.AnalogSensorOperations _irSensorPort = new
    irsensor.AnalogSensorOperations();
irsensor.AnalogSensorOperations _irNotifyPort = new
    irsensor.AnalogSensorOperations();
```

Verify that the service will compile. You can run it but it doesn't do much yet. The next step is to define a manifest that will run all the services you need and link them together properly.

## Creating a Manifest with the Manifest Editor

The Manifest Editor is a new tool provided with the 1.5 release of Microsoft Robotics Developer Studio. It is an easy alternative to hand-editing confusing XML manifests. Any tool that enables you to avoid hand-editing .xml files is a good thing.

You'll use it here to set up the manifest for your scenario because it consists of several services.

Run the Manifest Editor by clicking Start ⇨ Microsoft Robotics Developer Studio (1.5) ⇨ Microsoft Dss Manifest Editor. It starts up with an empty manifest. The left-most column contains a list of all services installed on the system. The textbox at the top can filter the services to make it easier to select the one you need. The center column is where you'll build the manifest. The right-most column contains properties and other information about items that are selected in the manifest. Follow these steps to build a manifest for the simulated Robo-Magellan scenario:

1. Click the SimMagellan service in the left column and drag it to the center column. Release the mouse button. The SimMagellan service is now listed as part of the manifest and its five partners are listed below it. You'll resolve those in just a minute.

2. Add the following services by dragging them into an empty area of the center column:

```
SimulatedQuadDifferentialDriveService
Simulated Webcam
Simulated Magellan Referee
Simulated IR Distance Sensor
```

3. Resolve the SimMagellan service's QuadDrive partner by selecting the SimulatedQuadDifferentialDriveService listed in the center column and then dragging it to the QuadDrive partner box under the SimMagellan service. This sets up the manifest so that the SimulatedQuadDifferentialDrive service is paired with the _quadDrivePort in the SimMagellan service.

**349**

**4.** Drag the SimulatedQuadDifferentialDrive service from the center column to the Drive partner under the SimMagellan service to associate this service with the `_drivePort` in the SimMagellan service.

**5.** Resolve the other three partners under the SimMagellan service by dragging the appropriate service in the center column to the partner box.

**6.** Now you need to specify a simulation entity partner for the services that need one. Select `SimulatedQuadDifferentialDriveService` by clicking it. Now right-click it to bring up a context menu and select Add a Partnership. In the dialog box that appears, enter **Entity** (without quotes) in the Name box. Enter the SimulationEngine service contract in the Namespace box: **http://schemas.microsoft.com/robotics/2006/04/simulation.html** and press OK. A new Partner box labeled Entity appears under the `SimulatedQuadDifferentialDriveService` box. Select this partner by clicking it, and a Simulation Entity box appears in the Properties column on the right side of the screen. Type the name of the simulation entity that partners with the QuadDrive service in the form of a URI: **http://localhost/Corobot**. That entity name appears in the Partner box under the `SimulatedQuadDifferentialDriveService` in the manifest.

**7.** Add an entity partner to the Simulated IR Distance Sensor service with the name `http://localhost/Corobot_frontIR`, just as in step 6.

**8.** Add an entity partner to the SimulatedWebcam service with the name `http://localhost/Corobot_cam`, just as you did in steps 6 and 7.

**9.** Save the manifest with the name MySimMagellan. The Manifest Editor will create a directory called `MySimMagellan` containing the manifest `MySimMagellan.manifest.xml`. Your manifest should look similar to the one shown in Figure 7-4.

**Figure 7-4**

10. Run the manifest by clicking Run ⇨ Run Manifest. A console window is displayed and the manifest is loaded. Verify that no errors are displayed in the console window except for initial state files that were not found. The simulation window should appear and display that SimMagellan environment.

## Adding a Windows Forms User Interface

It is very helpful to retrieve the state of any running service using a web browser, but sometimes you need a real user interface to debug or control a service. In this section, you'll add a Windows Forms user interface to display information about the SimMagellan service and to reset and start it. Robotics Tutorial 4, provided with the MRDS SDK, explains how to add a simple Windows Form to your service. This example shows more aspects of using a Windows Form in your service.

## Creating the Form

For this part of the project, you can either create the form yourself or copy the SimMagellanUI.* files from the Chapter7\SimMagellan directory into the MyChapter7\SimMagellan project. To create your user interface, follow these steps:

1. Add a new Windows Form file to your SimMagellan project called SimMagellanUI.cs.

2. Add a Reset button called _resetButton. Figure 7-5 shows the controls on the SimMagellan UI Form that you're creating.

3. Add a Start button called _startButton to the form.

4. Add a PictureBox control called _cameraImage to the form. Give it a size of 320 × 240.

5. Add a label control called _stateLabel.

6. Add a label control called _elapsedLabel.

7. Change the title of the main window to SRS Robo-Magellan and add an icon if you like.

**Figure 7-5**

# Connecting the Service to the Form

With the form created and the controls placed, you're ready to hook up the communication between the service and the form. Windows Forms controls may only be accessed by the thread that created them. This is a problem in the MRDS environment because there are no guarantees regarding which thread is running a particular piece of code.

Fortunately, MRDS provides a set of code that makes it easy for services to interact with Windows Forms. Add the following `using` statement to your `SimMagellan.cs` file along with a reference to the `Ccr.Adapters.WinForms` DLL:

```
using Microsoft.Ccr.Adapters.WinForms;
```

The first thing you'll do is establish a way for the form to pass events back to the service. This is easily done by defining a port that can receive messages from the form. The `FromWinformEvents` class defined in the following snippet defines a port that can receive a single message type, a `FromWinformMsg`. This message contains a command, an array of string parameters, and an arbitrary object reference. The meaning of the string parameters and the object depends on the command in the message.

```csharp
#region WinForms communication

public class FromWinformEvents : Port<FromWinformMsg>
{
}

public class FromWinformMsg
{
    public enum MsgEnum
    {
        Loaded,
        Reset,
        Test,
        Start,
    }

    private string[] _parameters;
    public string[] Parameters
    {
        get { return _parameters; }
        set { _parameters = value; }
    }

    private MsgEnum _command;
    public MsgEnum Command
    {
        get { return _command; }
        set { _command = value; }
    }

    private object _object;
    public object Object
    {
        get { return _object; }
        set { _object = value; }
```

```
        }

        public FromWinformMsg(MsgEnum command, string[] parameters)
        {
            _command = command;
            _parameters = parameters;
        }
        public FromWinformMsg(
            MsgEnum command, string[] parameters, object objectParam)
        {
            _command = command;
            _parameters = parameters;
            _object = objectParam;
        }
    }
    #endregion
```

Add the definitions of the `FromWinformEvents` class and the `FromWinformMsg` class to `SimMagellanTypes.cs`. Add a member variable to the `SimMagellanUI` class in `SimMagellan.cs` to receive messages from the Windows Form:

```
// This port receives events from the user interface
FromWinformEvents _fromWinformPort = new FromWinformEvents();
```

Now add a handler to the interleave to listen for messages from this new port in the `Start` method, just after the call to `base.Start`:

```
// Add the winform message handler to the interleave
MainPortInterleave.CombineWith(new Interleave(
    new TeardownReceiverGroup(),
    new ExclusiveReceiverGroup
    (
        Arbiter.Receive<FromWinformMsg>(
            true, _fromWinformPort, OnWinformMessageHandler)
    ),
    new ConcurrentReceiverGroup()
));
```

Instantiating a Windows Form is as simple as posting a message to the `WinFormsServicePort` defined by the CCR Winforms Adapter immediately after the interleave is set:

```
// Create the user interface form
WinFormsServicePort.Post(new RunForm(CreateForm));
```

`RunForm` is a class defined by the Winforms Adapter that takes a delegate as a parameter. The delegate is responsible for instantiating the form. The adapter code starts a thread that is dedicated to the form and then uses that thread to call the passed delegate. In this case, the delegate is a method on the `SimMagellan` class called `CreateForm`:

```
// Create the UI form
System.Windows.Forms.Form CreateForm()
{
    return new SimMagellanUI(_fromWinformPort);
}
```

**353**

You pass a reference to `_fromWinformPort` so that the form can use this port to pass back messages.

In the `SimMagellanUI.cs` file, you need to modify the `SimMagellanUI` class constructor to accept this parameter:

```
FromWinformEvents _fromWinformPort;

public SimMagellanUI(FromWinformEvents EventsPort)
{
    _fromWinformPort = EventsPort;
    InitializeComponent();
    _fromWinformPort.Post(
        new FromWinformMsg(FromWinformMsg.MsgEnum.Loaded, null, this));
}
```

The constructor keeps a reference to this port in `_fromWinformPort` and then initializes itself and sends a message back to the SimMagellan service to indicate that the form is now loaded.

Back in the `SimMagellanUI` class, add a handler for this message as follows:

```
SimMagellanUI _magellanUI = null;

// process messages from the UI Form
void OnWinformMessageHandler(FromWinformMsg msg)
{
    switch (msg.Command)
    {
        case FromWinformMsg.MsgEnum.Loaded:
            // the windows form is ready to go
            _magellanUI = (SimMagellanUI)msg.Object;
            break;

    }
}
```

When the `Loaded` message is received from the form, the service stores a reference to the form. This reference sends commands to the form.

Now you have established a way for the service to send commands and information to the form, and a way for the form to notify the service when user interface events occur.

You should take this opportunity to compile your SimMagellan service. Run it using the Manifest Editor as described in the section "Creating a Manifest with the Manifest Editor." The form associated with the SimMagellan service should appear when the service runs.

# How to Make a Robot Behave

The next task is to give the robot an objective and a way to accomplish that objective. In this case, the objective is fairly clear. Your robot needs to ask the referee service where the cones are, and then it needs to navigate to each one and touch it.

An common method for defining a behavior like this is to set up a state machine. A *state machine* defines several states that represent a current situation for the robot. Certain events from the outside world can change the state. For example, the robot could have a `Wander` state whereby it is moving about the environment trying to get closer to a cone. When the IR sensor detects an obstacle nearby, the state may change from `Wander` to `AvoidCollision`, and the SimMagellan service will issue commands to the Drive service to move the robot away from the obstacle. When the obstacle has been avoided, the robot state will change back to `Wander`.

In this section, you'll set up the state machine for the robot and define the actions it takes in each of these states. It should be stated again that this behavior algorithm is by no means optimal. You've done a lot of work in this chapter to set up the scenario and you'll have a basic behavior implemented to demonstrate it, but the whole purpose of the exercise is to provide a way to refine the behavioral algorithm and to prototype new ideas.

## Defining Behavior States

You will be defining the following states. Each state listed here includes a brief description of the associated robot behavior:

- **NotSpecified:** This is the initial state of the state machine. The robot does nothing at all in this state. It moves to the `Reset` state when the Reset button is pressed.

- **Reset:** In this state, the robot begins processing camera frames, displaying the elapsed time on the user interface. It retrieves a list of the cone positions from the Referee service and subscribes to the IR Sensor service for notifications. It enables the robot drive and moves the robot to its initial position. After all this is done, the state is changed to `Ready`.

- **Ready:** The `Ready` state is much like the `NotSpecified` state. The robot does nothing at all. It changes to the `Wander` state when the Start button is pressed.

- **Wander:** This is the main state of the robot while it navigates in the environment. In this state, the robot checks the last image received from the camera. If it detects a cone, then the state changes to `Approach`. Otherwise, the robot uses a method called `GetHappiness` to determine its next move. This method returns a "happiness" value that is calculated according to the robot's proximity to a cone that has not yet been touched. The idea is that the robot seeks higher and higher levels of happiness by navigating closer to a cone until it is close enough to see it and move directly toward it. There is some degree of randomness associated with the movement in this state because set patterns of movement can sometimes result in endless cycles. If the happiness level in the current position is greater than the happiness level in the previous position, the robot continues on its current course. Otherwise, it backs up to its previous position, adjusts its heading randomly, and then proceeds forward in the next direction. This behavior is repeated until the robot moves into the `Approach` state or the `AvoidCollision` state.

- **AvoidCollision:** This state is entered when the IR Sensor service notifies the SimMagellan service that the distance measurement has changed and is less than one meter. In this state, the robot backs away a quarter meter, turns 90 degrees to the left, and then moves forward one meter. The robot then returns to the `Wander` state.

- **Approach:** When the robot sees a cone, it enters the `Approach` state. In this state, the robot checks the distance from the IR sensor. If it is less than a meter, it goes to the `FinalApproach` state. Otherwise, it checks the position of the cone in the camera frame, adjusts its heading to head straight for the cone, and then moves forward a quarter meter and repeats the process.

❑ **FinalApproach:** By the time the robot reaches this state, it is closer than one meter to the cone and pointed more or less straight at the cone. In this mode, the robot no longer relies on the camera image but only on the IR distance data. It waits for another IR distance reading and then slowly moves that distance toward the cone. Then it marks the cone as having been touched, after which that cone is no longer considered in the GetHappiness method. After this is done, the robot enters the BackAway state.

❑ **BackAway:** The purpose of this state is to allow the robot to back away from the cone it just touched. It moves backward one meter and then turns 45 degrees to the left and resumes wandering. If all of the cones are touched, then the robot enters the Finished state.

❑ **Finished:** At this point, the robot has accomplished its objective. The elapsed time is no longer incremented on the user interface, and the robot spins in circles to show that it is victorious. It doesn't leave this state unless the Reset button is pushed to restart the scenario.

## Implementing the Behavior for Each State

Now that you understand the states and behaviors defined in the last section, you are ready to write the code to implement this state machine. The current robot state will be stored in the SimMagellan service state. Replace the definition of the SimMagellanState class in SimMagellanTypes.cs as follows:

```
[DataContract]
public enum ModeType
{
    NotSpecified = 0,
    Reset,
    Ready,
    Wander,
    AvoidCollision,
    Approach,
    FinalApproach,
    BackAway,
    Finished
}

/// <summary>
/// The SimMagellan State
/// </summary>
[DataContract()]
public class SimMagellanState
{
    [DataMember]
    public ModeType CurrentMode;
}
```

Add a BehaviorLoop method to the SimMagellan class:

```
// keep track of the time to finish
DateTime _startTime;
DateTime _endTime;

const float _rotateSpeed = 0.3f;
const float _driveSpeed = 0.5f;
```

```csharp
bool _iteratorsStarted = false;

IEnumerator<ITask> BehaviorLoop()
{
    bool driveCommandFailed = false;
    ModeType previousMode = ModeType.NotSpecified;
    while (true)
    {
        driveCommandFailed = false;

        if (previousMode != _state.CurrentMode)
        {
            // update the UI
            WinFormsServicePort.FormInvoke(
                delegate()
                {
                    _magellanUI.SetCurrentState(
                        _state.CurrentMode.ToString());
                }
            );
        }
        previousMode = _state.CurrentMode;
        switch (_state.CurrentMode)
        {
            case ModeType.NotSpecified:
            case ModeType.Ready:
                yield return Arbiter.Receive(
                    false,
                    TimeoutPort(100),
                    delegate(DateTime timeout) { });
                break;
        }
    }
}
```

For the previous block of code, the _driveSpeed and _rotateSpeed constants are used to set the default speed for the robot to drive and turn. _startTime and _endTime are used to keep track of how much time has elapsed since the Start button was pressed. The first thing the BehaviorLoop method does is to initialize the state to NotSpecified. It then enters an endless loop to continually process the robot's behavior. Each time through the loop, it checks whether the mode has changed. If it has, you call the SetCurrentState API on the SimMagellanUI class to display the current state on the user interface. Notice that you do this by posting a FormInvoke message on the WinFormsServicePort with a delegate containing the code you want to run. This code is always executed on the Windows Forms thread.

A switch statement is used to take a specific action based on the current state. You've already implemented a behavior for two states: NotSpecified and Ready. The behavior for these states is to wait 100 ms and then break out of the switch statement to go through the loop again. The only way to break out of one of these states is to click the Reset or Start buttons.

Add a SpawnIterator call at the end of the Start method so that the BehaviorLoop begins after the service has initialized:

```csharp
SpawnIterator(BehaviorLoop);
```

The next step is to add methods in the SimMagellanUI class that will process these button clicks and send a notification to the SimMagellan service. Add the following two event handlers to the _resetButton and _startButton controls in the SimMagellanUI class:

```
private void _startButton_Click(object sender, EventArgs e)
{
    _fromWinformPort.Post(
        new FromWinformMsg(FromWinformMsg.MsgEnum.Start, null));
}

private void _resetButton_Click(object sender, EventArgs e)
{
    _fromWinformPort.Post(
        new FromWinformMsg(FromWinformMsg.MsgEnum.Reset, null));
    _startButton.Enabled = true;
}
```

You also need to hook the Click event for each of the buttons in SimMagellanUI.designer.cs as shown here. If you double-click on the buttons in the designer, this is done for you:

```
this._resetButton.Click += new System.EventHandler(this._resetButton_Click);
this._startButton.Click +=new System.EventHandler(this._startButton_Click);
```

Each of these handlers posts a message to the _fromWinformPort to tell the SimMagellan service that a button was clicked. While you're modifying the SimMagellanUI class, you may as well add the methods that update the current state and the elapsed time labels on the form:

```
public void SetCurrentState(string state)
{
    _stateLabel.Text = "State: " + state;
}

public void SetElapsedTime(string time)
{
    _elapsedLabel.Text = time;
}
```

Back in the SimMagellan class, add the following two case statements to the OnWinformMessageHandler to process these messages from the user interface:

```
case FromWinformMsg.MsgEnum.Reset:
    _state.CurrentMode = ModeType.Reset;
    break;

case FromWinformMsg.MsgEnum.Start:
    if(_state.CurrentMode == ModeType.Ready)
    {
        _startTime = DateTime.Now;
        _state.CurrentMode = ModeType.Wander;
    }
    break;
```

At this point, you should be able to compile the SimMagellan service and run it. When you click the Reset button, the State label should change to Reset.

Now you're ready to implement the behavior for the Reset state. Add the following `case` statement to the `BehaviorLoop` method:

```
case ModeType.Reset:
    {
        if (!_iteratorsStarted)
        {
            // start processing camera frames
            SpawnIterator<DateTime>(DateTime.Now, ProcessFrame);

            // start displaying elapsed time
            SpawnIterator<DateTime>(DateTime.Now, UpdateElapsedTime);

            _irSensorPort.Subscribe(_irNotifyPort);

            MainPortInterleave.CombineWith(new Interleave(
                new TeardownReceiverGroup(),
                new ExclusiveReceiverGroup(),
                new ConcurrentReceiverGroup
                (
                    Arbiter.Receive<irsensor.Replace>(
                        true,
                        _irNotifyPort,
                        IRNotifyReplaceHandler)
                )
            ));

            // we don't want to start additional iterators if the reset
            // button is pushed to restart the scenario
            _iteratorsStarted = true;
        }

        yield return Arbiter.Choice(_refereePort.Get(),
            delegate(referee.MagellanRefereeState state)
                { SetWaypoints(state); },
            delegate(Fault f) { }
            );

        _drivePort.EnableDrive(true);
        _drivePort.SetDriveSpeed(0, 0);
        _quadDrivePort.SetPose(
            new quadDrive.SetPoseRequestType(
                new Microsoft.Robotics.PhysicalModel.Proxy.Pose()));
        _state.CurrentMode = ModeType.Ready;

        break;
    }
```

Several things are going on here to get our little robot ready to start. First, you start the iterator methods that process the camera frames and update the elapsed time. You'll look at those in just a bit.

Next, you subscribe to notifications from the IR Sensor service. This means that each time the value of the front IR sensor changes, the _irNotifyPort receives a message with the new IR sensor service state. You add a handler to the interleave to handle this message.

Next, you post a Get message to the Referee service to get a list of cone positions. You keep track of these positions in a list of waypoints. A *waypoint* holds the position of a cone and a Boolean indicating whether the cone has been visited.

Finally, the drive is enabled and a SetPose message is sent to the _quadDrivePort to move the robot back to the starting position. The default Pose that is sent has a position of (0,0,0) and an orientation with no rotations.

Let's look at the SetWaypoints method first because it is the simplest:

```
private void SetWaypoints(referee.MagellanRefereeState state)
{
    foreach (Vector3 location in state.Cones)
        _waypoints.Add(new Waypoint(location));
}
```

The _waypoints member is just a list of Waypoint objects, which are defined as follows:

```
class Waypoint
{
    public Vector3 Location;
    public bool Visited;

    public Waypoint(Vector3 location)
    {
        Location = location;
        Visited = false;
    }
}
```

Declare _waypoints in the SimMagellan class as follows:

```
List<Waypoint> _waypoints = new List<Waypoint>();
```

Now you need to add the UpdateElapsedTime method. This method runs in an endless loop to determine the elapsed time depending on the current state. If the string representing the time has changed since the last time it has been displayed, it is sent to the UI with a call to _magellanUI.SetElapsedTime:

```
private IEnumerator<ITask> UpdateElapsedTime(DateTime timeout)
{
    string previous = string.Empty;
    while (true)
    {
        string newString = string.Empty;

        switch (_state.CurrentMode)
```

```
            {
                case ModeType.NotSpecified:
                case ModeType.Ready:
                case ModeType.Reset:
                    newString = string.Format("Elapsed: {0}:{1:D2}", 0, 0);
                    break;

                case ModeType.Wander:
                case ModeType.Approach:
                case ModeType.FinalApproach:
                case ModeType.BackAway:
                    {
                        TimeSpan elapsed = DateTime.Now - _startTime;
                        newString = string.Format("Elapsed: {0}:{1:D2}",
                            elapsed.Minutes, elapsed.Seconds);
                        break;
                    }

                case ModeType.Finished:
                    {
                        TimeSpan elapsed = _endTime - _startTime;
                        newString = string.Format("Elapsed: {0}:{1:D2}",
                            elapsed.Minutes, elapsed.Seconds);
                        break;
                    }
            }

            if (newString != previous)
            {
                previous = newString;
                // update the UI
                WinFormsServicePort.FormInvoke(
                    delegate()
                    {
                        _magellanUI.SetElapsedTime(newString);
                    }
                );
            }

            yield return Arbiter.Receive(
                false, TimeoutPort(100), delegate { });
        }
    }
```

The handler for the IR Distance Sensor aborts any current drive requests by posting an `AllStop` message to the `_drivePort` if the distance reading is less than one meter. Any `DriveDistance` or `RotateDegrees` operations in progress will return a `Fault` response, which aborts the current operation and causes the state machine to begin processing the `AvoidCollision` state:

```
void IRNotifyReplaceHandler(irsensor.Replace replace)
{
    if (replace.Body.RawMeasurement < 1f)      // closer than 1 meter
    {
        if (_state.CurrentMode == ModeType.Wander)
```

*(continued)*

*(continued)*

```
            {
                // stop dead in our tracks, abort DriveDistance and
                // RotateDegrees in progress
                _drivePort.AllStop();
                _state.CurrentMode = ModeType.AvoidCollision;
            }
        }
    }
```

# Processing Camera Frames

The last method to implement to make the Reset state complete is ProcessFrame. Because this method is a little more complicated, we'll take a closer look at it.

The topic of robot vision and navigating using vision could fill many books, and it is still a largely unsolved problem. Fortunately, your requirements are simple, so the vision algorithm doesn't need to be very complicated.

Vision algorithms can be very difficult to debug, so it is useful to have the SimMagellan service provide some feedback about what it is doing. The ProcessFrame method is quite simple by itself. It sends a request to the SimulatedWebCam service each 200 milliseconds:

```
private IEnumerator<ITask> ProcessFrame(DateTime timeout)
{
    while (true)
    {

        yield return Arbiter.Choice(
            _cameraPort.QueryFrame(),
            ValidateFrameHandler,
            DefaultFaultHandler);

        yield return Arbiter.Receive(
            false, TimeoutPort(200), delegate { });
    }
}
```

The DefaultFaultHandler simply logs the fault in case the camera was unable to return a frame:

```
void DefaultFaultHandler(Fault fault)
{
    LogError(fault);
}
```

The camera frame response is handled by the ValidateFrameHandler. This handler checks the timestamp on the camera frame and discards the image if it is older than one second. This prevents the robot from making decisions based on stale data in case something delays the camera from sending frames. The frame is analyzed with the ProcessImage method and is then sent to the user interface form along with the analysis results. The analysis results are used to draw a rectangle around recognized objects:

```
private void ValidateFrameHandler(webcam.QueryFrameResponse cameraFrame)
{
    try
    {
        if (cameraFrame.Frame != null)
        {
            DateTime begin = DateTime.Now;
            double msFrame =
                begin.Subtract(cameraFrame.TimeStamp).TotalMilliseconds;

            // Ignore old images!
            if (msFrame < 1000.0)
            {
                _imageProcessResult = ProcessImage(
                    cameraFrame.Size.Width,
                    cameraFrame.Size.Height,
                    cameraFrame.Frame);

                WinFormsServicePort.FormInvoke(
                    delegate()
                    {
                        _magellanUI.SetCameraImage(
                            cameraFrame.Frame, _imageProcessResult);
                    }
                );
            }
        }
    }
    catch (Exception ex)
    {
        LogError(ex);
    }
}
```

Add an `ImageProcessResult` member variable to the SimMagellan service and add the definition for the `ImageProcessResult` class:

```
// latest image processing result
ImageProcessResult _imageProcessResult = null;
 [DataContract]
public class ImageProcessResult
{
    [DataMember]
    public int XMean;

    [DataMember]
    public float RightFromCenter;

    [DataMember]
    public int YMean;
```

*(continued)*

(continued)

```
      [DataMember]
      public float DownFromCenter;

      [DataMember]
      public int Area;

      [DataMember]
      public DateTime TimeStamp;

      [DataMember]
      public int AreaThreshold = 50*50;
  }
```

The definition of the `SetCameraImage` method on the `SimMagellanUI` class is fairly simple. It creates an empty bitmap and attaches it to the `PictureBox` control on the form if one doesn't already exist. It locks the bitmap and then copies the bits from the camera frame. After unlocking the bitmap, it draws a rectangle around the area identified by the `ImageProcessResult` if the area of the result is large enough. It also draws a small crosshair in the center of the area. The presence or absence of the white rectangle is a good way to know when the vision algorithm detects an object:

```
public void SetCameraImage(byte[] frame, ImageProcessResult result)
{
    if (_cameraImage.Image == null)
        _cameraImage.Image = new Bitmap(
            320, 240,
            System.Drawing.Imaging.PixelFormat.Format24bppRgb);

    Bitmap bmp = (Bitmap)_cameraImage.Image;
    System.Drawing.Imaging.BitmapData bmd = bmp.LockBits(
        new Rectangle(0, 0, bmp.Width, bmp.Height),
        System.Drawing.Imaging.ImageLockMode.ReadWrite,
        bmp.PixelFormat);

    System.Runtime.InteropServices.Marshal.Copy(
        frame, 0, bmd.Scan0, frame.Length);

    bmp.UnlockBits(bmd);

    if (result.Area > result.AreaThreshold)
    {
        int size = (int)Math.Sqrt(result.Area);
        int offset = size / 2;
        Graphics grfx = Graphics.FromImage(bmp);
        grfx.DrawRectangle(
            Pens.White,
            result.XMean - offset,
            result.YMean - offset, size, size);
        grfx.DrawLine(
            Pens.White,
            result.XMean - 5,
            result.YMean, result.XMean + 5, result.YMean);
        grfx.DrawLine(
            Pens.White,
            result.XMean,
```

```
                        result.YMean - 5, result.XMean, result.YMean + 5);
        }

        _cameraImage.Invalidate();
    }
```

The last method that should be explained is the `ProcessImage` function. Many vision algorithms do sophisticated edge detection or gradient processing. This simple algorithm is actually adapted from the vision algorithm shipped as part of the MRDS Sumo Competition package. It simply scans through the pixels and keeps track of how many pixels meet some criteria. It also keeps track of the average position of those pixels in the image:

```
private ImageProcessResult ProcessImage(
int width, int height, byte[] pixels)
{
    if (pixels == null || width < 1 || height < 1 || pixels.Length < 1)
        return null;

    int offset = 0;
    float threshold = 2.5f;
    int xMean = 0;
    int yMean = 0;
    int area = 0;

    // only process every fourth pixel
    for (int y = 0; y < height; y += 2)
    {
        offset = y * width * 3;
        for (int x = 0; x < width; x += 2)
        {
            int r, g, b;

            b = pixels[offset++];
            g = pixels[offset++];
            r = pixels[offset++];

            float compare = b * threshold;
            if((g > compare) && (r > compare))
            {
                // color is yellowish
                xMean += x;
                yMean += y;
                area++;

                // debug coloring
                pixels[offset - 3] = 255;
                pixels[offset - 2] = 0;
                pixels[offset - 1] = 255;
            }
            offset += 3;     // skip a pixel
        }
    }

    if (area > 0)
```

*(continued)*

*(continued)*

```
{
    xMean = xMean / area;
    yMean = yMean / area;
    area *= 4;
}

ImageProcessResult result = new ImageProcessResult();

result.XMean = xMean;
result.YMean = yMean;
result.Area = area;
result.RightFromCenter =
    (float)(xMean - (width / 2)) / (float)width;
result.DownFromCenter =
    (float)(yMean - (height / 2)) / (float)height;

return result;
}
```

In this case, you know that the cones are yellow and that there are few, if any, other yellow objects in the scene. Therefore, you look for pixels that have a red and green component much larger than their blue component. As an additional debug aid, the code colors the identified pixels magenta. This is a great debugging aid because it enables you to see exactly which pixels are being considered in the result.

To aid in performance, you need only analyze every other pixel in the image in width and height. The area is incremented by four for every pixel that is identified because each pixel represents four pixels in the image.

This is a good time to compile your service again and observe the behavior of the vision algorithm. Run your mySimMagellan manifest. When all the services have started, click the Reset button on the SimMagellan user interface form. The form should begin displaying the image from the Corobot camera updated five times per second. You can start the Simulation Editor by pressing F5 in the simulation window. Select the Corobot entity in the upper-left pane and select its position property in the lower-left pane. Drag the Corobot with the left mouse button while holding down the Ctrl key. Move the Corobot so that it is facing one of the street cones and verify that the image processing algorithm correctly identifies the object, and colors every other yellow pixel magenta, as in Figure 7-6.

Figure 7-6

## The Wander Behavior

Most of the time, the robot navigates and discovers the environment using the Wander state. The behavior in this state is intentionally left a little "fuzzy" and random. In part, this acknowledges that in the real world, positional data from GPS sensors has varying and limited accuracy. It is also an intentional way to deal with cyclical behavior. It is common to see a robot with a very specific objective get into a cycle in which it heads straight for its objective only to hit an obstacle, take a quick detour, and then head again straight for its objective and hit the same obstacle again. Without some random behavior and the ability to switch objectives, the robot may never complete the course. Once again, though, be aware that none of these behaviors is optimal and there is a lot of room for improvement and experimentation.

This is the code for the Wander case in the BehaviorLoop switch statement:

```
case ModeType.Wander:
    {
        // check to see if we have identified a cone
        if (_imageProcessResult != null)
        {
            if (_imageProcessResult.Area >
                _imageProcessResult.AreaThreshold)
            {
                // we have a cone in the crosshairs
                _state.CurrentMode = ModeType.Approach;
                continue;
            }
        }

        quadDrive.DriveDifferentialFourWheelState quadDriveState = null;
        yield return Arbiter.Choice(_quadDrivePort.Get(),
            delegate(quadDrive.DriveDifferentialFourWheelState state)
            {
                quadDriveState = state;
            },
            delegate(Fault f) { }
        );

        if (quadDriveState == null)
            continue;

        double distance;
        double degrees;
        float currentHappiness = GetHappiness(quadDriveState.Position);
        if (currentHappiness > _previousHappiness)
        {
            // keep going in the same direction
            distance = 0.25;
            degrees = 0;
        }
        else
        {
            // back up and try again
            if(!driveCommandFailed)
```

*(continued)*

*(continued)*

```
                    yield return Arbiter.Choice(
                        _drivePort.DriveDistance(-0.25, _driveSpeed),
                        delegate(DefaultUpdateResponseType response) { },
                        delegate(Fault f) { driveCommandFailed = true; }
                    );

                distance = 0.25;
                // choose an angle between -45 and -90.
                degrees = (float)_random.NextDouble() * -45f - 45f;
                // restore happiness level of previous position
                _previousHappiness = _previousPreviousHappiness;
            }
            if (degrees != 0)
            {
                if (!driveCommandFailed)
                    yield return Arbiter.Choice(
                        _drivePort.RotateDegrees(degrees, _rotateSpeed),
                        delegate(DefaultUpdateResponseType response) { },
                        delegate(Fault f) { driveCommandFailed = true; }
                    );
            }

            if (!driveCommandFailed)
                yield return Arbiter.Choice(
                    _drivePort.DriveDistance(distance, _driveSpeed),
                    delegate(DefaultUpdateResponseType response) { },
                    delegate(Fault f) { driveCommandFailed = true; }
                );

            _previousPreviousHappiness = _previousHappiness;
            _previousHappiness = currentHappiness;

            break;
    }
```

The first thing the robot does is check whether the image processing method has found a cone. If it has, it changes to the Approach state. If not, it continues wandering for another step.

Next, find out where the robot is by asking the SimulatedQuadDifferentialDrive for the current position. In the real world, this would be a GPS service and the coordinates returned would be latitude and longitude. Then, you call GetHappiness to find out how happy the robot is in this position. Being near a cone makes the robot very happy.

If the robot's happiness in this location is greater than the happiness in its previous location, it decides to continue moving in this direction for another quarter meter. Otherwise, it backs up to the previous position and picks a new direction. The new direction is a random angle between −45 and −90 degrees.

Notice that the code checks whether any previous drive command failed while processing this state. If it did, that could mean that the IR Distance Sensor service sent a notification that the robot is closer than one meter to an obstacle, terminating the current command by sending an AllStop message. In this case, you want to abort the current state processing and get back to the top of the switch statement because the state has probably changed to AvoidCollision.

As mentioned earlier, the GetHappiness method is a way for the robot to be motivated to move in the right direction. When it is moving toward a cone, happiness increases. Moving away from a cone decreases its happiness and causes it to rethink its life decisions. This is a fuzzy way of encouraging the robot to go in the right direction. The code is fairly simple:

```
float _previousHappiness = 0;
float _previousPreviousHappiness = 0;
Random _random = new Random();

float GetHappiness(Vector3 pos)
{
    float happiness = 0;
    for(int i=0; i<_waypoints.Count; i++)
    {
        if (_waypoints[i].Visited)
            continue;

        float dx = (pos.X - _waypoints[i].Location.X);
        float dz = (pos.Z - _waypoints[i].Location.Z);
        float distance = (float)Math.Sqrt(dx * dx + dz * dz);

        float proximity = 50f - Math.Min(distance, 50f);
        if(proximity > 40f)
            proximity = proximity * proximity;

        happiness += proximity;
    }
    return happiness;
}
```

The code only considers waypoints that have not been visited yet. As soon as a waypoint is visited, it no longer makes the robot happy and it moves on to other pursuits, not unlike some people. The happiness function is simply the sum of the proximity to each nonvisited cone, squared. If a cone is more than 50 meters away, then it doesn't contribute to the happiness function.

## The Approach State

When the vision processing algorithm detects that the robot is close enough to a cone to exceed the area threshold, the robot enters the Approach state, which has the following implementation:

```
case ModeType.Approach:
    {
        float IRDistance = -1f;

        // get the IR sensor reading
        irsensor.Get tmp = new irsensor.Get();
        _irSensorPort.Post(tmp);
        yield return Arbiter.Choice(tmp.ResponsePort,
            delegate(irsensor.AnalogSensorState state)
            {
                if (state.NormalizedMeasurement < 1f)
```

*(continued)*

*(continued)*

```
                      IRDistance = (float)state.RawMeasurement;
            },
            delegate(Fault f) { }
        );

        if (IRDistance >= 0)
        {
            // rely on the IR sensor for the final approach
            _state.CurrentMode = ModeType.FinalApproach;
            break;
        }

        if ((_imageProcessResult == null) ||
            (_imageProcessResult.Area <
             _imageProcessResult.AreaThreshold))
        {
            // lost it, go back to wander mode
            _state.CurrentMode = ModeType.Wander;
            continue;
        }
        float angle = _imageProcessResult.RightFromCenter * -2f * 5f;
        float distance = 0.25f;
        if (!driveCommandFailed)
            yield return Arbiter.Choice(
                _drivePort.RotateDegrees(angle, _rotateSpeed),
                delegate(DefaultUpdateResponseType response) { },
                delegate(Fault f) { driveCommandFailed = true; }
            );

        if (!driveCommandFailed)
            yield return Arbiter.Choice(
                _drivePort.DriveDistance(distance, _driveSpeed),
                delegate(DefaultUpdateResponseType response) { },
                delegate(Fault f) { driveCommandFailed = true; }
            );
        break;
    }
```

In the `Approach` state, the code first checks whether the robot is within one meter of an obstacle, assuming that the obstacle is a cone. If so, the robot moves to the `FinalApproach` state. If the robot no longer has the cone in its sights, it goes back to `Wander` mode. Otherwise, it needs to get closer.

You use the `_imageProcessResult.RightFromCenter` value to determine how far right or left the robot needs to turn in order to center the cone in the camera frame. The numbers used here were arrived at through experimentation. The code then adjusts the angle and moves forward a quarter meter. Eventually, the robot gets close enough to the cone to enter `FinalApproach` mode or it loses sight of the cone and goes back into `Wander` mode until the cone is sighted again.

## The Final Approach State

The FinalApproach state has the following implementation:

```
case ModeType.FinalApproach:
{
    // rely on the IR sensor for the remainder of the cone approach
    float IRDistance = -1f;

    // get the IR sensor reading
    irsensor.Get tmp = new irsensor.Get();
    _irSensorPort.Post(tmp);
    yield return Arbiter.Choice(tmp.ResponsePort,
        delegate(irsensor.AnalogSensorState state)
        {
            if (state.NormalizedMeasurement < 1f)
                IRDistance = (float)state.RawMeasurement;
        },
        delegate(Fault f) { }
    );

    if (IRDistance < 0)
    {
        // go back to the visual approach
        _state.CurrentMode = ModeType.Approach;
        break;
    }
    if (!driveCommandFailed)
        yield return Arbiter.Choice(
            _drivePort.DriveDistance(IRDistance, 0.2),
            delegate(DefaultUpdateResponseType response) { },
            delegate(Fault f) { driveCommandFailed = true; }
        );

    // mark the cone as touched
    quadDrive.DriveDifferentialFourWheelState quadDriveState = null;
    yield return Arbiter.Choice(_quadDrivePort.Get(),
        delegate(quadDrive.DriveDifferentialFourWheelState state)
        {
            quadDriveState = state;
        },
        delegate(Fault f) { }
    );

    if (quadDriveState != null)
    {
        for (int i = 0; i < _waypoints.Count; i++)
        {
            float dx = (quadDriveState.Position.X -
                        _waypoints[i].Location.X);
            float dz = (quadDriveState.Position.Z -
                        _waypoints[i].Location.Z);
            float distance = (float)Math.Sqrt(dx * dx + dz * dz);
            if (distance < 1f)
```

*(continued)*

(continued)

```
                              _waypoints[i].Visited = true;
                 }
         }

         _state.CurrentMode = ModeType.BackAway;

         break;
     }
```

In the `FinalApproach` state, you rely only on the IR Distance Sensor. The robot makes a reading of the distance to the cone and then moves exactly that far forward. In the real world, where IR distance readings are not always exact, it would probably be necessary to make smaller steps toward the cone. When the robot moves close enough to the cone to touch it, you ask the drive for the robot's position and then determine which cone it was that the robot touched, marking it as touched. Then the robot goes into the `BackAway` state that moves it away from the cone.

## The Back Away State

The `BackAway` state simply moves the robot away from the cone it just touched and points it in a different direction:

```
case ModeType.BackAway:
    {
        // back away and turn around
        if (!driveCommandFailed)
            yield return Arbiter.Choice(
                _drivePort.DriveDistance(-1, _driveSpeed),
                delegate(DefaultUpdateResponseType response) { },
                delegate(Fault f) { driveCommandFailed = true; }
            );

        if (!driveCommandFailed)
            yield return Arbiter.Choice(
                _drivePort.RotateDegrees(45, _rotateSpeed),
                delegate(DefaultUpdateResponseType response) { },
                delegate(Fault f) { driveCommandFailed = true; }
            );

        _state.CurrentMode = ModeType.Finished;
        for (int i = 0; i < _waypoints.Count; i++)
        {
            if (!_waypoints[i].Visited)
            {
                _state.CurrentMode = ModeType.Wander;
                _previousHappiness = 0;
                _previousPreviousHappiness = 0;
                break;
            }
        }

        if (_state.CurrentMode == ModeType.Finished)
```

```
        {
                // all the cones have been visited
                _endTime = DateTime.Now;
        }

        break;
    }
```

After that, it checks whether all the cones have been touched. If so, the robot enters the `Finished` state. Otherwise, it returns to the `Wander` state to find more cones.

## The Finished State

The `Finished` state simply causes the robot to spin around in wild celebration until you click the Reset button to try the scenario again:

```
case ModeType.Finished:
    if (!driveCommandFailed)
        yield return Arbiter.Choice(
            _drivePort.RotateDegrees(180, _rotateSpeed),
            delegate(DefaultUpdateResponseType response) { },
            delegate(Fault f) { driveCommandFailed = true; }
        );
    break;
```

## The Avoid Collision State

The only state that hasn't been described yet is the `AvoidCollision` state. Right now, this state does the same thing regardless of where the obstacle lies. It would be a great improvement if the robot did more discovery to determine the direction to drive to actually avoid the obstacle, rather than blunder around:

```
case ModeType.AvoidCollision:
    {
        // back away
        if (!driveCommandFailed)
            yield return Arbiter.Choice(
                _drivePort.DriveDistance(-0.25, _driveSpeed),
                delegate(DefaultUpdateResponseType response) { },
                delegate(Fault f) { driveCommandFailed = true; }
            );

        // turn left
        if (!driveCommandFailed)
            yield return Arbiter.Choice(
                _drivePort.RotateDegrees(90, _rotateSpeed),
                delegate(DefaultUpdateResponseType response) { },
                delegate(Fault f) { driveCommandFailed = true; }
            );

        _state.CurrentMode = ModeType.Wander;
```

*(continued)*

*(continued)*

```
            // move forward
            if (!driveCommandFailed)
                yield return Arbiter.Choice(
                    _drivePort.DriveDistance(1, _driveSpeed),
                    delegate(DefaultUpdateResponseType response) { },
                    delegate(Fault f) { driveCommandFailed = true; }
                );
            break;
    }
```

As you can see from the implementation, when the robot detects an impending collision, it backs up a quarter meter, turns 90 degrees to the left, changes to `Wander` mode, and then drives forward one meter.

Add the state implementations just discussed to your code, along with the `GetHappiness` method and its associated variables, and your SimMagellan service is complete.

# Using the SimMagellan Service

Once you click Reset and Start, the service runs itself. You will probably have to move the `MainCamera` around a bit to keep the Corobot in view, and you'll definitely want to be watching the SimMagellan user interface form to see the results of the image processing and the current state of the robot. The elapsed time gives you a good measure of your algorithm's efficiency, although the same algorithm can have radically different execution times as a result of very small variations in the robot's behavior. Each time you enhance the robot behavior, measure the elapsed time of a few runs to determine the amount of improvement you have achieved.

Don't forget that you can use the Simulation Editor to select the `Corobot` entity and move it around in the environment. This can come in handy if it happens to hit a wall on the edge and flip over or if you just want to see how it handles a particular approach to a cone.

# Future Directions

There are several things you can do with this simulation scenario to make it more like the actual SRS Robo-Magellan competition. For one thing, very few robot drives can report back on their actual position in the world. You should implement a simulated GPS entity and associated service that returns the position of the robot in latitude and longitude. Such a simulated service should probably also provide a way to model areas of poor reception in the world so that the navigation algorithm can handle these properly.

Another improvement to this simulation would be to change the happiness function to make the robot traverse the cones in a particular order. In the current simulation, the order in which the robot visits the cones is unimportant. One simple way to make the obstacle avoidance more efficient is to add waypoints with a negative happiness to the list each time an obstacle is encountered. This would keep the robot from moving toward an area where there is a known obstacle.

Finally, the obstacles in the environment are currently fairly simple. It is fairly easy to reconfigure the environment in the Referee service or to even have several canned environments that can be easily selected. A good navigation algorithm should be able to handle a variety of environments.

## Summary

Congratulations! Between this chapter and the last one, you have covered quite a bit of territory, not to mention nearly 3500 lines of code.

In this chapter, you built on the entities and services defined in the last chapter to create a complete scenario. A Referee service initialized the environment with trigger cones, barriers, and a single Corobot. Finally, you defined the Robo-Magellan orchestration service, complete with a Windows Forms user interface and a robot behavior model sufficient to navigate the course.

Having conquered the world of wheeled robots for the time being, you'll now move on to robotic arms and articulated robots in the next chapter.

# 8

# Simulating Articulated Entities

This chapter explains how to create articulated entities such as robotic arms and jointed robots in the simulation environment. Wheeled robots are fine for some applications, but sometimes you need a robot that can reach out and grab things.

The primary mechanism for defining articulated entities is the Joint object. This chapter begins by describing how to use a joint to join two entities and all of the options that are available. Then you will learn how to build a robotic arm with multiple joints.

## The Joint Class

A Joint is an object that joins two entities together. It has up to six degrees of freedom (three angular and three linear), which means it can be configured in a number of different ways. The simplest joint is a *revolute joint*, which has a single angular degree of freedom unlocked, and it behaves much like a door hinge. A joint with a single linear degree of freedom unlocked would operate somewhat like a shock absorber or a worm gear, with the joined entities able to move closer or farther along a single direction. A joint with all degrees of freedom unlocked can move in any direction and rotate and twist into any possible orientation.

The Joint class is defined in the Microsoft.Robotics.PhysicalModel namespace, which is implemented in the RoboticsCommon DLL. This Joint class is available for all services to use, not just simulation services. Its most important member is the JointProperties member called State. These properties specify the behavior of the joint when it is created in the physics engine.

A subclass of the Joint class called PhysicsJoint is defined in the Microsoft.Robotics .Simulation.Physics namespace, which is implemented in the PhysicsEngine DLL. This object contains some additional data that is used by the physics engine; it also includes some additional methods. This is the object you must use when interacting with the physics engine. It is simple to convert a `Joint` object to a `PhysicsJoint` object and back again, as you'll see in the example code in this chapter.

## The Joint Frame

Before you look in detail at the `JointProperties` class, it is important to understand the joint frame. A frame is a set of three axes that represent one coordinate system relative to another. This concept should be familiar to those readers who have worked with computer graphics. In a typical graphics scene, each object has its own frame relative to the world coordinates that determines its position and orientation. In graphics coordinates, we usually label the three axes in a frame as the X axis, the Y axis, and the Z axis.

Frames are used to specify the orientation of a joint as well. The three axes in a joint frame are called the local axis, the normal axis, and the binormal axis, as shown in Figure 8-1.

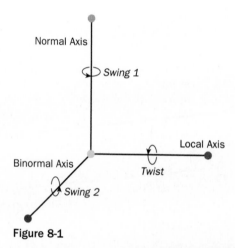

Figure 8-1

The joint frame follows the right-handed convention, meaning positive rotations around each axis are in the direction the fingers on your right hand curl when your thumb is pointed along the positive axis. The frame may be completely defined by specifying only two of the axes because the axes are always at right angles to each other and the third axis can be constructed by taking the cross product of the other two.

The joint names each angular degree of freedom (DOF), as shown in Figure 8-1. The Twist DOF is associated with rotations around the local axis, the Swing1 DOF is associated with rotations around the normal axis, and the Swing2 DOF is associated with rotations around the binormal axis.

## Joint Properties

The joint properties define how the joint behaves and how it attaches to the entities it joins. A joint may contain a reference to a `JointAngularProperties` class to unlock one or more angular degrees of freedom. It may also contain a reference to a `JointLinearProperties` class to unlock one or more linear degrees of freedom. It must always contain a two-element array of `EntityJointConnector` references to specify how it attaches to each entity.

The following sections describe the various members of the `JointAngularProperties` class, the `JointLinearProperties` class, the `EntityJointConnector` class, and other associated classes. These properties are explored in more detail in sample code following the descriptions.

## JointAngularProperties

The following table lists the members of the `JointAngularProperties` class:

| Member | Description |
|---|---|
| `TwistMode` | This member defines the mode for the Twist DOF, either locked, limited, or free. If locked, the joint is not free to move around the local axis. If limited, the rotation of the joint around the local axis is constrained by the `UpperTwistLimit` and the `LowerTwistLimit`. If free, the joint can move freely around the local axis. |
| `Swing1Mode` | Defines the mode for the `Swing1` DOF as above. If limited, the rotation of the joint around the normal axis is constrained by the `Swing1Limit`. |
| `Swing2Mode` | Defines the mode for the Swing2 DOF as above. If limited, the rotation of the joint around the binormal axis is constrained by the `Swing2Limit`. |
| `UpperTwistLimit` and `LowerTwistLimit` | `JointLimitProperties` that limit the motion of the joint around the local axis |
| `Swing1Limit` | `JointLimitProperties` that limit the motion of the joint around the normal axis |
| `Swing2Limit` | `JointLimitProperties` that limit the motion of the joint around the binormal axis |
| `TwistDrive` | `JointDriveProperties` that define how the joint is driven around the local axis |
| `SwingDrive` | `JointDriveProperties` that define how the joint is driven around the swing axis |
| `SlerpDrive` | A spherical linear interpolation `JointDriveProperties` that defines a special way to drive three angular DOF joints. |
| `GearRatio` | If this member is nonzero, then the angular velocity of the second entity is driven toward the angular velocity of the first entity multiplied by the `GearRatio`. |
| `DriveTargetOrientation` | The initial orientation of the joint |
| `DriveTargetVelocity` | The initial angular velocity of the joint |

## JointLimitProperties

The following table lists the members of the `JointLimitProperties` class:

| Member | Description |
| --- | --- |
| LimitThreshold | Specifies the angular or linear limit for the joint. Angular limits are expressed in radians. |
| Restitution | Floating-point value that specifies the "bounciness" of the joint as it moves against the limit |
| Spring | Spring properties that define a "soft limit." If the `SpringCoefficient` is nonzero, the joint can travel through the limit threshold but the spring brings it back to the threshold. The `DamperCoefficient` specifies how much the spring oscillates as it brings the joint back to the `LimitThreshold`. |

## JointDriveProperties

The following table lists the members of the `JointDriveProperties` class:

| Member | Description |
| --- | --- |
| Mode | Specifies the mode of driving this DOF. Position means that the joint is driven toward a specific position. Velocity means that the joint is driven toward a specific angular velocity. |
| Spring | Specifies the spring and damping coefficients for a spring that pulls the joint back to the target position |
| ForceLimit | Specifies the maximum force (torque) that the joint can exert to move to its target position |

## JointLinearProperties

The following table lists the members of the `JointLinearProperties` class:

| Member | Description |
| --- | --- |
| DriveTargetPosition, DriveTargetVelocity | The initial target position and velocity for the linear degrees of freedom for the joint |
| XMotionMode, YMotionMode, ZMotionMode | The DOF mode for each linear degree of freedom: locked, limited, or free. If limited, the motion of the joint along each axis is constrained by the `MotionLimit` member. The `XMotionMode` specifies movement along the local axis. The `YMotionMode` specifies movement along the normal axis, and the `ZMotionMode` specifies movement along the binormal axis. |

| Member | Description |
|---|---|
| XDrive, YDrive, ZDrive | The drive characteristics of the joint along each of the respective axes |
| MotionLimit | Only a single JointLimitProperties member is provided for all three linear degrees of freedom. The LimitThreshold of the MotionLimit member specifies the maximum of the absolute value of the motion allowed along each linear axis. For example, if all three linear DOFs are limited, then the LimitThreshold of the MotionLimit member defines a radius of a sphere in which the anchor of the second entity is constrained to move about the anchor of the first entity. |

## EntityJointConnector

The joint is anchored to each entity at a point and orientation specified by an EntityJointConnector, which has the following members:

| Member | Description |
|---|---|
| JointAxis | This is the vector that defines the local axis of the joint. It is often the vector (1,0,0) but it can be any vector. |
| JointNormal | This is the vector that defines the normal axis of the joint. It is often the vector (0,1,0) but it can be any vector that is perpendicular to theJointAxis vector. The binormal axis is the cross product of the JointAxis and the JointNormal. |
| JointConnectPoint | This is the point at which the joint attaches to the entity. It is in coordinates that are relative to the entity, not in world coordinates. The point doesn't actually have to be on the surface of the entity, it can be inside the entity or some distance away. |
| Entity | This is a reference to the entity that is connected. When the joint is serialized, this field cannot be serialized correctly. Prior to serialization, the name of this entity must be written to the EntityName property, which does serialize properly. After deserialization, this Entity reference must be restored by finding the entity with the specified name. |
| EntityName | This is the name of the entity referenced by the Entity property. Its value is only used after deserialization to restore the Entity reference. |

## *JointProperties*

The following table lists the members of the `JointProperties` class:

| Member | Description |
| --- | --- |
| Connectors | An array of two `EntityJointConnectors` that specifies the connection point and orientation for the first entity and the second entity |
| Angular | A reference to a `JointAngularProperties` class that specifies the angular properties for the joint. If this reference is null, then all angular degrees of freedom are considered locked. |
| Linear | A reference to a `JointLinearProperties` class that specifies the linear properties for the joint. If this reference is null, then all linear degrees of freedom are considered locked. |
| EnableCollisions | This Boolean property specifies whether collision detection checks should be performed between the entities connected by this joint |
| MaximumForce, MaximumTorque | This is the maximum force or torque that can be applied to the joint. If these limits are exceeded, the joint "breaks" and ceases to function. They should be set to similar values, and setting them to 0 ensures that the joint will never break. |
| Name | The name of the joint. Each joint should have a unique name. |
| Projection | The physics engine provides a way to correct large joint errors by projecting the joint back to a valid configuration. `Joint` errors occur when a joint's constrain is violated — for example, a joint is forced to move beyond its joint limit. There are normally some small joint errors due to the imprecise nature of floating-point math and numerical integrators, and the physics engine compensates for these by applying small corrective forces. If a joint error becomes very large, joint projection can allow the physics engine to directly change the position of the joint to reduce the error. If you wish to enable joint projection, the AGEIA documentation suggests that you set the `JointProjectionMode` to `PointMinimumDistance`, the `ProjectionAngleThreshold` to `0.0872`, and the `ProjectionDistanceThreshold` to `0.1`. |

# *A Joint TestBench*

The best way to gain a good understanding of these various joint properties is to build a sample program that creates a variety of joints with various properties to see how they work.

Open the Chapter 8 solution in `ProMRDS\Chapter8` and then open the `TestBench.cs` file in the TestBench project. This TestBench service is a basic simulation service much like Simulation Tutorial 1 or the Referee service from the previous chapter. Its purpose is to create a very long box called TestBenchEntity and then to create a number of test joints connected to SingleShapeEntities so that you can experiment with various joint parameters. It also creates a camera pointed at each joint so that it is very easy to switch to a view that shows the joint of interest.

## Subclassing Entity Types to Add Custom Joints

Before getting into the details about the joints created in the PopulateWorld method, it is important to define a new class, which you'll use for the segments of articulated entities. This class inherits from SingleShapeEntity and is called SingleShapeSegmentEntity. It adds an additional property called CustomJoint, which is of type Joint. The entire definition of this class is shown here:

```
/// <summary>
/// Defines a new entity type that overrides the ParentJoint with
/// custom joint properties.  It also handles serialization and
/// deserialization properly.
/// </summary>
[DataContract]
public class SingleShapeSegmentEntity : SingleShapeEntity
{
    private Joint _customJoint;

    [DataMember]
    public Joint CustomJoint
    {
        get { return _customJoint; }
        set { _customJoint = value; }
    }

    /// <summary>
    /// Default constructor
    /// </summary>
    public SingleShapeSegmentEntity() { }

    /// <summary>
    /// Initialization constructor
    /// </summary>
    /// <param name="shape"></param>
    /// <param name="initialPos"></param>
    public SingleShapeSegmentEntity(Shape shape, Vector3 initialPos)
        : base(shape, initialPos)
    {
    }

    public override void Initialize(
        Microsoft.Xna.Framework.Graphics.GraphicsDevice device,
        PhysicsEngine physicsEngine)
    {
        base.Initialize(device, physicsEngine);

        // update the parent joint to match our custom joint parameters
        if (_customJoint != null)
```

*(continued)*

*(continued)*

```
        {
            If(ParentJoint != null)
                PhysicsEngine.DeleteJoint((PhysicsJoint)ParentJoint);

            // restore the entity pointers in _customJoint after deserialization
            if (_customJoint.State.Connectors[0].Entity == null)
                _customJoint.State.Connectors[0].Entity = FindConnectedEntity(
                    _customJoint.State.Connectors[0].EntityName, this);

            if (_customJoint.State.Connectors[1].Entity == null)
                _customJoint.State.Connectors[1].Entity = FindConnectedEntity(
                    _customJoint.State.Connectors[1].EntityName, this);

            ParentJoint = _customJoint;
            PhysicsEngine.InsertJoint((PhysicsJoint)ParentJoint);
        }
    }

    VisualEntity FindConnectedEntity(string name, VisualEntity me)
    {
        // find the parent at the top of the hierarchy
        while (me.Parent != null)
            me = me.Parent;

        // now traverse the hierarchy looking for the name
        return FindConnectedEntityHelper(name, me);
    }

    VisualEntity FindConnectedEntityHelper(string name, VisualEntity me)
    {
        if (me.State.Name == name)
            return me;

        foreach (VisualEntity child in me.Children)
        {
            VisualEntity result = FindConnectedEntityHelper(name, child);
            if (result != null)
                return result;
        }

        return null;
    }

    /// <summary>
    /// Override the base PreSerialize method to properly serialize joints
    /// </summary>
    public override void PreSerialize()
    {
        base.PreSerialize();
        PrepareJointsForSerialization();
    }
}
```

This entity takes advantage of the fact that the simulation engine creates a fixed joint between a child and parent entity to join them together. In its override of the `Initialize` method, it first calls `base.Initialize`, which will, among other things, create the `ParentJoint`, which joins this entity to its parent entity if it is a child entity. If the `CustomJoint` has been defined, the `Initialize` method deletes the rigid `ParentJoint` that was automatically created and creates a new `ParentJoint` based on the properties of the `CustomJoint`. It is then very easy to define custom joints between child and parent entities simply by specifying the properties of `CustomJoint`.

This class also has some special code to handle proper serialization and deserialization of joints. A joint contains two connectors that each have a reference to the entity that is connected. When the `SingleShapeEntity` entity is serialized, either to be copied to disk or sent in a message to a service on another node, these entity references cannot be serialized; therefore, when a segment entity is deserialized, its entity references must be restored. This is done by overriding the `PreSerialize` method with a new method that calls `PrepareJointsForSerialization` after calling the base class `PreSerialize`.

`PrepareJointsForSerialization` is a method defined on `VisualEntity` that uses reflection to traverse all the fields in the entity to search for joints. When it finds a joint, it writes the name of the entity referenced by each `EntityJointConnector` into the `EntityName` field of each connector. This property is correctly serialized and it provides the information necessary to restore the entity connection after deserialization.

When a `SingleShapeSegmentEntity` is deserialized and then initialized, it eventually executes the code in the `Initialize` method override that checks whether either of the entity connectors is null. If so, the `FindConnectedEntity` method is used to traverse to the parent entity and then through all of the children entities to find an entity with the name contained in the `EntityName` property. When the entity is found, the reference to that entity is restored in the connector.

The `SingleShapeSegmentEntity` provides a `SingleShapeEntity` that supports custom `ParentJoints` and proper serialization and deserialization of joint objects within the entity. It is a simple matter to add the relevant code to other entity types by subclassing them to give them this same functionality.

## Adding Joints to the Test Bench

Now you are ready to define a test bench to experiment with the various joint properties. In the next few sections, you'll add five different types of joints to the test bench so that you can see how they work.

In the `PopulateWorld` method in `TestBench.cs`, you begin by creating sky and ground entities as usual. Then you create a box entity that is 50 meters long. You will attach other entities to this box at two-meter intervals.

## A Joint with One Angular Degree of Freedom

The first joint is the simplest possible revolute joint, with one degree of freedom:

```
// Simple single DOF joint
name = "1DOF";
segment = NewSegment(position + new Vector3(0, 0.5f, 0), name);
angular = new JointAngularProperties();
angular.TwistMode = JointDOFMode.Free;
angular.TwistDrive = new JointDriveProperties(JointDriveMode.Position,
    new SpringProperties(500000, 100000, 0), 100000000);
connectors = new EntityJointConnector[2];
connectors[0] = new EntityJointConnector(
    segment,
    new Vector3(0,1,0),
    new Vector3(1,0,0),
    new Vector3(0, -segment.CapsuleShape.State.Dimensions.Y/2 -
                    segment.CapsuleShape.State.Radius, 0));

connectors[1] = new EntityJointConnector(
    benchEntity,
    new Vector3(0,1,0),
    new Vector3(1,0,0),
    new Vector3(position.X, position.Y + 0.25f, position.Z));

segment.CustomJoint = new Joint();
segment.CustomJoint.State = new JointProperties(angular, connectors);
segment.CustomJoint.State.Name = name + "-twist";
benchEntity.InsertEntityGlobal(segment);
AddCamera(position, name);
position.X += 2;
```

The name variable is the base name for the segment, joint, and camera that will be created. You begin by creating a new SingleShapeSegmentEntity at the position defined by the position variable but raised up by half a meter so that its button rests on the top of the testbench.

Because this will be a revolute joint, you define a new JointAngularProperties variable and free the TwistMode degree of freedom. Recall from Figure 8-1 that the TwistMode degree of freedom is associated with the local axis.

Now you define the drive characteristics that will move the joint around this degree of freedom. The JointDriveProperties specify that you will be changing the position of the joint, rather than its velocity. They also specify the spring properties of the joint and the maximum torque that can be applied to the joint to force it to move to its target position. The spring properties enable a spring coefficient and a damping coefficient to be specified. The spring coefficient describes the amount of force that will be applied to the joint to move it toward its target position proportional to the distance from that target position. The damping coefficient roughly represents the friction of the joint to move. You'll examine the effect of lowering the damping coefficient in the second joint you create.

Next, the joint connectors are defined. The first connector attaches to the SingleShapeSegment entity you just created. In addition to the entity, the connector allows the local axis, the normal axis, and the connection point to be specified. In this case, you define the local axis to lie along the X axis (1,0,0) and the normal axis to lie along the Y axis (0,1,0). This leaves the binormal axis to lie along the Z axis (0,0,1) even though it is not specified here. The final parameter is the location on the segment entity where the joint is attached. The coordinates specified here are relative to the origin of the segment entity. You specify the bottom-most point of the capsule shape.

The next connector references the TestBenchEntity. It defines the normal and local axes the same way. Later you'll examine what happens if they are defined differently in the two connectors. The connection point is specified on the top surface of the testbench with X and Z coordinates according to the current position variable.

Now you create a Joint object and assign it to the CustomJoint property. You create a new JointProperties object using the angular joint properties and the joint connectors that you've already defined and assign it to CustomJoint.State. You give the joint a name and then insert the SingleShapeSegmentEntity as a child of the TestBenchEntity. You use InsertEntityGlobal because the coordinates you have given the segment entity are defined in world space, not relative to the TestBenchEntity.

Finally, create a camera that will look right at the joint you've just created. Run the TestBench manifest by typing the following in the MRDS command window. Alternatively, you can set the TestBench project as the StartUp project and then press F5 to run it.

```
C:\>cd "Microsoft Robotics Studio (1.5)"

C:\Microsoft Robotics Studio (1.5)>testbench
```

You should see a simulation window open with a scene similar to the one shown in Figure 8-2.

Figure 8-2

Press the F8 key once to switch the main camera to the 1DOF_cam camera to get a view of the segment entity you just created and attached. The TestBench manifest also runs the JointMover service, which can be used to move the joints you've specified. The user interface for this service is shown in Figure 8-3.

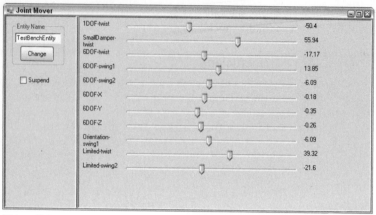

Figure 8-3

Each slider on the Joint Mover window corresponds to a degree of freedom in a joint on the entity specified in the Entity Name box. For scenes with multiple jointed entities, you can type a new entity name in the Entity Name box and press the change button to select the new entity. The sliders will reconfigure to match the joints in the new entity.

You can use the Suspend checkbox to set the top-level parent of the entity to kinematic and raise it off the ground a bit. This can be useful to test the joint movement of an entity that rests on the ground.

Slide the 1DOF-twist slider on the Joint Mover window to move the joint back and forth, as shown in Figure 8-4.

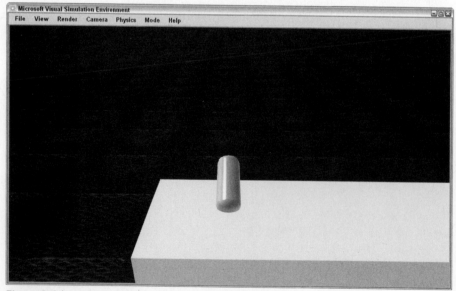

Figure 8-4

As you move the slider back and forth, the joint rotates around the X axis between $-180$ degrees and $180$ degrees. Note that if you move the slider quickly, the motion of the entity lags slightly behind.

## A Joint with a Smaller Damping Coefficient

For the next experiment, you'll create a joint with a much smaller damping coefficient on the drive to see how that affects the behavior of the joint. Here is the code for the "SmallDamper" joint:

```
// A joint with a very small damping coefficient on the drive
name = "SmallDamper";
segment = NewSegment(position + new Vector3(0, 0.5f, 0), name);
angular = new JointAngularProperties();
angular.TwistMode = JointDOFMode.Free;
angular.TwistDrive = new JointDriveProperties(JointDriveMode.Position,
    new SpringProperties(50000000, 100, 0), 100000000);
connectors = new EntityJointConnector[2];
connectors[0] = new EntityJointConnector(
    segment,
    new Vector3(0, 1, 0),
    new Vector3(1, 0, 0),
    new Vector3(0, -segment.CapsuleShape.State.Dimensions.Y / 2 -
                segment.CapsuleShape.State.Radius, 0));

connectors[1] = new EntityJointConnector(
    benchEntity,
    new Vector3(0, 1, 0),
    new Vector3(1, 0, 0),
    new Vector3(position.X, position.Y + 0.25f, position.Z));

segment.CustomJoint = new Joint();
segment.CustomJoint.State = new JointProperties(angular, connectors);
segment.CustomJoint.State.Name = name + "-twist";
benchEntity.InsertEntityGlobal(segment);
AddCamera(position, name);
position.X += 2;
```

This joint definition is identical to the previous 1DOF joint except for the damping coefficient specified in the `SpringProperties` of the `TwistDrive`. The damping coefficient used is 100 instead of the previous value of 100000.

Run the TestBench manifest again and select the "SmallDamper_cam" camera to look at the second joint. Move the SmallDamper-twist slider in the Joint Mover window to move the joint in the same way as you moved the first joint. You should notice that the joint is much more responsive and crisp in its movements. There is little or no lag between the movement of the slider and the movement of the joint. The damping coefficient can be a useful parameter to model the friction of the joint. Be careful not to make this value too large, though. A large damping coefficient relative to the spring coefficient can mean that the joint will never quite reach the target position when it is moved.

## A Six Degree of Freedom Joint

Next, you'll define a joint with all six degrees of freedom unlocked. The code for this joint is shown here:

```
// A joint with all 6 DOF free
name = "6DOF";
segment = NewSegment(position + new Vector3(0, 0.5f, 0), name);
angular = new JointAngularProperties();
angular.TwistMode = JointDOFMode.Free;
angular.Swing1Mode = JointDOFMode.Free;
angular.Swing2Mode = JointDOFMode.Free;
angular.TwistDrive = new JointDriveProperties(
    JointDriveMode.Position, new SpringProperties(50000000, 100, 0), 100000000);
angular.SwingDrive = new JointDriveProperties(
    JointDriveMode.Position, new SpringProperties(50000000, 100, 0), 100000000);

linear = new JointLinearProperties();
linear.XMotionMode = JointDOFMode.Free;
linear.XDrive = new JointDriveProperties(JointDriveMode.Position,
    new SpringProperties(50000000, 100, 0), 100000000);
linear.YMotionMode = JointDOFMode.Free;
linear.YDrive = new JointDriveProperties(JointDriveMode.Position,
    new SpringProperties(50000000, 100, 0), 100000000);
linear.ZMotionMode = JointDOFMode.Free;
linear.ZDrive = new JointDriveProperties(JointDriveMode.Position,
    new SpringProperties(50000000, 100, 0), 100000000);
connectors = new EntityJointConnector[2];
connectors[0] = new EntityJointConnector(
    segment,
    new Vector3(0, 1, 0),
    new Vector3(1, 0, 0),
    new Vector3(0, -segment.CapsuleShape.State.Dimensions.Y / 2 -
                    segment.CapsuleShape.State.Radius, 0));

connectors[1] = new EntityJointConnector(
    benchEntity,
    new Vector3(0, 1, 0),
    new Vector3(1, 0, 0),
    new Vector3(position.X, position.Y + 0.25f, position.Z));

segment.CustomJoint = new Joint();
segment.CustomJoint.State = new JointProperties(angular, connectors);
segment.CustomJoint.State.Linear = linear;
segment.CustomJoint.State.Name =
    name + "-twist" + ";" +
    name + "-swing1" + ";" +
    name + "-swing2" + ";" +
    name + "-X|-2|2|" + ";" +
    name + "-Y|-2|2|" + ";" +
    name + "-Z|-2|2|" + ";" ;
benchEntity.InsertEntityGlobal(segment);
AddCamera(position, name);
position.X += 2;
```

Some parts of this code require a little more explanation. When the angular properties are defined, all three angular degrees of freedom are freed but only two drives are defined: a `TwistDrive` and a `SwingDrive`. The AGEIA physics engine applies the `SwingDrive` characteristics to drive both the `Swing1` DOF and the `Swing2` DOF, even if they are both freed at the same time. This differs from the linear properties, for which a separate drive is allowed for each of the X, Y, and Z degrees of freedom.

When the `JointProperties` object is created, you pass the angular properties and the connectors as parameters and then assign the linear properties later.

The `JointMover` service determines how to map sliders to the joint by the name of the joint. This joint needs six different sliders, so six different names are provided in the joint name, separated by semicolons. There should be a name for each degree of freedom that is free or limited in the joint, in the order of Twist, Swing1, Swing2, X, Y, and Z degrees of freedom. Furthermore, the minimum and maximum values of each slider can be specified with two optional floating-point parameters following the name and separated with the " | " character. The default minimum and maximum values for each slider are –180 and 180, respectively. The only requirement that the simulation engine imposes on the joint name is that it must be unique relative to all the other joint names.

Run the TestBench manifest again and select the "6DOF_cam" camera to look at the third joint. Six different sliders appear in the Joint Mover window with the 6DOF name prefix to control each free degree of freedom. Experiment by moving the sliders, especially the linear DOF sliders. The 6DOF joint provides a good opportunity to understand the physics visualization for joints. Click Render ➪ Physics to put the simulator in physics visualization mode. Use the 6DOF-Y slider to move the `SingleShapeSegmentEntity` up off of the `TestBenchEntity`. You can see the frame for each connection point, with a line joining them that represents the linear positional offset, as shown in Figure 8-5.

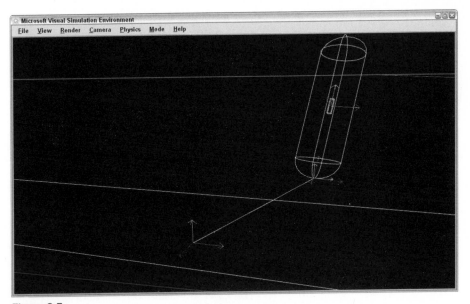

**Figure 8-5**

The frame displayed on the `TestBenchEntity` shows the point where the connection is made, as well as the orientation of the connection. When viewed on the screen, the red vector represents the local axis, the green vector represents the normal axis, and the blue vector represents the binormal axis. The frame displayed at the base of the capsule shape represents the connection point for the child entity. This frame is rotated by the angular rotation of the joint and offset by the linear position of the joint. If you look closely, you can see that there are actually two frames shown at the bottom of the capsule shape. One frame represents the target position, while the other frame represents the current position. Because the damping coefficient is set very low for this joint, the two frames are almost always aligned. With a higher damping coefficient, you would see one frame lag behind the other as the joint is moved.

## A Joint with Different Connector Orientations

Up until now, you've always specified the orientation of each connector to be the same. What happens if they are different? This sounds like a job for the `TestBench`! Here is the code for a single DOF joint that has a different orientation for each connector:

```
// Simple single DOF joint with differing connector orientations
name = "Orientation";
segment = NewSegment(position + new Vector3(0, 0.5f, 0), name);
angular = new JointAngularProperties();
angular.Swing1Mode = JointDOFMode.Free;
angular.SwingDrive = new JointDriveProperties(JointDriveMode.Position,
    new SpringProperties(500000, 100, 0), 100000000);
connectors = new EntityJointConnector[2];
connectors[0] = new EntityJointConnector(
    segment,
    new Vector3(0, 0, 1),
    new Vector3(0, 1, 0),
    new Vector3(0, -segment.CapsuleShape.State.Dimensions.Y / 2 -
                segment.CapsuleShape.State.Radius, 0));

connectors[1] = new EntityJointConnector(
    benchEntity,
    new Vector3(0, 1, 0),
    new Vector3(1, 0, 0),
    new Vector3(position.X, position.Y + 0.25f, position.Z));

segment.CustomJoint = new Joint();
segment.CustomJoint.State = new JointProperties(angular, connectors);
segment.CustomJoint.State.Name = name + "-swing1";
benchEntity.InsertEntityGlobal(segment);
AddCamera(position, name);
position.X += 2;
```

In this joint definition, you have specified the local axis of the child connector to align with the Y axis, and the normal axis to align with the Z axis. The parent connector follows the convention that the other joints have used, with the local axis aligned with the X axis and the normal axis aligned with the Y axis. Start the TestBench manifest and switch to the `Orientation_cam` camera to see what the result of this joint definition is. Switch to the physics view and you will see that the simulation engine has oriented the child entity so that the axes of each of the connectors line up. A difference in connector orientation means a difference in orientation between the joined entities.

## A Joint with Motion Limits

One more example: See what happens when you specify a joint limit. Consider the following joint definition:

```
// An angular joint with limited twist and swing2
name = "Limited";
segment = NewSegment(position + new Vector3(0, 0.5f, 0), name);
angular = new JointAngularProperties();
angular.TwistMode = JointDOFMode.Limited;
angular.Swing2Mode = JointDOFMode.Limited;
angular.TwistDrive = new JointDriveProperties(
    JointDriveMode.Position, new SpringProperties(5000, 100, 0), 100000000);
angular.SwingDrive = new JointDriveProperties(
    JointDriveMode.Position, new SpringProperties(5000, 100, 0), 100000000);

// specify a limit on twist and swing2 with an angle of PI/8, 0 restitution, and no
spring
angular.LowerTwistLimit = new JointLimitProperties((float)(-Math.PI / 8), 0,
    new SpringProperties(5000000, 1000, 0));
angular.UpperTwistLimit = new JointLimitProperties((float)(Math.PI / 8), 0,
    new SpringProperties(5000000, 1000, 0));
angular.Swing2Limit = new JointLimitProperties((float)(Math.PI / 8), 0,
    new SpringProperties(5000000, 1000, 0));

connectors = new EntityJointConnector[2];
connectors[0] = new EntityJointConnector(
    segment,
    new Vector3(0, 1, 0),
    new Vector3(1, 0, 0),
    new Vector3(0, -segment.CapsuleShape.State.Dimensions.Y / 2 -
               segment.CapsuleShape.State.Radius, 0));

connectors[1] = new EntityJointConnector(
    benchEntity,
    new Vector3(0, 1, 0),
    new Vector3(1, 0, 0),
    new Vector3(position.X, position.Y + 0.25f, position.Z));

segment.CustomJoint = new Joint();
segment.CustomJoint.State = new JointProperties(angular, connectors);

segment.CustomJoint.State.Name =
    name + "-twist" + ";" +
    name + "-swing2" + ";";
benchEntity.InsertEntityGlobal(segment);
AddCamera(position, name);
position.X += 2;
```

Here, you have specified joint limits on the Twist DOF and the Swing2 DOF of about 22.5 degrees in either direction. This should constrain the motion of the child entity to lie within a cone with a half-cone angle of 22.5 degrees. You've also specified a large spring coefficient on the JointLimitProperties, which applies a significant force to keep the joint within the joint limits.

Once again, run the TestBench manifest and switch the camera to Limited_cam to view this joint. Move the Limited-twist and Limited-swing2 sliders to move the joint and notice that it is constrained to the specified limits. As the sliders get significantly outside the limit angle, the physics solver can become less stable and cause the joint to oscillate. This can be improved somewhat by increasing the damping coefficient in the JointLimitProperties.

That concludes the joint experiments we'll cover in the TestBench example. It is hoped that these examples have explained some of the most important joint properties that affect joint behavior. The TestBench provides a great way to prototype a joint to see how it is going to behave in your application. In the following sections, you'll learn how to use joints to build a robotic arm.

# Building a Simulated Robotic Arm

Now you can use your newfound joint expertise to build a real-world object. In this section, you build a simulated robotic arm modeled after the Lynx 6 robotic arm. This hobbyist arm provides six independent degrees of freedom and is available at a reasonable price point. More information about this arm, shown in Figure 8-6, is available at www.lynxmotion.com/Category.aspx?CategoryID=25.

First, you'll learn how to write code to build the physics model of the arm by creating entities and connecting them with joints. Next, you'll learn how to add a visual model and then write a service to control the arm. Finally, you'll learn how to use inverse kinematics to drive the arm to a specified position.

Figure 8-6

As you can see from the figure, the arm consists of a base that swivels. The upper arm segment pivots at its connection to the base and the lower arm segment pivots at its connection to the upper arm. The wrist segment pivots at its connection to the lower arm and rotates. Finally, the gripper connects to the wrist and has the capability to open and close.

## The Physical Model of the Arm Entity

Refer to the SimulatedLynxL6Arm service in the Chapter 8 directory to see how the SimulatedLynxL6ArmEntity is defined. Figure 8-7, from the Lynxmotion website, was used to define the size and orientation of each arm segment.

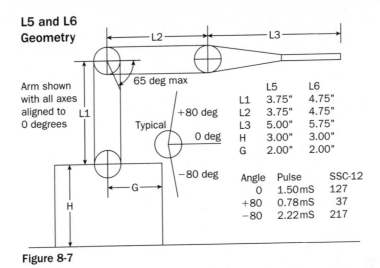

**Figure 8-7**

These dimensions are defined in `SimulatedLynxL6Arm.cs` at the top of the `SimulatedLynxL6Arm` class definition. Notice that this entity subclasses the `SingleShapeEntity`. This is because the arm entity itself only includes a single shape, which is the base of the arm. Each of the other segments in the arm is a separate entity joined to another entity in the arm as a child. This makes it easy to use the built-in `ParentJoint` that belongs to each entity to join the entities together.

The following code defines the dimensions of each part of the arm:

```
static float InchesToMeters(float inches) { return (float)(inches * 0.0254); }

// physical attributes of the arm
static float L1 = InchesToMeters(4.75f);
static float L2 = InchesToMeters(4.75f);
static float Grip = InchesToMeters(2.5f);
static float L3 = InchesToMeters(5.75f) - Grip;
static float L4 = 0.03f;
static float H = InchesToMeters(3f);
static float G = InchesToMeters(2f);
static float L1Radius = InchesToMeters(0.7f);
static float L2Radius = InchesToMeters(0.7f);
static float L3Radius = InchesToMeters(0.7f);
static float GripRadius = InchesToMeters(0.2f);
```

Because the dimensions are given on the diagram in inches but the units used by the simulator are meters, a simple utility function is used to convert the units. H specifies the height of the base, and G is the radius. The L1 entity corresponds to the upper arm, and the L2 entity is the lower arm. In Figure 8-7, the L3 segment includes both the wrist and the grippers. The entity divides this into the L3 entity, which includes the wrist to just before the grippers; an L4 entity, used to make the wrist rotate; and the Grip entity, which is 2.5 inches long. The GripRadius refers to the radius of each part of the gripper.

In addition to the size of each segment, you need to define the joints that will be used to join the segments. The following class defines the characteristics of each joint, as well as some run-time properties of the joint such as the target position, the current position, and the current speed of motion:

```
// This class holds a description of each of the joints in the arm.
class JointDesc
{
    public string Name;
    public float Min;   // minimum allowable angle
    public float Max;   // maximum allowable angle
    public PhysicsJoint Joint; // Phyics Joint
    public PhysicsJoint Joint2; // Alternate Physics Joint (used for gripper)
    public float Target;   // Target joint position
    public float Current;   // Current joint position
    public float Speed;   // Rate of moving toward the target position
    public JointDesc(string name, float min, float max)
    {
        Name = name; Min = min; Max = max;
        Joint = null;
        Joint2 = null;
        Current = Target = 0;
        Speed = 30;
    }

    // Returns true if the specified target is within the valid bounds
    public bool ValidTarget(float target)
    {
        return ((target >= Min) && (target <= Max));
    }

    // Returns true if the joint is not yet at the target position
    public bool NeedToMove(float epsilon)
    {
        if (Joint == null) return false;
        return (Math.Abs(Target - Current) > epsilon);
    }

    // Takes one step toward the target position based on the specified time
    public void UpdateCurrent(double time)
    {
        float delta = (float)(time * Speed);
        if (Target > Current)
            Current = Math.Min(Current + delta, Target);
        else
            Current = Math.Max(Current - delta, Target);
    }
}
```

The initialized properties of each joint are the name and the minimum and maximum angles that the joint supports. In the case of the grippers, you will be using linear joints, and the minimum and maximum values represent distances. ValidTarget provides a simple way to check a target joint position to determine whether it falls within the minimum and maximum for the joint. NeedToMove indicates whether the current joint position is within a certain distance of the target position.

`UpdateCurrent` calculates a new `Current` position based on the speed of movement and the time that has elapsed.

The joints for the arm are defined in an array of descriptors as follows:

```
// Initialize an array of descriptions for each joint in the arm
JointDesc[] _joints = new JointDesc[]
{
    new JointDesc("Base", -180, 180),
    new JointDesc("Shoulder", -90, 90),
    new JointDesc("Elbow", -65, 115),
    new JointDesc("Wrist", -90, 90),
    new JointDesc("WristRotate", -90, 90),
    new JointDesc("Gripper", 0, InchesToMeters(2))
};
```

Notice that the minimum and maximum for the elbow joint are a little unusual due to the physical constraints of the arm, and the minimum and maximum for the gripper joint are 0 inches and 2 inches, respectively. The gripper is actually modeled using two simulated joints, one for each of the parts of the gripper. From a logical standpoint, though, the gripper position is treated as a single joint that specifies the distance between the left and right gripper.

The segment dimensions and the joint descriptions are used in the initialization constructor to programmatically build the arm entity. In the `SimulatedLynxL6Arm` constructor, one of the first things you'll notice is that the physics shape for the base is not quite as tall as the actual base. This arises from a design decision to only enable one degree of freedom per joint. The point where the upper arm attaches to the base could be thought of as having two degrees of freedom: the base rotation and the shoulder joint swivel. Instead, we have defined another entity called `L0Entity` between the base and the `L1Entity`. The `ParentJoint` of the `L0Entity` is used for the base rotation, and the `ParentJoint` of the `L1Entity` is used for the shoulder joint. The height of the base is lowered to prevent the `L1Entity` from colliding with the base when the shoulder joint is moved.

The base state and its shape are initialized with the following code:

```
// The physics shape for the base is slightly lower than the actual base
// so that the upper-arm segment does not intersect it as it moves around.
float baseHeight = H - L1Radius - 0.001f;
State.Name = name;
State.Pose.Position = position;
State.Pose.Orientation = new Quaternion(0, 0, 0, 1);
State.Assets.Mesh = "L6_Base.obj";
MeshTranslation = new Vector3(0, 0.026f, 0);

// build the base
BoxShape = new BoxShape(new BoxShapeProperties(
    "Base",
    150, // mass
    new Pose(new Vector3(0, baseHeight / 2, 0), new Quaternion(0, 0, 0, 1)),
    new Vector3(G * 2, baseHeight, G * 2)));
```

Notice that the shape is a box, rather than a cylinder. The only way to specify a cylinder shape in the AGEIA engine is to use a mesh that is shaped like a cylinder with a `SimplifiedConvexMeshEnvironmentEntity`. The box is simpler and works fine.

# Part II: Simulations

In the spirit of full disclosure, it should be mentioned that there has been no attempt to specify accurate masses for each of the arm components. A mass of 150 Kg has been specified for the base, which is not accurate. If you wish to attach this arm to another entity in the environment, you will want to specify more accurate mass values to each entity in the arm. It is possible to make the base of the arm move as a result of the inertia of swinging the upper part of the arm quickly. The base of the arm can be made immobile by making it kinematic using the following:

```
State.Flags |= EntitySimulationModifiers.Kinematic;
```

The default pose specified for the box places it baseHeight/2 above the shape origin. This puts the origin of the base entity on the ground instead of baseHeight/2 above the ground, making it easier to position within the world. We set a value on the `MeshTranslation` property of the base entity, moving the mesh associated with the base up to correspond with the position of the shape.

The next part of the arm to be defined is the `L0Entity`. All of the remaining entities in the arm are `SingleShapeSegmentEntities`. As you will recall from the TestBench application earlier in this chapter, this type of entity allows its `ParentJoint` to be overridden. The `L0Entity` contains a single sphere shape, whose radius is the same as the `L1Entity`. The `ParentJoint` of this entity controls the rotation of the base. It is created as shown here:

```
// build and position L0 (top of the base)
SphereShape L0Sphere = new SphereShape(new SphereShapeProperties(
    "L0Sphere",
    50,
    new Pose(new Vector3(0, 0, 0), new Quaternion(0, 0, 0, 1)),
    L1Radius));

SingleShapeSegmentEntity L0Entity =
    new SingleShapeSegmentEntity(L0Sphere, position + new Vector3(0, H, 0));
L0Entity.State.Pose.Orientation = new Quaternion(0, 0, 0, 1);
L0Entity.State.Name = name + "_L0";
L0Entity.State.Assets.Mesh = "L6_L0.obj";
L0Entity.MeshTranslation = new Vector3(0, -0.02f, 0);
JointAngularProperties L0Angular = new JointAngularProperties();
L0Angular.Swing1Mode = JointDOFMode.Free;
L0Angular.SwingDrive = new JointDriveProperties(JointDriveMode.Position,
    new SpringProperties(50000000, 1000, 0), 100000000);
EntityJointConnector[] L0Connectors = new EntityJointConnector[2]
{
    new EntityJointConnector(L0Entity,
        new Vector3(0,1,0), new Vector3(1,0,0), new Vector3(0, 0, 0)),
    new EntityJointConnector(this,
        new Vector3(0,1,0), new Vector3(1,0,0), new Vector3(0, H, 0))
};
L0Entity.CustomJoint = new Joint();
L0Entity.CustomJoint.State = new JointProperties(L0Angular, L0Connectors);
L0Entity.CustomJoint.State.Name = "BaseJoint";

this.InsertEntityGlobal(L0Entity);
```

The `L0Entity` contains a single sphere. Its center is positioned H meters above the ground so that its center corresponds to the base's center of rotation. It is important that the initial position of the base and

`L0Entity` is very close to their position once they are connected with a joint. If the initial position is different, then the pieces will snap together violently during the first frame of the simulation and the arm will likely topple over. A joint is defined with the Swing1 degree of freedom unlocked so that the joint will rotate freely around the Y axis. The `L0Entity` is inserted as a child of the base entity using the `InsertEntityGlobal` method because the position of the `L0Entity` is defined in world coordinates, rather than coordinates relative to the base entity.

After all that work, you end up with the physics model shown in Figure 8-8.

**Figure 8-8**

It's not very impressive yet but it gets better. Next, you attach the upper arm segment (`L1Entity`) to the `L0Entity`:

```
// build and position L1 (upper arm)
CapsuleShape L1Capsule = new CapsuleShape(new CapsuleShapeProperties(
    "L1Capsule",
    2,
    new Pose(new Vector3(0, 0, 0), new Quaternion(0, 0, 0, 1)),
    L1Radius,
    L1));

SingleShapeSegmentEntity L1Entity =
    new SingleShapeSegmentEntity(L1Capsule, position + new Vector3(0, H, 0));
L1Entity.State.Pose.Orientation = new Quaternion(0, 0, 0, 1);
L1Entity.State.Name = name + "_L1";
L1Entity.State.Assets.Mesh = "L6_L1.obj";
JointAngularProperties L1Angular = new JointAngularProperties();
L1Angular.TwistMode = JointDOFMode.Free;
```

*(continued)*

*(continued)*

```
L1Angular.TwistDrive = new JointDriveProperties(JointDriveMode.Position,
    new SpringProperties(50000000, 1000, 0), 100000000);
EntityJointConnector[] L1Connectors = new EntityJointConnector[2]
{
    new EntityJointConnector(L1Entity,
        new Vector3(0,1,0), new Vector3(0,0,1), new Vector3(0, -L1/2, 0)),
    new EntityJointConnector(L0Entity,
        new Vector3(0,1,0), new Vector3(0,0,1), new Vector3(0, 0, 0))
};
L1Entity.CustomJoint = new Joint();
L1Entity.CustomJoint.State = new JointProperties(L1Angular, L1Connectors);
L1Entity.CustomJoint.State.Name = "Shoulder|-80|80|";

L0Entity.InsertEntityGlobal(L1Entity);
```

This is much like the code that was used to define the `L0Entity` except that you are now using a capsule shape for the segment. The entity is positioned exactly H meters above the ground so that its rounded end-cap coincides with the sphere in the `L0Entity`. Its center of rotation will correspond to the center of the `L0Entity` sphere.

When the joint position is specified for the `L1Entity` in `L1Connects[0]`, a position of `(0, -L1/2, 0)` is specified for the joint position for which this coordinate is relative to the `L1Entity`. A position of `(0,0,0)` is specified for the `L0Entity` because this joint will be located exactly at the sphere's center. When the joint is created, these two points on the two entities will be constrained to be in the same place. Notice that the name you give to the joint includes a minimum and maximum value that the JointMover service uses to scale its controls, as in the TestBench sample described previously.

Finally, the `L1Entity` is added as a child of the `L0Entity` and another segment of the arm is attached.

The `L2Entity` is created and joined to the `L1Entity` in much the same way. The only difference lies in the joint connectors, which are defined as follows:

```
EntityJointConnector[] L2Connectors = new EntityJointConnector[2]
{
    new EntityJointConnector(L2Entity,
        new Vector3(1,0,0), new Vector3(0,0,1), new Vector3(0, -L2/2, 0)),
    new EntityJointConnector(L1Entity,
        new Vector3(0,1,0), new Vector3(0,0,1), new Vector3(0, L1/2, 0))
};
```

Notice that the normal vector specified for the `L2Entity` connector is (1,0,0) and the normal vector specified for the `L1Entity` connector is (0,1,0). This joins the entities at right angles to each other.

The `L3Entity` and `L4Entity` are created and joined in much the same way. At this point the physical model looks like Figure 8-9.

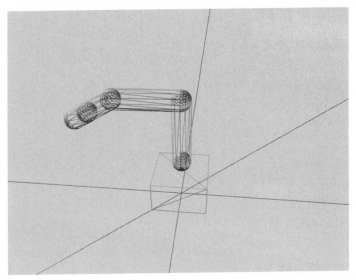

Figure 8-9

It is shown in wireframe view with the wrist joint slightly bent to better demonstrate how the entities are connected. The sphere in the middle of the wrist segment is really the end-caps of the L3Entity and the L4Entity.

All that remains is to add the two gripper entities. The gripper is modeled as two capsules that move together and apart using linear joints. Although each part of the gripper is joined to the L4Entity with its own joint, the gripper is treated logically as if it has one joint. The following code adds the left gripper to the entity:

```
// build and position LeftGrip
CapsuleShape LeftGripCapsule = new CapsuleShape(new CapsuleShapeProperties(
    "LeftGripCapsule",
    1f,
    new Pose(new Vector3(0, 0, 0), new Quaternion(0, 0, 0, 1)),
    GripRadius,
    Grip));
LeftGripCapsule.State.DiffuseColor = new Vector4(0, 0, 0, 1);

LeftGripEntity = new SingleShapeSegmentEntity(
    LeftGripCapsule, position + new Vector3(0, H, 0));
// position the entity close to its final position once the joint is connected
LeftGripEntity.Position = new xna.Vector3(-0.24f, 0.19f, 0.01f);
LeftGripEntity.Rotation = new xna.Vector3(179.94f, -176.91f, 89.67f);
LeftGripEntity.State.Name = name + "_LeftGrip";
// use a linear joint for the left grip
JointLinearProperties LeftGripLinear = new JointLinearProperties();
LeftGripLinear.XMotionMode = JointDOFMode.Free;
LeftGripLinear.XDrive = new JointDriveProperties(JointDriveMode.Position,
    new SpringProperties(5000000000, 1000, 0), 100000000);
EntityJointConnector[] LeftGripConnectors = new EntityJointConnector[2]
```

(continued)

*(continued)*

```
    {
        new EntityJointConnector(LeftGripEntity,
            new Vector3(1,0,0), new Vector3(0,0,1), new Vector3(0, -Grip/2, 0)),
        new EntityJointConnector(L4Entity,
            new Vector3(1,0,0), new Vector3(0,0,1), new Vector3(0, L4/2, GripRadius))
    };
    LeftGripEntity.CustomJoint = new Joint();
    LeftGripEntity.CustomJoint.State =
        new JointProperties(LeftGripLinear, LeftGripConnectors);
    LeftGripEntity.CustomJoint.State.Name = "LeftGripJoint|-0.0254|0|";

    L4Entity.InsertEntityGlobal(LeftGripEntity);
```

The main difference from the other joints is that this is a linear joint. The X degree of freedom is unlocked. The range of the gripper is 0 to 2 inches, so the range of this single half of the gripper is 0 to 1 inch. One other difference is that a `DiffuseColor` is specified for the capsule shape in this entity. This is because there is no custom mesh specified for this entity. The default capsule shape, colored dark black, is used for the visual representation. With the grippers added, the complete physics model looks like the image shown in Figure 8-10.

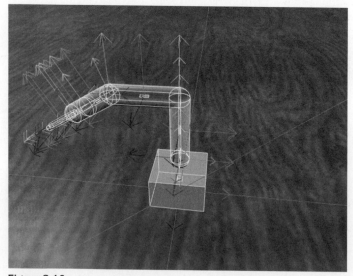

**Figure 8-10**

Finally, a camera entity is mounted just above the `L4Entity` so that you can get a close-up view of what the arm is manipulating:

```
// Add a camera to see what the gripper is gripping
AttachedCameraEntity gripCam = new AttachedCameraEntity();
gripCam.State.Name = "Arm Cam";

// move the camera above the L4entity and look at the grippers
gripCam.State.Pose = new Pose(new Vector3(0.05f, -0.01f, 0),
    Quaternion.FromAxisAngle(0, 1, 0, (float)(Math.PI / 2)) *
    Quaternion.FromAxisAngle(1, 0, 0, (float)(Math.PI / 3)));

// adjust the near plane so that we can see the grippers
gripCam.Near = 0.01f;

// the gripcam coordinates are relative to the L4Entity, don't use
InsertEntityGlobal
L4Entity.InsertEntity(gripCam);
```

The position and orientation of the camera is defined relative to the `L4Entity`, and `InsertEntity` is used rather than `InsertEntityGlobal`. Another interesting thing to note is that the near plane of the camera is adjusted to be 1 centimeter. All objects closer to the camera than the near plane are not displayed. The default value for the near plane is 10 centimeters, which clips the grippers from the scene. A closer near plane allows the camera to get closer to objects in the scene at the cost of reducing the depth buffer resolution for the scene. In some extreme circumstances, this can cause occlusion problems with some objects in the scene unless the far plane is brought in by the same factor as the near plane.

---

### The AttachedCameraEntity

The `CameraEntity` provided with the MRDS 1.5 SDK always attempts to keep the horizon level in the displayed scene. This works well for scene cameras but isn't really desirable for a camera attached to a robot or an arm. A different type of camera called `AttachedCamera`, defined in `SimulatedLynxL6Arm.cs`, enables the camera view to roll with the object to which it is attached. This is accomplished by overriding the `Initialize` method and setting the `private _frameworkCamera` field with a new `Camera` object called `AttachedCameraModel`. The code is interesting because it uses reflection to set a private field. You should use this type of camera whenever you want to attach a camera to an entity that may not always stay lined up with the horizon.

---

The physics model is perfectly adequate to model the motion and physical constraints of the arm, but it is pretty uninteresting visually. A few custom meshes make quite a difference to the simulation, as shown in Figure 8-11.

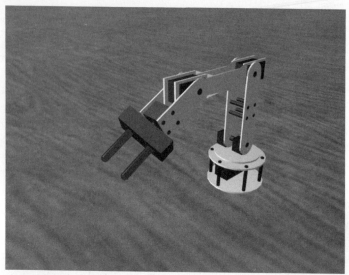

Figure 8-11

The custom meshes were modeled in Maya. The six Maya files for the models are included in the ProMRDS\mayamodels directory starting with 16_base.ma and ending with 16_14.ma. Corresponding .obj and .mtl files are included in the store\media directory for use by the simulator. The combined physics and visual view of the arm model is shown in Figure 8-12.

Figure 8-12

## Running the Arm Service

This is a good time to run the arm simulation and to manipulate the joints to get a feel for the range and capabilities of the arm. You can easily run the arm simulation from the command line using the provided .cmd file as follows:

```
C:\Microsoft Robotics Studio (1.5)>SimulatedLynxL6Arm
```

This starts a DSS node with the Lynx.L6Arm.Simulation manifest. This manifest runs the SimulatedLynxL6Arm service along with the JointMover service. You can use the JointMoverService to manipulate the arm joints and move the arm around. The SimulatedLynxL6Arm user interface is explained in a later section.

## Moving the Arm

The SimulatedLynxL6ArmEntity defines a MoveTo method that can be used to move the arm to a specific position. The method takes the following parameters, which completely specify the arm position:

| Parameter | Units | Description |
|-----------|-------|-------------|
| baseVal | Degrees | Rotation angle of the base joint |
| shoulderVal | Degrees | Pivot angle of the shoulder joint |
| elbowVal | Degrees | Pivot angle of the elbow joint |
| wristVal | Degrees | Pivot angle of the wrist joint |
| rotateVal | Degrees | Rotation angle of the wrist joint |
| gripperVal | Meters | Distance of separation between the grippers |
| Time | Seconds | Time to complete the motion |

This method returns a SuccessFailurePort after the motion has been initiated. A Success message is posted to the port when the motion successfully completes. Otherwise, an Exception message is posted if an error was encountered.

The arm entity has a private Boolean variable called _moveToActive that indicates whether a move operation is currently underway. If a call is made to MoveTo while a move operation is currently active, an exception message is posted to the response port.

Each parameter is checked against the bounds specified in the corresponding joint description. An invalid parameter results in an exception being posted to the response port, with a message indicating which parameter was bad:

```
SuccessFailurePort responsePort = new SuccessFailurePort();

if (_moveToActive)
{
    responsePort.Post(new Exception("Previous MoveTo still active."));
    return responsePort;
}

// check bounds.  If the target is invalid, post an exception message
// to the response port with a helpful error.
if(!_joints[0].ValidTarget(baseVal))
{
    responsePort.Post(new Exception(
        _joints[0].Name + "Joint set to invalid value: " + baseVal.ToString()));
    return responsePort;
}
```

After all of the parameters have been validated, each joint description `Target` value is set to the specified value. In addition, a speed value is calculated for each joint based on the distance between the current value and the target value and the overall time specified for completion of the motion. Each joint gets its own speed value to ensure that joints that have a large distance to move will move more quickly than joints that have a small distance to move so that all joints complete their motion at the same time:

```
// set the target values on the joint descriptors
_joints[0].Target = baseVal;
_joints[1].Target = shoulderVal;
_joints[2].Target = elbowVal;
_joints[3].Target = wristVal;
_joints[4].Target = rotateVal;
_joints[5].Target = gripperVal;

// calculate a speed value for each joint that will cause it to complete
// its motion in the specified time
for(int i=0; i<6; i++)
    _joints[i].Speed = Math.Abs(_joints[i].Target - _joints[i].Current) / time;
```

The `MoveTo` method then sets the _moveToActive flag to `true` and returns the response port. As you can see, the motion is set up in this method but the joint is actually not moved until its `update` method is called.

The entity's `Update` method is called once each frame, ideally about 60 times per second. Each time this method is called, the joint is moved a small amount based on how much time has elapsed since the last update.

The first time Update is called after the entity is initialized, it sets references to the newly created joints in the joint description array. It then follows the pattern shown in the following code to update each joint:

```
// update joints if necessary
if (_moveToActive)
{
    bool done = true;
    // Check each joint and update it if necessary.
    if (_joints[0].NeedToMove(_epsilon))
    {
        done = false;

        Vector3 normal = _joints[0].Joint.State.Connectors[0].JointNormal;
        _joints[0].UpdateCurrent(_prevTime);
        _joints[0].Joint.SetAngularDriveOrientation(
            Quaternion.FromAxisAngle(
                normal.X, normal.Y, normal.Z,
                DegreesToRadians(_joints[0].Current)));
    }
}
```

If a motion sequence is currently active, then each joint is evaluated to determine whether it still needs to be moved to hit its target value. If NeedToMove returns true, then UpdateCurrent is called to move the joint toward its target in a step that depends on the speed of the current movement and the amount of time that has elapsed since the last time Update was called. A new joint orientation is calculated from the new joint position and the joint is set to its new position.

If none of the joints need to be updated, then the motion sequence is finished. _moveToActive is set to false and a new SuccessResult message is posted to the response port.

This is a better way of controlling joint movement than the Simulation Tutorial 4 provided in the SDK, which relies on the damping coefficient of the joint drive to dictate the speed at which the joint moves. The method illustrated in this example provides much more control over the joint's rate of speed and allows for the possibility of setting a maximum motor movement speed according to the characteristics of the physical motors on the arm.

## A Fly in the Ointment

At this point, you have a great arm model that closely simulates the capabilities of the real arm. Unfortunately, the AGEIA physics engine appears to have a limitation that significantly affects this arm model. Although the gripper joints bring the grippers together and it is possible to close them on an object, the physics engine does not do a good job of simulating the friction between the grippers and the grasped object. Even with a high-friction material specified for the object and the grippers, the arm cannot pick up an object without it sliding away from the grippers. What do you do when the physics engine won't work properly? You cheat!

Because an arm really isn't very interesting if it is not able to pick up objects, you will define code that detects when the arm is closing on an object. A joint will be created on-the-fly to attach the object to the grippers, and then that object will follow the grippers as they move — just as if it had been grasped. You'll see how this works in detail because it provides a good example of using some additional simulator functionality.

Look at the following lines of code near the bottom of the MoveTo function that was described in the section "Moving the Arm":

```
if((_joints[5].Target > gripperVal) && (Payload == null))
{
    _attachPayload = true;
}
else if ((_joints[5].Target < gripperVal) && (Payload != null))
{
    _dropPayload = true;
}
```

_joints[5] represents the gripper joint. When the motion sequence is being set up, you detect whether the grippers are closing or opening. If they are closing, then you set _attachPayload to be true. If they are opening, then you set _dropPayload to be true. These flags are not used until the motion sequence is completed. This code is in the Update method:

```
// no joints needed to be updated, the movement is finished
if (_attachPayload)
{
    // we are attaching a payload object after the motion has completed
    if (_intersect == null)
    {
        // haven't yet cast the intersection ray, do it now
        _intersect = new IntersectRay(new Ray(Origin, Direction));
        SimulationEngine.GlobalInstancePort.Post(_intersect);
    }
    List<TriangleIntersectionRecord> results =
        _intersect.ResponsePort.Test<List<TriangleIntersectionRecord>>();

    if (results != null)
    {
        // we've received the intersection results,
        // see if we need to attach a payload
        AttachPayload(results);
        if (_payload != null)
        {
            // create a joint to hold the payload
            _payload.PhysicsEntity.UpdateState(false);
            L4Entity.PhysicsEntity.UpdateState(false);
            Vector3 jointLocation = TypeConversion.FromXNA(xna.Vector3.Transform(
                TypeConversion.ToXNA(_payload.State.Pose.Position),
                xna.Matrix.Invert(L4Entity.World)));

            Vector3 normal = new Vector3(0, 1, 0);
            Vector3 axis = new Vector3(1, 0, 0);

            // calculate a joint orientation that will preserve the orientation
            // relationship between L4Entity and the payload
            Vector3 parentNormal =
                Quaternion.Rotate(L4Entity.State.Pose.Orientation, normal);
            Vector3 parentAxis =
                Quaternion.Rotate(L4Entity.State.Pose.Orientation, axis);
```

```
                Vector3 thisNormal =
                    Quaternion.Rotate(_payload.State.Pose.Orientation, normal);
                Vector3 thisAxis =
                    Quaternion.Rotate(_payload.State.Pose.Orientation, axis);
                EntityJointConnector[] payloadConnectors = new EntityJointConnector[2]
                {
                    new EntityJointConnector(_payload,
                        thisNormal, thisAxis, new Vector3(0, 0, 0)),
                    new EntityJointConnector(L4Entity,
                        parentNormal, parentAxis, jointLocation)
                };
                _payloadJoint = PhysicsJoint.Create(
                    new JointProperties((JointAngularProperties)null,
                        payloadConnectors));
                _payloadJoint.State.Name = "Payload Joint";
                PhysicsEngine.InsertJoint(_payloadJoint);
                // the payload is now fixed to the L4Entity
            }
            // the payload attach is complete
            _attachPayload = false;
            _intersect = null;
            // the motion is also complete, send the completion message
            _moveToActive = false;
            _moveToResponsePort.Post(new SuccessResult());
        }
    }


    // once a ray has been cast into the environment,
    // this method interprets the results and
    // sets a payload in _payload if one is found within the grippers
    public void AttachPayload(List<TriangleIntersectionRecord> results)
    {
        foreach (TriangleIntersectionRecord candidate in results)
        {
            if (candidate.OwnerEntity.GetType() == typeof(SingleShapeSegmentEntity))
                continue;

            if (candidate.IntersectionDistance > Grip)
                break;

            _payload = candidate.OwnerEntity;
        }
    }
```

If _attachPayload is set after the motion is completed, a ray is constructed that originates in the center of the L4Entity and extends in the direction of the grippers. The ray is intersected with all entities in the environment. If a valid entity is found, that entity is attached to the L4Entity with a joint.

The first time this code is executed, _intersect is null and a new ray is constructed and cast into the environment by posting an IntersectRay message on the SimulationEngine port. This is the first time you've seen this method. It calculates intersections with the visual mesh of an object. The laser range finder discussed in Chapter 6 uses the PhysicsEngine.Raycast2D method to cast a ray into the environment that intersects with physics shapes. You can use either method depending on whether you desire an intersection with the visual mesh or an intersection with the physics shapes.

Each subsequent time the `Update` method is called, the `responseport` of the `IntersectRay` message is checked for results. When the results are available, they are passed to the `AttachPayload` method, which determines whether any of the intersection results represent an object that should be attached to the arm. If so, `_payload` is set to the entity. If a valid payload is found, then a joint is built that attaches the payload entity to the `L4Entity` while maintaining its relative position and orientation.

Once the results have been processed, a message is sent on the `MoveTo` response port to indicate that the motion sequence has been completed.

That covers the sequence of events when the payload is grasped. When the grippers are opened and a payload entity is currently attached, the `_dropPayload` Boolean is set to `true`. If this flag is set at the end of the `Update` method, the joint that attaches the payload is deleted and the payload entity is no longer fixed to the arm.

What are the implications of this terrible hack? It means that the arm simulation does a poor job of modeling how well an object is grasped by the grippers. In fact, if an object is within the grippers and they are closed even slightly, the object will be picked up. In the real world, the arm can only pick up objects if the grippers are set to an appropriate value for the size of the object. In addition, once the simulation arm grips an object, the object is firmly attached until it is released. With a real arm, moving the arm too fast may cause the object to dislodge from the grippers. These limitations should be taken into account when using the simulated arm to simulate algorithms intended for a real arm.

## Inverse Kinematics

Technically, you have everything you need at this point to move the arm around, but it isn't much fun to maneuver the arm to pick up an object by specifying each of the joint angles. It is much more convenient to simply specify the X,Y, Z coordinates of the grippers and their rotation and angle of approach, enabling the arm to move to that configuration. The process of calculating the joint angles from such a high-level specification is called *inverse kinematics*. The reverse process, calculating the gripper position from the joint angles, is called *forward kinematics*.

There are three solutions for the inverse kinematics for the Lynx 6 robotic arm, available on the Lynxmotion website at the following URLs:

```
www.lynxmotion.com/images/html/proj073.htm
www.lynxmotion.com/images/html/proj057.htm
www.lynxmotion.com/images/html/proj058.htm
```

The `SimulatedLynxL6Arm` service described in this chapter uses an inverse kinematics solution similar to the third link. The service implements a method called `MoveToPosition` that takes the parameters in the following table, calculates the joint positions, and then calls `MoveTo` on the arm entity to execute the motion:

| Parameter | Units | Description |
|-----------|-------|-------------|
| X | Meters | X position of the center of the tip of the gripper |
| Y | Meters | Y position of the center of the tip of the gripper |
| Z | Meters | Z position of the center of the tip of the gripper |
| P | Degrees | Angle of approach of the gripper (–90 is vertical with the gripper down) |
| W | Degrees | Rotation angle of the wrist joint |
| Grip | Meters | Distance of separation between the grippers |
| Time | Seconds | Amount of time to complete the motion |

The code is fairly straightforward even if the math isn't:

```
// This method calculates the joint angles necessary to place the arm into the
// specified position.  The arm position is specified by the X,Y,Z coordinates
// of the gripper tip as well as the angle of the grip, the rotation of the grip,
// and the open distance of the grip.  The motion is completed in the
// specified time.
public SuccessFailurePort MoveToPosition(
    float x, // x position
    float y, // y position
    float z, // z position
    float p, // angle of the grip
    float w, // rotation of the grip
    float grip, // distance the grip is open
    float time) // time to complete the movement
{
    // taken from Hoon Hong's ik2.xls IK method posted on the Lynx website

    // physical attributes of the arm
    float L1 = InchesToMeters(4.75f);
    float L2 = InchesToMeters(4.75f);
    float Grip = InchesToMeters(2.5f);
    float L3 = InchesToMeters(5.75f);
    float H = InchesToMeters(3f);   // height of the base
    float G = InchesToMeters(2f);   // radius of the base

    float r = (float)Math.Sqrt(x * x + z * z); // horizontal distance to the target
    float baseAngle = (float)Math.Atan2(-z, -x); // angle to the target

    float pRad = DegreesToRadians(p);
    float rb = (float)((r - L3 * Math.Cos(pRad)) / (2 * L1));
    float yb = (float)((y - H - L3 * Math.Sin(pRad)) / (2 * L1));
    float q = (float)(Math.Sqrt(1 / (rb * rb + yb * yb) - 1));
    float p1 = (float)(Math.Atan2(yb + q * rb, rb - q * yb));
    float p2 = (float)(Math.Atan2(yb - q * rb, rb + q * yb));
    float shoulder = p1 - DegreesToRadians(90);  // angle of the shoulder joint
```

*(continued)*

411

*(continued)*

```
    float elbow = p2 - shoulder;        // angle of the elbow joint
    float wrist = pRad - p2;            // angle of the wrist joint

    // Position the arm with the calculated joint angles.
    return _16Arm.MoveTo(
        RadiansToDegrees(baseAngle),
        RadiansToDegrees(shoulder),
        RadiansToDegrees(elbow),
        RadiansToDegrees(wrist),
        w,
        grip,
        time);
}
```

The X and Z coordinates are converted to a radius and an angle (cylindrical coordinates). The angle of approach (p) of the gripper, along with the radius and elevation (Y coordinate), is used to calculate joint angles from the gripper back to the base. The radian joint values are converted back to degrees and passed to the MoveTo method.

It is not possible for the arm to accommodate all combinations of gripper coordinates and angle-of-approach values. If an impossible position is requested, one of the joint angles will be out of bounds and the MoveTo method will post an exception message to the response port.

## Using the Arm User Interface

The SimulatedLynxL6Arm service implements a Windows Forms user interface using the same principles described in Chapter 7. The window is shown in Figure 8-13.

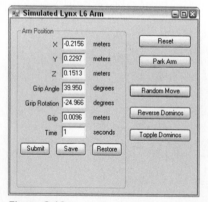

**Figure 8-13**

The seven parameters to the MoveToPosition method appear in this dialog. You can enter values using meters for distance parameters, degrees for angle parameters, and seconds for the time parameters. After the Submit button is pressed, the arm moves to the requested position if it is valid. Otherwise, an error message is displayed if the position is invalid.

Just to make things more interesting, 10 dominos have been added to the scene. These dominos are `SingleShapeEntities` consisting of a single box shape. Their visual mesh provides a texture map for each face to make them more interesting.

Pressing the Reset button causes the currently executing motion sequence to stop, resetting the position of the dominos and the arm to their initial position using the following code:

```
// Place the dominos back in their initial positions
public void ResetDominos()
{
    for (int i = 0; i < DominoCount; i++)
    {
        Dominos[i].State.Pose = new Pose(InitialDominoPosition(i));
        // Don't move the domino until its Update method is executing
        Dominos[i].DeferredTaskQueue.Post(new Task<int>(i, DeferredSetDominoPose));
    }
}

// Restore the arm to its initial position
public void ResetArmPosition()
{
    _16Arm.State.Pose = new Pose(new Vector3(0, 0, 0));
    // Don't set the pose until the Update method executes
    _16Arm.DeferredTaskQueue.Post(new Task(DeferredSetArmPose));
    // wait for the arm motion to settle before making it non-kinematic
    Activate(Arbiter.Receive(false, TimeoutPort(500), delegate(DateTime now)
    {
        _16Arm.DeferredTaskQueue.Post(new Task(ReleaseArm));
    }));
}

// Returns the initial position of the requested domino
Vector3 InitialDominoPosition(int i)
{
    return new Vector3(InchesToMeters(2.5f) + i * 0.011f, 0, 0);
}

// Create and place DominoCount dominos
public IEnumerator<ITask> CreateDominos()
{
    for(int i=0; i<DominoCount; i++)
    {
        Dominos[i] = new Domino(InitialDominoPosition(i), i);
        SimulationEngine.GlobalInstancePort.Insert(Dominos[i]);
    }

    yield break;
}

// This method sets the pose of a domino in its update method
void DeferredSetDominoPose(int i)
```

*(continued)*

413

*(continued)*

```
{
    Dominos[i].PhysicsEntity.SetPose(Dominos[i].State.Pose);
}

// Sets the arm pose in its update method
void DeferredSetArmPose()
{
    _16Arm.PhysicsEntity.IsKinematic = true;
    _16Arm.PhysicsEntity.SetPose(_16Arm.State.Pose);
}

// Restores the arm to non-kinematic
void ReleaseArm()
{
    _16Arm.PhysicsEntity.IsKinematic = false;
}
```

Notice that the dominos are moved back to their original positions by posting a task to each of their deferred task queues. The task will execute during their `Update` method when it is safe to call `PhysicsEntity.SetPose` to move them.

It is similar for the arm except you must also deal with the inertia of the arm. When the arm entity is moved quickly, its segments act like a pendulum, storing the inertia, and the physics engine causes the arm to continue moving during the next few frames. You can prevent this effect by setting the base of the arm as kinematic before you move it. After it is moved, the code waits for half a second before setting the arm back to its non-kinematic state.

Pressing the Park Arm button executes two calls to `MoveTo` as shown in the following code:

```
// This method puts the arm into its parked position
public IEnumerator<ITask> ParkArm()
{
    // Raise the grippers away from the working surface
    yield return Arbiter.Choice(_16Arm.MoveTo(0, 0, 80, 0, 0, 0.05f, 2),
        delegate(SuccessResult s) { },
        ShowError);

    // Move to the parked position
    yield return Arbiter.Choice(_16Arm.MoveTo(0, 80, -56, -75.2f, 0, 0.05f, 3),
        delegate(SuccessResult s) { },
        ShowError);
}
```

The sequence of events in this code snippet is as follows:

1. The first call to _16Arm.MoveTo is executed and returns a response port.

2. Arbiter.Choice returns a task that executes when a response message is posted to the response port.

3. The first delegate executes if the response message was a `SuccessResult`. Otherwise, the second delegate executes and prints the error information to the console.

4. The second call to `_16Arm.MoveTo` is executed and returns a response port.

5. `Arbiter.Choice` returns a task that executes when a response message is posted to the response port.

6. Finally, the first delegate executes if the response message was a `SuccessResult` or the error message is printed.

Stacking yield return statements in an iterator method like this guarantees that the arm motions will be executed one after another in the proper order with no wasteful spin-waiting on the CPU.

The first `MoveTo` command raises the gripper nearly vertical, and the second one puts it into the parked position shown in Figure 8-14.

**Figure 8-14**

Clicking the Random Move button causes the arm to be moved to a random position. Random values for each of the parameters are chosen without regard to what is valid and then passed to `MoveToPosition`. This continues until `MoveToPosition` returns a `SuccessResult`, meaning the position specified was valid. Whenever `MoveToPosition` is called, the parameters on the Winforms window are updated.

Both the Reverse Dominos button and the Topple Dominos button execute a sequence of arm moves. The first sequence causes the arm to pick up the dominos one by one and put them back down again on the other side of the arm. The Topple Dominos button causes the arm to pick up the dominos one by one and lay them out in a pattern. The dominos are then knocked down in classic fashion. The code for the Topple Dominos motion is shown here:

```
// This method executes a series of arm movements that cause the arm to pick up
// dominos from one side and put them down in a pattern on the other side and
// then the arm causes the dominos to fall.
public IEnumerator<ITask> ToppleDominos()
{
    _moveSequenceActive = true;
    for (int i = 0; i < DominoCount; i++)
    {
        Vector3 src = InitialDominoPosition(DominoCount - i - 1);

        // move the arm into position to grasp the domino
        yield return CheckAndMove(0.18f, 0.06f, src.Z, -80, 0, 0.04f, 3);
        yield return CheckAndMove(src.X, 0.025f, src.Z, -90, 0, 0.04f, 1f);
        // close the grip to grab the domino
        yield return CheckAndMove(src.X, 0.025f, src.Z, -90, 0, 0.025f, 0.5f);
        // move it out and up
        yield return CheckAndMove(0.17f, 0.2f, src.Z, 0, 0, 0.025f, 1);
        // move it to the other side

        Vector3 dst = ToppleDominoPosition[i];
        yield return CheckAndMove(dst.X - 0.02f, 0.2f, dst.Z, 0, 0, 0.025f, 2);
        // move it into position
        yield return CheckAndMove(dst.X - 0.02f, 0.03f, dst.Z, -90,
            AdjustWrist(0, dst.X, dst.Z), 0.025f, 1);
        yield return CheckAndMove(dst.X, 0.03f, dst.Z, -90,
            AdjustWrist(0, dst.X, dst.Z), 0.025f, 0.5f);
        // lower it
        yield return CheckAndMove(dst.X, 0.026f, dst.Z, -90,
            AdjustWrist(0, dst.X, dst.Z), 0.025f, 1);
        // release it
        yield return CheckAndMove(dst.X, 0.026f, dst.Z, -90,
            AdjustWrist(0, dst.X, dst.Z), 0.04f, 0.5f);
        // back away
        yield return CheckAndMove(dst.X, 0.06f, dst.Z, -80,
            AdjustWrist(0, dst.X, dst.Z), 0.04f, 1);
    }

    // knock them down
    Vector3 final = ToppleDominoPosition[9];    // the last domino position
    yield return CheckAndMove(final.X, 0.05f, final.Z - 0.04f, -80, 0, 0, 1);
    yield return CheckAndMove(final.X, 0.05f, final.Z - 0.04f, -80, 0, 0, 0.5f);

    // pause for dramatic effect
    yield return Arbiter.Receive(false, TimeoutPort(500), delegate(DateTime now)
        { });
    yield return CheckAndMove(final.X, 0.05f, final.Z + 0.06f, -80, 0, 0, 0.3f);
    yield break;
}

// This method calculates the wrist rotation angle that will orient the wrist to
// the specified angle relative to the X axis.
```

```
float AdjustWrist(float angle, float x, float z)
{
    float val = angle + RadiansToDegrees((float)(Math.Atan2(-z, -x))) + 90;
    while (val < -90)
        val += 180;

    while (val > 90)
        val -= 180;

    return val;
}


// check for early termination of the motion and call MoveToPosition
ITask CheckAndMove(
    float x, float y, float z, float gripAngle,
    float gripRotation, float grip, float time)
{
    if (_moveSequenceActive)
    {
        return Arbiter.Choice(
            MoveToPosition(x, y, z, gripAngle, gripRotation, grip, time),
            delegate(SuccessResult s) { },
            ShowError);
    }
    else
    {
        return new Task(DoNothing);
    }
}


void DoNothing()
{
}


// The destination positions for the dominos in the topple movement
Vector3[] ToppleDominoPosition = new Vector3[]
{
    new Vector3(-0.11f,0,0.125f),
    new Vector3(-0.12f,0,0.10f),
    new Vector3(-0.125f,0,0.075f),
    new Vector3(-0.12f,0,0.05f),
    new Vector3(-0.11f,0,0.025f),
    new Vector3(-0.10f,0,0.00f),
    new Vector3(-0.095f,0,-0.025f),
    new Vector3(-0.10f,0,-0.05f),
    new Vector3(-0.11f,0,-0.075f),
    new Vector3(-0.12f,0,-0.10f)
};
```

The CheckAndMove method ensures that the sequence is still active. If the Reset button is clicked, _moveSequenceActive becomes false and a null task is returned instead of the MoveToPosition task. The ToppleDominoPosition array specifies where the arm will place each domino (see Figure 8-15).

Look at the calls to the `AdjustWrist` method. This method calculates the angle that the gripper will make relative to the X axis and then adjusts that angle by rotating the wrist so that the domino is placed at the specified angle.

Have some fun putting the arm through its paces. You can change the pattern of the dominos by editing the `ToppleDominoPosition` array or you can define entirely new motion sequences. As you debug a new motion sequence, watch the console output for messages indicating that a requested motion is invalid or set a breakpoint on the `ShowError` method.

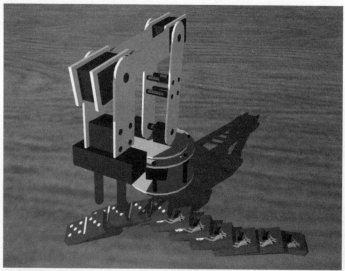

Figure 8-15

# Summary

This chapter explored the world of joints and articulated entities. You learned how to join two entities together with a joint and became familiar with the many properties of a joint. The TestBench sample service provided a way to examine the behavior of different joint types and to prototype new joints.

The SimulatedLynxL6Arm service showed how to build a robotic arm entity with six degrees of freedom. The arm service included a user interface and inverse kinematics code that made it easy to define new motion sequences for the arm to perform.

Chapter 9 includes additional examples of articulated entities in the simulation environment.

# 9

# Adventures in Simulation

The previous chapters in this part have covered what you need to know to make your own simulation scenarios, complete with custom robots, simulation services, and orchestration services. Unfortunately, all this information is a little like being handed a toolbox and a pile of lumber and being told to build a house. It helps to have a few examples to work from. That's what this chapter is all about.

Several examples are presented that demonstrate how to use the simulator in a variety of ways. You'll see how to define a new sumo player service using the Simulated Sumo Competition package. You'll see how to take articulated robotic arms to the next level by building a simulated walking hexapod. You'll also see how to implement a soccer strategy with a team of Corobots in the Simulated Soccer Competition package. Finally, you'll have an opportunity to build a complete simulation environment that can be explored by a simulated robot and built from a floor-plan image.

## Simulating a Sumo Competition

In the Spring of 2007, the Microsoft Robotics Team sponsored a Sumo Robotics competition at the Mobile and Embedded Developers Conference (MEDC) in Las Vegas. Figure 9-1 shows one of the final matches of that competition. The instructions to build a sumo robot and the software needed to drive the robot are provided in a package available in a separate download from the MRDS SDK.

Figure 9-1

The Sumo package is a great place to get started with writing and modifying MRDS services because most of the infrastructure has already been provided and a complete sumo strategy can be implemented by making just a few changes to the existing code.

This section explains how to use the simulated sumo services and how to implement a new sumo player service with a unique strategy.

## SimulatedSumo

In a sumo competition, two robots are placed within a circular sumo ring where they attempt to push each other outside the ring. Each sumo match consists of three rounds that last, at most, one minute each. When a robot leaves the sumo ring, either under its own power or by being pushed out, the round ends and the remaining robot scores one point. If neither robot leaves the ring after one minute, the round is a tie and neither robot scores a point. At the end of three rounds, the robot with the most points wins the match.

The sumobot uses its front bumpers to sense when it has made contact with its opponent, and it uses the webcam to locate its opponent. The infrared sensors on the underside are used to determine when the sumobot is getting close to the edge of the sumo ring. Note in Figure 9-1 that on the outside edge of the circular ring is a band that is painted a different color and has a different reflectivity than the interior of the ring. The IR sensors on each robot can detect when the robot is over this region, and the robot can adjust its course so that it doesn't drive out of the ring unless it is being pushed by the other sumobot.

The Sumo package is available from the Microsoft Robotics website (`www.microsoft.com/robotics`) under Downloads. When you download and install the package, it installs the source code for the SimulatedIRobotLite service and the SumoPlayer service to `samples\simulation\competitions\SimulatedSumoServices`. The binaries for these services are installed in the `bin` directory along with the binaries for the SumoReferee service.

The Sumo robot used with this package is built with an iRobot Create, an onboard computer running Windows CE, and a webcam. You can find the instructions for building this platform at `http://msdn2.microsoft.com/en-us/robotics/bb403184.aspx`.

Figure 9-2 shows the complete hardware platform alongside its simulation model.

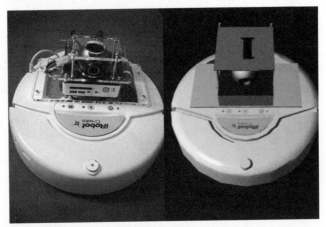

**Figure 9-2**

You can see that the sumobot uses an iRobot Create for its motor platform. The Create has two powered wheels and uses a differential drive to steer. In addition, there are left-front and right-front bumper sensors and four IR sensors on the underside. The ICOP eBox 2300 processor that runs Windows CE is mounted on the transparent platform just above the payload bay. A webcam is mounted just above that to provide vision capability for the sumobot.

## The Simulated Sumo Referee

The SimulatedSumoReferee service constructs the simulated sumo ring environment, places the competitors in the sumo ring, and then runs the match and determines the winner. The user interface for the SimulatedSumoReferee service is shown in Figure 9-3. (Never mind that the skinny cartoon referee looks more like a soccer referee than a sumo referee.)

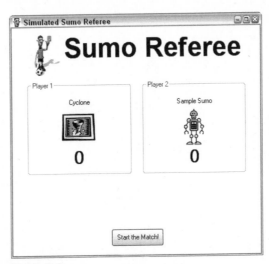

**Figure 9-3**

You can run the SimulatedSumoReferee service by typing **sumoreferee** from the c:\Microsoft Robotics Studio (1.5) directory. This runs the sumoreferee.cmd file in the bin directory, installed along with the software in this book. This command file starts the simulatedsumoreferee manifest installed with the Sumo package described previously.

When the SimulatedSumoReferee service runs, it starts the simulation engine and builds the sumo ring environment, which consists of a sumo ring, a ground plane, and an overhead light. No sky is used in this simulation environment. The Sumo Referee user interface displays a drop-down box listing all the available sumo players on the system, identified by their contracts. The sample sumo player provided with the sumo package is called "samplesumo," and the sumo player that you develop in this section and which is provided with the book software is called "cyclone." In addition to the sumo players, the following options are available:

❑ **Refresh:** This causes the list of players to refresh in case a new player has been installed on the system.

❑ **Continuous:** Selecting this option causes the referee to pick two players at random and start the match. When the first match is complete, two more players are selected and another match starts.

❑ **None:** This prevents a player from being selected. This is handy when you want to observe the behavior of your sumobot without the annoyance of another crazed robot trying to push it out of the ring. Of course, a match of None against None is not very exciting to watch.

❑ **Qualify:** The qualifying round was used in the MEDC competition to enable developers to show that they had a functioning sumobot service prior to being admitted to the hardware round. A sumobot is specified for the first player and "Qualify" is selected for the second player. The referee then starts a match with a sumobot opponent that does not move but patiently waits to be pushed out of the ring.

Try playing a match by selecting two players and then clicking the Start Match button. The Referee service takes care of placing the robots and starting each round. It also keeps track of the time and the score. At the end it declares a winner or a tie. A sumo round in progress is shown in Figure 9-4.

Figure 9-4

Each player is identified by the number on its top platform.

From a simulation standpoint, one of the interesting things about this simulation environment is the way it detects when a sumobot has left the ring. The ring itself is several centimeters high and it is placed in the middle of a trigger volume that is slightly lower than the ring. In Figure 9-5, rendering of the trigger entity has been enabled so that you can see it.

Figure 9-5

When a sumobot leaves the ring, it intersects with the trigger shape, the Referee service is notified, and the round is ended.

## Communicating with the Sumo Referee Service

Most of the examples in this book have a manifest that starts a number of services, and those services continue to run as long as the node is running. In the case of the sumo competition, sumo player services are started and ended dynamically.

When the sumo player service runs, it starts up an instance of the SimulatedIRobotLite service. This service is analogous to the hardware service that would be running on an actual hardware robot. It provides an interface between the sumo player service and the hardware.

The sample sumo player service provided with the sumo package is `\Microsoft Robotics Studio (1.5)\samples\simulation\competitions\simulatedsumoservices\sumoplayer\sumoplayer.csproj`. You can examine the `ConnectWithRobot` method in `sumoplayer.cs` to see how the sumo player service configures and connects to the SimulatedIRobotLite service.

When the SimulatedIRobotLite service receives the `Connect` message, it sends a message to the SimulatedSumoReferee service announcing that a new player is about to join the match. The Referee service adds a new sumobot entity to the simulation environment and sets up a `ServiceForwarder` to send messages back to the SimulatedIRobotLite service.

This method of sending an announcement message to another service so that it can send messages back is a pattern that is often used in systems with dynamic services, so we'll look at it in a little more detail. In the SimulatedIRobotLite service, the AnnounceToReferee method is called in response to a Connect message from the sumo player service. The method first verifies that the sumo Referee service is running and that it has a valid port to send messages to that service.

Then a PlayerAnnounceRequest is constructed to send to the Referee service. This class contains the sumo player name, an image that represents the sumo player service, and the ServiceInfo.Service string that identifies the SimulatedIRobotLite service. It also contains the name of the entity specified as a partner to the SimulatedIRobotLite service. The referee uses this entity name to name the player entity that is created in the simulation environment:

```
// let the referee know that we would like to join the match.
referee.PlayerAnnounce announce = new referee.PlayerAnnounce();
announce.Body = new referee.PlayerAnnounceRequest(ServiceInfo.Service);
announce.Body.Name = _state.Name;
announce.Body.Image = _state.RobotImage;

// our entity instance name is passed as a partner
PartnerType entityPartner =
    Dss.ServiceModel.DsspServiceBase.DsspServiceBase.FindPartner(
    new System.Xml.XmlQualifiedName(Microsoft.Robotics.Simulation.Partners.Entity,
        Microsoft.Robotics.Simulation.Contract.Identifier),
        ServiceInfo.PartnerList);

if (entityPartner == null)
{
    LogError(
        "Invalid entity name specified as a partner to the iRobotLite Service");
}
else
    announce.Body.EntityName = entityPartner.Service.ToString();

// find the alternate contract service info for the refereeannounce service
foreach (ServiceInfoType altService in AlternateContractServiceInfo)
{
    if (altService.Contract.Contains("simulatedsumorefereeannounce"))
    {
        announce.Body.AnnounceURI = altService.Service;
        break;
    }
}
_refereePort.Post(announce);

referee.PlayerAnnounceResponse announceRsp = null;
Fault failure = null;
yield return Arbiter.Choice(announce.ResponsePort,
    delegate(referee.PlayerAnnounceResponse rsp)
    {
        announceRsp = rsp;
    },
```

```
delegate(Fault f)
{
    failure = f;
    LogError("Announce failed, shutting down", f);
}
);
```

When the referee receives the announce request, it sets up a port to send messages back to the SimulatedIRobotLite service as follows:

```
refereeannounce.SimulatedSumoRefereeAnnounceOperations port =
    ServiceForwarder<refereeannounce.SimulatedSumoRefereeAnnounceOperations>
        (announce.Body.AnnounceURI);
```

When both players have been connected and the match is started, the Referee service sends a `PlayerButtonPress` message to the SimulatedIRobotLite service associated with each sumobot, indicating that the Play button has been clicked, and the sumobot begins running its strategy for the round.

When the match has concluded and a new match begins, the referee sends a `Drop` message to the sumo player service, as well as its associated SimulatedIRobotLite service, and they terminate.

The code for the sample sumo player and for the SimulatedIRobotLite services is included with the MRDS Sumo package. Unfortunately, the source code for the SimulatedSumoReferee service has not been publicly released yet. Look for it in a future MRDS release.

## The Sample Sumo Behavior

Before you learn how to make your own sumo behavior, it pays to study the behavior implemented in the sample sumo player. The behavior implemented in the sample sumo relies on a state machine with the following states:

❑ **Uninitialized:** This is the initial state. No state information has been received from the robot yet.

❑ **Initializing:** A state packet has been received and the sumo player service is initializing the hardware.

❑ **Ready:** The hardware has been initialized and the sumobot is ready to begin a round.

❑ **Pending:** The Play button has been pressed and the sumobot is waiting five seconds before entering the Wander state and starting to move.

❑ **Contact:** One of the bumper sensors indicated that the sumobot is in contact with an opponent.

❑ **AvoidBoundary:** One of the infrared sensors on the bottom of the sumobot detected that the sumobot is too close to the edge of the ring.

❑ **Tracking:** The opponent has been spotted; move toward it.

❑ **Wander:** The sumobot doesn't know what else to do and wanders around the sumo ring until something interesting happens.

❑ **Blind:** The sumobot has not received a sensor packet for some time. Stop movement and spin slowly to avoid leaving the sumo ring.

Most of the processing for this state machine takes place in the `TimerHandler` method in `sumoplayer.cs`, which is called at regular intervals. Three times out of four, the `TimerHandler` begins by requesting sensor state from the hardware. On the fourth time, an image is requested from the webcam. Both the sensor handler method (`IRobotStateHandler`) and the image handler method (`ValidateFrameHandler`) execute asynchronously when the data is returned from the hardware.

The sumo player service starts in the `Uninitialized` state. Once it has made a request to initialize the hardware, it moves to the `Initializing` state and then to the `Ready` state when the hardware confirms that it is properly initialized. At that point, the robot beeps once to indicate that it is ready for action. On the real hardware, booting the hardware and making it to the `Ready` state can take a minute or so. In the simulator, it is very fast.

When the hardware robot is in the `Ready` state, it is placed in the sumo ring, and the Play button is clicked when the referee starts the round. In the simulator, the referee places the simulator model in the ring and then passes a message to the SimulatedIRobotLite service associated with each player, which causes the Play button to be clicked. The next time the sumo player service retrieves the hardware state, it detects that the button has been pressed and enters the `Pending` state. The `Pending` state lasts for five seconds during which the robot can't move. In the real world, this gives the robot owner time to get out of the ring before the robots start moving.

The sumo player service sets the state to `Pending` in the `RobotUpdateButtonsHandler` following code. It then enforces the five-second wait with a call to `InternalDrivingMilliseconds`, which specifies a wheel speed of 0 for both the left and right wheels and a duration of 5000 milliseconds. The `InternalDrivingMilliseconds` method is important because it schedules the movement of the robot for the next period of time. If the sumo ring border or an opponent is detected, the duration of the movement can be preempted. In this case, you are instructing the robot to stay still for the next five seconds. It can still gather webcam images and do processing during this time, so if it is facing its opponent it will be ready to attack immediately after the time limit expires:

```
public void RobotUpdateButtonsHandler(irobot.ReturnPose pose)
{

    // Ignore buttons before the iRobot is initialized
    if (_state.SumoMode < SumoMode.Ready)
        return;

    if ((pose.ButtonsCreate & create.ButtonsCreate.Play) ==
        create.ButtonsCreate.Play && _state.SumoMode < SumoMode.Pending)
    {
        _state.SumoMode = SumoMode.Pending;
        LogVerbose(LogGroups.Console, "Sumo Mode: Pending");

        // play tones to indicate that the competition has begun
        _robotPort.RoombaPlaySong(3);

        // wait 5 seconds before moving
        InternalDrivingMilliseconds(0, 0, 5000.0);

        // Set Advanced LED On and Power LED to bright Green.
        _robotPort.RoombaSetLeds(irobot.RoombaLedBits.CreatePlay, 0, 255);

    }
}
```

The `TimerHandler` runs about every 12.5 milliseconds, and each time it runs, it requests new sensor state or image state. If the time has expired from the previous `InternalDrivingMilliseconds` command, it plots its next move. If it is in the `Pending` state, it moves to the `Wander` state and makes a call to `SetWanderDrive`, which causes the robot to drive forward with a slight curve to the right for the next 250 milliseconds. The robot remains in this `Wander` state until something interesting happens.

In the `Wander` state, the robot eventually crosses the outer band in the sumo ring. When this happens, the next state packet returned from the hardware will be processed by the `RobotUpdateFloorSensorsHandler` method. This method takes immediate action to preempt the current movement. It wouldn't be wise to wait until the duration of the last movement completed because by then the robot could be out of the ring and the round would be lost. This method immediately calls `InternalDrivingMilliseconds` to spin the robot in a direction away from the edge of the ring. This call cancels the previous driving command. When the spin command expires, the `TimerHandler` method drives the robot away from the edge and puts it back into the `Wander` state.

This way of handling drive commands works well because it enables the robot to choose its movement by setting the motor drive speeds for a period of time, but it also allows for important events to interrupt this movement when necessary.

Another event that can interrupt movement commands is when the robot detects the presence of an opponent. This occurs when the `ProcessImage` method detects a large white area in the webcam image. It determines whether the center of the white region is to the left or the right, and then adjusts the wheel speed to move the robot toward its opponent and changes the state to `Tracking`.

When the sumobot makes contact with its opponent, the `RobotUpdateBumpersHandler` changes the state to `Contact` and pushes against the opponent with full power.

In the `Tracking` state or the `Contact` state, if the robot loses its opponent, it returns to the `Wander` state.

The only state we haven't covered is the `Blind` state. The robot enters this state if it receives a sensor packet that is older than a certain time threshold. This indicates that there is a problem transferring the sensor data or the robot hardware is running slowly and the sensor data is too old to be trusted. In this case, the robot enters the `Blind` state and stops moving forward until it begins to receive sensor state data that is not stale. This prevents the robot from relying on outdated sensor state data to make its decisions, which in turn prevents it from going out of the ring.

That covers the behavior that is provided with the sample sumo player. Of course, you are free to implement anything you like for your winning strategy. Take some time to open the sumo player project, run it, and set some breakpoints so that you understand how the robot makes decisions and handles important events such as crossing the outer border and detecting its opponent.

## Creating a New Sumo Player

Now that you are familiar with the sample sumo player service, you can build your own. Microsoft has provided a utility called MakeSumoPlayer that makes this process simple. From the Microsoft Robotics Developer Studio Command Prompt window, type the following: **MakeSumoPlayer /name:Cyclone**.

This utility takes the source code for the sample sumo player service in \Microsoft Robotics Studio (1.5)\samples\simulation\competitions\simulatedsumoservices and copies it to a new sumo player service with the specified name in the same directory. All of the necessary variables and namespaces are updated to make a new and completely independent service.

Compile the new sumo player service that you have created and run the sumo referee by typing **sumoreferee** at the MRDS command prompt. A new sumo player called Cyclone should now show up in the drop-down box, and you can pair it with the old sample sumo player for a match. At this point, the behavior of the new sumo player is the same as the sample sumo player. Let's make some changes to make things interesting.

You can customize your new sumo player by giving it a new name and an image that will be used in the sumo referee user interface. Change the initial value of the _name field to define a new name for your sumobot. You can change the image by editing the PlayerImage.bmp file in the Resources subdirectory. Be sure to keep the image at 64 by 64 pixels.

A copy of the Cyclone sumo player has been provided in the Chapter9 directory with a few modifications to the out-of-bounds handling. Open this project and look at the RobotUpdateFloorSensorsHandler method. A minor change or two has created a major change in the behavior of the robot.

This is the old code, which handled the case when the robot first encounters the colored band at the edge of the ring:

```
if (_state.Sensors.LineDetected)
{
    _state.SumoMode = SumoMode.AvoidBoundary;
    if (_state.Sensors.LineLeft && !_state.Sensors.LineRight &&
        !_state.Sensors.LineFrontRight)
        InternalDrivingMilliseconds(300, -300, 100.0);
    else if (_state.Sensors.LineRight && !_state.Sensors.LineLeft &&
            !_state.Sensors.LineFrontLeft)
        InternalDrivingMilliseconds(-300, 300, 100.0);
    else
        InternalDrivingMilliseconds(-200, 200, 400.0);
}
```

Sensors.LineDetected is true if any of the floor sensors have detected the outer boundary of the ring. In this case, the old behavior was to spin the robot for a period of time and then move it away from the boundary. The new behavior is shown here:

```
if (_state.Sensors.LineDetected || _state.SumoMode == SumoMode.FollowLine)
{
    bool lineRight = _state.Sensors.LineRight || _state.Sensors.LineFrontRight;
    bool lineLeft = _state.Sensors.LineLeft || _state.Sensors.LineFrontLeft;
    bool lineMiddle = !lineLeft && lineRight;
    if (_state.SumoMode != SumoMode.FollowLine)
```

```
    {
        if (!lineMiddle)
        {
            // spin until our orientation is right to follow the line
            InternalDrivingMilliseconds(-100, 120, 1000.0);
        }
        else
        {
            _state.SumoMode = SumoMode.FollowLine;
            InternalDrivingMilliseconds(100, 200, 1000);
        }
    }
    else
    {
        if (lineMiddle)
        {
            // right on the line, continue by curving slightly to the left
            InternalDrivingMilliseconds(100, 200, 1000);
        }
        else if (!lineRight)
        {
            // curve to the right
            InternalDrivingMilliseconds(200, 50, 1000);
        }
        else if (lineLeft)
        {
            // curve to the left
            InternalDrivingMilliseconds(50, 250, 1000);
        }
    }
}
```

A new state called FollowLine has been added. If the robot is not in the FollowLine state when it encounters the edge of the ring, it turns until the right side of the robot is over the colored outer ring and the left side of the robot is not. At that point, it enters the FollowLine state. In this state, it follows along the inside edge of the outer boundary. If it detects that it is still over the boundary, it moves with a slight curve to the left to roughly follow the contour of the ring. If the right sensors move left of the boundary, the robot curves to the right to re-acquire the line. If the left sensors move right of the boundary, then the robot curves to the left to re-acquire the line.

This causes the robot to circle the ring looking for its prey. The new FollowLine state is defined in the CycloneTypes.cs file. Compile and run the new cyclone service either by pressing F5 to debug or by running the sumo Referee service and selecting the Cyclone player. You should see a very noticeable difference in the behavior of the two opponents as one wanders randomly across the ring while the other follows the outer boundary, as shown in Figure 9-6.

Figure 9-6

## Where to Go from Here

Is the new strategy better or worse than the previous strategy? Well, the jury is still out on that one, but you now understand how to build your own sumobot and implement your own strategy. You have everything you need to organize your own simulated sumo competition whereby each participant creates a unique sumo player and competes.

You can also take it to the next level by following the instructions on the Microsoft Robotics Developer Studio website to build a hardware version of the sumobot. After you compile your service for Windows CE, you will be able to see it running your strategy in a real sumo ring.

# Building a Six-Legged Walker

If you've ever dreamed about building a giant mechanical cockroach, then this project is for you. You'll build on the previous chapter that discussed articulated entities in the simulation environment to learn how to build a robotic leg with three degrees of freedom and replicate it five times to build a hexapod that can walk in any direction.

In this section, you'll learn how to define the hexapod entity and then how to define a service that enables the hexapod to walk. The code for this project is in the ProMRDS\Chapter9 directory under HexapodEntity and HexapodDifferentialDrive. We'll look at HexapodEntity first. Figure 9-7 shows what our hexapod entity will look like.

Figure 9-7

# The Hexapod Entity

The HexapodEntity service was initially created by copying and renaming the files from SimulationTutorial1. All references to SimulationTutorial1 were replaced with references to HexapodEntity, and the namespace was changed appropriately.

This service takes a data-driven approach to define the hexapod. An array called ShapeDescriptors is initialized with information about all of the shapes that make up the entity. Then another array called Relationships is initialized with all of the information about the joints that join the shapes together. The method that builds the entity consists of fairly generic code that reads the information from the arrays and puts the entity together. One advantage of this approach is that it is much easier to change the shapes and joints without having to modify any code.

## Building the Main Body and Legs

The following code shows how the shapes in the main body and the front right leg are defined:

```
public struct HexapodShapeDescriptor
{
    public Shapes ShapeID;
    public string Name;
    public double xPosition;
    public double yPosition;
    public double zPosition;
    public double xSize;
    public double ySize;
    public double zSize;
    public double radius;
    public double xRotation;
    public double yRotation;
    public double zRotation;
    public double mass;
    public string mesh;
```

*(continued)*

(continued)

```
        public HexapodShapeDescriptor(
            Shapes _ShapeID,
            string _Name,
            double _xPosition,
            double _yPosition,
            double _zPosition,
            double _xSize,
            double _ySize,
            double _zSize,
            double _radius,
            double _xRotation,
            double _yRotation,
            double _zRotation,
            double _mass,
            string _mesh)
        {
            ShapeID = _ShapeID;
            Name = _Name;
            xPosition = _xPosition;
            yPosition = _yPosition;
            zPosition = _zPosition;
            xSize = _xSize;
            ySize = _ySize;
            zSize = _zSize;
            radius = _radius;
            xRotation = _xRotation;
            yRotation = _yRotation;
            zRotation = _zRotation;
            mass = _mass;
            mesh = _mesh;
        }
    }

    // units are in meters
    // Adjustable parameters:
    const float _bodyLength = 0.3f;
    const float _bodyWidth = 0.2f;
    const float _bodyHeight = 0.08f;
    const float _upperLegRadius = 0.03f;
    const float _lowerLegRadius = 0.025f;
    const float _upperLegLength = 0.1f;
    const float _lowerLegLength = 0.15f;

    HexapodShapeDescriptor[] ShapeDescriptors = new HexapodShapeDescriptor[]
    {
        new HexapodShapeDescriptor(Shapes.Box, "Hexapod",
            0,
            _upperLegLength + _lowerLegLength,
            0,
            _bodyWidth, _bodyHeight, _bodyLength, 0,
            0, 0, 0, 1, "HexapodBody.obj"),

        new HexapodShapeDescriptor(Shapes.Sphere, "FRShoulder",
```

```
        _bodyWidth/2 + _upperLegRadius * 2,
        _bodyWidth/2 + _upperLegRadius,
        _bodyLength/2 - _upperLegRadius,
        0, 0, 0, _upperLegRadius,
        0, 0, 0, 1, "HexapodShoulder.obj"),

new HexapodShapeDescriptor(Shapes.Capsule, "FRUpper",
        _bodyWidth/2 + _upperLegRadius * 5,
        _upperLegLength/2 + _lowerLegLength + _lowerLegRadius,
        _bodyLength/2 - _upperLegRadius,
        0, _upperLegLength, 0, _upperLegRadius,
        0, 0, 0, 1, "HexapodUpperLeg.obj"),

new HexapodShapeDescriptor(Shapes.Box, "FRLower",
        _bodyWidth/2 + _upperLegRadius * 5,
        _lowerLegLength/2 + _lowerLegRadius,
        _bodyLength/2 - _upperLegRadius,
        _lowerLegRadius*2, _lowerLegLength + _lowerLegRadius*2,
        _lowerLegRadius*2, _lowerLegRadius,
        0, 0, 0, 1, "HexapodLowerLeg.obj"),
};
```

The HexapodShapeDescriptor class holds information about each shape, such as its type, its name, its position, its size, its rotation, its mass, and whether it has a visual mesh associated with it. The size of the body and the length and radius of each part of each leg are specified, and then the ShapeDescriptors array is initialized with all of the data. The body consists of a single box shape. Each leg consists of three shapes: a sphere, called the shoulder, a capsule for the upper leg, and a box for the lower leg. The positions and sizes of each shape are defined according to the size variables so that it is easy to change the size of the various parts of the entity. The physics representation for the front right leg and body is shown in Figure 9-8.

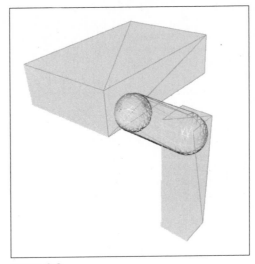

**Figure 9-8**

The sphere attached to the side of the hexapod body is the shoulder shape. It is there so you can define a vertical shoulder joint and an additional horizontal shoulder joint and have them operate independently of each other.

Five additional legs are specified similarly to the front right leg defined earlier. The shapes are placed in approximately the same position and orientation that they will end up with the joints attached. That prevents an abrupt motion in the first frame after the entity is initialized as all of the shapes snap together.

## Defining the Joints

The class holding the joint information is called `ParentChild` because it defines the joint that joins a child entity to its parent. This class and the joint descriptors are shown in the following code:

```
public class ParentChild
{
    public string Parent;
    public string Child;
    public string JointName;
    public Vector3 ChildConnect;
    public Vector3 ParentConnect;
    public Vector3 ParentNormal;
    public Vector3 ParentAxis;
    public Vector3 ChildNormal;
    public Vector3 ChildAxis;
    public ParentChild(string child, string parent, string jointName,
                    Vector3 childConnect, Vector3 parentConnect,
                    Vector3 childNormal, Vector3 childAxis,
                    Vector3 parentNormal, Vector3 parentAxis)
    {
        Child = child;
        Parent = parent;
        JointName = jointName;
        ChildConnect = childConnect;
        ParentConnect = parentConnect;
        ParentNormal = parentNormal;
        ParentAxis = parentAxis;
        ChildNormal = childNormal;
        ChildAxis = childAxis;
    }
}

static Vector3 X = new Vector3(1, 0, 0);
static Vector3 Y = new Vector3(0, 1, 0);
static Vector3 Z = new Vector3(0, 0, 1);
static Vector3 nX = new Vector3(-1, 0, 0);
static Vector3 nY = new Vector3(0, -1, 0);
static Vector3 nZ = new Vector3(0, 0, -1);

ParentChild[] Relationships = new ParentChild[]
{
    new ParentChild("FRShoulder", "Hexapod", "FRShoulder",
        new Vector3(0, 0, _upperLegRadius),
        new Vector3(_bodyWidth/2, 0, _bodyLength/2 - _upperLegRadius),
        Y, X, -X, Y),
```

```
        new ParentChild("FRUpper", "FRShoulder", "FRShoulder2",
            new Vector3(0, _upperLegLength/2, 0),
            new Vector3(0,0,0),
            Y, X, Z, -Y),
        new ParentChild("FRLower", "FRUpper", "FRElbow",
            new Vector3(0, _lowerLegLength/2, 0),
            new Vector3(0,-_upperLegLength/2,0),
            Y, X, Y, X)
    }
```

The ParentChild class specifies the name of the child entity, the name of the parent entity, the name of the joint; the position of the joint relative to the child, the position of the joint relative to the parent; and the normal and axis vectors of the child, and the normal and axis vectors of the parent. This information enables each joint to be fully specified.

Just like the shapes, the joint positions are specified using the size and radius variables defined previously. This makes it easy to change the size of the legs or body without having to redefine all of the shapes and joints.

Notice that the joints are defined with a different orientation for the parent connector and the child connector. This causes the shapes to be joined at right angles and ensures that the axis of each joint is correct. The Shoulder joint moves up and down relative to the hexapod body, while the Shoulder2 joint moves backward and forward.

## Building the Hexapod

Now that all the hard work has been done in defining the shapes and joints, the code that actually builds the hexapod is very simple. The first part of the method creates the shapes and wraps them with SegmentEntities:

```
string _parentName = "Hexapod";

HexapodEntity CreateHexapod(string name, Vector3 initialPosition)
{
    Dictionary<string, VisualEntity> hexapodShapes =
        new Dictionary<string, VisualEntity>();
    string prefix = name + "_";

    foreach(HexapodShapeDescriptor desc in ShapeDescriptors)
    {
        Shape newShape = null;
        switch(desc.ShapeID)
        {
            case Shapes.Box:
                newShape = new BoxShape(new BoxShapeProperties(
                    desc.Name + " Shape", (float)desc.mass,
                    new Pose(), new Vector3((float)desc.xSize,
                    (float)desc.ySize, (float)desc.zSize)));
                break;

            case Shapes.Capsule:
```

(continued)

*(continued)*

```
                        newShape = new CapsuleShape(new CapsuleShapeProperties(
                            desc.Name + " Shape", (float)desc.mass, new Pose(),
                            (float)desc.radius, (float)desc.ySize));
                        break;

                    case Shapes.Sphere:
                        newShape = new SphereShape(new SphereShapeProperties(
                            desc.Name + " Shape", (float)desc.mass, new Pose(),
                            (float)desc.radius));
                        break;
                }
                SingleShapeEntity shapeEntity = null;
                if(desc.Name == _parentName)
                {
                    shapeEntity = new HexapodEntity(newShape, new Vector3(
                        (float)desc.xPosition + initialPosition.X,
                        (float)desc.yPosition + initialPosition.Y,
                        (float)desc.zPosition + initialPosition.Z));
                }
                else
                {
                    shapeEntity = new SegmentEntity(newShape, new Vector3(
                        (float)desc.xPosition + initialPosition.X,
                        (float)desc.yPosition + initialPosition.Y,
                        (float)desc.zPosition + initialPosition.Z));
                }

                shapeEntity.State.Name = prefix + desc.Name;
                shapeEntity.State.Pose.Orientation = UIMath.EulerToQuaternion(
                    new xna.Vector3((float)desc.xRotation, (float)desc.yRotation,
                                    (float)desc.zRotation));
                if (!string.IsNullOrEmpty(desc.mesh))
                    shapeEntity.State.Assets.Mesh = desc.mesh;

                hexapodShapes.Add(shapeEntity.State.Name, shapeEntity);
            }
```

`_parentName` is used to hold the name of the shape that becomes the top-level parent. The code goes through each defined shape and creates the shape with the specified parameters and then creates an entity to hold the shape. If the shape is the parent shape, it becomes a `HexapodEntity`; otherwise, it becomes a `SegmentEntity`. The entity is given a name based on the top-level name specified as a parameter to the `CreateHexapod` method, and a visual mesh is assigned if one was specified for the shape. Each entity is stored in the `hexapodShapes` dictionary for later retrieval.

## Creating and Attaching the Joints

Once the entities have been created, joints can be created and attached to them. The following code accomplishes this task:

```
        // now set up the Parent/Child relationships
        foreach (ParentChild rel in Relationships)
```

```
    {
        string[] names = rel.JointName.Split(';');
        JointAngularProperties angular = new JointAngularProperties();

        if ((names.Length > 0) && (names[0] != string.Empty))
        {
            angular.TwistMode = JointDOFMode.Free;
            angular.TwistDrive = new JointDriveProperties(
                JointDriveMode.Position,
                new SpringProperties(500000, 1000, 0), 1000000);
        }

        if ((names.Length > 1) && (names[1] != string.Empty))
        {
            angular.Swing1Mode = JointDOFMode.Free;
            angular.SwingDrive = new JointDriveProperties(
                JointDriveMode.Position,
                new SpringProperties(500000, 1000, 0), 1000000);
        }

        if ((names.Length > 2) && (names[2] != string.Empty))
        {
            angular.Swing2Mode = JointDOFMode.Free;
            if(angular.SwingDrive == null)
                angular.SwingDrive = new JointDriveProperties(
                    JointDriveMode.Position,
                    new SpringProperties(500000, 1000, 0), 1000000);
        }

        EntityJointConnector[] connectors = new EntityJointConnector[]
        {
            new EntityJointConnector(hexapodShapes[prefix + rel.Child],
                rel.ChildNormal, rel.ChildAxis, rel.ChildConnect),
            new EntityJointConnector(hexapodShapes[prefix + rel.Parent],
                rel.ParentNormal, rel.ParentAxis, rel.ParentConnect)
        };

        SegmentEntity child = (SegmentEntity)hexapodShapes[prefix + rel.Child];
        child.CustomJoint = new Joint();
        child.CustomJoint.State = new JointProperties(angular, connectors);
        child.CustomJoint.State.Name = rel.JointName;
        hexapodShapes[prefix + rel.Parent].InsertEntityGlobal(
            hexapodShapes[prefix + rel.Child]);
    }
```

This code attempts to configure the joint based on the joint name. If it contains more than one string separated by a semicolon, then multiple degrees of freedom are unlocked. In this example, each joint has only a single degree of freedom unlocked.

After the joint properties have been defined, the child entity is retrieved from the dictionary and its CustomJoint field is set to the newly created joint. The child entity is then inserted as a child into the parent entity using InsertEntityGlobal. When the SegmentEntity is initialized, it will replace its parent joint properties with the properties of its custom joint.

## Attaching a Camera

The last task is to attach a camera to the front of the hexapod so that you can see the world from its perspective. You use the same `AttachedCameraEntity` that was used with the Lynx 6 Robotic Arm simulation:

```
HexapodEntity retValue = (HexapodEntity)hexapodShapes[prefix + _parentName];

// add a camera
AttachedCameraEntity cam = new AttachedCameraEntity();
cam.State.Name = name + "_cam";

// move the camera to the front of the hexapod
cam.State.Pose = new Pose(new Vector3(0, 0, -_bodyLength / 2 + 0.02f));

// adjust the near plane so that we can see objects that are close
cam.Near = 0.01f;

// use a visual model with the camera
cam.State.Assets.Mesh = "WebCam.obj";

// the camera coordinates are relative to the hexapod
retValue.InsertEntity(cam);


retValue.State.Flags |= EntitySimulationModifiers.Kinematic;
retValue.State.Name = name;
return retValue;
}
```

That's the entire hexapod! It has a total of 18 degrees of freedom with three joints in each of its six legs. The hexapod consists of a `HexapodEntity`, which is the top-level parent entity, and multiple child entities, which define the camera and the legs. The `HexapodEntity` has a `MoveTo` method much like the `MoveTo` method defined for the Lynx 6 Robotic Arm in the previous chapter. It enables the position of each joint to be specified along with a time value, and it will cause the joints to be moved to that position over the specified time.

The HexapodEntity service creates an entire simulation environment, including sky and ground and two hexapod entities. It uses the terrain from Simulation Tutorial 5 to give the hexapods something interesting to walk around. Any other service that needs to use a hexapod entity can simply add this service as a reference, just as was done with the Corobot entity in Chapter 7.

# The Hexapod Differential Drive Service

At this point, you could simply run the HexapodEntity service and see a complete simulation environment with two hexapods, but you wouldn't have a convenient way to make them walk around. You could start the JointMover service from Chapter 8, but that really only enables you to move one joint at a time, making it fairly tedious to control 18 joints.

The HexapodDifferentialDrive service comes to the rescue. This service implements the `GenericDifferentialDrive` contract, so it will work with the SimpleDashboard service and you will be able to control the motion of the hexapods just as if they were differential-drive robots.

This service was generated with the `DssNewService` utility, much like the SimulatedQuadDifferentialDrive in Chapter 6. The command line was as follows:

```
dssnewservice /s:HexapodDifferentialDrive
/i:"\Microsoft Robotics Studio (1.5)\bin\RoboticsCommon.dll"
/alt:"http://schemas.microsoft.com/robotics/2006/05/drive.html"
```

This service uses the usual strategy of subscribing to the simulation engine for an entity specified as its partner. When it receives a notification that the entity has been initialized, it spawns an iterator that is used to control the legs. Two different leg movements have been included with this service. If the entity name has a "2" in it, the second leg movement method is used.

The `SetDrivePowerHandler` method is executed when a `SetDrivePower` message is received. It sets the `_left` and `_right` member variables to indicate the current power assigned to each motor in the differential drive. These power values range from −1 to 1 and they are used by the leg movement methods to control how far the legs move during each cycle.

## The MoveLeg Method

Let's look at the `MoveLegs` method first. It uses the default `ShoulderAngle`, `ElbowAngle`, `StepAngle`, and `LiftAngle` defined by the `HexapodEntity` to define the angles used to make the legs walk. The first thing it does is extend the legs straight out so that the entity will be suspended above the ground. The `HexapodEntity` is initially created as a kinematic object so that it won't be moved by the physics engine until it is fully initialized. This prevents any abrupt movements from moving the entity as all the pieces come together in the first frame after it is initialized.

After 200 milliseconds, the `HexapodEntity` is changed from a kinematic entity to a dynamic entity and it falls to the ground. Then the legs are pulled into the neutral position to enable the hexapod to stand:

```
const float minDrive = 0.05f;

public IEnumerator<ITask> MoveLegs()
{
    float shoulderAngle = _hexapodEntity.ShoulderAngle;
    float elbowAngle = _hexapodEntity.ElbowAngle;
    float stepAngle = _hexapodEntity.StepAngle;
    float liftAngle = _hexapodEntity.LiftAngle;

    // legs straight out
    yield return Arbiter.Choice(_hexapodEntity.MoveTo(
        90, 0, 0, 90, 0, 0, 90, 0, 0,
        90, 0, 0, 90, 0, 0, 90, 0, 0, 1),
        delegate(SuccessResult s) { },
        ShowError);

    // wait for the leg movement inertia to settle
    yield return Arbiter.Receive(
        false, TimeoutPort(200), delegate(DateTime dt) { });

    _hexapodEntity.DeferredTaskQueue.Post(
        new Task<VisualEntity>(_hexapodEntity, DeferredDrop));
```

*(continued)*

*(continued)*

```
        // pull the legs in to the default position
        yield return Arbiter.Choice(_hexapodEntity.MoveTo(
            90, shoulderAngle, elbowAngle,
            90, shoulderAngle, elbowAngle,
            90, shoulderAngle, elbowAngle,
            90, shoulderAngle, elbowAngle,
            90, shoulderAngle, elbowAngle,
            90, shoulderAngle, elbowAngle, 2),
            delegate(SuccessResult s) { },
            ShowError);
```

Next, the method goes into an endless loop to move the legs according to the current values of `_left` and `_right`. The `rightStepAngle` is the maximum step angle multiplied by the `_right` motor power. As the right motor power is increased, the angle that the legs are moved forward and backward also increases and the speed of the entity increases. Each side of the hexapod is independently controlled, just as a differential drive would enable the hexapod to be steered in any direction and even driven in reverse.

If both `_right` and `_left` are near 0, the entity doesn't need to move its legs at all, and the method waits 100 milliseconds and then tries again. If the legs do need to move, then a four-phase leg movement is started. First, the `FrontRight`, `RearRight`, and `MiddleLeft` legs are lifted and moved forward or backward according to the step angle. Next, these legs are dropped back to the ground in their new position. Next, the `MiddleRight`, `FrontLeft`, and `RearLeft` legs are lifted and moved forward or backward according to the step angle. At the same time, the previous three legs are drawn back to their original position. It is this motion that moves the entity relative to the ground. In the last phase, the second set of three legs is again brought back to the ground. When phase one starts again, these legs are brought back to their original position and the entity moves again:

```
        // endless loop to make the legs move
        while (_hexapodEntity != null)
        {
            float stepTime = 0.1f;

            shoulderAngle = _hexapodEntity.ShoulderAngle;
            elbowAngle = _hexapodEntity.ElbowAngle;
            stepAngle = _hexapodEntity.StepAngle;
            liftAngle = _hexapodEntity.LiftAngle;
            float rightStepAngle = stepAngle * _right;
            float leftStepAngle = stepAngle * _left;

            // no movement necessary
            if ((Math.Abs(rightStepAngle) < minDrive) &&
                (Math.Abs(leftStepAngle) < minDrive))
            {
                yield return Arbiter.Receive(
                    false, TimeoutPort(100), delegate(DateTime now) { });
                continue;
            }

            // Raise FR, RR, ML
```

```
yield return Arbiter.Choice(_hexapodEntity.MoveTo(
    90 + rightStepAngle, shoulderAngle + liftAngle, elbowAngle,
    90, shoulderAngle, elbowAngle,
    90 + rightStepAngle, shoulderAngle + liftAngle, elbowAngle,
    90, shoulderAngle, elbowAngle,
    90 + leftStepAngle, shoulderAngle + liftAngle, elbowAngle,
    90, shoulderAngle, elbowAngle, stepTime),
    delegate(SuccessResult s) { },
    ShowError);


// Lower FR, RR, ML
yield return Arbiter.Choice(_hexapodEntity.MoveTo(
    90 + rightStepAngle, shoulderAngle, elbowAngle,
    90, shoulderAngle, elbowAngle,
    90 + rightStepAngle, shoulderAngle, elbowAngle,
    90, shoulderAngle, elbowAngle,
    90 + leftStepAngle, shoulderAngle, elbowAngle,
    90, shoulderAngle, elbowAngle, stepTime),
    delegate(SuccessResult s) { },
    ShowError);


// Raise MR, FL, FR
yield return Arbiter.Choice(_hexapodEntity.MoveTo(
    90, shoulderAngle, elbowAngle,
    90 + rightStepAngle, shoulderAngle + liftAngle, elbowAngle,
    90, shoulderAngle, elbowAngle,
    90 + leftStepAngle, shoulderAngle + liftAngle, elbowAngle,
    90, shoulderAngle, elbowAngle,
    90 + leftStepAngle, shoulderAngle + liftAngle, elbowAngle, stepTime),
    delegate(SuccessResult s) { },
    ShowError);


// Lower MR, FL, FR
yield return Arbiter.Choice(_hexapodEntity.MoveTo(
    90, shoulderAngle, elbowAngle,
    90 + rightStepAngle, shoulderAngle, elbowAngle,
    90, shoulderAngle, elbowAngle,
    90 + leftStepAngle, shoulderAngle, elbowAngle,
    90, shoulderAngle, elbowAngle,
    90 + leftStepAngle, shoulderAngle, elbowAngle, stepTime),
    delegate(SuccessResult s) { },
    ShowError);
    }
}
```

This is a relatively efficient walk algorithm and it works well with the differential-drive model. One problem is that sometimes only three legs are on the ground, which can reduce the stability and friction of the entity.

## The MoveLegs2 Method

Another leg movement algorithm is shown in the `MoveLegs2` method. This method follows the same initialization sequence as the first method except that the hexapod is never brought to a standing position. It lies on the ground with its legs extended. It uses a three-phase movement to move along the ground like an animal with flippers. It first raises its legs above the ground in unison and moves them forward or backward according to the values of `_left` and `_right`. Then it lowers its legs back to the ground in their new positions. Finally, it draws its legs back to their original position, and the body of the hexapod is dragged across the ground:

```
// Raise all legs
yield return Arbiter.Choice(_hexapodEntity.MoveTo(
    90 + rightStepAngle, shoulderAngle - liftAngle, elbowAngle,
    90 + rightStepAngle, shoulderAngle - liftAngle, elbowAngle,
    90 + rightStepAngle, shoulderAngle - liftAngle, elbowAngle,
    90 + leftStepAngle, shoulderAngle - liftAngle, elbowAngle,
    90 + leftStepAngle, shoulderAngle - liftAngle, elbowAngle,
    90 + leftStepAngle, shoulderAngle - liftAngle, elbowAngle, stepTime),
    delegate(SuccessResult s) { },
    ShowError);

// Lower all legs
yield return Arbiter.Choice(_hexapodEntity.MoveTo(
    90 + rightStepAngle, shoulderAngle, elbowAngle,
    90 + rightStepAngle, shoulderAngle, elbowAngle,
    90 + rightStepAngle, shoulderAngle, elbowAngle,
    90 + leftStepAngle, shoulderAngle, elbowAngle,
    90 + leftStepAngle, shoulderAngle, elbowAngle,
    90 + leftStepAngle, shoulderAngle, elbowAngle, stepTime),
    delegate(SuccessResult s) { },
    ShowError);

// Back to neutral position
yield return Arbiter.Choice(_hexapodEntity.MoveTo(
    90, shoulderAngle, elbowAngle,
    90, shoulderAngle, elbowAngle,
    90, shoulderAngle, elbowAngle,
    90, shoulderAngle, elbowAngle,
    90, shoulderAngle, elbowAngle,
    90, shoulderAngle, elbowAngle, stepTime),
    delegate(SuccessResult s) { },
    ShowError);
```

This algorithm is surprisingly efficient in the simulator, although it would probably result in a much shorter life span for a real-world hexapod because it is constantly banged against the ground.

You can run the Hexapod simulation environment by typing **Hexapod** in the \Microsoft Robotics Studio (1.5) directory. It starts the Hexapod manifest, which in turn starts the HexapodEntity service and two copies of the HexapodDifferentialDrive service, along with the SimpleDashboard service. The Hexapod simulation environment is shown in Figure 9-9.

**Figure 9-9**

When the Dashboard form appears, type **localhost** into the Machine textbox and click the Connect button. Two HexapodDifferentialDrive services will be displayed in the Service Directory list. Double-click one of them and click the Drive button in the upper-left group box. You should then be able to drag the directional control to control the movement of the selected hexapod. Double-click the other drive service to see the other robot move with a different walk algorithm.

## Where to Go from Here

The DifferentialDrive model does not tap the full potential of hexapod movement. With three degrees of freedom per leg, it is possible for the hexapod to directly move in any direction without having to turn in that direction first, much like an omni-wheel. Of course, this would require another hexapod drive service that implements something other than the differential drive generic contract, and a user-interface to control it.

You can modify the basic hexapod entity to make it an octoped or a quadruped or even a biped. You should be able to use the same basic code with different shape and joint data. The walk algorithm is somewhat more complicated because the robot goes through motions where it is not perfectly balanced.

Another step you can take is to define an XML file format that contains information about joints and legs. This could enable you to prototype any number of exotic articulated entities simply by defining them in an XML document.

It might also be interesting to explore swarm algorithms with this type of robot. Now that we've conjured the mental picture of dozens of mechanical cockroaches skittering across the simulated landscape, we'll move on to the next project.

# Implementing a Soccer Strategy

In the spring of 2007, the Microsoft Robotics Developer Studio Team sponsored a simulated soccer competition as part of RoboCup 2007 in Atlanta. They developed a simulated soccer environment that works with robot models provided by third parties. Robosoft provided a simulated quadruped called the *Robudog* for the competition.

In this section, you'll add your own robots to the soccer simulation and then create your own soccer player and put it up against the sample soccer player that Microsoft provides.

## The MRDS Simulated Soccer Environment

The Soccer package is available from the Microsoft Robotics website (www.microsoft.com/robotics) under Downloads. When you download and install the package, it installs the source code for the SimpleSoccerPlayer service in samples\simulation\competitions\SimulatedSoccerServices. The binaries for this service are installed in the bin directory along with the binaries for the SimulatedSoccerReferee service. Unfortunately, the source code for the Referee service is not provided.

Several manifests are provided to run the soccer simulation. You can start the simulation with two players on each of the red and blue teams using the following command line:

```
Dsshost -p:50000 -t:50001
-m:"samples\config\simulatedsoccer.legonxt.fourplayers.manifest.xml"
```

When the simulation engine starts up, it displays the soccer field with four LEGO NXT robots. Each team, red and blue, has a goalkeeper and a field player. The Referee service also displays a window that shows the user interface for controlling the match. The manifest starts the simulation engine with the SimulatedSoccer.legonxt.fourplayers.state.xml state file. This file contains the soccer scene, including the soccer field, the lights and ground, the goalposts, and the player robots, as shown in Figure 9-10.

Figure 9-10

When you click the Start button in the Referee window, the players and ball are reset to their kickoff position and the match begins. While the match is running, a rough ASCII representation of the soccer field displays with a single letter for each player and the ball. You can drag these representations with the left mouse button to move the players or the ball on the field. Right-clicking a player moves it back to its default kickoff position, and middle-clicking a player puts it into a 30-second penalty during which it cannot move. This is shown in Figure 9-11.

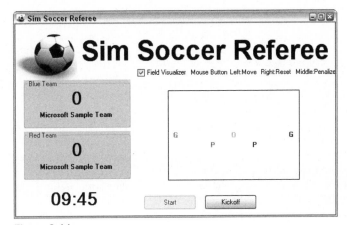

**Figure 9-11**

Each match lasts 10 minutes but the referee does not automatically stop the player action when the time expires. A point is scored for the blue team each time the ball enters the blue goal, and the red team scores when the ball enters the yellow goal. At any time, you can click the Kickoff button to return the players to their default position and orientation. This is useful if one or more of the robots gets stuck. If a single robot tips over, it can be reset by left- or right-clicking it.

Each soccer player has a camera mounted on it so you can see the game from each player's perspective; you can press the F8 key to switch between cameras or you can select the appropriate camera from the Camera menu.

Each player has a SimulatedDifferentialDrive service associated with it to move it around the field. Each player also has a SimulatedWebcam service associated with it to retrieve images from the onboard camera for processing. The orchestration service, which controls the behavior of the soccer players, is called SimpleSoccerPlayer. This service is designed to work with both the field player and the goalkeeper. In the manifest referenced above, this service controls both teams, so the behavior of both teams is identical.

The soccer player robot entities are identified by the Referee service according to their names. Press F5 while the soccer simulation is running to see the name of each entity. The name of the blue team field player, for example, is blueteam/field/2/simulatedsoccerplayer/robotmotioncontrol. It has a child entity, which is a real-time camera, with the name of blueteam/field/2/simulatedsoccerplayer/robocam.

Open the service directory for the DSS node by pointing your browser to http://localhost:50000/ directory while the soccer simulation is running, and you will see something similar to Figure 9-12.

| Service Instance |
|---|
| /blueteam/field/2 |
| /blueteam/field/2/simulatedsoccerplayer |
| /blueteam/field/2/simulatedsoccerplayer/robocam |
| /blueteam/field/2/simulatedsoccerplayer/robocam/webcamservice |
| /blueteam/field/2/simulatedsoccerplayer/robotmotioncontrol |
| /blueteam/field/2/simulatedsoccerplayer/robotmotioncontrol/drive |
| /blueteam/goalkeeper |
| /blueteam/goalkeeper/simulatedsoccerplayer |
| /blueteam/goalkeeper/simulatedsoccerplayer/robocam |
| /blueteam/goalkeeper/simulatedsoccerplayer/robocam/webcamservice |
| /blueteam/goalkeeper/simulatedsoccerplayer/robotmotioncontrol |
| /blueteam/goalkeeper/simulatedsoccerplayer/robotmotioncontrol/drive |

**Figure 9-12**

These are the services associated with the blue team players. The /blueteam/field/2 service is the SimpleSoccerPlayer orchestration service that controls the behavior of this player. It implements the simulatedsoccerplayer contract as an alternate contract so that the Referee service can communicate with it. The /blueteam/field/2/simulatedsoccerplayer/robocam service is a SimulatedWebcam service associated with the camera on this player. It implements the webcamservice contract as an alternate contract. The /blueteam/field/2/simulatedsoccerplayer/robotmotioncontrol service is a SimulatedDifferentialDrive service that implements the GenericDifferentialDrive contract as an alternate contract. Notice that the names of the services correspond to the names of the entities they are paired with in the simulation environment.

When the SimpleSoccerPlayer service starts, it sends a message to the Referee service announcing its presence. The Referee service checks the name of the player and finds running services with matching names, passing information about these services back to the player. The player uses the services to control its entity in the simulation environment. The SimpleSoccerPlayer never interacts directly with its entity in the simulator, only with the DifferentialDrive and Webcam services attached to the entity. This makes it possible for the player services to be running on a different node or even a different computer from the simulator simply by changing the manifest.

## Changing the Players on the Field

The soccer environment in the Microsoft package is configured to use LEGO NXT robots on the field. These robots are fine but they tend to tip over easily. In this section, you'll change the environment to use the simulated Corobot entity defined in Chapter 6.

As previously described, everything in the soccer simulation environment is defined in the simulatedsoccer.legonxt.fourplayers.state.xml file. It is easy to remove the LEGO robots from the environment by deleting their XML descriptions from the state file.

A new manifest has been provided in the ProMRDS\config directory: simulatedsoccer.corobot.fourplayers.manifest.xml. This manifest uses the state file called simulatedsoccer.corobot.fourplayers.state.xml, which does not contain the LEGO robots. If the simulation environment does not contain four properly named entities, then their corresponding DifferentialDrive services will not appear in the service directory and the referee will be unable to start the match. In this case, the Start button is disabled.

You can add your own player entities to the environment by adding their serialized state to the simulator state file or by creating a service that inserts them into the simulation environment programmatically. You will find the CorobotSoccerPlayers service in the `Chapter9` directory, which inserts the entities into the environment using the following code:

```
corobot.CorobotEntity blueGoalKeeper;
corobot.CorobotEntity blueField;

protected override void Start()
{
    base.Start();
    // create four Corobot Soccer Players
    blueGoalKeeper = new corobot.CorobotEntity(
        "blueteam/goalkeeper/simulatedsoccerplayer/robotmotioncontrol",
        new Vector3(0.5f, 0, 0));
    RenameCamera(blueGoalKeeper,
        "blueteam/goalkeeper/simulatedsoccerplayer/robocam");
    blueGoalKeeper.InsertEntity(new TeamIDEntity(new Vector4(0, 0, 1, 1),
        blueGoalKeeper.State.Name + "_ID"));
    SimulationEngine.GlobalInstancePort.Insert(blueGoalKeeper);

    blueField = new corobot.CorobotEntity(
        "blueteam/field/2/simulatedsoccerplayer/robotmotioncontrol",
        new Vector3(2f, 0, 0));
    RenameCamera(blueField, "blueteam/field/2/simulatedsoccerplayer/robocam");
    blueField.InsertEntity(new TeamIDEntity(new Vector4(0, 0, 1, 1),
        blueField.State.Name + "_ID"));
    SimulationEngine.GlobalInstancePort.Insert(blueField);
```

Each player entity is created with the proper name. The camera child entity is renamed to match the soccer environment naming convention. An additional child entity is added consisting of a colored sphere to identify the team affiliation of each robot. This customization makes it much easier to tell which robot is on which team.

The CorobotSoccerPlayers service is started in the manifest along with the Players, DifferentialDrive, and Webcam services and the SimulatedSoccerReferee service. When this service adds the entities to the simulation environment, the simulator notifies the services that have requested notifications. These services then insert themselves into the service directory, and when the Players services announce themselves, the Referee service can find the proper services and the Start button is enabled.

The CorobotSoccerPlayers service does a couple of other tricky things. It subscribes to the simulation engine for a notification when the player entities are inserted into the environment:

```
// add a handler for the notification from the simulation engine
MainPortInterleave.CombineWith(
    Arbiter.Interleave(
        new TeardownReceiverGroup(),
        new ExclusiveReceiverGroup
        (
            Arbiter.Receive<InsertSimulationEntity>(false, _notificationTarget,
                InsertEntityNotificationHandler),
            Arbiter.Receive(true, _timerPort, TimerHandler)
        ),
```

*(continued)*

*(continued)*

```
            new ConcurrentReceiverGroup()
        )
    );

    // request a notification when a player is inserted into the sim engine
    EntitySubscribeRequestType esrt = new EntitySubscribeRequestType();
    esrt.Name = blueField.State.Name;
    SimulationEngine.GlobalInstancePort.Subscribe(esrt, _notificationTarget);
```

When the service receives this notification, it initializes the main camera view to a top-down view that is more convenient than the default view provided by the Referee service:

```
    // override the viewpoint set up by the referee by waiting until
    // the first player is inserted into the simulator and then setting
    // the main camera view.
    void InsertEntityNotificationHandler(InsertSimulationEntity ins)
    {
        // Set up initial view
        CameraView view = new CameraView();
        view.EyePosition = new Vector3(0.02f, 3.66f, 3.42f);
        view.LookAtPoint = new Vector3(0.02f, 2.87f, 2.81f);
        SimulationEngine.GlobalInstancePort.Update(view);
    }
```

The CorobotSoccerPlayers service keeps track of the player entities once they have been created and inserted to ensure that they remain upright. It is not uncommon for the robots to push each other over as they jockey for position on the field. Once a robot is on its side, it must be manually reset using the referee user interface.

The CorobotSoccerPlayers service starts a `TimerHandler` method that is executed periodically. It examines the pose of each of the players. If it detects that a player has been knocked over, it rights it:

```
        Activate(Arbiter.Receive(false, TimeoutPort(10000), _timerPort.Post));
    }

    Port<DateTime> _timerPort = new Port<DateTime>();

    void TimerHandler(DateTime time)
    {
        AdjustIfNecessary(redGoalKeeper);
        AdjustIfNecessary(redField);
        AdjustIfNecessary(blueGoalKeeper);
        AdjustIfNecessary(blueField);

        Activate(Arbiter.Receive(false, TimeoutPort(500), _timerPort.Post));
    }

    // right a robot that has fallen over
    void AdjustIfNecessary(VisualEntity entity)
    {
        xna.Vector3 rotation = entity.Rotation;
```

```
        if ((Math.Abs(rotation.X) > 75) || (Math.Abs(rotation.Z) > 75))
        {
            entity.Rotation = new xna.Vector3(0, 0, 0);
        }
    }
```

Although it is not very realistic to have the robots magically right themselves, it does make the soccer matches more interesting to watch. Not everything done in the simulator has to be a reflection of the real world.

You can run the modified soccer environment using the ProMRDSSoccer.cmd file as follows:

```
C:\Microsoft Robotics Studio (1.5)>ProMRDSSoccer
```

You should see something similar to the environment shown in Figure 9-13.

Figure 9-13

## Building a Better Soccer Player

The behavior of the SimpleSoccerPlayer shipped with the soccer package is very simple. The robot spins slowly, looking for the soccer ball. When it has located the ball, it drives toward it. When it loses sight of the ball, it repeats this behavior.

Even this very simple behavior can result in some interesting behavior on the field. The robot players do occasionally manage to make a goal, although there is no guarantee that it will be in the proper goal box.

A ProMRDSsimplesoccerplayer service has been provided in the Chapter9 directory, which is essentially the same as the Microsoft SimpleSoccerPlayer service with a few minor modifications to make it work better with the Corobot entity. The most significant change is in the code that creates a port to talk to the DifferentialDrive service. It must append the string "/drive" to the service name so that the port talks to the GenericDrive contract of the SimulatedQuadDifferentialDrive service. Some of the drive power factors have also been scaled to better match the gearing of the Corobot model. Other than this, the behavior of SimpleSoccerPlayer has not been changed.

## Improved Image Processing

The ProMRDSbettersoccerplayer service in the `Chapter9` directory implements several enhancements that improve the soccer player behavior substantially. One of the most significant changes is in the vision processing code. The original soccer player only had the ability to recognize the soccer ball. An improved soccer player must be able to recognize both the yellow and blue goals as well as the ball.

The original image processing code, as well as the new and improved code, uses the HSV color space to identify objects in the image. HSV is an alternate way of representing colors that is different from RGB. The letters stand for the following:

❑ **H:** Stands for *hue* and it represents the base color. Its value ranges from 0, which represents red, to 360, which represents violet.

❑ **S:** Stands for *saturation* and is a measure of how "washed out" the color is. High values of saturation result in colors that are closer to white, like pastels, whereas low values of saturation yield deep, rich colors.

❑ **V:** Stands for *value* and it scales the intensity of the color. Low V values yield colors that are nearly black.

One reason why HSV is a good way to classify colors in an image is because it is possible to ignore the V component of a color to take into account the variations in color intensity due to lighting and shadows. For example, the image processing code in the original SimpleSoccerPlayer service searches for colors in the image that have a very low hue and a reasonably high saturation. This selects pixels in the image that are red regardless of whether they are dark red or bright red.

The new and improved image processing code enables hue and saturation ranges to be specified for each object that needs to be identified. You can find it in the ProMRDSBetterSoccerPlayer project in the `Chapter9` directory in `VisionProcessing.cs`:

```
public class ImageSegment
{
    public float Saturation;
    public float SaturationEpsilon;
    public float Hue;
    public float HueEpsilon;
    public float Area;
    public float X;
    public int XMin;
    public int XMax;
    public int XSpan;
}

public static class ImageProcessing
{
    /// <summary>
    /// Find objects in the image, using the segment definitions provided
    /// </summary>
    public static void ProcessFrameHandler(
```

```
            float width,
            float height,
            ImageSegment [] segments,
            byte[] rgbValues)
{
    // initialize the results
    foreach (ImageSegment segment in segments)
    {
        segment.Area = segment.X = 0;
        segment.XMax = 0;
        segment.XMin = (int)(width + 1);
    }

    // scan the image
    int offset = 0;

    for (int y = 0; y < height; y++)
    {
        for (int x = 0; x < width; x++)
        {
            int r, g, b;

            b = rgbValues[offset++];
            g = rgbValues[offset++];
            r = rgbValues[offset++];

            Color color = Color.FromArgb(r, g, b);

            float hue = color.GetHue();
            float saturation = color.GetSaturation();

            foreach (ImageSegment segment in segments)
            {
                if ((Math.Abs(hue - segment.Hue) < segment.HueEpsilon) &&
        (Math.Abs(saturation - segment.Saturation) < segment.SaturationEpsilon))
                {
                    segment.Area++;
                    segment.X += x;
                    segment.XMax = Math.Max(segment.XMax, x);
                    segment.XMin = Math.Min(segment.XMin, x);
                }
            }
        }
    }

    foreach (ImageSegment segment in segments)
    {
        if (segment.Area > 0)
        {
            segment.X = segment.X / segment.Area;
            segment.XSpan = segment.XMax - segment.XMin;
        }
    }
}
```

The `ProcessFrameHandler` method steps through each pixel in the image and calculates the hue and saturation values. These values are compared to the ranges specified in each `ImageSegment` structure that is passed. If it matches a particular segment, then the count for that segment is calculated and the minimum X and maximum X values for the segment are updated. After all of the pixels have been classified, the average X value and the XSpan are calculated for each segment.

The call to `ProcessFrameHandler` is in the `ProcessFrameAndDetermineBehavior` method in `BetterSoccerPlayer.cs`:

```
if (_segments[0] == null)
{
    _segments[0] = new ImageSegment();
    _segments[0].Hue = 0.1f;
    _segments[0].HueEpsilon = 0.1f;
    _segments[0].Saturation = 0.75f;
    _segments[0].SaturationEpsilon = 0.5f;
    _segments[1] = new ImageSegment();
    _segments[1].Hue = 180f;
    _segments[1].HueEpsilon = 5f;
    _segments[1].Saturation = 0.75f;
    _segments[1].SaturationEpsilon = 0.5f;
    _segments[2] = new ImageSegment();
    _segments[2].Hue = 60f;
    _segments[2].HueEpsilon = 5f;
    _segments[2].Saturation = 0.75f;
    _segments[2].SaturationEpsilon = 0.5f;
}

try
{
    ImageProcessing.ProcessFrameHandler(
        _queryFrameRequest.Size.X,
        _queryFrameRequest.Size.Y,
        _segments,
        _lastFrame);
}
catch (Exception ex)
{
    LogError(ex);
}
```

`_segments[0]` identifies the soccer ball and selects all pixels that have a hue between 0 and 0.2 and a saturation between 0.25 and 1.0. `_segments[1]` defines hue and saturation values for the blue goal, and `_segments[2]` defines hue and saturation values for the yellow goal. These values were determined experimentally by running the soccer simulation and then using the Simulation Editor to select each object and display its hue and saturation.

Select the object you want to measure by holding down the Ctrl key and right-clicking it. Expand the `Meshes` property and then the `Rendering Materials` property. Select the desired `Rendering Material` in the object and click the ellipse that appears to display the MaterialEditor. This dialog shows the HSV and RGB values for the ambient, diffuse, and specular colors of the object (see Figure 9-14).

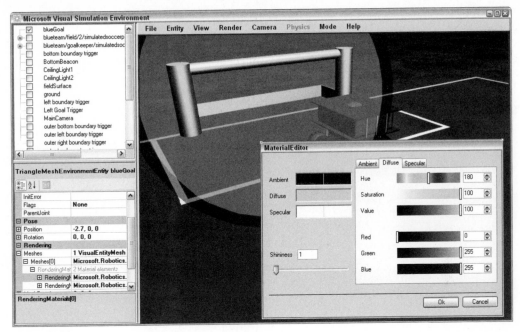

**Figure 9-14**

Now you have the tools you need to develop even more sophisticated vision and image processing algorithms in the simulation environment.

## Improved Behavior

Now that the soccer players can identify the most important objects on the playing field, it is possible to implement a more intelligent player algorithm. The new code is in the `ProcessFrameAndDetermineBehavior` method in `BetterSoccerPlayer.cs`. The SoccerPlayer service requests frames from the webcam periodically, and this method is called each time a frame is received.

The soccer player already keeps track of its state, whether it is currently waiting to play, playing, or penalized. A new `Mode` property has been added to the state to keep track of the current behavior while in the playing state. These modes are as follows:

❑ **FindBall:** The ball is not currently in the image. Search for it by slowly spinning.

❑ **ReAlign:** The ball is in the image but it is not lined up with the goal. Attempt to reposition.

❑ **Kick:** The ball was being tracked but it has abruptly disappeared. This could mean that you are so close to it that it doesn't show up on the camera. Move forward quickly to push it toward the goal.

❑ **Approach:** The ball is in sight but too far away to do anything. Move closer.

Before you can make a decision about the next behavior to implement, you need to determine which objects have been found using the following code:

```
bool foundBall = _segments[0].Area > 10;
bool foundBlueGoal = (_segments[1].Area > 500) && (_segments[1].XSpan > 75);
bool foundYellowGoal = _segments[2].Area > 500 && (_segments[1].XSpan > 75);
```

If the webcam managed to find more than 10 pixels that are soccer-ball-colored, then the ball has been located. It is a little more complicated with the goals because you want to identify the goal but not the colored posts at midfield. In this case, you must see at least 500 pixels that are the right color and they must cover a horizontal span greater than 75 pixels. This effectively eliminates the midfield posts while still allowing the goals to be seen from across the soccer field.

The soccer player starts out in the FindBall mode and executes the following code:

```
switch (_state.Mode)
{
    case SoccerPlayerMode.FindBall:
    case SoccerPlayerMode.Approach:
        {
            if (!foundBall)
            {
                if (_previousArea > 4000)
                {
                    // we are right on top of the ball so we can't see it.  Kick!
                    _previousArea = 0;
                    _mainPort.Post(new ExecuteCommand(
                        new ExecuteCommandRequest(
                            SimpleSoccerPlayerCommand.Kick)));
                }
                else
                {
                    _mainPort.Post(new ExecuteCommand(
                        new ExecuteCommandRequest(
                            SimpleSoccerPlayerCommand.FindBall)));
                    _previousArea = 0;
                }
            }
        }
```

If the ball isn't in sight, then you check to see whether it was previously very big. When the robot gets very close to the ball, the camera actually penetrates the ball and it disappears in the Webcam view. This is analogous to a real-world scenario in which a camera gets very close to the ball and no longer recognizes it because it is suddenly very dark. In this case, you send a command to the main port to instruct the player to move forward quickly to kick the ball. This is often more effective than attempting to push the ball closer to the goal because it gives the other player less time to locate and steal the ball.

If the ball was not previously very large, then you are out of luck and must continue searching for it. A FindBall command is sent to the port:

```
else
{
    if (((_state.Team == referee.TeamIdentifier.BlueTeam) &&
        foundBlueGoal) ||
```

```
                    (_state.Team == referee.TeamIdentifier.RedTeam &&
                    foundYellowGoal) ||
                    (_segments[0].Area < 2000) || (_segments[0].Area > 4000))
           {
                    // either the goal is lined up for a shot or the ball is
                    // too far away or too close to decide
                    _state.Mode = SoccerPlayerMode.Approach;
                    _previousArea = (int)_segments[0].Area;

                    // calculate an offset from center that will be used to
                    // calculate left/right drive power
                    _state.BallOffset = 0.5f *
                        (_segments[0].X - 0.5f * _queryFrameRequest.Size.X) /
                        _queryFrameRequest.Size.X;
                    _mainPort.Post(new ExecuteCommand(
                        new ExecuteCommandRequest(
                            SimpleSoccerPlayerCommand.ApproachBall)));

           }
           else
           {
                    _state.Mode = SoccerPlayerMode.ReAlign;
                    // we found the ball but we're not lined up properly
                    // for a shot, realign
                    _mainPort.Post(new ExecuteCommand(
                        new ExecuteCommandRequest(
                            SimpleSoccerPlayerCommand.Realign)));

           }
       }
       break;
   }
}
```

If the ball is in sight, then you check whether your target goal is also in sight. If it is or if the ball is very far away or very close, you continue to approach the ball. This gets you very close to the ball until you eventually kick it.

If the ball is in sight but the goal is not, then you attempt to reposition the player until the goal is in sight. This is done by sending the Realign command to the main port.

The implementation of each of these commands is in the ExecuteCommand handler. The ApproachBall and FindBall commands are fairly straightforward. FindBall sets the motor power so that the robot slowly spins. ApproachBall uses the position of the ball in the frame to set the left and right motor power to cause the player to move directly toward the ball.

The Kick command is handled by spawning an iterator to execute a more complicated move. The KickBall iterator is shown here:

```
IEnumerator<ITask> KickBall()
{
    yield return Arbiter.Choice(
        _robotDrive.DriveDistance(0.1, 0.8),
        delegate(DefaultUpdateResponseType response) { },
```

(continued)

*(continued)*

```
        delegate(Fault f) { }
    );

    _state.Mode = SoccerPlayerMode.FindBall;
}
```

This iterator sends a `DriveDistance` command to the differential drive to drive forward 0.1 meters at a fast speed. The statement that sets the player mode back to `FindBall` doesn't execute until the `DriveDistance` command has been completed. The player remains in the `Kick` mode until the kick motion has finished.

The `Realign` command is also implemented with an iterator:

```
IEnumerator<ITask> RealignApproach()
{
    yield return Arbiter.Choice(
        _robotDrive.RotateDegrees(-45, 0.5),
        delegate(DefaultUpdateResponseType response) { },
        delegate(Fault f) { }
    );

    yield return Arbiter.Choice(
        _robotDrive.DriveDistance(0.3, 0.5),
        delegate(DefaultUpdateResponseType response) { },
        delegate(Fault f) { }
    );

    yield return Arbiter.Choice(
        _robotDrive.RotateDegrees(70, 0.5),
        delegate(DefaultUpdateResponseType response) { },
        delegate(Fault f) { }
    );

    _state.Mode = SoccerPlayerMode.FindBall;
}
```

This is very similar to the `KickBall` iterator except that more motions are used. First the player is rotated 45 degrees to the left. Then it drives forward 0.3 meters and turns to the right by 70 degrees. This enables the player to circle around the ball while drawing closer to it. The `ProcessFrameAndDetermineBehavior` method does not change the mode of the player until this motion sequence has completed and the mode is set back to `FindBall`.

The most significant change in behavior is that the player no longer moves the ball in a random direction. It actually attempts to align itself with the appropriate goal before advancing the ball. Nor does it just keep pushing the ball, but instead attempts to kick it in the right direction.

The manifest `simulatedsoccer.corobot.fourplayers.manifest.xml` starts the soccer environment with four Corobot players. The red team runs the ProMRDSSimpleSoccerPlayer service, which is very similar to the sample service that Microsoft ships in the soccer package. The blue team runs the ProMRDSBetterSoccerPlayer service, which has the new code just described. In this scenario, the blue team consistently outscores the red team by a factor of 4 to 1, as shown in Figure 9-15.

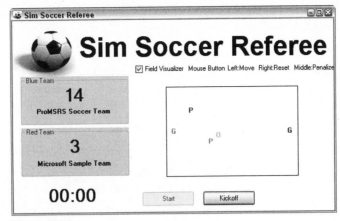

**Figure 9-15**

# Where to Go from Here

We've barely scratched the surface of a good soccer algorithm. Much more sophisticated behavior can be implemented using the same basic concepts illustrated. None of the players is aware of the position or movement of any of the other players. The vision algorithm can be enhanced to identify other players, and the player algorithm can take them into account.

In addition, no attempt is really made to find the position of the player other than to determine whether the target goal is in view. The vision algorithm can be enhanced to detect the posts at midfield, and their position and size in the frame can be used to determine the player's position on the field. This enables the player to make more intelligent decisions about how it will approach the ball and line itself up with the goal.

Finally, we haven't done anything at all with the goalkeepers. They can do much more than just sit in front of the goal by constantly tracking the ball and taking action when it gets too close to the goal.

The ProMRDSBetterSoccerPlayer service represents an improvement over the basic service in the soccer package, but it is very primitive compared to what can be done.

The Corobot soccer services can provide a way for robotics clubs to organize a simulation-only competition to see who can come up with the best soccer player strategy.

In addition, the Corobot is available commercially and runs Microsoft Robotics Developer Studio Services, so with enough time and money it is even possible to build a real soccer field with real robots to run these algorithms; and then you're well on your way to hosting your own RoboCup competition.

# Exploring Virtual Worlds

The Holy Grail for many robotics researchers is to build a fully autonomous robot that can be set loose to explore its environment and build a map as it goes. It sounds simple enough, and you might even believe that this problem has already been solved based on what you have seen on television and in the movies. In fact, we are possibly going to perpetuate this myth with the example in this section. Unfortunately, *Simultaneous Localization and Mapping (SLAM)* is still an active area of research, and many real-world problems are yet to be solved.

The ExplorerSim application illustrates several aspects of autonomous exploration. Figure 9-16 shows a screenshot of the ExplorerSim in action as the robot wanders around building a map. The simulated Pioneer 3DX robot is at the center of the simulation window. The top two windows show the map (so far) and the view from the robot's camera.

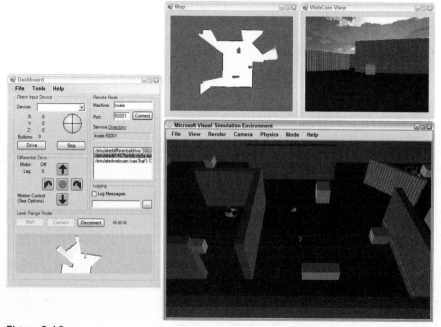

**Figure 9-16**

## *Background*

In the original version of MRDS, Microsoft supplied an example of an exploration service in `samples\Misc\Explorer`. This uses a laser range finder (LRF) to wander around at random. However, most people do not have a robot equipped with an LRF because they are quite expensive. The obvious solution was to use the simulator. (Do you have the impression yet that we like the simulator?)

The Explorer code uses the `RotateDegrees` and `DriveDistance` operations in the generic `DifferentialDrive` contract, but the original SimulatedDifferentialDrive didn't support these. Another important piece of the puzzle was obtaining the robot's pose — its position and orientation. A modified version of the SimulatedDifferentialDrive service solves both of these problems.

None of the Simulation Tutorials included with MRDS provide environments that are similar to a typical home or office. This is where the Maze Simulator comes in, as it enables you to easily build a virtual world for the robot to explore simply by creating a floor plan.

With the updated DifferentialDrive service and the Maze Simulator, you have the necessary components to begin exploring and mapping using the simulator.

## The Modified DifferentialDrive Service

There isn't a lot to say about the modified version of the SimulatedDifferentialDrive service other than that it is required to run ExplorerSim. You can view the source code in `ProMRDS\Chapter9\DifferentialDriveTT`.

If you are recompiling ExplorerSim, you must compile the services in the following order because of the dependencies:

1.  DifferentialDriveTT
2.  MazeSimulator
3.  ExplorerSim

The original reason for creating this service was to implement the `RotateDegrees` and `DriveDistance` operations. Although these operations were added to MRDS V1.5, there is a problem when the robot is driven at high speed because it sometimes misses the stopping point and keeps on going. Therefore, we still use our own version of the service.

The modified Drive service also supplies the current pose of the robot, allowing the mapping to work. This is clearly cheating! In real life, a robot rarely knows where it is. In fact, this is one of the problems that SLAM is supposed to solve (the "localization" step).

> You might immediately think of using a GPS, but this isn't a suitable solution. GPS receivers don't work indoors; and even if they did, they are only accurate to the nearest 5 or 10 meters. In an office environment, a 5-meter error could put you in the office next door, or even outside the building.

Changing a pre-defined entity in the simulator is a little complicated. Although the source for all the entities is supplied in `samples\simulation\Entities\Entities.cs`, you cannot just edit this file and recompile it because it is built into the simulator.

Consequently, a new entity has to be defined with a different name. In `SimDiffDriveEntities.cs` you will find the new differential drive entity:

```
[Category("DifferentialDrive")]
[BrowsableAttribute(false)] // prevent from being displayed in NewEntity dialog
public class TTDifferentialDriveEntity : VisualEntity
```

The `Update` method for this entity is responsible for implementing the `RotateDegrees` and `DriveDistance` operations, as shown previously in Chapter 6, so it is not covered here.

Changing the name of this entity has a ripple effect, which forces you to create new robot entities as well because they are built on the differential drive:

```
[DataContract]
[DataMemberConstructor]
```

*(continued)*

*(continued)*

```
[CLSCompliant(true)]
// A new entity is required because it uses a new
// version of the Simulated Differential Drive
public class TTPioneer3DX : TTDifferentialDriveEntity
```

The final important change is in the `UpdateStateFromSimulation` method in `SimulatedDifferentialDrive.cs`. The current position of the robot is inserted into the left and right wheel motor states. This is a hack, but it works. Note that the same pose is inserted for both wheels, but in fact the wheels are on either side of the robot's center by half of the wheelbase. Because you don't care about the actual wheel positions, the position of the robot's center is good enough:

```
void UpdateStateFromSimulation()
{
    if (_entity != null)
    {
        _state.TimeStamp = DateTime.Now;
        _state.LeftWheel.MotorState.CurrentPower =
            _entity.LeftWheel.Wheel.MotorTorque;
        _state.RightWheel.MotorState.CurrentPower =
            _entity.RightWheel.Wheel.MotorTorque;

        // Construct a new pose
        Pose p = new Pose();
        Vector3 posn = new Vector3(
            _entity.Position.X, _entity.Position.Y, _entity.Position.Z);
        // We need to create a quaternion for the orientation
        Quaternion q = new Quaternion();

        // Rotations are about the Y axis because we are moving in 2D
        // which happens to be the X-Z plane. The Y axis is (0,1,0).
        // Figure out the orientation angle in Radians
        Double angle = _entity.Rotation.Y * Math.PI / 180;
        // Now create the quaternion
        q.X = 0;
        q.Y = (float) Math.Sin(angle/2);
        q.Z = 0;
        q.W = (float) Math.Cos(angle/2);
        // Finally, insert the values into the pose
        p.Position = posn;
        p.Orientation = q;

        // NOTE: This is NOT quite correct!!!
        // The actual pose of the wheels should differ by the wheelbase.
        _state.LeftWheel.MotorState.Pose = p;
        _state.RightWheel.MotorState.Pose = p;

    }
}
```

Now it is a simple matter of getting the robot's state whenever you want to know its pose.

# The Maze Simulator

The Maze Simulator was written in the early days of the MRDS V1.0 Community Technology Previews. It enables fairly complex environments to be created, such as the one shown in Figure 9-17. These are much more interesting to drive around in than the Simulation Tutorials.

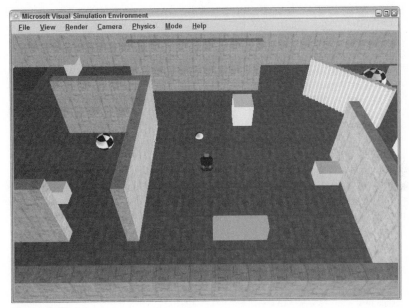

Figure 9-17

To make a new environment, you simply create a bitmap image with the floor plan, something like what is shown Figure 9-18, and then edit the `config` file for the Maze Simulator to specify the new maze image.

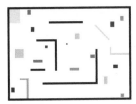

Figure 9-18

The `config` file is called `ProMRDS\Config\MazeSimulator.config.xml` and an example is shown here, followed by a bit of explanation:

```xml
<?xml version="1.0" encoding="utf-8"?>
<MazeSimulatorState
xmlns:s="http://www.w3.org/2003/05/soap-envelope"
xmlns:wsa="http://schemas.xmlsoap.org/ws/2004/08/addressing"
xmlns:d="http://schemas.microsoft.com/xw/2004/10/dssp.html"
```

(continued)

*(continued)*

```
xmlns="http://schemas.tempuri.org/2006/08/mazesimulator.html">
    <Maze>ModelLarge.bmp</Maze>
    <GroundTexture>cellfloor.jpg</GroundTexture>
```

The `Maze` tag specifies the filename for the maze bitmap. The file itself can be created as a Windows bitmap (.BMP file) using Windows Paint. It can actually be any image format, but only GIF and BMP guarantee that no information is lost, so don't use JPG format because it uses lossy compression.

The example configuration file contains just a filename with no path. If you want to specify an explicit path, just begin the filename with a backslash (or a slash), e.g., `\ProMRDS\Config\MyMaze.bmp`. If there is no leading slash, then the path will default to the `store\media\Maze_Textures` folder.

You can also specify a texture for the floor. This texture bitmap must reside in the `store\media` directory (or a subfolder of it).

Next in the configuration file are the Wall Textures and Wall Colors. The maze bitmap is examined by the Maze Simulator and used to create blocks in the simulated world based on the colors of the pixels in the bitmap. (This should be obvious from Figure 9-19, although you can't see the colors in the black-and-white screenshot.) In other words, the walls are color-coded and you can have different types of walls.

The Maze Simulator divides the RGB color space into 16 different colors, giving you 15 usable block types because one color has to be the floor. The list of colors recognized by the Maze Simulator is shown in the following example. (If you want to see the standard 16 colors, you can look at `ColorPalette16.bmp` in the Maze Simulator source directory.) The values from 0 to 7 are the fully saturated colors; 8 to 15 are their darker equivalents. The Maze Simulator matches pixel values in the maze bitmap to the closest of these 16 colors in order to select one of the 16 block types.

```
public enum BasicColor : byte
{
    Black       = 0,
    Red         = 1,
    Lime        = 2,
    Yellow      = 3,
    Blue        = 4,
    Magenta     = 5,
    Cyan        = 6,
    White       = 7,
    DarkGrey    = 8,
    Maroon      = 9,
    Green       = 10,
    Olive       = 11,
    Navy        = 12,
    Purple      = 13,
    Cobalt      = 14,
    Grey        = 15
}
```

The floor color is always taken to be the value of the top-left pixel. In Figure 9-19, the top-left pixel is white because there is a white border around the outside wall, which is black. It is a good idea to have a wall enclosing the entire environment so that your robot can't run away!

For each of the 16 pixel colors, you can specify either a texture file or a color as an RGB value for the corresponding wall blocks. Textures always override colors. To use a color, leave the corresponding texture empty, i.e., <string />.

The values in the following config file equate to cyan (light blue) for block type 0 (which is black pixels in the maze bitmap) and red for block type 1 (which is red pixels). Ignore the fact that the configuration file says XYZ — that is just a side-effect of using a Vector3. In this case, however, the cyan color is overridden by the water.jpg texture. Cyan is only there in case the texture file is missing.

```
<WallTextures>
  <string>Maze_Textures/water.jpg</string>
  <string />
  ... (Up to 16 strings)
</WallTextures>
<WallColors>
  <Vector3>
<X xmlns="http://schemas.microsoft.com/robotics/2006/07/physicalmodel.html">0</X>
<Y xmlns="http://schemas.microsoft.com/robotics/2006/07/physicalmodel.html">255</Y>
<Z xmlns="http://schemas.microsoft.com/robotics/2006/07/physicalmodel.html">255</Z>
  </Vector3>
  <Vector3>
<X xmlns="http://schemas.microsoft.com/robotics/2006/07/physicalmodel.html">255</X>
<Y xmlns="http://schemas.microsoft.com/robotics/2006/07/physicalmodel.html">0</Y>
<Z xmlns="http://schemas.microsoft.com/robotics/2006/07/physicalmodel.html">0</Z>
  </Vector3>
  ... (Up to 16 Vector3s)
</WallColors>
```

Note that there is no requirement to use the same wall colors as the pixel colors in the maze bitmap, although you will probably confuse yourself if you make them different!

Each of the block types has an associated height in the HeightMap (which is scaled as explained below) and a mass (in kilograms) in the MassMap. If you specify the mass as zero, then it means infinite, and the robot will never be able to move blocks of this type. It is fun to create blocks that can be pushed around.

```
<HeightMap>
  <float>0.01</float>
  <float>1.5</float>
  ... (Up to 16 floats)
</HeightMap>
<MassMap>
  <float>0</float>
  <float>2</float>
  ... (Up to 16 floats)
</MassMap>
```

The next few properties control the geometry. The UseSphere flags, one per block type, enable you to create balls instead of blocks. Note in the following example that the second block type is a sphere, and from the MassMap just shown it has a mass of 2kg and a size in the HeightMap of 1.5 meters (possibly scaled). You can see some balls in Figure 9-18 with a checkerboard texture applied to them. When you run the Maze Simulator you can push them around.

```
<UseSphere>
   <boolean>false</boolean>
   <boolean>true</boolean>
   ... (Up to 16 booleans)
</UseSphere>
<WallBoxSize>0.98</WallBoxSize>
<GridSpacing>0.1</GridSpacing>
<HeightScale>0.1</HeightScale>
<DefaultHeight>10</DefaultHeight>
<SphereScale>1</SphereScale>
```

The remaining parameters are used to scale the blocks so that you can get a realistic size without having to create an enormous maze bitmap.

*The WallBoxSize is deliberately a little less than the GridSpacing so that different blocks do not overlap, although adjacent blocks of the same type are combined. The simulator frame rate decreases as more objects are added to the simulation. It is very easy to create hundreds or thousands of blocks using the Maze Simulator. It tries its best to combine blocks, but there might still be a serious impact on performance. The final number of blocks is displayed in the console window when the Maze Simulator starts. If it is running very slowly, try to simplify your environment using only horizontal and vertical lines (East-West and North-South) in the maze bitmap. Lines that are at an angle or curved are approximated by a lot of separate blocks. In the top-right corner of Figure 9-18, you can see this in the light-colored wall, which is at 45 degrees.*

Finally, you can select either the Pioneer 3DX robot or the LEGO NXT robot using the RobotType and specify where it should start in terms of grid cells using RobotStartCellRow and RobotStartCellCol. The position (0, 0) is at the center of the maze:

```
<RobotStartCellRow>0</RobotStartCellRow>
<RobotStartCellCol>0</RobotStartCellCol>
<RobotType>Pioneer3DX</RobotType>
</MazeSimulatorState>
```

That's it! You have defined all of the necessary parameters to create a virtual maze-like environment. All you need to do is use the Dashboard to drive your robot around and see whether you like your new construction. You can use ProMRDS\chapter9\mazesimulator\mazesimulator.cmd to start the Maze Simulator and the Dashboard together.

## The ExplorerSim Application

With a maze already built, you can now run the ExplorerSim program to let the robot explore and build a map. The batch file to run ExplorerSim is ProMRDS\chapter9\explorersim\explorersim.cmd.

When ExplorerSim starts up, it also starts a Dashboard. You don't need to use the Dashboard unless the robot gets stuck. The exploration algorithm is not much more than a random walk, and sometimes the robot will oscillate backward and forward as though it can't make up its mind. In that case, you can drive it somewhere else, although you will find that you are fighting with ExplorerSim!

The main reason for running the Dashboard is that it can be used to display the LRF data and show the view from the webcam on the robot. However, ExplorerSim does not use the webcam images, only the LRF data.

ExplorerSim also creates a Map window, which, after all, is the main purpose of exploring. An example is shown in Figure 9-19 for a partially completed map that corresponds to the maze bitmap in Figure 9-18. Compare the two and see whether you can see the similarities.

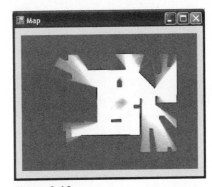

**Figure 9-19**

ExplorerSim can create three different types of maps, which use different algorithms for drawing new data into the map. It is beyond the scope of this book to explain mapping algorithms, but a brief introduction might be useful.

The map in Figure 9-19 is what is known as an *occupancy grid*. This is a type of metric map in which the world is broken up into a set of grid cells of equal size. In the case of Figure 9-19, each pixel represents a five-centimeter square. The values in the cells represent the probability that the cell is vacant, and the common convention is that black means occupied and white means empty. Shades of gray, therefore, indicate how uncertain the robot is of the cell's occupancy. Initially, the whole map is set to gray (a 50% probability of being occupied).

The map dimensions can be set by editing the `config` file `ProMRDS\Config\ExplorerSim.config .xml`, shown here:

```xml
<?xml version="1.0" encoding="utf-8"?>
<State xmlns:s="http://www.w3.org/2003/05/soap-envelope"
xmlns:wsa="http://schemas.xmlsoap.org/ws/2004/08/addressing"
xmlns:d="http://schemas.microsoft.com/xw/2004/10/dssp.html"
xmlns="http://schemas.microsoft.com/robotics/2007/06/explorersim.html">
  <Countdown>0</Countdown>
  <LogicalState>Unknown</LogicalState>
  <NewHeading>0</NewHeading>
```

*(continued)*

*(continued)*

```
<Velocity>0</Velocity>
<Mapped>false</Mapped>
<MostRecentLaser>2007-07-14T18:25:32.2205888+10:00</MostRecentLaser>
<X>0</X>
<Y>0</Y>
<Theta>0</Theta>
<DrawMap>true</DrawMap>
<DrawingMode>BayesRule</DrawingMode>
<MapWidth>24</MapWidth>
<MapHeight>18</MapHeight>
<MapResolution>0.05</MapResolution>
<MapMaxRange>7.99</MapMaxRange>
<BayesVacant>160</BayesVacant>
<BayesObstacle>96</BayesObstacle>
</State>
```

Notice that you can set the `MapWidth` and `MapHeight` (in meters) as well as the `Resolution` (which is the grid cell size).

Three drawing modes can be specified using the `DrawingMode` parameter:

❑ **Overwrite:** This simply draws a ray in the map from the robot's current position out to the point where the laser hits an obstacle. It does this for all range scans, which are typically one degree apart with a field of view of 180 degrees. As the robot moves, the map is overwritten with new data. Note that if the robot lost track of its position, then it would start drawing map information in the wrong place and potentially destroy a perfectly good map! The overwrite method only works when you have near-perfect pose information, such as in simulation.

❑ **Counting:** This is a little more robust. Each time the robot processes a laser scan, it increments the cell values along the laser ray and decrements the value at the point where the ray hits an obstacle. Eventually, if the robot sees the same things often enough, the map will "develop," just like developing a film — the obstacles become darker and the free space becomes lighter.

❑ **Bayes Rule:** Bayes Rule is based on probabilities. It enables the probability value in a cell to be updated based on new information and a belief in the accuracy of the new information. For this particular exercise, there is not much difference between Bayes Rule and the Counting method. However, in the real world the robot would have a Measurement model that assigned probability values to new range-scan data and a Motion model that estimated the true position after each move. For more information about this field, see *Probabilistic Robotics* by Thrun, Burgard, and Fox (MIT Press, 2005).

The last point to make about ExplorerSim, which will become quite obvious if you watch it for a while, is that it has a very poor exploration algorithm. It takes a very long time for the robot to explore the entire maze because it keeps covering ground that it has already seen.

## Where to Go from Here

There are literally dozens of different directions that you could follow from here, limited only by your imagination.

The robot has a webcam, but the exploration is not using vision. However, using a camera introduces complications with perspective, and it is not an easy task to measure distances accurately. One way to process images is using the RoboRealm software (www.roborealm.com), available free and compatible with MRDS.

You could also try using different sensors other than the LRF. In Chapter 6, there is a Simulated IR sensor, but you would need several of these for map building. Bear in mind that most LRFs supply hundreds of data points in each scan, which you can use to add new information to the map. Some of the information is redundant, but it is not possible to get the same density of information with IR.

Raúl Arrabales Moreno has written a simulated sonar sensor service for the Pioneer 3DX and adapted the ExplorerSim to use it. The code is available on the Conscious Robots website (www.conscious-robots.com).

The exploration algorithm in the ExplorerSim is not good (it was inherited from the original Explorer sample). Exploration algorithms are another active research area. Maze solving is often taught in classes on artificial intelligence, and it is a form of exploration. You could definitely improve the exploration process by coming up with an approach that tries to direct the robot toward fresh, unexplored areas. You just need to realize that sometimes you will find a wall in your way!

Real robots don't always know their position, so there are opportunities for SLAM researchers to develop algorithms *without* using the pose information from the simulator. This is a difficult task.

Lastly, we have not dealt with moving obstacles at all. People wandering around a building might get in the way of robots. It doesn't make any sense to put an obstacle in the map if it is a person, because it probably won't be there the next time you come back.

# Summary

This chapter has provided several examples of simulation projects. The sumo and soccer environments provide an opportunity to immediately code behavior for a robot without having to define the entire environment. The hexapod project shows how to implement a walking, articulated robot that could be evolved into a quadruped or biped walking robot. The Explorer project shows how to create a complete simulation environment that can be navigated by a robot model.

It is hoped that these projects have provided some inspiration for your own simulation ideas and some concrete examples that will aid you in your development.

# Part III

# Visual Programming Language

# 10

# Microsoft Visual Programming Language Basics

The Microsoft Visual Programming Language (VPL) is a new application development environment designed specifically to work with DSS services. Programs are defined graphically in data flow diagrams rather than the typical sequence of commands and instructions found in other programming languages. VPL provides an easy way to quickly define how data flows between services. It is useful for beginning programmers because they can quickly specify their intent, but it is also well suited to help with prototyping and code generation for more experienced programmers. It is also useful for specifying robotic orchestration services within the context of the Microsoft Robotics Developer Studio SDK, but it can be used outside of robotics as well.

This chapter covers what it means to work with a data flow language, how to specify a data flow diagram using the basic activities provided by VPL, and how to debug and run it. You won't find any robots in this chapter but you will learn how to use VPL to control robots in Chapter 11.

## What Is a Data Flow Language?

Many people learn how to program using an imperative language such as BASIC or C++. Such a language uses control statements to modify program state. This matches the underlying hardware implementation of the CPU well because it is built to execute machine code statements that modify memory.

When a program is data- or event-driven and has different parts that execute asynchronously, the imperative programming model can become inefficient and difficult to use. The following example shows why this might sometimes be true.

Imagine a program as a city grid. Each intersection represents some processing that must be done, and cars on the roads between intersections represent data, as shown in Figure 10-1.

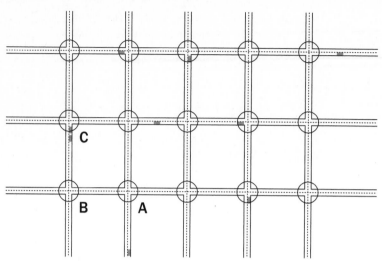

**Figure 10-1**

A typical imperative program might take the following form:

```
For each intersection
     While not enough cars are waiting
          Sleep for a time.
     Do some calculations and call the next intersection routine with the results
```

The calculations done at each intersection can be arbitrarily complex and may depend on the state of multiple cars waiting at the intersection. This approach can lead to inefficiency because the instructions associated with each intersection must periodically run to determine whether enough cars are waiting. In addition, because cars may arrive at each intersection at an arbitrary time, race conditions can arise if the processing instructions at each intersection are not carefully defined.

In a data flow language, the emphasis is on following the cars rather than processing the intersections. A data flow diagram for the city example above would look much like the city grid, with the roads representing data connections and the intersections representing processing activities. None of the intersection code runs until it is presented with data.

Let's say that data enters the diagram at the bottom and approaches intersection A. The code for intersection A runs to process the data and puts the resulting state on the way to intersection B. The code for intersection B then runs and sends the modified state to intersection C. Intersection C is programmed to wait until data is available on at least two of its roads before passing the resulting data off to the left and out of the grid. Now that there is no data in the grid, no code runs until more data is available.

Multiple cars in the grid represent multiple data packets being processed simultaneously. Because the calculations at each intersection depend only on the cars waiting for it, they can run at the same time as another intersection.

It is simpler to write a program for each intersection than it is to write instructions for managing the entire grid. Similarly, it is simpler and more efficient to write small segments of code that respond to data inputs than it is to write code that attempts to control every intersection simultaneously.

VPL is a data flow language that enables you to specify one or more MRDS services to process data in parallel and to define the data connections between those services. In a robotics context, it is useful for writing high-level orchestration services that control robot behavior. In the rest of this chapter, you will learn how to implement algorithms using VPL.

# The VPL Development Environment

Start the VPL development environment by clicking Start ⇨ Visual Programming Language. A window similar to the one shown in Figure 10-2 is displayed.

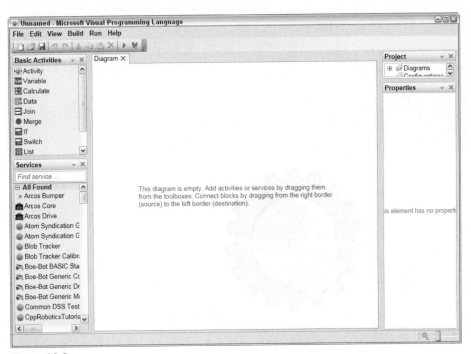

**Figure 10-2**

Note the following about the environment:

❑ The main commands are accessible from the main menu bar and the most common commands are on the toolbar.

❑ The work area consists of a central diagram area surrounded by four toolboxes. A diagram consists of one or more activity boxes that are connected together. You drag activity boxes from the various toolboxes to the central area. These toolboxes include the following:

❑ The Basic Activities toolbox contains blocks that control data flow and create data and variables.

❑ The Service toolbox shows all of the services available to VPL. This list includes all of the services that are present in the MRDS `bin` directory.

❑    The Project toolbox shows the diagrams and service configuration files included in your project.

❑    The Properties toolbox shows the properties for the currently selected item so that you can view and edit the settings exposed by a service or activity.

# A Simple VPL Diagram

VPL programs are called diagrams because they are defined by placing graphical icons in a diagram and connecting them with data flow connections in the VPL development environment. Once the diagram has been defined, it can be executed within the VPL development environment or you can generate an MRDS service from the diagram and run that.

The easiest way to see how a diagram works is to actually build one. The following sections describe how to create a new diagram and run it.

## Creating a VPL Diagram

To create a program, follow these steps:

1.    Start with a fresh diagram by clicking File ➪ New. The environment should look like Figure 10-2. Drag a Data activity from the Basic Activities toolbox into the diagram space.

2.    Click the data type drop-down arrow at the bottom of the activity and select string for the type of data. Type a string such as, **Hello world, I'm VPL!**, in the input field of the Data activity. The activity should look like Figure 10-3.

Figure 10-3

3.    Now drag the TexttoSpeechTTS activity from the Services toolbox. The Services toolbox contains a lot of entries so it often saves time to type part of the name of the service that you need in the Search box above the list of services. This narrows the list down to only those services that contain the search text. Position the TexttoSpeechTTS activity to the right of the Data activity.

---

### Search Groups in the Services Toolbox

The Services toolbox contains many services, so the Search box at the top is very helpful. You will probably find that you do some of the same searches repeatedly. The Services toolbox provides a way for you to save common searches. Type **dialog** into the Search box and a few services are displayed. Now click the plus (+) symbol next to All Found. This creates another heading in the services list that contains all the services that this search finds. You can expand or contract the list using the icon to the left of the dialog heading. All the other services are found under the All Found heading. When you no longer want to keep the dialog search results, you can click the X symbol to the right of this heading and it is deleted.

---

**4.** Place the mouse on the right output pin of the `Data` activity and drag the mouse to the `TexttoSpeechTTS` activity. Notice that the corners of the `TexttoSpeechTTS` activity turn green, indicating that this connection is allowed. When you release the mouse button, the Connection dialog appears. This dialog enables you to select which message is being output from the source activity and which message accepts the input in the destination activity. In this case, select DataValue in the From: box and SayText in the To: box and press OK.

**5.** Now the Data Connections dialog box opens. This enables you to map a particular data value in the message output from the source to a target value in the message input to the destination. Because the data value being passed is a string, it has a value and a length. Select value in the Value box and SpeechText in the Target box. A connection should appear between the two activities in the diagram as shown in Figure 10-4.

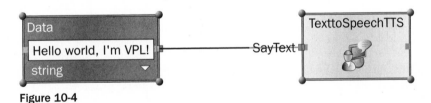

**Figure 10-4**

Congratulations! You have just written your first VPL diagram. Save it by clicking File ⇨ Save As and select an appropriate directory and project name. If you have any trouble creating this diagram, you can find it in the `Chapter10` directory in the `1-Hello` folder.

## Running a VPL Diagram

To run a diagram, first click File ⇨ Open to load a previously saved diagram. Then follow these steps:

**1.** Click Run ⇨ Start and press F5, or press the green forward arrow in the toolbar to run your program. A dialog box appears that shows the console output as the program is executed. After a few seconds, you should hear your computer speak the words in the `Data` activity box. The console output window is shown in Figure 10-5.

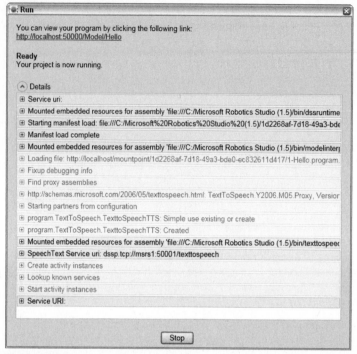

**Figure 10-5**

**2.** After the program has finished its execution, press the Stop button to dismiss the Run dialog.

---

### If Your Computer Won't Talk

If you run the Hello diagram and you aren't able to hear the computer speak the words, check the connections to your speakers and the volume. Select the Speech item in the system control panel and verify that the speech engine is configured properly. If you still cannot hear the computer speak, you can substitute the `SimpleDialog` activity for the `TexttoSpeechTTS` activity in your diagram to display the words in a dialog box instead. The `2-Hello` VPL project in the `Chapter10` directory shows how this is done.

---

Let's take a look at exactly what happened when this basic little diagram in Figure 10-4 was executed. The `Data` activity has a very simple job. When it receives a message on its input pin, it sends a message on its output pin containing the data specified. Where did the input message come from? Its input pin is unconnected. It turns out that VPL automatically sends an input message to all top-level `Data` activities when the program starts.

Up until now, we've been speaking about messages on wires in pretty general terms, but it's time to get more specific. You can think of a message much like a C# or C++ structure. It has one or more fields and each has a name and a type. When you specify values in the Data Connections dialog, you are actually selecting which field of the structure to send to a target value in the destination activity.

Go back to the 1-Hello project and save it to a different project name. This is important because when you make a modification to a project and then press F5 to run it, VPL automatically saves it. If you make modifications to 1-Hello and then run it, you will modify the project that was installed as part of the book software.

Select the connection between the Data activity and the TexttoSpeechTTS activity. The properties of the connection are shown in the Properties toolbox. The properties closely resemble the contents of the Data Connections dialog. Change the Value property to Length as shown in Figure 10-6.

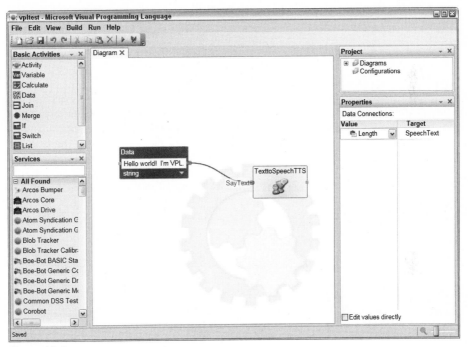

**Figure 10-6**

Now, when you execute the diagram, the engine says "22" because the length of the data string was passed to the speech activity instead of its value.

Now you know that the Data activity builds a message that contains a value, which is a string and a value associated with the string, which is its length. This message is passed to the TexttoSpeechTTS activity, where, in the original diagram, the value of the string is assigned to the SpeechText input. When the TexttoSpeechTTS activity receives a message on its SayText input pin, it speaks the text passed to the SpeechText input and then sends a success or failure message on its output pin. Because nothing is connected to this pin, the message drops on the floor. Not to worry — VPL has a great little floor sweeper that gathers dropped messages and disposes of them properly.

# Inputs, Outputs, and Notifications

Each activity can have one or more inputs shown as square pins on the left side of the activity icon and one or more outputs shown as square pins on the right side of the icon. When there are multiple inputs or outputs, only a single pin is shown and the desired input or output is selected with the Connections dialog when a connection is made.

Some activities have an additional output called a *notification*. A single response output message is generated on an output pin as the result of a single input message on an input pin. Notification outputs can also send a message as a result of an input message but they more typically send a message as the result of a change in internal state. As an example, consider the `DirectionDialog` activity shown in Figure 10-7.

**Figure 10-7**

When a message is received on the Action Input Pin, a different message is sent on the Result Output Pin, usually indicating success or failure. If the input message changes the internal state of the activity, it typically generates a notification message as well. Sometimes an external action, such as a user input event, changes the state of an activity and it spontaneously generates a notification without any input message at all.

The 3-Notifications diagram in the `Chapter10` directory illustrates this behavior. It is shown in Figure 10-8.

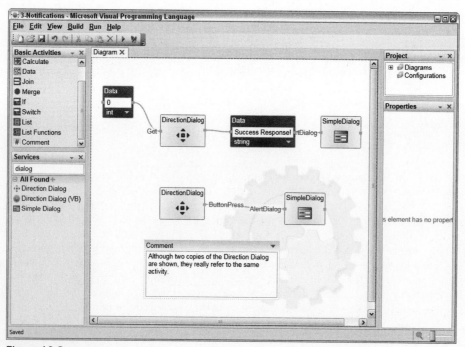

**Figure 10-8**

In the top row of activities, a message is initially generated by the Data activity on the left, which is sent to the Get input of the DirectionDialog activity. It sends a response message containing its state to the second Data activity, which then sends an output message containing the string "Success Response!" to the SimpleDialog activity, where it is displayed in an AlertDialog. So far you have used the Action Input Pin and the Response Output Pin of the DirectionDialog activity.

On the second row of activities, another DirectionDialog activity is shown on the left. However, this icon actually represents the same activity as the DirectionDialog in the top row. You can tell because they have the same name. When you drag onto the diagram an activity that is already present, VPL asks you whether you want a completely new activity or just another reference to the existing activity. If you specify that you want a new activity, the new activity is given a different name than the original activity. While this might initially seem confusing, it really makes the diagrams simpler because it reduces the rat's nest of connections coming from a single activity when multiple inputs and outputs are used.

The output from the DirectionDialog activity that is used in the second row is the notification output. This output sends a message to the SimpleDialog activity whenever the internal state of the DirectionDialog changes. Just like the DirectionDialog, the SimpleDialog on the second row is really just another reference to the same DirectionDialog activity shown on the first row.

This diagram also shows the use of the Comment activity. This activity has no effect on the functionality of the diagram but it provides a way to document what is going on in the diagram.

Run the diagram and notice that the response is immediately displayed as a result of the data input message on the top row of the diagram. Click the buttons on the DirectionDialog and you will see messages displayed as a result of the notifications generated by the DirectionDialog activity on the second row.

# Basic Activities

The following activities can be found in the Basic Activities toolbox. These basic activities are used to build the data connections between the activity blocks that represent services. Unlike the service activities, they do not have a unique name.

They are defined briefly in the following table and discussed in more detail in the following sections.

| Activity | Description |
| --- | --- |
| Variable | This activity is used to get or set the value of a state variable. |
| Calculate | This activity is used to calculate a value from one or more input values. |
| Data | This activity is used to define static data that is used as an input to another activity. |

*(Continued)*

| Activity | Description |
| --- | --- |
| Join | This activity waits to send an output message until it has received a message on all of its inputs. |
| Merge | This activity forwards a message to its output from any of its inputs. |
| If | This activity forwards a message on the first output that has an expression that evaluates to true. |
| Switch | This activity forwards a message on the first output that matches the input value. |
| List | This activity holds a list of data. |
| List Functions | This activity performs various operations on a list of data. |
| Comment | This activity displays a comment in the diagram. |
| Activity | This activity can be used to define a custom activity that, in turn, contains other activities. |

# VPL Variables

In a data flow diagram, the program state is mostly encapsulated in the messages passed between activities. Sometimes it is necessary to define state that belongs to the diagram. This state is analogous to the state in a service. Activities can set or retrieve this state. This is useful because activities do not, by default, pass the same data to their output that they receive on their input.

You use the Variable activity to define and set or get the value of a variable. When you drag a Variable activity from the Basic Activities toolbox onto the diagram, it is initially blank. Click the ellipse on the bottom of the activity to bring up the Define Variables dialog box, which enables you to add and delete variables. Any Variable activity in the diagram can represent any defined variable.

Load the 4-Variables diagram in the Chapter10 directory. It is shown in Figure 10-9.

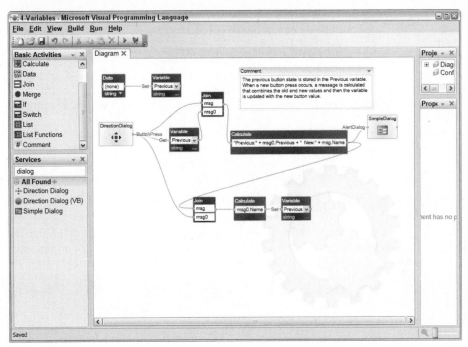

**Figure 10-9**

A string variable called `Previous` is defined and initialized using the `Data` and `Variable` activities in the upper-left corner. When a button is pressed in the `DirectionDialog`, a notification is sent to three different activities. The first activity to take action is the `Variable` activity, which gets the value of the variable and sends it to the `Join` activity connected to its output.

This is the first time we've used a `Join` activity. Its function is very simple. It waits for an input message on both of its input pins and then combines the messages into a single message and sends it on its output pin. In Figure 10-9, when the `Join` activity receives the output from the `Variable` activity and the notification from the `DirectionDialog`, it combines these two messages into one and sends it to the `Calculate` activity, which performs arithmetic, logical, or string operations on its input values according to the expression entered in the activity. These operations are shown the following table:

| Type of Operation | Operation |
|---|---|
| Arithmetic | +, −, *, / and % for modulo |
| Logical | &&, ‖, and ! |

The plus operator (+) can also be used to concatenate strings. Parentheses can be used to group operations.

The operation defined in the `Calculate` activity in Figure 10-9 is `"Previous:" + msg0.Previous + " New:" + msg.Name`. This builds a string that contains literals combined with the string values of `msg0.Previous` and `msg.Name`. The message is displayed using the `SimpleDialog` activity.

**481**

All that remains is to update the value of the `Previous` variable. This is done with the `Join` activity on the bottom row. It waits until the string has been built and then combines this message with the original notification message from the `DirectionDialog` activity. The `Calculate` activity extracts the `Name` string from the original `DirectionDialog` notification and passes it to a variable activity to be stored. The `Join` at the bottom of the diagram is necessary because you must be certain that the string has been constructed before you change the value of the variable. Without this `Join`, there is a race condition between constructing the string and updating the variable that can cause unpredictable results.

Run the diagram and notice that the message displayed contains both the current button press as well as the previous button press.

# Looping and Conditionals in VPL

No language is complete without the ability to make logical decisions and iterate through loops. In an imperative language, you use an `If` statement to make a decision and a `For` or a `While` statement to iterate. In VPL, decisions are made with an `If` activity, and iteration is done using a data flow loop.

You might expect to find a basic `ForLoop` activity in the Basic Activities toolbox, but it isn't there. That's because a `ForLoop` cannot usually be implemented in VPL with a single activity box. As you will see in the next few examples, you must take special care to ensure that the `ForLoop` executes in the proper order.

The diagram in Figure 10-10 shows how a simple `ForLoop` can be implemented in VPL. You can find it in the `Chapter10` directory under "5-Loops."

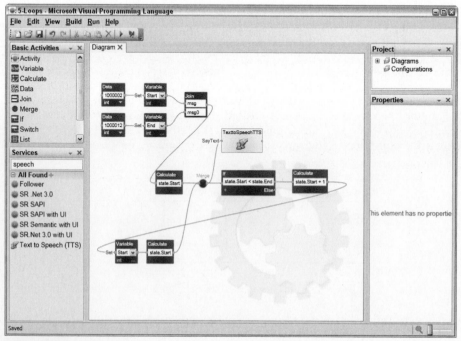

Figure 10-10

Two variables are defined: Start and End. They are given initial values by the Data activities in the upper left. When both Variable activities have been initialized, the Join activity immediately after them passes a message to the Calculate activity, which then sends a message that contains the value of the Start variable. This message is passed to the TexttoSpeechTTS activity and to an If activity. If the value of the Start variable is less than the End variable, then a message is passed along to the Calculate activity on the far right, which calculates the value of Start+1. This value is set back into the Start variable and the value of the Start variable is placed again on the wire to traverse the loop again. This continues until the Start variable is equal to the End variable, at which point no message is passed on to the Calculate block and execution stops.

There are a few interesting things to note in this diagram. This is the first time that the Merge activity has been used. When it receives an input message on any of its inputs, it simply passes that message to its output. Unlike the Join activity, it doesn't wait until all of its inputs have received a message. The Merge activity will sometimes display a yellow exclamation point with a warning that indicates a loop has been detected. VPL is trying to gently encourage you to implement loops with recursion, rather than use the Merge activity. This keeps the diagrams simpler and more readable. Implementing loops with recursion is covered in the section "Refining a Custom Activity."

The Calculate boxes that use the values of the Start and End variables don't use any of the values that are on the wire. They only use the variables stored in the diagram state. The two Calculate boxes that place the value of state.Start on the wire prior to the Merge activity are simply there to ensure that the same message is on both connections leading into the Merge activity. The Merge activity requires the messages it receives on all of its inputs to be of the same type. These messages are not actually used by any activity.

In VPL, loops are implemented as literal loops in the diagram with an If activity or other conditional to ensure that the loop doesn't continue forever.

Run the diagram to hear the beautiful sound of the numbers from 1,000,002 to 1,000,012 being read aloud. Listen carefully and you may hear one or more of the numbers read out of sequence. This occurs because there are two possible routes for messages to take after the Merge activity. VPL schedules processing for each route asynchronously. Occasionally, a message makes it all the way through the If activity and back to the Merge before the first message has been processed on its way to the TexttoSpeechTTS activity. This can cause a later message to be posted to the TexttoSpeech activity before an earlier one is posted, causing the out-of-order number reading.

The diagram in Figure 10-11 solves this problem. You can load it from the Chapter10 directory in 6-Loops (in order).

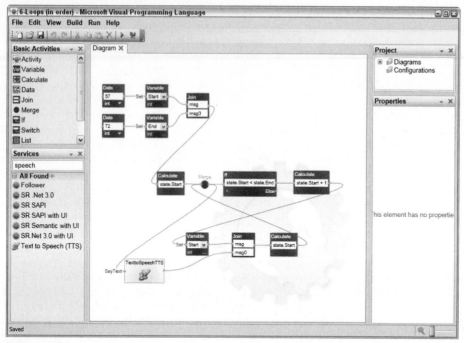

**Figure 10-11**

This diagram adds a Join activity after the Variable activity where the new Start variable value is saved. The Join activity will wait until the TexttoSpeechTTS activity has completed saying the previous phrase before passing an output message to the Calculate activity in the lower-right corner. This keeps the rest of the loop executing at the same speed as the TexttoSpeechTTS activity and prevents multiple SayText messages from being queued at the TexttoSpeechTTS input.

# The VPL Execution Model

The 6-Loop (in order) diagram shown in Figure 10-11 provides a good opportunity to discuss the way in which VPL executes a diagram. It helps to understand how VPL evaluates activities and propagates messages when debugging a diagram.

A diagram is inactive until a message enters it. The message can originate from a notification on an activity or from a message that is injected into the inputs of a top-level Data activity. As long as a message is actively propagating in the diagram, VPL continues its execution.

You can see how VPL evaluates a diagram by manually tracing messages through the diagram. For the diagram shown in Figure 10-11, the diagram is inactive until VPL injects an input message into each of the upper-left Data activities. Because two messages are now active, VPL schedules a parallel thread of execution to evaluate each of them. One message propagates through the topmost Data activity and its output message propagates to the topmost Variable activity. It sets the value of the Start variable and its output message propagates to the Join activity. The Join activity has not yet received a message on its msg0 input so that evaluation thread terminates because it can no longer propagate a message.

Meanwhile, the second message propagates through the lower Data activity and to the lower Variable activity where the End variable is set, and then the message propagates to the Join activity. Because the Join activity has now received a message on both of its inputs, an output message is able to propagate and this thread continues to execute.

The message moves to the Calculate activity and through the Merge activity. The message is then sent to two activities: If and TexttoSpeechTTS. VPL spawns another thread to evaluate the message that goes to the TextoSpeechTTS activity and the previous thread continues to evaluate the diagram at the If activity. If state.Start < state.End, the If activity output message moves to the Calculate activity, through the Variable activity, and then it gets to the Join activity. Because the Join activity has not received a message at its msg0 input, no further message can be propagated and the thread is terminated.

At the same time, the TexttoSpeechTTS thread continues to execute. After this activity has completed speaking the text that was passed to it, it sends an output message to the msg0 input of the Join activity. Because a message was already received at the msg input, the Join activity propagates a message to the Calculate activity, which passes a message to the Merge activity and the loop is evaluated once again. When the message has been around the loop enough times that state.Start is equal to state.End, the output message from the Merge is sent to the If activity and, once again, another thread is spawned to evaluate the message in the TexttoSpeechTTS activity. This time, no message is propagated from the output of the If activity because its condition is false. This causes the original thread to terminate, leaving only the thread, which is running the TexttoSpeechTTS activity. When this activity completes, it sends an output message to the msg0 input of the Join activity. Because the Join activity has not received a message on its msg input, this message cannot propagate and the last thread is terminated. Because all threads have now terminated, the diagram is again inactive. There are no other sources of messages in the diagram and so it remains inactive until it is stopped.

# Debugging a VPL Diagram

It is useful to be able to manually trace messages through a diagram but it is much nicer to let the computer do all the busy work of this process. That is what the VPL debugger is for. You can run a diagram in debug mode by clicking Run ⇨ Debug Start. This starts VPL in single-step mode. The debugger output is shown in a web page that is launched when the diagram starts execution.

*The debugger is known to work with Internet Explorer but it may not work with other browsers. You should set your default browser to Internet Explorer prior to debugging VPL diagrams.*

The debugger page is divided into four sections:

❑ **Diagram State:** This is the topmost section, shown in Figure 10-12. It shows values of all of the state variables in the diagram.

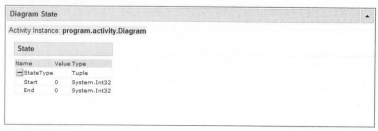

**Figure 10-12**

❑ **Current Node:** The next section is a graphical representation of the diagram, much like the one shown in VPL. The difference is that this diagram is annotated to show the currently executing activity, activities that have been scheduled for execution, and breakpoints. As shown in Figure 10-13 (without the color), the currently executing activity is outlined with bright red, while other activities that are scheduled for execution are outlined with dark red. A red circle is shown next to activities that have a breakpoint set.

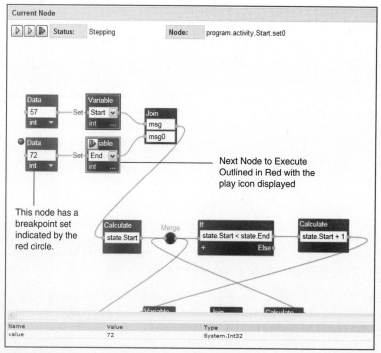

**Figure 10-13**

If the button on the active activity is clicked, VPL will step to the next activity. This can also be accomplished by pressing the Step button at the top of this section. The other two buttons can be used to run the diagram at slow speed or regular speed. The slow-speed execution is useful to Vtrace through large diagrams without having to continually press the Step button.

At the bottom of this section, the value of the message on the input pin of the current activity is shown.

❑ **Breakpoints:** The next section shows the currently set breakpoints, as shown in Figure 10-14. If a breakpoint has been set on an activity, execution will stop when it becomes the active activity. Breakpoints can be cleared, enabled, and disabled in this section.

**Figure 10-14**

❑ **Pending Nodes:** The fourth section shows all activities scheduled to be executed, as shown in Figure 10-15. As described in the previous section, multiple activities can be executing at the same time on multiple threads. A breakpoint can be set on any of the scheduled activities by pressing the corresponding SetBP button.

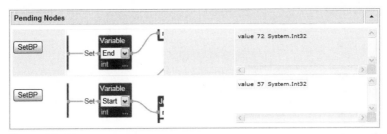

**Figure 10-15**

The diagram can be terminated by pressing the Stop button on the Run dialog, just as when the diagram is not running in debugger mode.

# Creating Custom Activities

As shown in Figure 10-11, simple functions can be composed of many different activity blocks. Real-world diagrams can quickly become unmanageable. Fortunately, VPL enables custom activities to be defined that are similar to subroutines in imperative languages.

The ForLoop function defined in Figure 10-11 is a good candidate to become a custom activity because it is a function that might be used multiple times in a diagram and it is self-contained. This section starts by showing you how to define a custom activity to implement a ForLoop. Later you will improve that custom activity to make the ForLoop execute in order.

## Defining a ForLoop Custom Activity

Defining a custom activity is very similar to defining a diagram except that the input and output pins must be defined and connected. The following steps show how to define a custom activity that implements the ForLoop functionality. You can see the final result in the 7-CustomActivity project in the Chapter10 directory.

1. Start out with a new project by clicking File ⇨ New.

2. Drag an Activity block from the Basic Activities toolbox.

3. Give the activity a name such as **ForLoop** by typing it in the Name: box in the Properties toolbox. Give it a friendly name and a description as well.

4. Double-click the ForLoop activity to open its diagram. The diagram page looks similar to the top-level diagram except that now input and output pins are present.

5. Define the actions that this activity will support by clicking Edit ⇨ Actions and Notifications. The Actions and Notifications dialog is displayed.

6. Add a new action by pressing the Add button in the Actions column. Name the action by typing **Initialize** in the box. Add two input values by pressing the Add button in the Input values column. Name the input values StartValue and EndValue and make them both of type int. A properly configured dialog is shown in Figure 10-16.

**Figure 10-16**

**7.** Select the Notifications tab on the Actions and Notifications dialog. Add a notification by clicking the Add button in the Notifications column. Name the new notification "CountNotification." Add a notification value by clicking the Add button in the Notification values column. Name the value "Count" and give it the type of `integer`. A properly configured Notifications tab is shown in Figure 10-17.

**Figure 10-17**

**8.** The `ForLoop` activity now supports an `Initialize` action that expects an input message with two integer values, `StartValue` and `EndValue`, as well as a notification output that contains a single integer called `Count`. Make sure that the `Initialize` action is selected on the Action drop-down menu at the top of the `ForLoop` diagram.

**9.** Add actions to the diagram to match the diagram in Figure 10-18. The upper `Calculate` activity extracts the `StartValue` value from the input message and then waits for the `EndValue` to be processed before starting the loop. The lower `Calculate` activity extracts the `EndValue` value from the input message and then sets it in the `End` state variable. This implementation of the `ForLoop` does not store the incremented value in a state variable as the last implementation did. The state is maintained in the messages that are passed between activities. When both input values have been processed, a response message is sent to the outside world via the result pin. Because we didn't specify any specific types for output messages, any message will do. The `Join` activity passes a message that once again contains both values. The `Calculate` activity extracts the `StartValue` value and passes it to the `If` activity, which only passes along a message if the value on the wire is less than the `End` state variable. This message is passed back to the outside world as a `CountNotification` that contains the count value as an integer. The value on the wire is incremented and passed back to the `If` activity to be checked again.

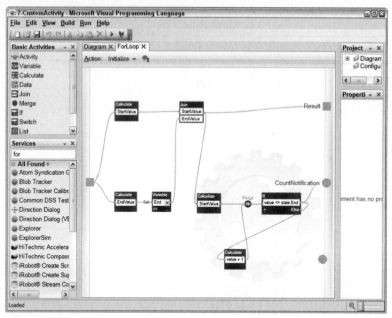

**Figure 10-18**

10. Select the main diagram tab. Add actions to the diagram to match Figure 10-19. The two `Data` activities are combined into a single message with the `Join` activity and sent to the `Initialize` action on the `ForLoop` activity. Copy and Paste the `ForLoop` activity to create another instance of it and connect its `CountNotification` output to the `SayText` action of the `TextoSpeechTTS` activity.

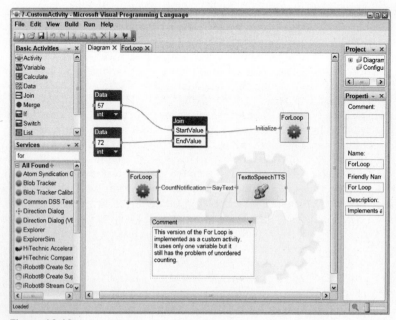

**Figure 10-19**

The top-level diagram is now much simpler because much of the complexity has been hidden in the ForLoop activity. It is important to design VPL diagrams in this hierarchical fashion to keep complex diagrams from becoming unwieldy.

## Improving the ForLoop Custom Activity

In some ways, this diagram is a step backward from the previous one because your iteration loop has no feedback to ensure that the count is always done in the right order. In this section, you'll see how to improve the ForLoop custom activity to provide this feedback

The 8-CustomActivity (in order) project in the Chapter10 directory improves upon the diagram in the previous section by introducing a custom activity with multiple actions. It also uses fewer total activities.

The top-level diagram is shown in Figure 10-20.

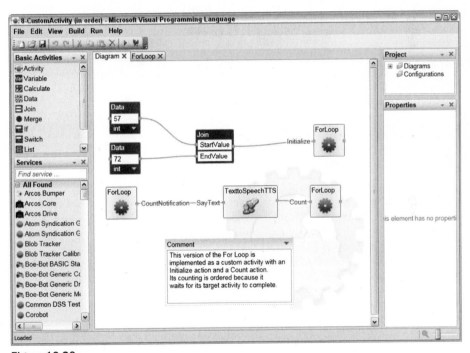

**Figure 10-20**

The ForLoop activity appears three times in this diagram but each reference is to the same activity. The diagram begins by combining the Start and End values into a single message that is passed to the Initialize action of the ForLoop activity. The ForLoop then sends a CountNotification containing the first count to the TexttoSpeechTTS activity. When this activity has completed, it sends a message back to the Count action of the ForLoop, which causes another CountNotification. This continues until the count reaches the End value and then the ForLoop issues no more CountNotifications. This is a clean way to use the ForLoop activity and it provides the needed synchronization to keep the count from getting out of order.

The new `Initialize` action in the `ForLoop` activity is shown in Figure 10-21.

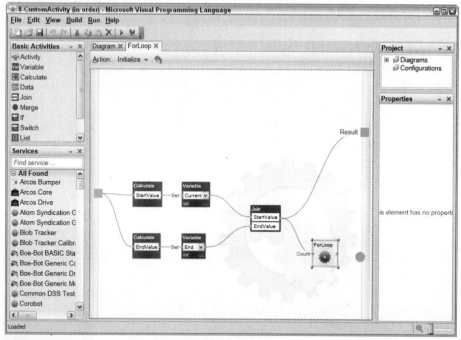

**Figure 10-21**

You can specify multiple actions in an activity by creating them in the Actions and Notification dialog. You then specify inputs and outputs for each action. The Action: drop-down menu at the top of the diagram enables you to select the action you are working with. Each action is independent from all the others except that they all refer to the same state variables.

In the `Initialize` action, the `StartValue` and `EndValue` parts of the input message are separated and stored in their respective state variables. When both values have been stored, a message is sent to the result pin to indicate that the operation has completed. You may have noticed that the `ForLoop` activity appears in its own action definition. This is called *recursion*. It is perfectly legal to pass a message from one action to the input of another action by referring to the activity in this way. Simply use the Edit menu to copy the `ForLoop` activity from the main diagram and then paste it into this diagram. The output from the `Join` activity passes a message to the `Count` action on the `ForLoop` activity. The definition for this action is shown in Figure 10-22.

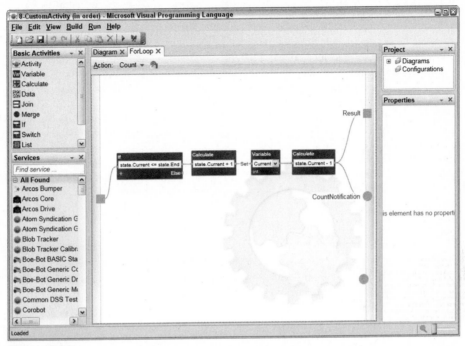

**Figure 10-22**

This diagram is significantly simpler than previous implementations of the `ForLoop`. When a message is passed to the input, the `If` activity determines whether any counts are remaining. If so, it passes a message to the `Calculate` activity, which calculates the value of the `Current` variable plus one and then sets it back into the `Current` variable. The final `Calculate` is used to put the previous value of the `Current` variable on the wire to be output as a `CountNotification` and a result message.

Execute this diagram to verify that it does count in the proper order and then take the time to step through it using the debugger if the execution order is not immediately clear.

Another possible way to improve the `ForLoop` custom activity is to add an `End` notification that sends a message when the loop has completed. You'll see how to add this in Example 9 in the next section.

# Using Lists and Switch Activities

The only activities left to describe in the Basic Activities toolbox are the `List` activities, the `List Functions` activities, and the `Switch` activities.

The `Switch` activity is much like the `If` activity. Instead of conditions, the `Switch` activity expects data values to be specified, like case statements associated with a switch statement in an imperative language. The data values are compared to the data value on the wire and the input message is propagated to the first case that matches. If none of the cases matches, the message is propagated to the `Else` output.

The `Switch` activity is useful for taking an action based on a specific data value on the wire. This is often done when implementing a state machine. In this case, the values in the `Switch` activity represent various possible states, and the action that the state machine takes depends on the value of the state variable.

List and ListFunctions activities are used together to manage and operate on collections of typed data. A List activity generates an empty list of a specific type. Select the type of the list from the drop-down menu at the bottom of the activity icon. All of the basic VPL types are supported.

A list may be stored in a variable like any other data value. Simply create a new variable with the desired list type and use the Set and Get actions on the Variable action to set and retrieve the list.

The List Functions activity performs one of several possible operations on a list. Its inputs and outputs depend on the function that is selected from the drop-down menu on the bottom of the List Functions activity icon. The possible functions are listed in the following table:

| Function | Inputs |
| --- | --- |
| Append | List: A list<br>Item: A data item of the same type as the list. The item on the Item input is appended to the end of the list and the resulting list is present on the output pin. |
| Concatenate | List1: A list<br>List2: A list of the same type<br><br>List2 is concatenated on the end of List1 and the resulting list is present on the output pin. |
| Reverse | List: A list. A new list is constructed with the items in reverse order from the input List. The new list is present on the output pin. |
| Sort | List: A list. A new list is constructed with the items from the input List sorted in ascending order. |
| RemoveItem | List: A list<br>Index: An integer that represents the position of the item to be removed. Index 0 refers to the first item in the list.<br><br>The resulting list is present on the output pin. |
| InsertItem | List: A list<br>Item: A data item of the same type as the list<br>Index: An integer representing the position where the item is to be inserted. Index 0 refers to the first item in the list.<br><br>The resulting list is present on the output pin. |
| GetIndex | List: A list<br>Item: A data item of the same type as the list. If the data item is present in the list, an integer representing its position is present on the output pin. As with the other functions, index 0 refers to the first item in the list. |

# An Example Using List Activities

The 9-Lists project in the `Chapter10` directory provides several examples showing how these activities can be used, including the `Switch` activity and a slightly enhanced version of the `ForLoop` activity from the previous section.

This diagram, shown in Figure 10-23, does text processing to transform input text into output text by replacing keywords with random words, similar to *Mad Libs*. In *Mad Libs*, a story is constructed by substituting new words into a story framework. A list of plural nouns, a list of adjectives, a list of professions, and a list of source phrases are all initialized and then manipulated by the diagram. The source phrases are replaced, if necessary, with random words from the other list and then concatenated into a final output, which is displayed in a dialog and spoken by the `TexttoSpeechTTS` activity.

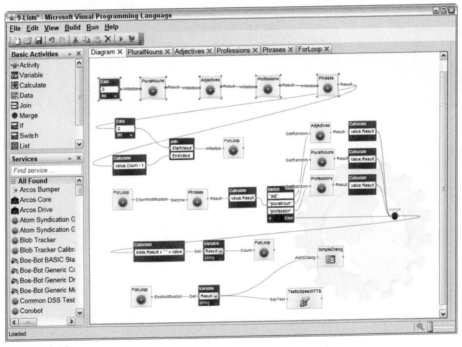

**Figure 10-23**

The various lists are initialized at the top of the diagram. The `Data` activity in the upper left passes a message to the `Initialize` action of the `PluralNouns` activity. When it has initialized itself, it passes a message from its output to the `Initialize` action of the `Adjectives` activity. The `Professions` activity is initialized next and then the `Phrases` activity. The `Phrases` activity outputs a message with an integer `Count` value that represents the number of phrases in the list. This value is used to initialize the `ForLoop` activity in the second row of the diagram, much like what you have seen in previous diagrams.

The `ForLoop` activity begins counting and passes a `CountNotification` message to the `GetOne` action input on the `Phrases` activity. The phrase identified by that index is passed to a `Calculate` activity where the value.Result string is placed on the input to the Switch activity. The Switch activity checks whether the input string is either "adj," "pluralnoun," or "profession." If it matches any of those cases, it replaces that word with a Random entry from the appropriate list by passing a message to the GetRandom action input of the associated activity. The random string is passed to the Merge activity. If none of these keywords is matched, the string is passed unmodified to the MERGE.

The updated phrase is passed to a Calculate activity that concatenates the current Result with a space and the value of the phrase. The result is stored in the Result variable and a message is passed back to the ForLoop activity telling it to generate another CountNotification.

After all of the phrases in the Phrases activity have been processed, the ForLoop sends an EndNotification message to the Variable activity at the bottom of the diagram. The value of the Result variable is retrieved and sent to an AlertDialog as well as the TexttoSpeechTTS activity.

The source text was taken from a paragraph in the MRDS documentation, which describes potential users of VPL:

> VPL is targeted for beginner programmers with a basic understanding of concepts like variables and logic. However, VPL is not limited to novices. The compositional nature of the programming language may appeal to more advanced programmers for rapid prototyping or code development. As a result, VPL may appeal to a wide audience of users including students, enthusiasts/hobbyists, as well as possibly web developers and professional programmers.

Some of the words in the text were replaced with one of the three keywords described above:

> VPL is targeted for **adj profession** with a basic understanding of concepts like **pluralnoun** and **pluralnoun**. However, VPL is not limited to **pluralnoun**. The **adj** nature of the programming language may appeal to **adj pluralnoun** for rapid prototyping or code development. As a result, VPL may appeal to a wide audience of **pluralnoun** including **profession**, **profession**, as well as possibly **profession** and **adj profession**.

Each time the diagram is executed, a new paragraph is generated with different word substitutions. Go ahead and run the diagram a few times to see and hear its output.

## The PluralNouns Activity

The PluralNouns, Adjectives, and Professions activities are essentially identical to each other except that they maintain different lists. The Initialize action of the PluralNouns activity is shown in Figure 10-24.

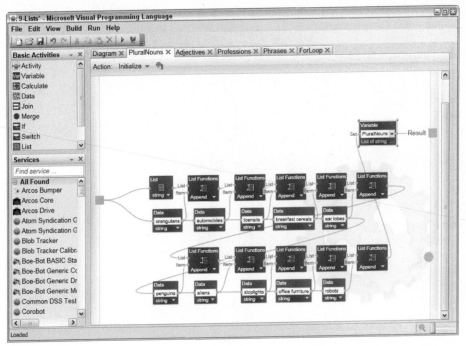

**Figure 10-24**

The `List` activity in the upper right builds a new list and passes it to the first `List Functions` activity when a message is received on the input pin. The first `Data` activity also generates a single string of data and passes it to the first `List Functions` activity, which then adds it to the end of the list. Subsequent `List Functions` and `Data` activities are then executed until 10 plural nouns have been added to the list. At that point, the list is stored to the `PluralNouns` variable associated with this activity and a response message is sent to the output pin. The `Initialize` actions are nearly identical for the `Adjectives` and `Professions` activities.

## The Phrases Activity

The `Phrases` activity passes the number of items in its list as an output to be used by the main diagram. Figure 10-25 shows the `Initialize` action of the `Phrases` activity. Notice how the source text has been broken up into discrete phrases so that the keywords are isolated and can be replaced before the output text is reconstructed.

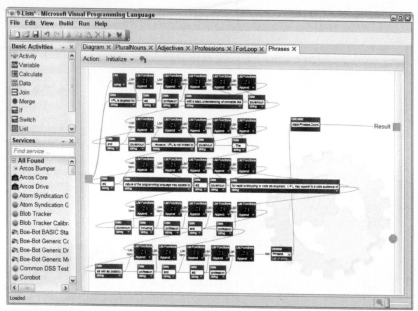

Figure 10-25

## The GetRandom Action

The `PluralNouns`, `Adjectives`, and `Professions` activities support a `GetRandom` action that chooses a random element from each of their respective lists and places it on the output pin. Figure 10-26 shows the `GetRandom` action from the `PluralNouns` activity.

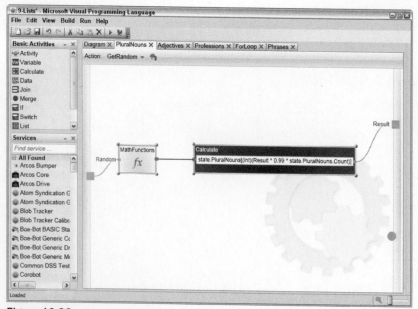

Figure 10-26

Here, the `MathFunctions` activity from the Services toolbox is used to generate a random floating-point number between 0.0 and 1.0, inclusive. The `MathFunctions` activity supports a number of math functions, including random, trigonometry, exponent and logarithm, round and truncate, power, square root, and conversion functions between radians and degrees.

The `Calculate` activity calculates an index into the `PluralNouns` list with the following expression:

```
(int)(Result * 0.99 * state.PluralNouns.Count)
```

The resulting index ranges from 0 to the index of the last element in the list, inclusive. The 0.99 factor is to slightly scale the random number so that if a 1.0 value is input, the index will not reference past the end of the list.

The syntax used to retrieve an element from the list is similar to array indexing in C#:

```
state.PluralNouns[(int)(Result * 0.99 * state.PluralNouns.Count)]
```

The string retrieved from the list is passed to the output pin.

The `GetOne` action on the `Phrases` activity is very similar except that the input integer value is used to index the list instead of a random number.

## A Final ForLoop Improvement

The `ForLoop` activity includes one small enhancement from the previous section. In the `Count` action, an `EndNotification` has been added. This notification is triggered once when the count has exceeded its maximum. The diagram for this action is shown in Figure 10-27.

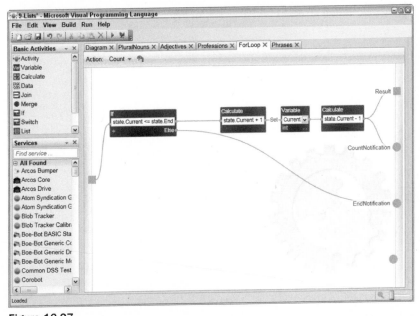

Figure 10-27

499

This diagram is guaranteed to produce endless hours of entertainment for you and your coworkers, and with only slight modifications it can be used to generate haiku poetry for fun and profit.

# Summary

This chapter has shown how to implement basic diagrams in VPL using the built-in basic activities along with a handful of services such as the `TexttoSpeechTTS` activity and the `SimpleDialog` activity. The VPL debugger was also demonstrated as a way to understand and debug diagram execution.

The next chapter demonstrates how to use VPL to control simulated and real-world robots.

# 11

# Visually Programming Robots

The Microsoft Visual Programming Language (VPL) can be used to implement general-purpose programs that have nothing to do with robotics, but because it is part of the Microsoft Robotics Developer Studio SDK, this chapter focuses on robotic applications.

First, the relationship between VPL activities and DSS services is explained, and you'll learn how to compile a diagram into a C# service implementation. Next you'll learn how to associate an activity with a specific service. Simple examples are provided that demonstrate how to control actuators and read data from sensors.

More complicated examples are presented in Chapter 12.

## VPL Activities and Services

A VPL activity can represent a DSS service. As explained in Chapter 3, a service implements a specific contract by supporting a number of operations on its main port. A service may also implement an alternate contract and its associated operations on a separate port. The VPL activity that represents the service has inputs for all of the operations on the main port and any alternate ports.

You know from the previous chapter that an activity may have more than one action associated with it. Each action has its own set of inputs and outputs. Each action associated with an activity corresponds to an operation supported by the service. For example, when you connect a wire to the input of the SimpleDialog service, three possible actions are supported: AlertDialog, PromptDialog, and ConfirmDialog. If you look at the implementation of the simple dialog in samples\Misc\Utility\Dialog\DialogTypes.cs, the DialogOperations class has operations defined for DsspDefaultLookup, DsspDefaultDrop, Alert, Prompt, and Confirm. The DisplayName attribute for the Alert class is "AlertDialog," and this Alert operation corresponds to the Alert action in the SimpleDialog activity. The inputs and outputs for this action are also defined by the `Alert` class. The input message for this operation is an `AlertRequest` class that contains a single `DataMember` string with the `DisplayName` attribute of "AlertText." The output or response message is a `DefaultSubmitResponseType` or a `Fault`.

When you define an activity in a VPL diagram, you are really defining a DSS port. Each action in the activity corresponds to an operation on that port. The inputs and outputs of each action correspond to the input and response messages of that operation. The logic implemented in the diagram for a particular action defines what happens in the handler for that particular input message. VPL is really just a graphical way of describing the DSS model of service ports and messages.

As you might imagine, the notification outputs of an action correspond to notifications from a service, and a connection from a notification output to an input pin on a service corresponds to that service subscribing to the notifications. Not only do activities within a diagram correspond to ports within services, the top-level diagram itself represents a single service.

Note that VPL doesn't enable any service functionality that is not already achievable using C# or some other .NET language and programming directly to the DSS model. However, nearly anyone who has written code to define services and pass messages between them will agree that VPL is a much simpler and more concise way of specifying the same functionality.

# Compiling Diagrams to Services

Now that you know that VPL diagrams represent the DSS model of services, ports, and messages, you might wonder if it is possible to convert services to diagrams and whether it is also possible to convert a diagram to a service. Of course, the answer is yes; otherwise, this would be a very short section.

In the case of converting services to diagrams, you have already seen how this works in Chapter 10. All of the activities in the Services toolbox represent services that are present on the current machine. Each time one of these services is dragged onto the diagram, an activity icon is displayed that represents the service. All of the information about the service is available on the input and output pins showing the various actions (operations) supported by the service port and the data passed in the input and response messages of each operation.

The more interesting case is converting a diagram to a service. When you press F5 or F10 to run a service, VPL has an interpreter, which executes the diagram. Another way to execute the diagram is to compile it into a service and then run that service.

You can compile a diagram into a service by clicking Build ⇨ Compile as a Service. The first time you select this option for a particular diagram, VPL prompts for the destination directory for the service.

## Service Compilation Options

A number of other properties affect how the service will be compiled. You can display these properties by clicking the Diagrams item in the Project toolbox. The properties are displayed in the Properties toolbox under the CompileSetting heading. The compile properties for the 8-CustomActivity (in order) VPL diagram are shown in Figure 11-1.

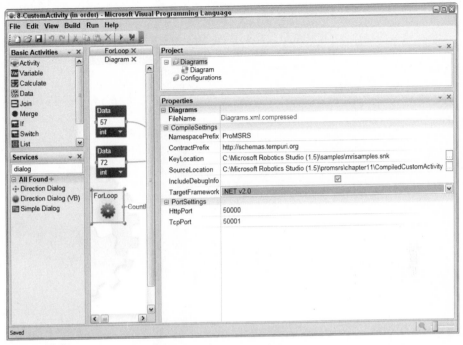

**Figure 11-1**

The following properties are supported:

❏ **NamespacePrefix:** This prefix is prepended to every namespace in the project. It enables you to use namespaces in the generated service that correspond to your particular organization or group.

❏ **ContractPrefix:** This prefix is used as the first part of the contract identifier for the service. For example, the contract identifier constructed for this service is derived from the ContractPrefix, the current date, and the name of the main diagram: http://schemas.tempuri.org/ 2007/12/customactivityorder/customactivityinorder.html.

❏ **KeyLocation:** This is the location of the key file used for signing the service code.

❏ **SourceLocation:** This is the directory where the source code for the service will reside.

❏ **IncludeDebugInfo:** This determines whether the service is compiled with debug information present.

❏ **TargetFramework:** This specifies whether the service targets .NET V2.0 or the .NET Compact Framework V2.0.

## A Service Compilation Example

Open the 8-CustomActivity (in order) project in the chapter10 directory and compile it by selecting the item on the Build menu. A compiled version of this diagram is in ProMRDS\chapter11\ CompiledCustomActivity. The source diagram is shown in Figure 11-2.

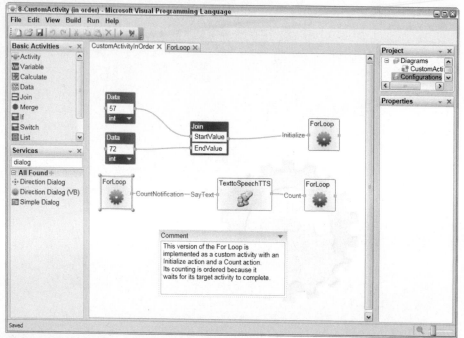

Figure 11-2

The files generated by the VPL compilation are as follows:

```
CustomActivityInOrderService.cs
CustomActivityInOrderTypes.cs
CustomActivityorder.csproj
CustomActivityorder.csproj.user
CustomActivityorder.manifest.xml
ForLoopService.cs
ForLoopTypes.cs
```

VPL has taken the .mvpl filename and converted it to a valid project name: CustomActivityorder
.csproj. The associated .user file specifies a reference path and debug actions that cause the manifest
to be executed when F5 is pressed. This project contains everything needed to build the target service.

VPL has also generated a manifest that can be used to run the target service. The source code for the
manifest shows all of the services that are executed:

```
<?xml version="1.0"?>
<Manifest xmlns:customactivityinorder=
"http://schemas.tempuri.org/2007/12/customactivityorder/customactivityinorder.html"
xmlns:texttospeech="http://schemas.microsoft.com/2006/05/texttospeech.html"
xmlns:this="urn:uuid:996d98f0-d63b-40a1-bb40-312c8cb2fac3"
xmlns:dssp="http://schemas.microsoft.com/xw/2004/10/dssp.html"
xmlns:forloop="http://schemas.tempuri.org/2007/12/customactivityorder/forloop.html"
xmlns="http://schemas.microsoft.com/xw/2004/10/manifest.html">
```

```
<CreateServiceList>
  <ServiceRecordType>
    <dssp:Contract>
 http://schemas.tempuri.org/2007/12/customactivityorder/customactivityinorder.html
    </dssp:Contract>
    <dssp:PartnerList>
      <dssp:Partner>
        <dssp:Contract>
          http://schemas.microsoft.com/2006/05/texttospeech.html
        </dssp:Contract>
        <dssp:PartnerList />
        <dssp:Name>customactivityinorder:TexttoSpeechTTS</dssp:Name>
        <dssp:ServiceName>this:TexttoSpeechTTS</dssp:ServiceName>
      </dssp:Partner>
      <dssp:Partner>
        <dssp:Contract>
          http://schemas.tempuri.org/2007/12/customactivityorder/forloop.html
        </dssp:Contract>
        <dssp:PartnerList />
        <dssp:Name>customactivityinorder:ForLoop</dssp:Name>
        <dssp:ServiceName>this:ForLoop</dssp:ServiceName>
      </dssp:Partner>
    </dssp:PartnerList>
    <Name>this:CustomActivityInOrder</Name>
  </ServiceRecordType>
  <ServiceRecordType>
    <dssp:Contract>
      http://schemas.microsoft.com/2006/05/texttospeech.html
    </dssp:Contract>
    <dssp:PartnerList />
    <Name>this:TexttoSpeechTTS</Name>
  </ServiceRecordType>
  <ServiceRecordType>
    <dssp:Contract>
      http://schemas.tempuri.org/2007/12/customactivityorder/forloop.html
    </dssp:Contract>
    <dssp:PartnerList />
    <Name>this:ForLoop</Name>
  </ServiceRecordType>
</CreateServiceList>
</Manifest>
```

Three services are started in this manifest that are identified by their contract identifiers:

```
http://schemas.tempuri.org/2007/12/customactivityorder/customactivityinorder.html
http://schemas.microsoft.com/2006/05/texttospeech.html
http://schemas.tempuri.org/2007/12/customactivityorder/forloop.html
```

The texttospeech service corresponds to the TexttoSpeechTTS activity in the top-level diagram. The forloop service corresponds to the ForLoop custom activity, and the customactivityinorder service corresponds to the top-level diagram. Both the texttospeech and forloop services are specified as partners to the customactivityinorder service.

Open the `CustomActivityorder.csproj` project. There are four source files. The `CustomActivityInOrderService.cs` and `CustomActivityInOrderTypes.cs` files implement the `customactivityinorder` service, while the `ForLoopService.cs` and `ForLoopTypes.cs` files implement the `forloop` service. Both of these services are contained in the target assembly called `CustomActivityorder.Y2007.M12.dll`.

It is worthwhile to take some time to look at the generated code to better understand how the VPL diagram maps to actual service code. The first few lines of code in `CustomActivityInOrderService.cs` look very familiar because they are nearly identical to many of the other services you have looked at in previous chapters:

```
namespace ProMRDS.Vpltest.CustomActivityInOrder
{
    [DisplayName("Vpltest")]
    [Description("A diagram that illustrates a custom activity.")]
    [dssa.Contract(Contract.Identifier)]
    public class CustomActivityInOrderService : dssm.DsspServiceBase
    {
        // Service state
        [dssa.InitialStatePartner(Optional = true)]
        private CustomActivityInOrderState _state;

        // Service operations port
        [dssa.ServicePort("/Vpltest", AllowMultipleInstances = false)]
        private CustomActivityInOrderOperations _mainPort =
            new CustomActivityInOrderOperations();
```

The state for the service is declared with the type `CustomActivityInOrderState`, which has the following definition. This class has no members because the top-level diagram has no state variables.

```
namespace ProMRDS.Vpltest.CustomActivityInOrder
{
    static class Contract
    {
        public const string Identifier =
            "http://schemas.tempuri.org/2007/12/vpltest/customactivityinorder.html";
    }

    [dssa.DataContract]
    public class CustomActivityInOrderState
    {
    }
```

The `CustomActivityInOrderOperations` class contains fairly standard operations, including `HttpGet`, `Get`, `Replace`, `Subscribe`, and a generic operation called `Action`. The handler for this operation does nothing but return an empty response message.

In the `Start` method, the state is initialized and the `DoStart` iterator is started to begin the diagram execution. In this iterator, subscriptions to service notifications are set up. Because there is only one case where a notification output is used in this diagram (`CountNotification`), only a single subscription is initialized.

Now the `RunHandler` is executed to take care of the initial message propagation in the diagram due to top-level data blocks. The relevant code is shown here:

```
public IEnumerator<ccr.ITask> RunHandler()
{
    Increment();

    JoinAlpha a = new JoinAlpha();
    a.EndValue = 72;
    a.StartValue = 57;
    ForLoop.InitializeRequest request = new ForLoop.InitializeRequest();
    request.EndValue = a.EndValue;
    request.StartValue = a.StartValue;

    Increment();
    Activate(
        ccr.Arbiter.Choice(
            ForLoopPort.Initialize(request),
            OnInitializeSuccess,
            delegate(soap.Fault fault)
            {
                base.FaultHandler(fault, @"ForLoopPort.Initialize(request)");
                Decrement();
            }
        )
    );

    Decrement();

    yield return WaitUntilComplete();
}
```

The `JoinAlpha` class represents the `Join` activity in the top-level diagram. Its two input values are initialized with the values from the `Data` activities and then a `ForLoop.InitializeRequest` object is created and initialized with the values and posted to the `ForLoopPort`. If the response from the `ForLoop` is success, nothing is done and the `RunHandler` waits until it receives a drop request.

This behavior fully implements the functionality specified on the top line of the main diagram that initializes the `ForLoop` activity and then waits for a notification from that activity with the first count value.

The second line of activities in the main diagram begins executing when the `ForLoop` service sends a `CountNotification`. This is handled in the `_joinAlphaHandler` method, shown here:

```
void _joinAlphaHandler(object[] args)
{
    JoinAlpha message = new JoinAlpha(args);

    Increment();
    Activate(
        ccr.Arbiter.Choice(
            TexttoSpeechTTSPort.SayText((texttospeech.SayTextRequest)message),
```

*(continued)*

*(continued)*

```
                OnSayTextSuccess,
                delegate(soap.Fault fault)
                {
                    base.FaultHandler(fault,
@"TexttoSpeechTTSPort.SayText((texttospeech.SayTextRequest)message)");
                    Decrement();
                }
            )
        );

        Decrement(args.Length);
    }

    void OnSayTextSuccess(dssp.DefaultUpdateResponseType response)
    {
        ForLoopPort.Count(new ForLoop.CountRequest());
        Decrement();
    }
```

The message received in the notification is posted to the `TexttoSpeechTTSPort` in a `SayText` message. On success, the `OnSayTextSuccess` handler executes and posts a new `CountRequest` message to the `ForLoopPort` to start the next count.

The `RunLoopService` code is similar. The implementation of the `CountAction` is in the `RunHandler` method of the `CountMessageHandler` class, and the implementation of the `Initialize Action` is in the `RunHandler` method of the `InitializeMessageHandler` class.

Occasionally, errors may be generated when a diagram is compiled that don't show up when it is executed with the interpreter. The interpreter tends to be more forgiving than the C# compiler. Check the errors reported by the compiler to determine what action needs to be taken to fix the service. Sometimes it requires defining the data members of input and output messages that weren't previously defined.

Diagram compilation is a relatively new feature in version 1.5 of MRDS, so it is possible that a bug in the code generator may prevent the diagram from compiling correctly. In this case, it may be possible to manually modify the generated code to fix the compilation errors. (Be sure to report errors in diagram compilation on the MRDS forum so that they can be fixed.)

Now that this diagram has been compiled, it is possible to use the custom activities in the diagram in other diagrams. The next time you run VPL, the `CustomActivityorderForLoop` service will appear in the Services toolbox and it can be used as a standalone activity in any other diagram. Although the CustomActivityorder service, which represents the top-level diagram, also appears in the list, it doesn't provide any useful inputs or outputs to make it useful in another diagram.

# Configuring Activities

Now that you know that a VPL activity can represent a service, all you need to do to make VPL work with the many robotics services available is to associate a particular activity with a particular service. In many cases, the association is implicit. For example, when you used the `TexttoSpeechTTS` activity in the previous chapter, VPL automatically started the TexttoSpeech service. In some cases, the service in the Services toolbox represents a generic contract containing the definition of a port but no implementation. In this case, you need to explicitly associate the activity with a service that implements that generic contract.

Another reason to configure a diagram is that sometimes a number of services need to run together and not all of them are specified in the diagram. In this case, you associate a manifest with the diagram, and all of the services in the manifest are started together.

## Setting the Configuration for a Diagram

Let's try an example to see how this works. Start up a fresh copy of VPL:

1.  Drag the `GenericContactSensors` activity onto the diagram and connect its `ContactSensorUpdate` notification to the `SayText` input of a `TexttoSpeechTTS` activity. In the DataConnections dialog, connect the Pressed Value to the `SpeechText` input as shown in Figure 11-3.

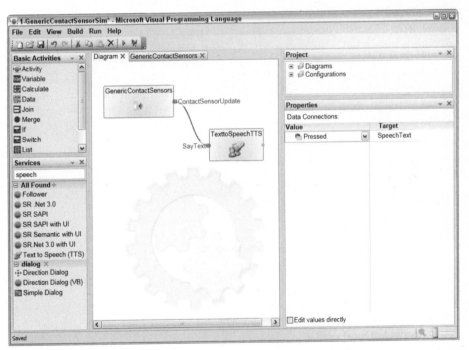

Figure 11-3

2.  Press F5 to run the diagram. VPL is unable to start the service associated with the `GenericContactSensors` activity because it refers to a generic contract that has not been implemented. The error shown in Figure 11-4 is displayed in the browser.

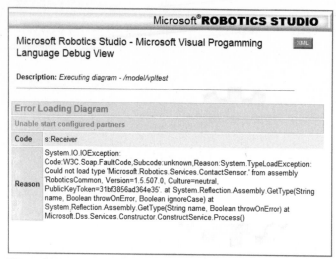

**Figure 11-4**

3. To make this work properly, you must tell VPL the service to run that implements the GenericContactSensor contract. The simplest way to do this is to associate an existing manifest with the diagram. You will use a simulation manifest to keep things simple for now. Right-click the GenericContactSensors activity and select Set Configuration. In the configuration screen, select Use a Manifest and then click the Import Manifest button. A list of all the manifests under \Microsoft Robotics Studio (1.5), which contains a service that implements the GenericContactSensors contract, is shown in Figure 11-5.

**Figure 11-5**

4. Select the `LEGO.NXT.Tribot.Simulation.manifest.xml` manifest and click OK. VPL now copies this manifest into the VPL project folder along with any associated configuration files. It associates the `GenericContactSensors` activity with the SimulatedGenericContactSensors service started in the manifest.

5. Now press F5 and the results are quite different, as shown in Figure 11-6.

Figure 11-6

This manifest starts a number of services, including SimulationEngine. The configuration file for the SimulationEngine service contains a definition for the scene shown in Figure 11-6. Now you just need to find a way to press the simulated LEGO touch sensor to test the diagram. The easiest way to do this is to drive the LEGO Tribot into one of the obstacles in the scene, but you don't have any way to drive it because the manifest didn't start the SimpleDashboard service.

## Starting the SimpleDashboard Service Upon Execution

You could easily start the SimpleDashboard service at this point by opening a browser window, typing `http://localhost:50000` in the address window, selecting ControlPanel and then starting the SimpleDashboard service. However, it is much nicer to configure the diagram so that the SimpleDashboard service is always launched when this diagram is executed. This is as simple as dragging the `SimpleDashboard` activity from the Services toolbox onto the diagram. This adds the SimpleDashboard service as a partner to the main diagram, which causes it to start when the diagram is executed.

You can load the completed diagram from 1-GenericContactSensorSim in the `chapter11` directory:

1.  Run the diagram by pressing F5.

2.  When the SimpleDashboard window appears, type **localhost** in the Machine: textbox and then double-click the (LEGONXTMotorBase) service in the Service Directory list and verify that the Motor is on in the Differential Drive group box.

3.  Press the Drive button and then drag the directional control to drive the LEGO NXT around the scene. The touch sensor is on the front of the Tribot, so try crashing it into one of the obstacles in the scene. You should hear the TexttoSpeechTTS service announce the state of the touch sensor each time it changes.

Congratulations! You have successfully used VPL to read a sensor from a simulated robot. Next you will see how easy it is to modify the diagram to use a real robot.

## Modifying the Diagram to Use a Real Robot

This example requires a LEGO NXT robot. If you don't have one, you may be able to use a different manifest that supports another robot. Even if you don't have the hardware, it is a good idea to follow along with the example to learn how you can switch a VPL diagram from working with a simulated robot to a real one.

To modify the diagram, follow these steps:

1.  Expand the Configurations heading in the Project toolbox.

2.  Delete the manifest that is listed there by right-clicking on it and selecting Delete.

3.  Right-click the `GenericContactSensor` activity and select Set Configuration.

4.  In the Set Configuration screen, select Use a Manifest and click the Import Manifest button.

5.  Select the `LEGO.NXT.Tribot.Manifest.xml` manifest. This is similar to the simulated Tribot manifest but intended for the real LEGO hardware. Actually, at this point, you can select any manifest that is listed that works with the hardware you have.

*You can load a diagram that has these changes from the 2-GenericContactSensorNXT project in the* `chapter11` *directory.*

Make sure that you have set up your hardware according to the instructions in the MRDS documentation under the "Setting Up Your Hardware" topic of the Robotics Tutorials section. Press F5 to run the diagram. This time, instead of seeing the simulation environment, you should be presented with a web page that enables you to configure your LEGO NXT connection. After you have done so and clicked Connect at the bottom, the web page should indicate that the LEGO is connected. Press the touch sensor on the Tribot and you should hear the TexttoSpeechTTS service say the state of the touch sensor.

Notice that all you needed to change was the manifest associated with the diagram. In many cases, it is possible to define an algorithm using a VPL diagram that can run equally well in simulation and on a real robot simply by changing the manifest associated with the diagram.

# VPL in the Driver's Seat

You may have noticed that there is a GenericDifferentialDrive service in the Services toolbox. This is the generic contract to control two motors together to implement a differential drive. This activity can be used to control the motion of a differential drive robot. Consider the diagram shown in Figure 11-7, which you can find in the 3-DriveInCirclesSim project in the `chapter11` directory.

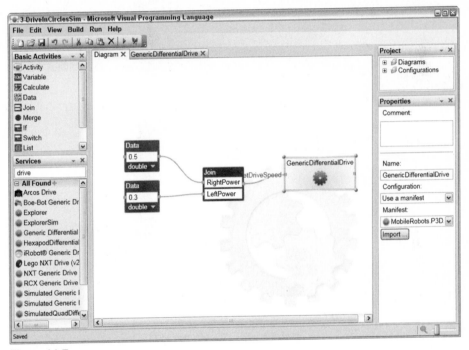

Figure 11-7

A differential drive allows a different power or speed to be specified for both the left and right drive motors. In this very simple example, the right wheel is given a drive speed of 0.5, and the left wheel drives at 0.3. Because the right wheel moves faster, the robot turns to the left at a constant rate and therefore travels in a circle, as shown in the overlay image in Figure 11-8.

**Figure 11-8**

You can also specify a nice spiral with a Tribot, as shown in the 4-Spiral project in Figure 11-9.

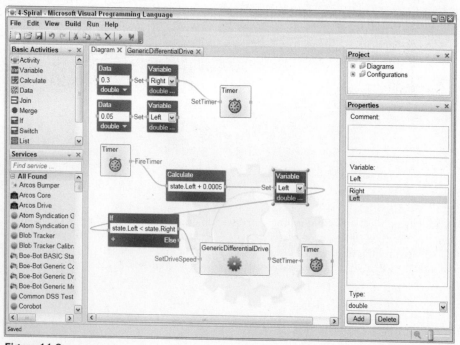

**Figure 11-9**

This is the first time the `Timer` activity has been used. A timer interval in milliseconds is passed to the `SetTimer` input, and after that length of time the `FireTimer` notification sends a message. The `Right` and `Left` variables start out at 0.3 and 0.05, respectively, which causes the Tribot to start in a tight turn. The radius of the turn is gradually increased as 0.0005 is added to the Left speed 10 times per second.

The manifest used for this example is the `Lego.NXT.Tribot.Simulation.Manifest.xml` manifest. Because a copy is made of the manifest and its associated configuration files when it is imported, the main camera `Location` and `LookAt` values were changed in the VPL project to change the initial viewpoint of the camera by editing the XML configuration file: `lego.nxt.tribot.simulationenginestate (LEGO.NXT.Tribot.Simulation.Manifest).xml`. The drive path is shown in Figure 11-10.

**Figure 11-10**

It is possible to control the movement of a robot more precisely by using the `DriveDistance` and `RotateDegrees` inputs of the differential drive, as illustrated in the 5-DriveDistance diagram shown in Figure 11-11.

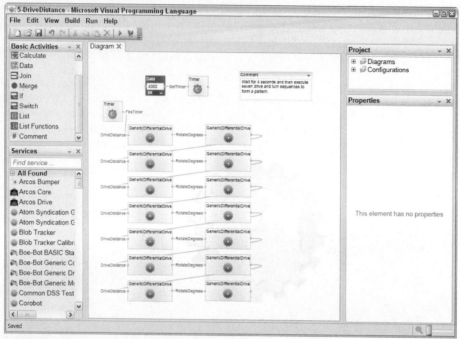

**Figure 11-11**

In this diagram, you use the `DriveDistance` and `RotateDegrees` inputs of the GenericDifferentialDrive to move forward a precise distance at a specified speed and then to turn a precise number of degrees at a specified speed. The Corobot manifest developed in Chapter 6 has been used for the configuration. The GenericDifferentialDrive does not send an output message until the requested motion has been completed, so the output of one motion can be fed directly to the input of another motion. The `DriveDistance` and `RotateDegrees` inputs can be used to drive a robot in a search pattern or to navigate to a specific location, as shown in Figure 11-12.

**Figure 11-12**

# Using VPL to Read Sensors

Any real robot control algorithm needs to be able to read values from multiple sensors and then control actuators such as drive motors or joint servos to move the robot. The next two examples show how this can be done together.

The diagram in the 6-BackAndForth project shows how to use a sensor reading to control the differential drive. You use the Corobot manifest again because it has an IR distance sensor mounted on its front and back. The IR distance sensor implements the `GenericAnalogSensor` contract so that is the activity used in the diagram.

The first thing the diagram does is rotate the robot by 180 degrees so that its front is facing the large box. The `Ready` variable is used to discard the sensor notifications until this move has been completed. After the `Ready` variable is set to `true`, the notifications from the distance sensor are used to set the motor speed. The `Normalized` distance value is used, which varies between 0 and 1. If the distance is larger than 0.9, then the robot is driven forward to get closer to the obstacle. If it is less than 0.2, then the motors are reversed to back away from the obstacle. The robot approaches the obstacle and then reverses before hitting it, backs away to a sufficient distance, and then approaches again. The diagram is shown in Figure 11-13.

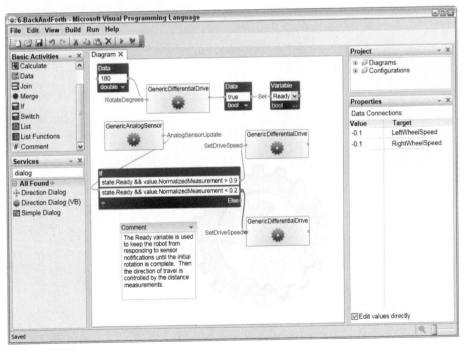

Figure 11-13

This diagram uses two generic contract activities, and both of them must be properly configured or VPL will report an error as it is initializing.

It can be difficult to debug diagrams that have a high frequency of notifications. The debugger tends to be unresponsive due to the notification processing that is going on. In this case, it is sometimes easier to debug a diagram by inserting `SimpleDialog` activities at strategic points to print the values on the wire or the values of state variables. It's just like inserting `printf`s in the old days.

# The Laser Range Finder

The Laser Range Finder sensor presents more of a challenge than the `GenericAnalogSensor` because it returns so much data. It casts 360 rays and reports the distance to intersection of each of them. It is up to the VPL diagram to interpret all of that data and determine the distance to an important object.

The 7-LaserRangeFinder project in the `chapter11` directory shows one way to do this. The top-level diagram is shown in Figure 11-14; the implementation of the `Evaluate` action in the `EvaluateRanges` activity is shown in Figure 11-15.

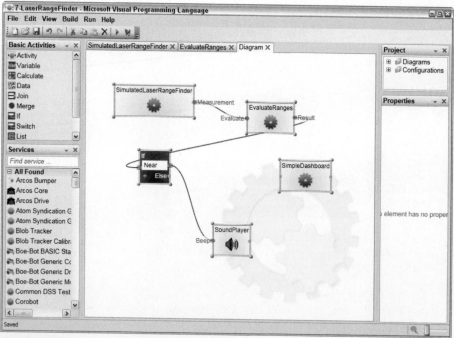

**Figure 11-14**

The `SimulatedRangeFinder` activity is configured to use the `MobileRobots.P3DX.Simulation.Manifest.xml` manifest. At periodic intervals, this activity sends a Measurement notification, which consists of 360 distance measurements in a list of integers along with some other data. In the main diagram shown in Figure 11-14, the data is passed to the `EvaluateRanges` activity along with a distance threshold. This activity then sends a response message containing a Boolean called Near, which is true if any of the distance measurements are less than the threshold. When the value is true, the SoundPlayer activity sounds a single beep. When you drive the robot so that an obstacle is nearer to the robot than the threshold distance, the diagram beeps. You can try this by running the diagram and maneuvering the robot close to one of the obstacles in the scene When you turn the robot so that the laser no longer intersects any obstacles, the beeping stops.

Let's take a closer look at how the laser range finder data is processed in the EvaluateRanges activity. The implementation of this activity is shown in Figure 11-15.

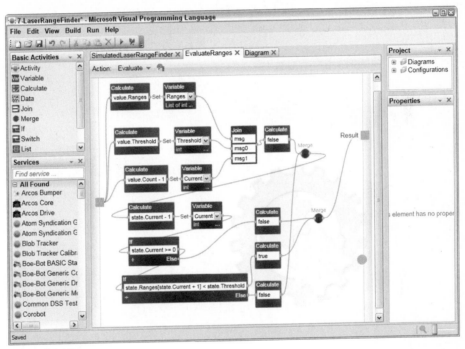

**Figure 11-15**

Three values are extracted from the input message and stored in state variables. The list of integers is stored in the Ranges variable, the Threshold value is stored in the Threshold variable, and the count of data values –1 is stored in the Current variable. This variable is used to count backwards through the list of data.

When all of the state variables have been initialized, the diagram enters a loop in which it steps down from the value of the Current variable to 0. If any of the ranges are less than the Threshold value, the loop takes an early exit and passes true to the output. If all of the ranges are traversed without a single one falling below the threshold, then a value of false is passed to the output.

Sometimes it is possible to implement a diagram without using any state variables at all. For example, the diagram in Figure 11-15 can be simplified by removing the Threshold, Current, and Ranges variables. Instead of being stored in variables, these values are passed around on the wires between the activities. Optimization of this diagram is left as an exercise for you.

# Controlling Multiple Robots

All of the examples you have seen in this chapter so far have dealt with only a single robot or a single sensor. It has been fairly easy to configure each activity because there has been only one service to choose from in each manifest that implements the port represented by the activity. It is fairly common to have multiple sensors on a single robot that each have an associated service of the same type. In addition, it is desirable to be able to control multiple robots at the same time with a single diagram.

When there are multiple services of the same type in a manifest, you must give each service in the manifest a unique name, and then you must assign each activity to the appropriate service in the manifest.

Consider the 8-TwoRobots project in the chapter11 directory. This is a very simple diagram consisting of two different GenericDifferentialDrive activities and the data to initialize the motors to unique values. The diagram is shown in Figure 11-16.

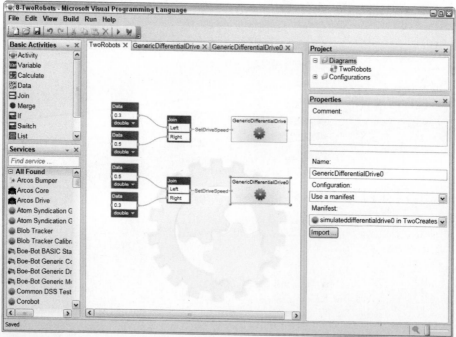

**Figure 11-16**

Notice that the two GenericDifferentialDrive activities have different names. This is because when the second one was dragged from the Services toolbox, VPL asked whether to create a new activity or make it refer to the existing GenericDifferentialDrive activity. A new activity was created and so it has a different name. Each of these GenericDifferentialDrive activities must be configured.

The manifest for this project was created by running the iRobot.Create.Simulation.Manifest.xml manifest and then using the Simulation Editor to duplicate the iRobot Create entity and give it a unique name. Refer to Chapter 5 for more examples of the Simulation Editor.

The manifest that runs the services assigns a different name to each of the SimulatedDifferentialDrive services, as is shown in the excerpt below. (The contracts should appear on a single line. They are shown on two lines here due to formatting constraints).

```
<ServiceRecordType>
  <dssp:Contract>
http://schemas.microsoft.com/robotics/simulation/services/2006/05/
simulateddifferentialdrive.html
  </dssp:Contract>
  <dssp:PartnerList>
    <dssp:Partner>
      <dssp:Service>http://localhost/IRobotCreateMotorBase</dssp:Service>
      <dssp:PartnerList />
      <dssp:Name>simulation:Entity</dssp:Name>
    </dssp:Partner>
  </dssp:PartnerList>
  <Name>this:simulateddifferentialdrive</Name>
</ServiceRecordType>
<ServiceRecordType>
  <dssp:Contract>
http://schemas.microsoft.com/robotics/simulation/services/2006/05/
simulateddifferentialdrive.html
  </dssp:Contract>
  <dssp:PartnerList>
    <dssp:Partner>
      <dssp:Service>http://localhost/IRobotCreateMotorBase0</dssp:Service>
      <dssp:PartnerList />
      <dssp:Name>simulation:Entity</dssp:Name>
    </dssp:Partner>
  </dssp:PartnerList>
  <Name>this:simulateddifferentialdrive0</Name>
</ServiceRecordType>
```

The first `ServiceRecord` runs the SimulatedDifferentialDrive service and associates it with the IRobotCreateMotorBase entity. This service is given the name this:simulateddifferentialdrive. The second `ServiceRecord` runs the SimulatedDifferentialDrive service and associates it with the IRobotCreateMotorBase0 entity. This service is given the name this:simulateddifferentialdrive0.

Now, back in VPL, right-click the `GenericDifferentialDrive` activity and select Set Configuration. Select Use a Manifest and select the `TwoCreates` manifest. In the Manifest drop-down menu in the Properties toolbox, two entries appear that correspond to the names of the differentialdrive services in the manifest. Associate the simulateddifferentialdrive service with this activity. Configure the `GenericDifferentialDrive0` activity in the same way, but select the simulateddifferentialdrive0 service with this activity.

Press F5 to run the diagram. Unfortunately, a bug in the 1.5 release of MRDS prevents the VPL model interpreter from running the diagram correctly; only one of the robots moves. Fortunately, you have another way to execute the diagram: compile it. A compiled version of the diagram has already been provided in the `chapter11\CompiledTwoRobots` directory. You can also compile the diagram to another directory of your choosing. Now click Run ⇨ Run Compiled Services and you should see both robots happily driving in circles, as shown in Figure 11-17.

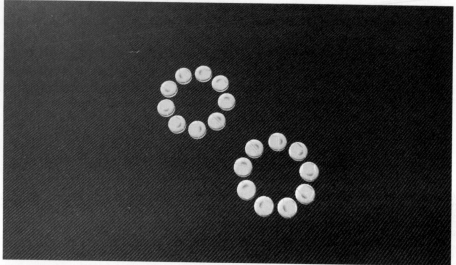

Figure 11-17

Use this same strategy anytime you must deal with two services of exactly the same type in a single manifest. Make sure that the services are uniquely named and then verify that each activity is paired with the proper service.

## Summary

In this chapter, you've seen several simple examples of VPL diagrams that read sensor data from simulated and actual robots. In addition, you've seen several different ways to control a differential drive. The last example showed how to control multiple robots or multiple sensors of the same type from a single diagram.

The next chapter provides several more examples of VPL diagrams.

# 12

# Visual Programming Examples

The previous two chapters explained the basics of the Microsoft Visual Programming Language (VPL) and showed how VPL can be used to read robot sensors and control robot actuators. Be sure that you are familiar with the material in Chapters 10 and 11 before reading this one. This chapter offers a few more examples that show how VPL can be used to solve a variety of problems. The examples use the Simulation Environment instead of actual robot hardware so that they are accessible to everyone. In most cases, they can be converted to run with actual hardware just by changing the manifest that is used.

Each example is independent of the others so feel free to dive in and devour a few or just take a bite from all of them.

## VPL Explorer

The ExplorerSim service presented in Chapter 9 showed an example of a simulation scenario in which a robot explores its environment and avoids obstacles using its sensors. This type of orchestration service can also be implemented using VPL.

The VPLExplorer project in the `chapter12` directory shows an example of how this can be done with a simulated Pioneer3DX robot. This robot is the same one used in the Chapter 9 ExplorerSim service. It is popular for this type of algorithm because of its built-in laser range finder (LRF), which provides accurate distance measurements to obstacles around the robot. The main challenge in this scenario is processing all of the data coming from the LRF at the rate it is produced.

This VPL example was first posted by Paul Roberts on the MSDN Channel 9 Forum in January, 2007. You can see his original post at http://channel9.msdn.com/ShowPost.aspx?PostID=270510.

This section describes several modifications that have been made to Paul's original diagram. When you run the diagram by pressing F5, the Pioneer3DX robot should be displayed in a Simulation Environment along with several obstacles. The robot should begin moving around on its own, turning as necessary to avoid the obstacles. If you have a wired Xbox controller, you can also use the controller to override the movement of the robot. The robot goes into Manual mode when the Right Shoulder button is held down. In Manual mode, the left thumbstick turns the robot, while the left trigger moves the robot backward and the right trigger moves it forward. In this mode, the controller vibration is used to indicate when the robot is close to an obstacle.

When the VPLExplorer diagram is loaded, it should look like the diagram shown in Figure 12-1.

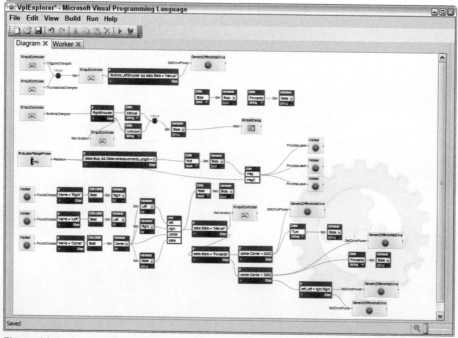

Figure 12-1

# Interfacing with the Xbox Controller

It is often easiest to understand a complicated diagram like this one when it is broken apart into pieces. The activities associated with the Xbox controller are shown in Figure 12-2.

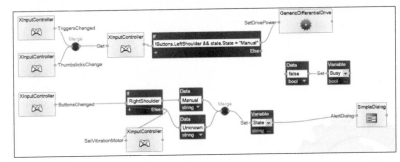

**Figure 12-2**

As you can see from the diagram, three different notifications are used from the `XInputController` activity: `TriggersChanged`, `ThumbsticksChanged`, and `ButtonsChanged`. When either the triggers or the thumbsticks change, a `Get` message is sent to the `XInputController` activity, which responds with the current state of the controller. If the LeftShoulder button is not pressed and the value of the `state.State` variable is Manual, then the positions of the triggers are used to set the `DrivePower` on the robot.

The state variable is set to Manual when the RightShoulder button is pressed. The second row of the diagram shows how this happens. When a `ButtonsChanged` notification is sent, the `State` variable is set to Manual if the RightShoulder button is pressed and to Unknown if it is not pressed. In addition, the vibration of the controller is set to 0 if it is not pressed. An `AlertDialog` displays the new state whenever it changes due to a RightShoulder button press.

## Reading the LRF Data

The activities used to process the LRF data are shown in Figure 12-3.

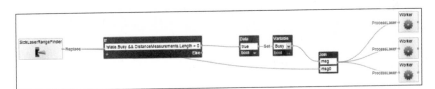

**Figure 12-3**

One interesting aspect of this diagram is the configuration of the `SickLaserRangeFinder` activity. This activity represents a service for the hardware LRF. Unfortunately, not everyone has a robot with a $10,000 LRF, so we have used simulation for this example. In the past, we have used a generic contract activity and configured it by attaching it to a real or simulated service in a manifest. Because there is no LRF generic contract defined in MRDS, we have configured this activity differently. Instead of selecting Use a Manifest on the configuration page, we specified Use Another Service and then the `SimulatedLaserRangeFinder` service was dragged onto the page. When VPL starts up all the services, it associates the `SickLaserRangeFinder` activity with the SimulatedLRF service, which was started in the `demoscene.manifest.xml` manifest associated with the diagram.

Each time the `LaserRangeFinder` activity generates a new set of distance measurements, it passes a message from its measurement notification. The simulated LRF sends measurement notifications four times per second. On some computers, it is difficult for the VPL diagram to process that much range data while also displaying the simulation engine window. In this case, the `Busy` variable is used to throttle the data back. If the `Busy` variable is true or if the array of measurements is empty, then the data is discarded; otherwise, it is passed along and the `Busy` variable is set to `true`.

At this point, the data is passed on to three separate `Worker` activity blocks. Each of these activity blocks represents the same `Worker` activity but the three messages are processed in parallel. Look carefully at the `Offset` and `Limit` values passed to each worker. The topmost activity is used to process the first 3/8 of the data, the middle activity processes the middle 2/8, and the bottom activity processes the remaining 3/8. Each of the activities is passed a `Name` string that identifies the range it is processing.

The purpose of the `Worker` activity is to scan through all of the range data and to return the smallest range that is not 0. The implementation of this activity is shown in Figure 12-4.

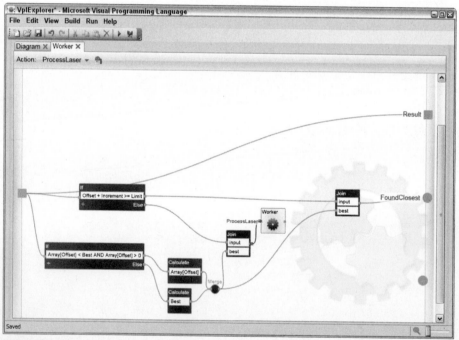

**Figure 12-4**

The LaserRangeFinder project in Chapter 11 showed how to iterate through the range data using a loop construct. The `Worker` activity uses recursion to step through the data.

If the input `Offset + Increment` value is larger than the input `Limit`, then the currently calculated nearest range is sent to the `FoundClosest` notification. This is the recursion limit. Otherwise, the current array value is compared to the previously nearest distance, and the `Offset` is incremented and passed back to the Worker `ProcessLaser` input. The `Worker` activity continues to pass messages to its own input until the `Limit` is reached and then it will send a notification. Recursion is often a cleaner way to iterate over data in VPL because it doesn't require any state variables. The state is maintained in the messages being passed.

Back in the main diagram, you see what happens to the `Worker` activity notifications in Figure 12-5.

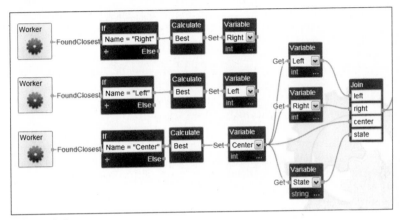

**Figure 12-5**

The notifications are fed into three different `If` activities. Each `If` activity determines which result was calculated by matching the name that was passed into the `Worker` activity. Keep in mind that all three of these calculations are occurring in parallel and there is no way to know which one will finish first. When one finishes, it is identified and its output is saved in an appropriate variable, either `Right`, `Left`, or `Center`. When the `Center` notification is received, a message is constructed from all three variables and the `State` variable and passed on to the part of the diagram that acts on the results, as shown in Figure 12-6.

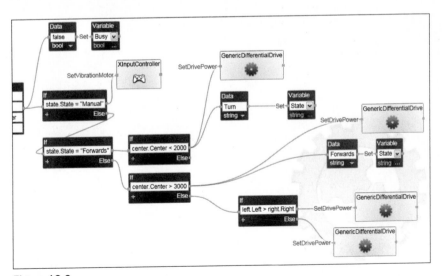

**Figure 12-6**

When the processing has completed, the Busy variable is set back to false so that another set of range data can be processed. The first If block checks whether the current state is Manual. If so, it sets the left vibration motor according to the center distance and the right vibration motor according to a combination of the right and left distances. This provides tactile feedback to the controller when the robot is running in Manual mode.

If the robot is not running Manual mode, then you start looking more closely at the distance data. If the robot state is Forwards and the closest center value is less then 2000, then the robot is approaching an obstacle head-on and is relatively close. The robot is stopped to prevent a collision with the obstacle and the State is set to Turn.

If the robot is in some state other than Manual or Forwards, then the diagram checks to see if there are no obstacles in the center closer than 3000. If that is the case, then the drive motor is set to move the robot forward and the state is set to Forwards. Finally, if the way ahead isn't clear, then the robot is turned to the left or right depending on which value is closest.

This diagram incorporates many elements that are used in other diagrams, so it is worthy of some study until you fully understand what is going on. Don't forget to click the connections between the activities so that you can see how the outputs from one activity are modified before being assigned to the inputs of another activity. Often, the calculations done between activities are relatively invisible but critical parts of the overall algorithm.

When you run the activity, you should see the robot move randomly through the environment, while, it is hoped, avoiding all obstacles, as shown in Figure 12-7.

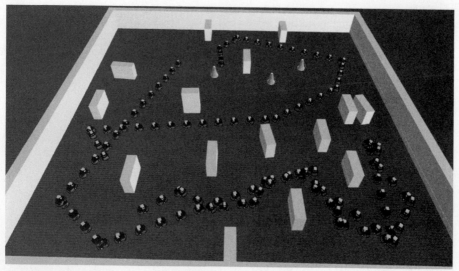

Figure 12-7

# VPL Sumo

Here is another example showing how you can take a complicated C# orchestration service and distill the robot behavior into a reasonably simple VPL diagram. In this case, parts of the sumo player algorithm implemented in the sample sumo player service shipped in the MRDS Sumo package have been implemented in a VPL diagram.

Like the VPLExplorer diagram in the previous section, the VPLSumo diagram shown here was written by Paul Roberts and posted to the MSDN Channel 9 forum. If you like this project, you should check out his posting to see the two other related diagrams he posted: http://channel9.msdn.com/ShowPost.aspx?PostID=341708.

Open the VPLSumoOne project in the `chapter12` directory. You may be surprised at the simplicity of the main diagram and even more surprised when you see the `Manouver` action on the `PlayerOne` activity. Most of the functionality in the diagram is in the `Start` action on the `PlayerOne` activity.

## The VPL Sumo Main Diagram

The main diagram is shown in Figure 12-8.

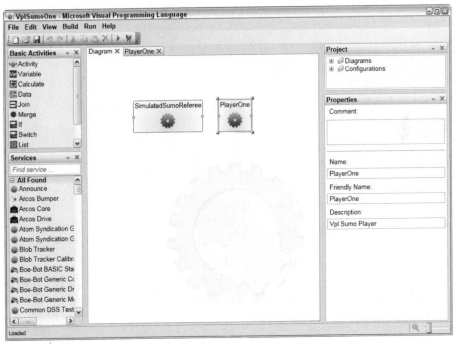

Figure 12-8

There are just two activities in the diagram and no connections at all. All of the sumo player stuff is in the `PlayerOne` activity. The `SimulatedSumoReferee` is there so that its service is started along with the sumo player service.

## The PlayerOne Start Action

To understand the `PlayerOne` sumo player activity, you'll have to look at the `Start` action. The first part of this action is shown in Figure 12-9.

**Figure 12-9**

The behavior of the sumo player is determined by its current mode, which is stored in the `SumoMode` variable. Here, it is initialized to the Initial mode. Then the state of the `SimPlayerOne` activity is retrieved with a `Get` Message. The `SimPlayerOne` activity represents the SimulatedIRobotLite service, which is the basic service that controls the simulated sumo robots. Its state is modified by setting a new name: "VPLSumo." It is then passed back to the Configure input on the `SimPlayerOne` activity to set the new state. Finally, a message is passed to the Connect input of the `SimPlayerOne` activity, which causes it to announce itself to the referee service. At that point, the referee creates a Player1 sumobot in the Simulation Environment and allows the SimPlayerOne service to connect to it. The process of initializing a sumo player service through the sumo referee is explained in more detail in Chapter 9. Once the SimPlayerOne service is connected to the simulated sumobot, it can read the sensors on the robot and control the simulated robot's drive.

The next part of the `Start` action is shown in Figure 12-10.

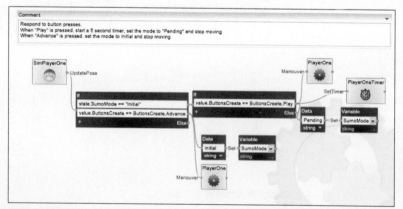

**Figure 12-10**

This part of the logic handles the button pushes on the top of the iRobot Create. When a sumobot is placed in the sumo ring, its Play button is pressed and it begins moving exactly five seconds later. When its Advance button is pressed, it becomes inactive.

In the simulated sumo ring, the referee service sends a message to the SimulatedIRobotLite service associated with the simulated player, telling it that one of its buttons has just been pushed. The SimulatedIRobotLite service responds by sending an `UpdatePose` notification to our service.

If the current `SumoMode` is Initial and the Play button has been pushed, then three things happen:

1. A timer is initialized with an interval of 5000 milliseconds.

2. The `SumoMode` is set to Pending, indicating that when the timer completes, play can begin.

3. The sumobot is stopped by sending a left and right drive value of 0 to the `Manouver` input of the `PlayerOne` activity.

If the `SumoMode` is not Initial, then the robot must already be executing its strategy. In this case, you check whether the Advance button was pressed. If so, the `SumoMode` is set back to Initial and the robot is stopped by setting its drive values to 0.

The logic that handles the timer completion in Pending mode is shown in Figure 12-11.

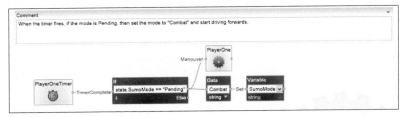

**Figure 12-11**

If the `SumoMode` is still Pending when the timer completes its five-second wait, then the `SumoMode` is set to Combat and the sumobot starts moving forward because the left and right drive values of the `Manouver` input are set to 250.

The logic that keeps the sumobot inside the sumo ring is shown in Figure 12-12.

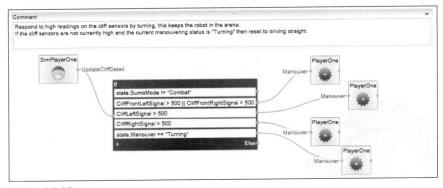

**Figure 12-12**

531

# Reading Data from the Infrared Sensors

The iRobot Create has four infrared floor sensors called *cliff sensors*. These sensors are identical to the cliff sensors used by the Roomba robot to avoid stairs as it vacuums. Two of the sensors are positioned near the front of the robot just left and right of center. The other two sensors are a little farther back on the left and right sides. When one of the floor sensors changes its value, the SimPlayerOne activity sends an UpdateCliffDetail notification. There are five possible cases to process.

1. If the current SumoMode is anything other than "Combat," the UpdateCliffDetail notification is ignored.

2. If one of the front sensors detects the edge of the ring, drastic action must be taken to keep the robot from traveling outside the ring because that is the direction in which it is headed. In this case, a speed value of 300 is applied to the left wheel and a value of -400 is applied to the right wheel to cause the robot to make a sharp right turn.

3. If neither of the front sensors is over the outer ring but the left cliff sensor is, then the robot is traveling parallel to the edge of the ring. In this case, a speed value of 500 is applied to the left wheel and a value of 100 is applied to the right, causing the robot to make a more gradual right turn.

4. Like case 3, if the right cliff sensor is over the outer ring, the robot makes a gradual left turn.

5. The last case occurs when none of the cliff sensors is over the outer ring. In this case, the robot moves straight ahead with a speed value of 250 applied to each wheel.

These five cases specify the motion of the robot as it moves around the sumo ring. The only case not covered is when the front bumpers are engaged, and that is covered by the part of the Start action shown in Figure 12-13.

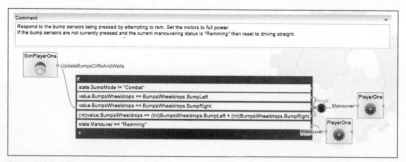

Figure 12-13

# Reading the Create Bumper Sensors

It would be nice if our little sumobot were a bit more aggressive when it comes in contact with its opponent, and that's what this logic does. The iRobot Create has a front-right bumper and a front-left bumper. This logic is fairly straightforward. When one of the bumpers state changes, an

UpdateBumpsCliffsAndWalls notification is sent. If the SumoMode is not Combat, then the notification is ignored. If either or both of the bumpers are pressed, then the sumobot pushes ahead with all its might with a wheel speed of 500 set on both wheels. Finally, if neither of the bumpers is pressed and the current Manouver is "Ramming," the robot calms down and goes back to its normal straight Wander mode.

## The Manouver Message

The only part of the diagram we haven't examined is what exactly happens when a Manouver message is sent to the PlayerOne activity. This action is shown in Figure 12-14.

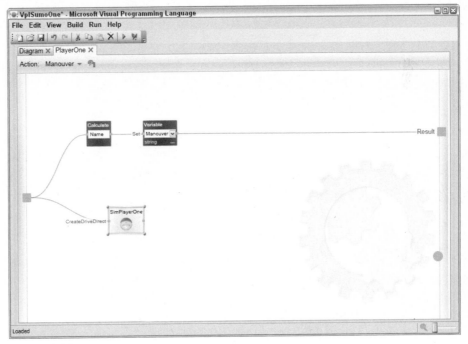

Figure 12-14

As you can see, the implementation of this action is very simple. Three values are passed in: the name of the maneuver, the left wheel speed value, and the right wheel speed value. The name is stored in the Manouver variable for later reference and the left and right values are applied to the robot drive.

This diagram gives our little VPLSumo robot the basic behaviors of a sumobot but it is usually at a disadvantage to the sample sumo player shipped by Microsoft because it does not use its webcam to locate its opponent. You'll examine how to do vision processing later in this chapter.

# Arm Mover

In Chapter 8, you developed a simulated robotic arm based on the Lynxmotion L6 arm, and in Chapter 15 you'll see how to write services for a real L6 arm. In this section, you'll learn how to write a VPL diagram that can control the movements of the arm.

The project is appropriately titled MoveArm and you can find it, as you are certainly aware by now, in the ProMRDS\chapter12 directory. The idea is to use an Xbox controller to control a piece of hardware (or simulated hardware) to make the arm interactive or to be able to teach it movements.

You need an actual Xbox 360 controller connected to a USB port on your computer to make this diagram work. This controller, shown in Figure 12-15, has a number of controls on it that make it useful for controlling robots.

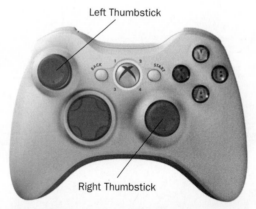

Figure 12-15

Both of the thumbstick controls provide X and Y offsets. In this application, you'll use the left thumbstick horizontal direction to control the movement of the arm in the X direction. Pulling the left thumbstick to the left decreases the X position of the gripper, while pulling it to the right increases the X position. The vertical direction of the left thumbstick controls the Z position of the arm in the same way. You can think of the 2D thumbstick control as moving the arm in the 2D X-Z plane. The vertical direction of the right thumbstick controls the vertical or Y position of the arm. Finally, the angle of the wrist is controlled by the yellow Y button and the green A button.

It is recommended that you use a wired Xbox 360 controller rather than a wireless one because it is easier to configure it and get it working properly on a PC.

The main diagram for this project is shown in Figure 12-16.

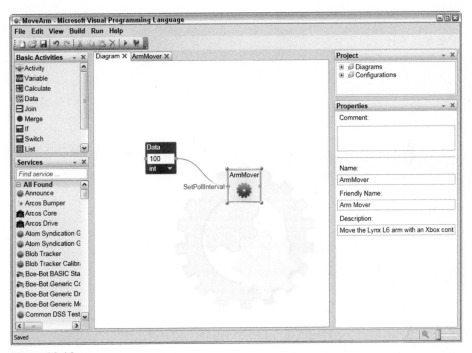

Figure 12-16

## The ArmMover Activity

Like the preceding project, it's not much to look at. It consists of a single custom activity called ArmMover, which is initialized with an integer data value of 100. Obviously, all of the logic is in the ArmMover activity.

The Start action of the ArmMover activity is shown in Figure 12-17.

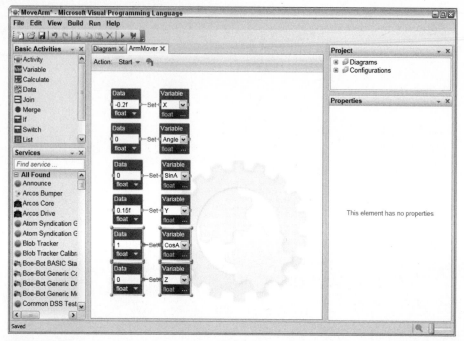

**Figure 12-17**

In this action, each of the variables that keeps track of the position of the arm is set to a default value. The diagram keeps track of the position of the arm gripper with the X, Y, and Z variables, which default to the position (-0.2, 0.15, 0). The Angle variable keeps track of the angle of the wrist and is initialized to 0. The SinA and CosA variables are used to store the sine and cosine of the wrist angle to make the conversion of the wrist pose to a quaternion easier; and they are initialized to the sine and cosine of the default wrist angle.

## The SetPollInterval Action

The diagram for the SetPollInterval action is much larger because it contains all of the logic to move the arm so it will be shown in sections. Figure 12-18 contains the polling part of the action.

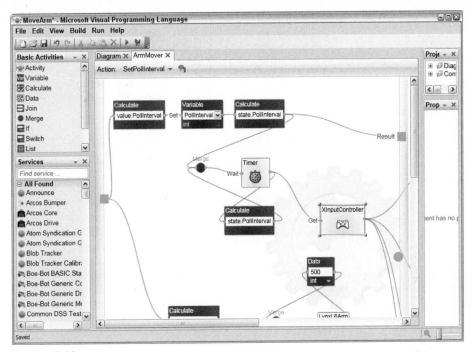

**Figure 12-18**

When an integer `PollInterval` value is set on the input, the `PollInterval` variable is initialized with it and a message is passed to the result pin to indicate that the `SetPollInterval` action message was successfully processed. The action never really completes, however, because it runs in a continuous loop, polling the Xbox controller.

The `PollInterval` value is passed to the `Timer` activity, which waits that number of milliseconds before sending a message to its output, which does two things:

❑ It sends a message back around to the timer input so that the timer is fired on a regular cycle determined by the value of `PollInterval`.

❑ It sends a `Get` message to the `XInputController` activity, which retrieves the current state of the controller. The action taken based on the movement of the left thumbstick is shown in Figure 12-19.

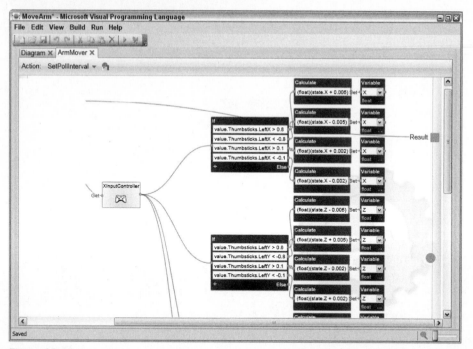

**Figure 12-19**

The thumbstick sends values between -1.0 and 1.0 in both the X and Y directions. The topmost `If` activity takes an action based on the horizontal position of the left thumbstick. If the thumbstick is pulled all the way to the right (> 0.8), then the X position of the arm gripper is increased by five millimeters. If it is pulled only partially to the right (> 0.1), then the X position is increased by only two millimeters. This enables the arm to be moved quickly but it also provides a finer degree of control for exact positioning. Similar actions are taken for negative values of the thumbstick.

The second `If` activity does the same thing but it uses the Y position of the thumbstick to set the Z position of the arm.

Figure 12-20 shows how the right thumbstick and buttons are handled.

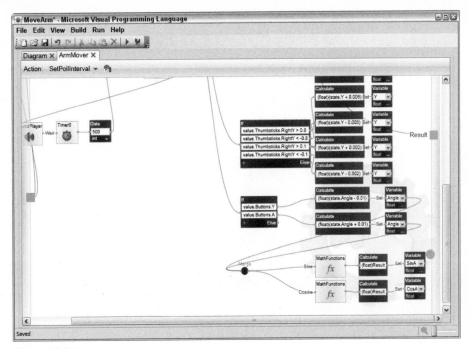

**Figure 12-20**

The Y position of the right thumbstick is used to change the Y position of the arm, much like the Y position of the left thumbstick is used to change the Z position.

The Y and A buttons are handled in the bottom If activity. They decrease and increase the wrist angle by 1/100 of a radian, respectively. After the Angle variable has been updated with the new angle, the MathFunctions activity calculates the sine and cosine of the new angle and these are stored in the appropriate variables for later use.

## Moving the Arm

The final section of the diagram to discuss is the part that actually moves the arm based on the values in the variables. This is shown in Figure 12-21.

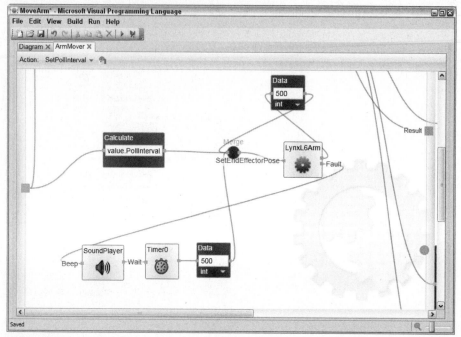

Figure 12-21

This part of the diagram consists of only seven activity blocks. Their purpose is to continually present the position of the arm grip to the LynxL6Arm service using the SetEndEffectorPose message so that as the position variables are updated, the arm moves.

The PollInterval value is extracted from the input message and this begins the execution of the loop. This message propagates to the Merge activity, which then sends a SetEndEffectorPose message to the LynxL6Arm activity with the inputs shown in the Properties Toolbox in Figure 12-22.

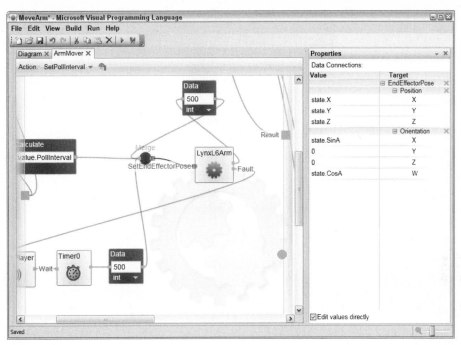

**Figure 12-22**

The X, Y, and Z variables feed the X, Y, and Z inputs of the EndEffectorPose message as you would expect. However, the angle of the wrist is specified in this message using a quaternion.

Don't worry; there is no need to wade through pages and pages of explanations about quaternions. Suffice it to say that the angle of the wrist can be converted to a quaternion by specifying the sine of the angle in the X position and the cosine of the angle in the W position. These values were calculated and stored in variables when the wrist angle was updated.

When the LynxL6Arm activity has completed moving the arm to the specified position, it sends a success message, which feeds back to a Data block that is used to place an integer value on the wire so that it matches the other inputs to the Merge activity. This message is eventually fed back into the LynxL6Arm activity to position the arm again.

If the arm is unable to move to the position specified due to mechanical constraints, the LynxL6Arm activity sends a Fault message instead. To indicate to the user that the arm is in an invalid position, the SoundPlayer activity is used to sound a system beep and the Timer0 activity is used to wait for one second to give the beep time to complete before a new EndEffectorPose is presented to the LynxL6Arm activity.

When you run the diagram, you might hear a couple of beeps because the diagram sends messages to the Arm service before the Arm entity has been initialized in the Simulation Environment. As soon as the simulator window appears, the arm should reset to its default position and then you should be able to move it using the controller. When you attempt to move the arm to an invalid position, you will hear periodic beeps until you manipulate the controls to bring the arm back to a valid position.

The `SetPollInterval` action of the `ArmMover` activity has two loops that execute simultaneously. An action may have any number of internal loops or instances of recursion running at the same time.

This activity controls a simulated arm but you've done nothing that is specific to the simulated arm so it is possible to run it with an actual hardware arm. If you have a Lynxmotion L6 arm, you can run the diagram in the `MoveArmHW` folder to control it in the same way.

Two parameters can be set on the arm that the diagram does not handle: the wrist rotation and the grip open/close position. While this functionality is left as an exercise for you (don't you just hate that?), here are some hints about how you might go about it.

You can set up another variable to store the wrist rotation with its own sine and cosine values, like the `Angle` variable. The quaternion passed to the `SetEndEffectorPose` input of the `LynxL6Arm` activity needs to include this angle information. The following code, which converts Euler angles to a quaternion, can be used for guidance:

```
double c1 = Math.Cos(heading / 2);
double c2 = Math.Cos(pitch / 2);
double c3 = Math.Cos(roll / 2);
double s1 = Math.Sin(heading / 2);
double s2 = Math.Sin(pitch / 2);
double s3 = Math.Sin(roll / 2);
quat.W = (float)(c1 * c2 * c3 - s1 * s2 * s3);
quat.X = (float)(s1 * s2 * c3 + c1 * c2 * s3);
quat.Y = (float)(s1 * c2 * c3 + c1 * s2 * s3);
quat.Z = (float)(c1 * s2 * c3 - s1 * c2 * s3);
```

The wrist angle is pitch and the wrist rotation angle is roll.

The gripper open and closed position can be set by using the `SetJointTargetPose` input of the `LynxL6Arm` and specifying the joint name **Grip** along with a joint position. Perhaps you could use the X and B keys on the controller to open and close the grip.

# Line Follower

One of the classic basic robotics projects is to build a line-following robot. Typically, the robot has two color or brightness sensors mounted on the left and right side of the robot, facing the ground. The signals from the sensors control the motor power so that the robot follows a line on the floor. When the robot strays from the line, the left or right side motor is driven faster to correct the course.

The MRDS 1.5 release includes a line-follower diagram for the iRobot Create but it is not very handy if you don't happen to have an iRobot Create, so this section presents a slightly different line-following algorithm that works with the Simulation Environment, along with a utility that can be used to generate line geometry on the floor of the Simulation Environment.

## The LineMesh Utility

Every visible object in the Simulation Environment is represented by a 3D mesh of triangles. 2D images can be painted on these 3D meshes using texture maps, but texture maps tend to get very fuzzy when viewed edge-on. Objects rendered using triangles stay much sharper, so the authors decided to build a utility that converts a 2D line drawing into a 3D mesh that can be inserted into the environment.

You can skip this section if you are only interested in the VPL implementation of the line-following algorithm.

## Using the DSS Command-Line Tools

The source code for this utility is in ProMRDS\Chapter12\LineMesh. This utility shows how to use the DSS command-line tools to build console applications with a consistent command-line parameter style. In the file LineMesh.cs, a class called CommandArguments is defined that contains all of the variables to be filled in by the command-line parameters. Each member variable has an attribute that contains information about how the parameter can appear on the command line, whether it is required, whether it may appear at most once or many times, and some help text for the parameter. The long strings in the following code are spread across several lines due to formatting constraints:

```
class CommandArguments
{
    internal CommandArguments()
    {
    }

    [Argument(ArgumentType.Required, ShortName = "i", LongName = "input",
        HelpText = "image file to convert to a .obj mesh.  The wildcard characters
* and ? are allowed.")]
    public string inFilename;

    [Argument(ArgumentType.AtMostOnce, ShortName = "o", LongName = "output",
        HelpText = "Optional output filename.  A .obj extension is not required.
If this parameter is not specified, the output file will default to the same name
as the input file but with a .obj extension.")]
    public string outFilename;

    [Argument(ArgumentType.AtMostOnce, ShortName = "x", LongName = "xOffset",
        HelpText = "Optional X offset for the mesh")]
    public string xOffset;

    [Argument(ArgumentType.AtMostOnce, ShortName = "z", LongName = "zOffset",
        HelpText = "Optional Z offset for the mesh")]
    public string zOffset;

    [Argument(ArgumentType.AtMostOnce, ShortName = "s", LongName = "scale",
        HelpText = "Optional scale multiplier for the mesh.  The default size is
10 cm per pixel.")]
    public string scale;

    [Argument(ArgumentType.AtMostOnce, ShortName = "h", LongName = "height",
        HelpText = "Optional height multiplier for the mesh.  The default height is
10 cm.")]
    public string height;
}
```

You can see from the preceding code that this utility supports a required input filename; all of the remaining parameters are optional. The output filename can be used to name the output mesh and material files. The X and Z offsets are used to specify where the mesh appears relative to the center of the entity with which it is associated. The `scale` parameter is used to define how large each pixel in the image appears in the Simulation Environment. Finally, the `height` parameter scales how tall the line geometry is. How tall, you ask? The pixels in the input image are converted to boxes in the Simulation Environment, rather than just flat polygons, to make it easier to position the lines above the ground without a gap between the top of the line and the ground.

The command-line parameters are processed in the `ParseCommandArgs` method after the default values are set in the call to `Parser.ParseArgumentsWithUsage`. This method assigns values to each of the `CommandArguments` member variables that are referenced on the command line:

```
private static CommandArguments ParseCommandArgs(string[] args)
{
    CommandArguments parsedArgs = new CommandArguments();
    parsedArgs.inFilename = null;
    parsedArgs.outFilename = null;
    parsedArgs.xOffset = "0";
    parsedArgs.zOffset = "0";
    parsedArgs.scale = "1";
    parsedArgs.height = "1";

    if (!Parser.ParseArgumentsWithUsage(args, parsedArgs))
    {
        return null;
    }

    return parsedArgs;
}
```

## Generating a Mesh from an Image File

The image file is converted in the `ConvertFile` method. The input file is read into a bitmap object. The utility supports .BMP, .GIF, .JPG, .EXIF, .PNG, and .TIFF file formats because that is what the .NET library bitmap object supports. It is better to use a lossless format such as .BMP or .TIFF because lossy formats like .JPG introduce pixels that have varying colors due to compression artifacts, and more unique colors means that more materials need to be generated, which can make the output files very large.

The code steps through each pixel in the bitmap. It checks whether it has seen that color before and if not, generates a material definition for that color pixel that goes in the .mtl output file. The cube geometry for that pixel is then generated and output to the .obj file, as shown in the following code:

```
for (int z = 0; z < input.Height; z++)
{
    for (int x = 0; x < input.Width; x++)
    {
        Color pixel = input.GetPixel(x, z);

        if ((pixel.R == 0) && (pixel.G == 0) && (pixel.B == 0))
            continue;

        string matName = GetMaterialName(pixel);
```

```
        if (!colors.ContainsKey(matName))
        {
            colors.Add(matName, pixel);
            mtlWriter.WriteLine("newmtl " + matName);
            mtlWriter.WriteLine(string.Format("Kd {0:0.00} {1:0.00} {2:0.00}",
                pixel.R / 255.0, pixel.G / 255.0, pixel.B / 255.0));
            mtlWriter.WriteLine("Ka 0.00 0.00 0.00");
        }

        // find the largest rectangular area covered by this color
        int xr = 1;
        int zr = 1;
        FindRectangle(x, z, input, out xr, out zr);
        ClearRectangle(x, z, xr, zr, input);

        objWriter.WriteLine(string.Format("v {0} {1} {2}",x * pixelSize + xOffset,
            yOffset, zr * pixelSize + zOffset));
        objWriter.WriteLine(string.Format("v {0} {1} {2}",xr * pixelSize + xOffset,
            yOffset, zr * pixelSize + zOffset));
        objWriter.WriteLine(string.Format("v {0} {1} {2}",x * pixelSize + xOffset,
            -yOffset, zr * pixelSize + zOffset));
        objWriter.WriteLine(string.Format("v {0} {1} {2}",xr * pixelSize + xOffset,
            -yOffset, zr * pixelSize + zOffset));
        objWriter.WriteLine(string.Format("v {0} {1} {2}",x * pixelSize + xOffset,
            -yOffset, z * pixelSize + zOffset));
        objWriter.WriteLine(string.Format("v {0} {1} {2}",xr * pixelSize + xOffset,
            -yOffset, z * pixelSize + zOffset));
        objWriter.WriteLine(string.Format("v {0} {1} {2}",x * pixelSize + xOffset,
            yOffset, z * pixelSize + zOffset));
        objWriter.WriteLine(string.Format("v {0} {1} {2}",xr * pixelSize + xOffset,
            yOffset, z * pixelSize + zOffset));

        objWriter.WriteLine("usemtl " + matName);
        objWriter.WriteLine(string.Format("f {0}/1/1 {1}/2/2 {2}/4/3 {3}/3/4",
            1 + vertexOffset, 2 + vertexOffset,
            4 + vertexOffset, 3 + vertexOffset));
        objWriter.WriteLine(string.Format("f {0}/3/5 {1}/4/6 {2}/6/7 {3}/5/8",
            3 + vertexOffset, 4 + vertexOffset,
            6 + vertexOffset, 5 + vertexOffset));
        objWriter.WriteLine(string.Format("f {0}/5/9 {1}/6/10 {2}/8/11 {3}/7/12",
            5 + vertexOffset, 6 + vertexOffset,
            8 + vertexOffset, 7 + vertexOffset));
        objWriter.WriteLine(string.Format("f {0}/7/13 {1}/8/14 {2}/10/15 {3}/9/16",
            7 + vertexOffset, 8 + vertexOffset,
            2 + vertexOffset, 1 + vertexOffset));
        objWriter.WriteLine(
            string.Format("f {0}/2/17 {1}/11/18 {2}/12/19 {3}/4/20",
            2 + vertexOffset, 8 + vertexOffset,
            6 + vertexOffset, 4 + vertexOffset));
        objWriter.WriteLine(
            string.Format("f {0}/13/21 {1}/1/22 {2}/3/23 {3}/14/24",
            7 + vertexOffset, 1 + vertexOffset,
            3 + vertexOffset, 5 + vertexOffset));
        vertexOffset += 8;
    }
}
```

Rather than generate a cube object for each pixel, the code tries to be smart in identifying rectangular regions that have the same color, and a cube is generated to cover the region. The pixels in the region are cleared to black so that they are not processed again later. No geometry is generated for black pixels.

Figure 12-23 shows a bitmap that you could use to generate lines in the Simulation Environment.

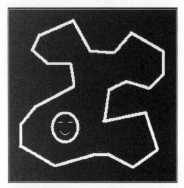

Figure 12-23

White is used for the line to be followed because it provides a high brightness contrast with a dark floor material. A typical command line to generate a mesh from such a bitmap would be as follows:

```
Linemesh /i:line.bmp /s:0.2
```

This command generates two output files called line.obj and line.mtl. The default size for each pixel is 10 cm cubed. The /s parameter scales this by 0.2 so that each pixel is only 2 cm in each dimension.

The resulting mesh is shown in the Simulation Environment in Figure 12-24.

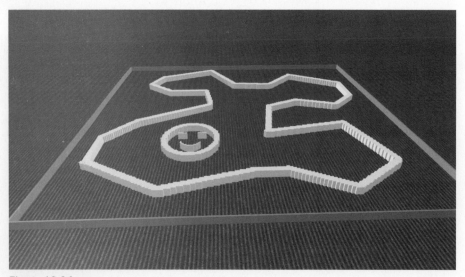

Figure 12-24

The mesh can be positioned vertically so that it barely protrudes from the ground by either adjusting the mesh transform for the entity or by moving the entity lower. The result is shown in Figure 12-25.

Figure 12-25

# Building a Brightness Sensor

The next big problem that needs to be solved is that no simulated brightness sensor is shipped with the MRDS SDK. That means you need to implement your own!

You can find the source code for this sensor in the ProMRDS\Chapter12\SimPhotoCell directory. It was very easy to implement this sensor by making some minor modifications to the SimulatedIR sensor presented in Chapter 6.

The service was changed to work with a CameraEntity, rather than a CorobotIREntity. The service periodically retrieves frames from the camera, converts the pixels in the image to luminance values, and then finds the average luminance for the entire image. When the service receives a Get message, it simply returns the most recently calculated brightness (luminance) value.

The code to retrieve the camera frame and calculate the brightness value is shown here:

```
private IEnumerator<ITask> CheckForStateChange(DateTime timeout)
{
    while (true)
    {
        if (_entity != null)
        {
            // get the image from the CameraEntity
```

(continued)

*(continued)*

```
                    PortSet<Bitmap, Exception> result = new PortSet<Bitmap, Exception>();
                    _entity.CaptureScene(
                        System.Drawing.Imaging.ImageFormat.MemoryBmp, result);
                    double brightness = 0;
                    yield return Arbiter.Choice(result,
                        delegate(Bitmap b)
                        {
                            int count = 0;
                            for (int y = 0; y < b.Height; y++)
                            {
                                for (int x = 0; x < b.Width; x++)
                                {
                                    Color c = b.GetPixel(x, y);
                                    brightness += (c.R * 0.3) +
                                            (c.G * 0.59) + (c.B * 0.11);
                                    count++;
                                }
                            }
                            // calculate the average brightness, scale it to (0-1) range
                            brightness = brightness / (count * 255.0);
                        },
                        delegate(Exception ex)
                        {
                        }
                    );
                    if (Math.Abs(brightness - _previousBrightness) > 0.05)
                    {
                        // send notification of state change
                        _state.RawMeasurement = brightness;
                        _state.NormalizedMeasurement =
                            _state.RawMeasurement / _state.RawMeasurementRange;
                        _state.TimeStamp = DateTime.Now;
                        _state.Pose = _entity.State.Pose;
                        base.SendNotification<Replace>(_submgrPort, _state);
                        _previousBrightness = (float)brightness;
                    }
                }
                yield return Arbiter.Receive(false, TimeoutPort(200), delegate { });
            }
        }
```

This is implemented with an iterator that runs in an endless loop. If the camera entity that this service is partnered with has been initialized in the Simulation Environment, then _entity will be non-null. In this case, CaptureScene is called on the entity to retrieve the latest rendered frame. A CameraEntity used with this service will typically have a very small number of pixels and a small field of view to simulate the limited field of view of a typical photo sensor.

When the image has been successfully retrieved, the code steps through the image calculating the brightness value for each pixel. When all of the brightness values have been summed, the average is calculated and scaled to lie within the range 0.0 to 1.0.

If the brightness has changed substantially since the previous brightness was calculated, then the service state is updated with the new brightness value and a notification is sent to other services. Finally, the handler waits for 200 milliseconds and then repeats the whole process.

Two different services are actually built from the same `SimPhotoCell.cs` file: one for the left sensor and one for the right sensor. Each service has a unique contract identifier even though it uses the same code. This is to work around a bug in the 1.5 version of VPL mentioned in the previous chapter whereby VPL does not work properly with two services of the same type specified in the same manifest.

## Building the Simulation Environment

This section goes into more detail about how the line follower Simulation Environment was built. If you are only interested in the VPL line-follower algorithm, you can skip this section.

The line follower Simulation Environment began life as the Simulation Tutorial 1 environment, shown in Figure 12-26.

Figure 12-26

Using the simulation editor, the sphere was deleted and the kinematic flag was set in the state of the box. This enables us to position the cube without regard to forces acting on it from the physics engine. The cube was moved below the ground plane so that it wouldn't interfere with our line-following robot. Then a custom mesh called `line.obj` was specified in its state, which caused our line mesh to be loaded and associated with the box. The transform on the mesh was modified to raise the mesh until it just showed above the ground. Finally, the texture map applied to the ground was changed to `Carpet_gray.jpg` to provide more contrast between the line and the ground. The combined physics and visual view of the resulting scene is shown in Figure 12-27.

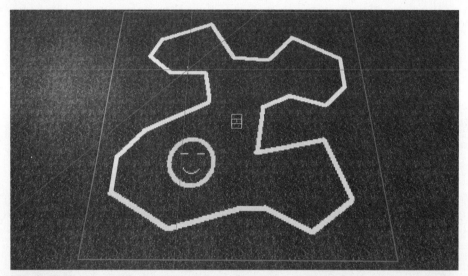

**Figure 12-27**

To insert a robot in the environment, you save the scene and then load the `iRobot.Create.Simulation.Manifest.xml` manifest into the simulator. Again, using the Simulation Editor, the iRobot Create entity is cut from the scene. The previous scene is reloaded and the iRobot Create pasted and placed in the proper position.

Now you have a robot but it doesn't have any photocell sensors on it. The SimPhotoCell service is designed to run with a `CameraEntity` so a `CameraEntity` named `LeftCam` is added to the environment as a child of the iRobot Create Entity. This camera is given a resolution of only four pixels by four pixels and its near plane is adjusted to be 0.001 meters, while the far plane is set to 50 meters. The field of view is set to only five degrees.

The camera is positioned 7 cm left of the robot's center line and 14 cm forward of the center. The camera is copied and then pasted as a child of the iRobot Create. The copy is given the name "RightCam" and positioned the same distance forward and 7 cm right of the robot's center line. Figure 12-28 shows the position of the cameras in the iRobot Create.

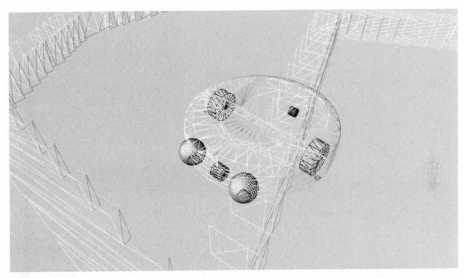

**Figure 12-28**

The cameras are rotated to point down at the ground and are set as real-time cameras. Finally, the ServiceContract field of each of the cameras is set to the contract identifier of the SimPhotoCell service that would be partnered with it. The entire scene is then saved. The simulation state file is called LineScene.xml and the manifest that runs all of the services is called LineScene.manifest.xml.

Now the simulation scene is ready to be controlled by a VPL diagram.

## The Line-Follower VPL Diagram

This diagram is much simpler than the previous few examples covered in this chapter. It can be found in the ProMRDS\Chapter12\FollowLine directory. It is shown in its entirety in Figure 12-29.

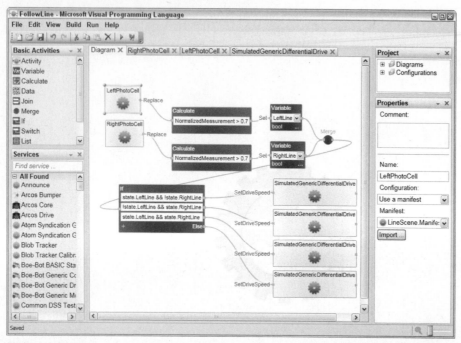

Figure 12-29

The `LeftPhotoCell` and `RightPhotoCell` activities represent the services partnered with the sensors on the simulated iRobot Create. When they send a notification indicating that the brightness value has changed, the `LeftLine` or `RightLine` state variables are set to `true` if the brightness value is greater than 0.7. This value was determined experimentally by driving the robot over the line and looking at the `SimPhotoCell` sensor state using a browser.

Each time a notification is received, the diagram must determine how to set the motor speed for the left and right wheels. The robot wants to have its right sensor over the line, with its left sensor not over the line. If this is the case, the diagram sends a `SetDriveSpeed` message to the generic differential drive, which causes it to move straight ahead. If both sensors are over the line, then you assume that the robot has strayed to the right too far and you move it back left by setting the left speed to 0.05 and the right speed to 0.2. If neither sensor is over the line, then you assume that the robot has strayed too far to the left and you move it back by setting the right speed to 0.05 and the left speed to 0.2. Finally, if the left sensor is over the line but the right sensor is not, the robot is too far to the right and you move it back to the left by setting the left wheel speed to 0.05 and the right wheel speed to 0.2.

As long as sensor notifications continue to arrive in a timely manner, this is all that is required to keep the robot traveling along the line. In some cases, the line might turn so abruptly that the robot overshoots it. In this case, the robot does what it normally does when neither sensor is over the line: it curves to the right (and usually re-acquires the line). This problem can be minimized by making the robot move more slowly or by defining a line that doesn't turn so abruptly.

This diagram does nothing to force the robot to traverse the line in any particular direction and it is possible for the robot to get confused and turn around and move in the other direction along the line.

This is a fairly basic line-following algorithm. There is a lot of opportunity to make it more sophisticated so that the robot can navigate abrupt changes in the line more consistently.

Figure 12-30 shows the robot making its way along the line from the perspective of a passenger in the Create's payload bay. Figure 12-31 shows a time-lapse image of the robot as it traverses the entire course.

Figure 12-30

Figure 12-31

# Ball Follower

Now you'll make that all-important transition from line-following to ball-following, and you'll use vision processing to do it. Robots of the future must be able to see and interpret the world around them. In this project, we'll show you how to design a VPL diagram that enables a robot to track and follow a ball. Vision-based autonomous navigation is left as an exercise for you.

Like the previous project, you'll implement a Simulation Environment conducive to ball-following, along with any needed services, and then implement a VPL diagram to orchestrate the behavior of the robot.

## *Building the Simulation Environment*

In this project, you use a service to initialize the Simulation Environment. You can find the source code in Chapter12\BallCourt. This service is fairly simple. The PopulateWorld method in the BallCourt.cs file does the usual work of creating sky and ground entities. In this case, the "ground" is a slab rather than an infinite plane but it has essentially the same function.

The BuildArena method builds a ball court with four walls as follows:

```
MaterialProperties bouncyMaterial =
    new MaterialProperties("Bouncy", 1.0f, 0.5f, 0.6f);
BoxShape[] boxShapes = new BoxShape[]
{
    new BoxShape(new BoxShapeProperties(0,
        new Pose(new Vector3(0, 0.5f*_state.Scale, -4*_state.Scale)),
        new Vector3(12*_state.Scale, 1*_state.Scale, 0.25f*_state.Scale))),
    new BoxShape(new BoxShapeProperties(0,
        new Pose(new Vector3(0, 0.5f*_state.Scale, 4*_state.Scale)),
        new Vector3(12*_state.Scale, 1*_state.Scale, 0.25f*_state.Scale))),
    new BoxShape(new BoxShapeProperties(0,
        new Pose(new Vector3(-(6f - 0.125f)*_state.Scale, 0.5f*_state.Scale, 0)),
        new Vector3(0.25f*_state.Scale, 1*_state.Scale, 8f*_state.Scale))),
    new BoxShape(new BoxShapeProperties(0,
        new Pose(new Vector3( (6f - 0.125f)*_state.Scale, 0.5f*_state.Scale, 0)),
        new Vector3(0.25f*_state.Scale, 1*_state.Scale, 8f*_state.Scale))),
};
boxShapes[0].State.Material = bouncyMaterial;
boxShapes[1].State.Material = bouncyMaterial;
boxShapes[2].State.Material = bouncyMaterial;
boxShapes[3].State.Material = bouncyMaterial;

MultiShapeEntity arena = new MultiShapeEntity(boxShapes, null);
arena.State.Name = "BallCourt";
arena.State.Assets.DefaultTexture = "BrickWall.dds";
SimulationEngine.GlobalInstancePort.Insert(arena);
```

Notice that a new material called Bouncy is defined with a restitution set to 1.0f. We also make the walls bouncy so that the ball continues to bounce from wall to wall within the court, enabling our robot to follow it. Each of the walls is defined with a size that can be scaled with the _state.Scale variable. A MultiShapeEntity is created to hold the wall shapes and then inserted into the Simulation Environment.

The `AddBall` method creates a ball entity that is both bouncy and slick:

```
void AddBall(DateTime now)
{
    MaterialProperties SlickMaterial =
        new MaterialProperties("Slick", 1.0f, 0.0f, 0.0f);
    SphereShape ballShape = new SphereShape(
        new SphereShapeProperties(0.1f, new Pose(), 0.1f * _state.Scale));
    ballShape.State.Material = SlickMaterial;
    ballShape.State.DiffuseColor = new Vector4(1, 0, 0, 1);
    SingleShapeEntity Ball = new SingleShapeEntity(
        ballShape,
        new Vector3());
    Ball.State.Name = "Ball";
    Ball.State.Velocity = new Vector3(_state.BallSpeed * _state.Scale, 0,
                                      -_state.BallSpeed * _state.Scale);
    Ball.State.Pose.Position = new Vector3(0, 0, 0.5f);
    SimulationEngine.GlobalInstancePort.Insert(Ball);
}
```

The slick material has 0 static and dynamic friction and 100% restitution. With this material, the ball will bounce from wall to wall and slide along the ground endlessly without losing any speed. The ball is given a `DiffuseColor` of red so that the vision algorithm can separate it from the rest of the scene. It is also given an initial velocity based on the `BallSpeed` variable in the state. This variable can be set in a configuration file.

After the `BallCourt` is added to the environment, a `Corobot` entity is also inserted. Five seconds later, the ball is added. The complete `BallCourt` environment is shown in Figure 12-32.

Figure 12-32

One last interesting thing to notice in the BallCourt service is the `SetBounceThreshold` method. This method waits for the physics engine to be initialized and then sets the `BounceThreshold` variable to −0.25. This variable is new in the 1.5 Refresh release. The physics engine bounce threshold governs the minimum speed at which objects bounce off of each other. If an object is traveling slower than the bounce threshold, then it will not bounce. The default value is 2.0 meters per second but we would like our ball to travel more slowly than that and still bounce off the walls, so it is set to −0.25 meters per second. The value is negative because it represents the speed of one object with respect to another.

## The BallFollower VPL Diagram

With the Simulation Environment in place, the VPL diagram used to analyze the video signal and control the robot is remarkably simple. It is shown in Figure 12-33.

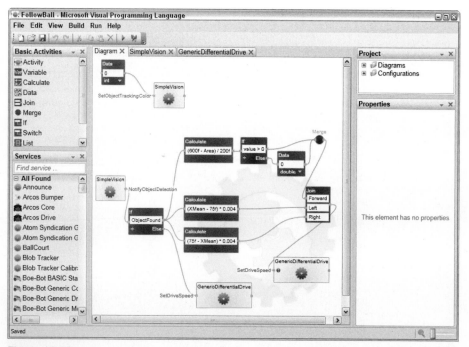

**Figure 12-33**

The `SimpleVision` activity is used to analyze the video from the camera on the Corobot. The `Data` activity in the upper left of the diagram sends an initialization message to the `SetObjectTrackingColor` input of the `SimpleVision` activity. It sets the target color to red with a similarity measure of 0.9. Only pixels that are closer to red than that threshold will be counted by the SimpleVision service.

This service sends periodic notifications to indicate whether it is currently tracking an object. If the `ObjectFound` output is false, then a message is sent to the simulated Differential Drive service to stop the robot from moving. If the robot doesn't see the ball, it waits where it is.

If the service indicates that the ball is visible, then the three Calculate activities determine values to drive the robot towards the ball but not too close. Figure 12-34 shows a graph of the equation used to calculate the Forward value.

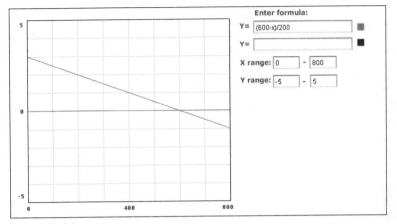

**Figure 12-34**

The horizontal axis represents the area of the recognized object and the vertical axis represents the forward speed of the robot. You can see that when the object is very far away (its area approaches 0), the robot moves forward quickly. As the object gets closer and its area approaches 600 pixels, the robot's forward velocity approaches 0. If the object is so close that its area is greater than 600 pixels, the If and Data activities clamp the Forward value at 0 so that the robot does not move backwards. The equations for the Left and Right values are shown in Figure 12-35.

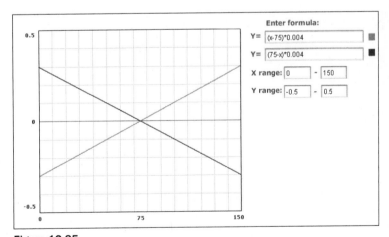

**Figure 12-35**

In this graph, the horizontal axis represents the X coordinate of the center of the object in the camera frame, and the vertical axis represents the forward motion of each wheel. The line with a negative slope represents the right wheel and the other line represents the left wheel.

When the object is in the exact horizontal center of the frame, both wheels have a speed of 0 and the robot doesn't turn. When the object is very near the left side of the frame, the left wheel turns backward and the right wheel turns forward, which turns the robot to the left, toward the object. Likewise, when the object is very near the right side of the frame, the wheels are driven so that the robot turns to the right to face the object.

When all three values have been calculated, a message is passed to the GenericDifferentialDrive activity with the results:

```
LeftWheelSpeed = value.Left + value.Forward
RightWheelSpeed = value.Right + value.Forward
```

The values are added together rather than multiplied so that the two different behaviors can act independently. If they were multiplied, the robot would not be able to turn to face the ball if the ball were larger in the frame than 600 pixels.

Run the diagram by pressing F5 and the ball court should appear along with the Corobot. After a few seconds, the ball should roll into the scene and the Corobot should notice it and accelerate toward it. As the ball rebounds off a wall, the robot will momentarily get too close to the ball and slow down or stop its forward motion. It should still turn to track the ball and it should again move forward when the ball is far enough away.

The rate of notifications coming from the SimpleVision service depends on the frame rate of the simulation engine. If you are running on a slower computer, the following behavior may not work correctly. You can slow down the ball to give the robot more time to react by changing the BallSpeed parameter in the ProMRDS\config\BallCourt.config.xml file to something smaller. You can also experiment with making the ball move faster by increasing this value.

Figure 12-36 shows the motion of the robot as it follows the ball around the court.

Figure 12-36

# Summary

The examples in this chapter have shown you a variety of ways to use VPL to write orchestration services. In each case, one or more C# services were used to interface with the hardware (or the simulation equivalent) or to perform complicated processing such as in the case of the Simple Vision service. VPL was used to quickly and simply join these basic services together to define how the whole system works.

There is nothing that can be done in VPL that can't also be done by writing a custom service but it is often easier and faster to write the behavior using a VPL diagram instead of a code editor.

That concludes the section of this book dealing with VPL. You should now have a good understanding of VPL's capabilities and how it can help you to quickly and easily orchestrate services and prototype new algorithms.

# Part IV

# Robotics Hardware

# 13

# Using MRDS with Robotics Hardware

Welcome to the hardware part of the book. So far in this book you have learned a lot about Microsoft Robotics Developer Studio (MRDS), mostly using the simulator, a key feature of which is that you can use the same code to control a simulated robot as a real robot. Fortunately, this means that you already know how to drive a real robot provided that the low-level communication is taken care of for you.

This chapter provides an overview of the hardware chapters as well as some key concepts in robotics; it does not introduce new code. Subsequent chapters explain how to use a variety of real robots with MRDS. Several different types of robots are covered, including wheeled robots and a robotic arm. If you have experience with robots, you might want to skim quickly through this material. Conversely, if you have not used robots before, then you should find the chapter useful for gaining a basic understanding of the field.

Some material from previous parts of the book is discussed again in this part of the book. This is intended to make the individual parts independent of each other, although repetition is also good when you are learning new material.

Because this book is designed to introduce programmers to Microsoft Robotics Developer Studio, you should view it as a supplement to the online documentation, not a replacement. In particular, you should regularly visit the MRDS home page and check the discussion forum: http://forums.microsoft.com/MSDN/default.aspx?ForumGroupID=383&SiteID=1.

It is not assumed that you know much about robotics. There are countless textbooks on the subject of robotics and even entire degree programs with titles like "Robotics Engineering" and "Mechatronics." If you want to learn more about robotics, you should spend some time looking at your favorite online bookstore or in your local bookstore. In addition, most robot manufacturers have active discussion forums, and you can learn a lot from the experiences of other people. You might also be able to find a robotics club in your local area.

Industrial robots are not covered in this book, and it does not teach you how to build a robot starting from spare parts. It is definitely not suitable for designing and programming robots that might be used for handling hazardous materials or that operate in medical environments, especially if the application is life-critical. You should consider the robots in this book to be educational toys or research tools.

# Safety Considerations Before Starting

Before you get started, be aware that working with real robots can be dangerous. There are many different safety issues, but the most important ones are as follows:

❑ **Your work environment:** This is the most important. Make sure that it is well lit and properly ventilated, and that there is plenty of clearance around the robot before you start operating it.

❑ **Keep away from pets and children:** As they say in the movies, you should never work with children or animals. Toddlers and pets might be fascinated by your robot and want to play with it. Robots often have sharp corners and protrusions and can do unexpected things if you interfere with them, especially if you rip their sensors off!

❑ **Keep a safe distance:** Robots sometimes move without warning. Even adults can trip over and fall down if a robot drives in front of them at the wrong moment. This could result in serious injury to both the human and the robot. Similarly, robotic arms can move quite fast and easily poke you in the eye if you are not careful. Therefore, make sure that you give your robots plenty of room to move.

❑ **Wear safety goggles:** If you are building a robot, you will have a lot of small components. Apart from the obvious risk to children from swallowing them, these small items might fly off suddenly if a tool slips. You should therefore wear eye protection whenever you assemble (or disassemble) a robot. Soldering presents a similar risk from molten solder flying through the air if you flick the tip of the soldering iron, as well as the chance of burns from the hot iron.

❑ **Read the manufacturer's instructions:** Always read the manufacturer's instructions for a robot and follow any safety warnings. Never try to push it beyond its limits or use it for purposes for which it was not intended.

That should be enough of the doom and gloom. If you act sensibly, you should have an enjoyable and educational experience. It might even be entertaining.

# Types of Robots

There are several different types of robots, but there are no universally recognized categories. Robots can be classified according to their purpose, abilities, and basic structure. Typically they are divided into categories such as mobile or stationary, autonomous or tethered, wheeled or legged, industrial versus domestic, and so on.

A common distinction is whether the robot is *mobile* or not. Robotic arms are usually stationary, although they can be attached to a mobile robot. Most hobby robots, however, have wheels and can run around and frighten your cat.

MRDS deals primarily with mobile robots — specifically, robots with two independently driven wheels, which are referred to as *Differential Drive robots*. Steering a differential drive robot is very simple because there are only two wheels. A vehicle with tracks or treads is much harder to steer because it relies on the tracks slipping in order to turn. A car (with four wheels) has what is called *Ackermann steering* and relies on turning only the front wheels, while the back wheels just follow along behind. A car, therefore, cannot turn on the spot, whereas a differential drive can.

One type of robot that is becoming popular is the *humanoid robot*. These robots are designed to look and operate similarly to a human. In other words, they have two legs (and usually arms and a head). However, they are a subset of the general category of *legged robots*, which include dogs (with 4 legs) or even spider-like robots (with 6 or 8 legs). One of the main areas of development for legged robots is *walking gaits*. As humans, we take it for granted that we can walk around on two feet, but teaching a robot to do the same and not fall over is quite difficult.

Although the Kondo KHR-1 Humanoid robot is one of the supported platforms in MRDS V1.5, it is only one of many different humanoid robots on the market. The walking abilities of these various robots vary dramatically. Many of them cannot get up when they fall over.

Articulated robotic arms use servo motors to move the joints and in some ways are similar to legged robots. No robotic arms are supported in MRDS V1.5, but Chapter 15 shows you how to use a Lynx 6 robotic arm using modified services from Lynxmotion.

> *Neither humanoid robots nor robots using Ackermann steering are covered in this book. The CoroBot simulation in Chapter 6 has four wheels but uses what is called* skid steering.

## Tethered Robots

Many robots have to be directly connected to a PC via a *tether* or *umbilical cable*. These robots are controlled by the PC and have little or no onboard intelligence. You do not download a program to the robot in this case because all of the control is performed on the PC. Robotics Tutorial 6 refers to these as *remotely connected robots*.

You might also hear the term *teleoperation* in relation to robots that use a tether. This means that a human remotely operates the robot. Using the Dashboard in MRDS or the Drive-by-Wire sample in Robotics Tutorial 4 are examples of teleoperation. In this case, the service on the PC does not perform any of the control logic — it is all left to the human operator.

Note that the tether does not have to be a physical connection between the PC and the robot. A wide variety of wireless solutions are available, such as Bluetooth, WiFi, and ZigBee. It could also be infrared, similar to a TV remote control or even a radio modem.

Note that most of the robots supported by MRDS are remotely controlled by a PC running MRDS services. When people first come across MRDS, they want to know how to use it to write programs to download into a robot. This is the way that the LEGO Mindstorms software works. For example, you can use the Mindstorms software to write code in a GUI environment that is similar to VPL and then download the code into the LEGO NXT brick for execution. The BASIC Stamp Editor is the equivalent software for the Boe-Bot, although you have to write programs in PBASIC.

At the risk of being repetitive, MRDS does not work this way. It cannot generate code for loading into an ARM processor, a BASIC Stamp, or a PIC microcontroller. Instead, you (or the vendor) need to write a simple *monitor program* to run inside the robot. (The term "monitor" reflects the fact that it is not a full operating system but instead a small run-time environment with very basic functionality.) Then you write a service under MRDS that communicates with this monitor program to send commands to the actuators and receive input from the sensors.

Figure 13-1 shows this arrangement diagrammatically. Notice that the monitor program has direct access to the sensors and actuators, but the MRDS service does not. It must send and receive information via the Comms Link, which can become a bottleneck. Moreover, if the Comms Link is unreliable, or the connection keeps breaking, then the robot might become uncontrollable.

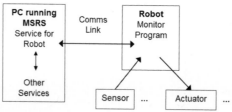

**Figure 13-1**

## The Monitor Program

In Robotics Tutorial 6, the monitor program is referred to as an Onboard Remote Communication Interface. Microsoft has written the monitor programs for you (if necessary) for the supported robots, or the robots themselves already have appropriate onboard firmware:

❑ **LEGO NXT:** When you first make a connection using MRDS, it automatically downloads and runs the necessary monitor program. Chapter 14 has examples for the NXT.

❑ **Boe-Bot:** You have to load the monitor program (supplied by Parallax) into the BASIC Stamp yourself. The Boe-Bot is used in the examples in Chapter 14 and also discussed below.

❑ **The Lynx 6 robotic arm:** This includes a small board called the SSC-32, which can control up to 32 servos. The firmware is already loaded into the onboard microcontroller when you get it. Chapter 15 covers the L6 arm.

❑ **Stinger:** The Stinger robots have a Serializer board that controls all of the I/O and accepts commands via a serial port. This robot is discussed in Chapter 16.

❑ **Hemisson:** These robots already have firmware on them that can be used to control the robot, as demonstrated in Chapter 17.

For the purposes of illustration, a simple monitor program is developed in Chapter 17 for the Integrator robot.

The monitor program must be written so that it is very responsive to requests from the MRDS service. If the monitor does a lot of "busy waits" as a way to introduce timing delays (in other words, it runs in tight loops doing nothing), then it might not be able to keep up with requests from MRDS.

## Boe-Bot Monitor

As an example, consider the Boe-Bot monitor program. The BASIC Stamp used in the Boe-Bot is single-threaded. To generate the appropriate signals for the servo motors that drive the wheels, it has to output pulses of the correct duration and repeat them at a rate that is close to 50 Hz. The protocol that has been defined for the Boe-Bot, therefore, requires the PC to continuously send a command until it is acknowledged by the robot because the robot is mostly busy sending pulses to the motors. (A full explanation of the Boe-Bot protocol is available in the documentation on the Parallax website.)

The following code is the main loop in the Boe-Bot monitor program, which is called BoeBotControlFormsrs.bs2 and is included in the software available from Parallax. The code is written in PBASIC, the native language of the BASIC Stamp, but it is quite easy to understand.

If you have installed the Parallax software, then you can double-click the filename in Windows Explorer and the BASIC Stamp Editor, shown in Figure 13-2, should start up, because there is a file association with the bs2 file type.

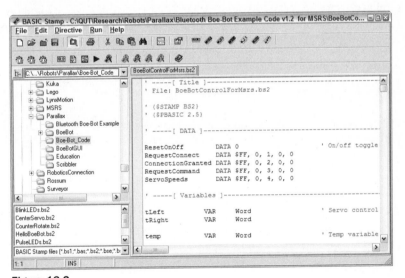

Figure 13-2

You must scroll down through the code to find the `Main Routine`, shown here:

```
' -----[ Main Routine ]-------------------------------------------------------

DO
  Resume:                              ' If Message not rcvd, try again
  IF IN5 = 0 THEN Reset                ' EB500 disconnected?

  PULSOUT 13, tLeft                    ' Servo control pulses
  PULSOUT 12, tRight

  SERIN 0, 84, 5, Resume,              ' Get next command
      [WAITSTR buffer \ 2, buffer2,
      buffer3, buffer4]

  PULSOUT 13, tLeft                    ' Servo control pulses again
  PULSOUT 12, tRight
```

The main loop checks whether the Bluetooth adapter has disconnected, and if so it jumps to `Reset`. Then it issues timed pulses on the two motor control pins 13 and 12 using the `PULSOUT` command. The variables `tLeft` and `tRight` are the durations of the pulses and can be changed by sending a `Set_Servo_Speed` command to the robot. In between pulsing, the code reads from the serial port with a timeout using the `SERIN` command. If this command times out, it jumps back to the label `Resume` at the top of the loop. However, if it succeeds, the program pulses the motors once more and then continues.

Note that the Boe-Bot uses continuous rotation servo motors. The pulse width durations range from 650 (1.3ms), which is full-speed clockwise, to 850 (1.7ms), which is full-speed counterclockwise. A value of 750 (1.5ms) results in the servo staying still. (If no pulses are received, the servo also stops.)

This requirement to continuously send pulses is a significant limitation of the BASIC Stamp. In contrast, many of the PIC microcontrollers on the market can generate PWM (pulse width modulation) signals automatically — all you have to do is load the relevant registers with the timing parameters.

The following section of code checks for a restart command from the PC. If this isn't found, then it determines which of the various command codes was received from the PC by using the `LOOKDOWN` statement, and then executes a computed `GOSUB` to call the corresponding subroutine. The names of the subroutines are self-explanatory:

```
IF buffer2 = 192 THEN                  ' Handle restart req from PC
  msgIndex = 0
  GOTO Request_Connect
ENDIF

LOOKDOWN buffer2,[32, 33,
                  64, 65,
                  96, 97, 98,
                  128, 129, 130,
                  160, 161, 162], routine

ON routine GOSUB  Set_Servo_Speed, Maneuver,
                  Get_Ir, Get_Whiskers,
                  Speaker_Tone, Set_Pins, Delay,
```

```
                      Enable_Digital_Sensors, Enable_Ir, Enable_Whiskers,
                      Disable_Digital_Sensors, Disable_Ir, Disable_Whiskers.
     ' Load digital sensor values into buffer byte 3 for return message.
     IF flagDigSens = 1 THEN GOSUB Digital_Sensors

     ' Increment message index for reply.  Next message from PC has to use
     ' reply's buf[1].
     msgIndex = msgIndex + 1
     buffer1 = msgIndex
     SEROUT 1, 84, [STR buffer \5]

   LOOP
```

The last part of the loop reads the digital sensors if they are enabled, and then it constructs a reply packet to be sent to the PC. This is an acknowledgment and includes a sequence number (msgIndex) so that the PC can verify that no packets have been dropped. Finally, it loops back to the top again, which immediately pulses the motors to keep them operating.

Ideally, the monitor program should use interrupts, but this is highly dependent on the hardware architecture. PIC microcontrollers and more sophisticated processors such as the ARM can handle interrupts. When an interrupt occurs as a result of an external signal, the processor saves the context (register values and processor state) and jumps to an interrupt handler. The handler does whatever is necessary and then restores the processor context so that the previously executing program can resume. This enables a primitive form of multi-tasking to be implemented. The main loop does not need to poll continuously for characters arriving on the serial port, for example, because the interrupt handler can assemble packets so that they only have to be processed once a complete packet is received.

There are two fundamentally different ways to handle sensor data: Either the monitor can periodically send updates or the MRDS service can poll the robot whenever it needs data. There are arguments both for and against each of these options, so it becomes the developer's decision.

In the preceding Boe-Bot code, you can see that the values of the digital sensors, which in this case are the whiskers and the IR sensors, can be sent back to the PC as part of every command acknowledgment. This way, the PC is usually current with the latest sensor information.

Clearly, it is not reasonable for the MRDS service running on the host computer to expect to obtain sensor values instantaneously. There is some transmission delay while a request is sent to the robot, and then the monitor has to read the sensor value(s) and send back the data. Depending on the type of connection and the speed of the CPU in the robot, this can take a few milliseconds or even tenths of a second. This means that there is an upper limit on the number of times per second that you can expect to get updates to sensor data. Stale data is a common problem in robotics.

## Boe-Bot Evasive Behavior

Another issue to be aware of is that controlling the movements of a robot precisely from a remotely connected PC is difficult due to the delays involved in transmitting commands and getting feedback on what the robot is doing. Therefore, the Boe-Bot has some pre-programmed behaviors that are designed specifically to get it out of trouble if it encounters an obstacle, e.g., a whisker is depressed.

The following code implements these evasive behaviors by using timed loops to move certain distances. The PAUSE command makes the program wait for a specified number of milliseconds. Notice that the delay is 20 ms, which means that the loops execute at 50 Hz. The first loop causes the Boe-Bot to move backward. The rest of the CASE statements handle the different turn directions:

```
' -----[ Subroutine - Maneuver ]-----------------------------------------------

' Preprogrammed maneuver example.  Current setup allows:
' "U" - Back up then U-turn
' "R" - Back up then right turn
' "L" - Back up then left turn

Maneuver:

  FOR temp = 0 TO 35
    PULSOUT 13, 650
    PULSOUT 12, 850
    PAUSE 20
  NEXT

  SELECT buffer3
    CASE "U"
      FOR temp = 0 TO 40
        PULSOUT 13, 650
        PULSOUT 12, 650
        PAUSE 20
      NEXT
    CASE "R"
      FOR temp = 0 TO 20
        PULSOUT 13, 850
        PULSOUT 12, 850
        PAUSE 20
      NEXT
    CASE "L"
      FOR temp = 0 TO 20
        PULSOUT 13, 650
        PULSOUT 12, 650
        PAUSE 20
      NEXT
  ENDSELECT

  RETURN
```

One of the side-effects of these tight loops is that the Boe-Bot does not read from the serial port while they are being executed. This means that it is effectively "deaf" for the duration of a behavior. This is obviously not desirable, but at the same time it does make it much easier to program simple wandering behaviors.

## The Communications Link

The Comms Link between the PC (MRDS service) and the robot (monitor program) is an important factor in the overall performance of the system. This can be a serial connection in the case of a tethered robot with an umbilical cable, or a wireless connection. Common types of wireless (radio) connections include IEEE 802.11b or 802.11g (also called WiFi), Bluetooth, or ZigBee. Even with a wireless connection, the usual approach is to simulate a serial (COM port) connection.

The LEGO NXT brick has built-in Bluetooth. For the Boe-Bot it is available as an optional extra circuit board that plugs into the robot. The Scribbler robot has a serial port into which you can plug a Serial-to-Bluetooth device (called a *dongle*). See Chapters 16 and 17 for examples using a Sena dongle.

In fact, any robot with a serial port can quickly and easily be converted to wireless operation. The software on the robot doesn't even need to know that Bluetooth is involved because the Bluetooth device establishes a virtual serial port connection.

Most PCs these days, especially laptops and PDAs, have built-in Bluetooth. Even if your PC does not have Bluetooth, there are many USB-to-Bluetooth dongles on the market and all PCs now have USB ports. The range of Bluetooth is quite good, and certainly adequate for experimenting with robots. For example, a LEGO NXT Tribot can be controlled up to 30 meters away using a laptop.

Setting up a Bluetooth connection, however, can be full of trials and tribulations. With the large number of different manufacturers making Bluetooth hardware, incompatibilities are possible. Look for advice on the Internet about which brands of Bluetooth devices are known to work with your robot before buying a device and wasting a lot of time trying to get it to work. Unfortunately, in some cases you just have to "suck it and see."

> The quality of Bluetooth devices on the market varies considerably. Many people have had problems getting particular brands of Bluetooth devices to talk to each other. This shouldn't happen if the devices conform to the Bluetooth specification, but it does. Several different brands of Bluetooth devices are used in this book and are therefore known to work.

If you purchase a Surveyor SRV-1, one of your options is ZigBee. ZigBee radio is a fairly new technology. It is similar to Bluetooth in a lot of ways, but not compatible with Bluetooth. Usually you install the drivers on your PC, plug in a USB-to-ZigBee device, and you are ready to start talking to the robot. You might need to select a channel, but for this you must read the instructions that came with the robot.

ZigBee is designed to support more nodes than Bluetooth (a maximum of eight nodes), and ZigBee can even be used in mesh networks. The ZigBee standard is set by a consortium, but it is basically the same as IEEE 802.15.4. It can operate in three different frequency bands, but it commonly uses the 2.4 GHz band, like Bluetooth, and has a similar range. Because it offers fewer functions (called *profiles* in Bluetooth), the ZigBee stack is much simpler than Bluetooth. The intended purpose of ZigBee is for low-data-rate, low-power applications such as devices that are always on. Bluetooth has a slightly better data rate, but it is still slow compared to WiFi. The main point in favor of Bluetooth over ZigBee is that it is available in so many devices these days, such as laptops and PDAs.

To set up a WiFi link, you must either use a so-called ad-hoc or peer-to-peer connection with fixed IP addresses or you need a Wireless Access Point to operate in that is called *infrastructure mode*. A Wireless Access Point (WAP) can issue addresses via DHCP. However, if the robot gets different addresses each time it is turned on, this makes communicating with it more complicated. It is far preferable to hard-code the IP address on the robot.

If you are trying to operate in a corporate environment, you might run into quite restrictive policies regarding connecting to the building's wireless network and you might not be allowed to hard-code an address. Access might be limited to known MAC addresses, or you might be required to install special software that establishes a VPN (virtual private network) in order to use encryption to ensure security. You might find it easier to buy your own wireless access point, although the "network police" might find you because you might interfere with the corporate network. Of course, if you are working at home, then you will need to set up your own wireless network anyway.

If you have a robot with an onboard PC, the setup is similar to your desktop or laptop. However, for the embedded WiFi cards used on some robots, you have to follow the manufacturer's instructions and it simply might not be possible to configure the robot to work with a corporate network. A discussion about how to set up wireless networks, and advice on the wide range of scenarios, is outside the scope of this book.

## Autonomous Robots

The most important distinction for this book is whether a robot is *autonomous* or not. An autonomous robot has an onboard computer capable of running Microsoft Robotics Developer Studio. The computer might be running Windows CE, Embedded XP, or even Vista. Typical hardware ranges from SBCs (single-board computers) to PDAs (personal digital assistants) and laptops.

SBCs come in a variety of formats. One of the earliest was PC104, an industry standard specification for a small embedded PC. However, manufacturers such as VIA Technologies have developed their own formats for small computers, including nano-ITX and pico-ITX. The pico-ITX boards are less than 10 cm square (about 4 inches) and have low power requirements, but they are still full x86-based PCs.

Programming under Windows CE requires additional skills beyond simply programming under Windows XP or Vista. The CE environment only supports a subset of .NET called the Compact Framework (CF). In addition, you have to transfer your programs from your development machine to the target PC on the robot. See Chapter 16 for more information.

Even though the PC is located directly on the robot, it still needs an interface board to control the actuators and read from the sensors. This controller board might contain a PIC microcontroller and have its own communications protocol. Therefore, a monitor program might still be required.

> *The Sumo robots used in a competition sponsored by Microsoft at the MEDC conference in 2007 were autonomous robots. They were based on the iRobot Create with an eBox-2300 PC mounted on top and a Logitech webcam. The onboard PC talks directly to the robot using a serial port.*

The ICOP eBox is available in a Jump Start Kit that runs Windows CE. See www.icoptech.com/ products_detail.asp?ProductID=272. This same PC is used on the Stinger CE robot discussed in Chapter 16.

It is possible to monitor the operations of an autonomous robot remotely, or even to give it *supervisory* instructions. In this case, the onboard computer directly controls the robot, but higher-level strategies are implemented in a separate computer. Communication is usually via WiFi, which has become quite popular due to its low cost and good range. A WiFi card can be purchased as an optional extra for the eBox-2300.

*In order to run MRDS, the onboard PC on the robot must have .NET V2.0 installed. You don't run Visual Studio on this PC (unless, of course, it is a laptop, in which case you can run whatever you like). Instead, you develop the services on a desktop PC and download them to the robot's PC.*

In addition, it is not necessary to install the full version of MRDS on the robot because a utility called `DssDeploy` can be used to package up a set of services. (`DssDeploy` is discussed in Chapter 3 for desktop PCs, and again in Chapter 16 in relation to installing a package on a PDA or embedded PC.) It automatically determines the dependencies and includes the necessary components to make a run-time version of MRDS. This saves memory, which is often at a premium on an embedded PC or PDA.

The procedure you use to deploy your services to the robot varies according to the type of PC that you are using on the robot and the communications hardware that it has installed. Common approaches include using USB flash drives, connecting via WiFi, and even downloading via a serial port, although this should be avoided because it is extremely slow.

# Buying Your First Robot

The type of robot you choose to work with depends on several factors, including the intended purpose and your level of expertise with hardware. (It is assumed that you are competent with software!) As previously stated, this is an introductory book and it does not cover industrial robots, but the principles that apply to educational robots also apply to industrial robots. For example, KUKA is a major manufacturer of industrial arms, and they have developed MRDS services and a set of tutorials, available from `www.kuka.com.en.products/software/educational_framework`.

Several robots have been selected for use in the examples in this book and are outlined briefly below. They cover a range of different types and prices, although the objective has been to keep them within a price range that a student or hobbyist can afford. If you are a teacher, then it is important to keep the price down because you need multiple robots in the classroom (as well as PCs).

*Information on prices is deliberately vague because prices are constantly changing.*

Conversely, if you are a researcher, then you might have access to some very expensive robots (such as the MobileRobots Pioneer 3DX) and devices (such as a SICK Laser Range Finder) that are supported by MRDS. The principles in this book translate to research robots as well, and MRDS has already been used successfully to control a Segway with a robotic arm balanced on top, a car for the DARPA Urban Challenge, an unmanned underwater vehicle, and even a huggable robot. For the latest examples, see the Microsoft Robotics Developer Studio Community web page at `http://msdn2.microsoft.com/en-au/robotics/aa731519.aspx`.

It is clearly not possible to cover every available model of robot. This is especially true for home-brewed robots. Questions are often posted to the Microsoft Robotics Developer Studio Discussion Forum at `http://forums.microsoft.com/msdn/default.aspx?forumgroupid=383&siteid=1` from new users who are hoping to build their own robots. Some of these people have unrealistic expectations about what their robots will do (probably because of what they have seen in science-fiction movies) and how easy it will be to build and program them.

## General Recommendations

If you have no experience with robots, then it would be best for you to select a robot kit or even a pre-assembled robot as your first choice. With that in mind, here are some general recommendations:

❑ **MRDS uses Windows platform code only.** A common type of question is "How do I program my brand xxx robot with an onboard model xxx PIC microcontroller using MRDS?" The answer is, you don't! MRDS does not generate code for anything other than Windows platforms. (Actually, it generates .NET code, which is platform-independent, but that basically limits MRDS to the various flavors of Windows.)

❑ **If your robot has some sort of onboard CPU and you want to program it directly, then you should investigate the options available from the manufacturer.** Many robots, including the LEGO Mindstorms NXT, Parallax Boe-Bot, RoboticsConnection Stinger, and so on, have programming tools available for them that enable you to run programs directly on the robot. In some cases, these are GUI tools similar to VPL. In other cases, they might use C-like languages or BASIC.

❑ **Use a wireless connection between your PC and your robot.** This eliminates the need to attach the robot to the PC via a serial cable, which severely limits the robot's freedom to move around. Many of the robots on the market have a Bluetooth module, although it is sometimes an optional extra. This requires Bluetooth hardware to be installed in your PC; otherwise, you have to purchase a USB-to-Bluetooth device. For testing robots within a typical office or laboratory environment, Bluetooth provides adequate range and is usually reliable.

An alternative to Bluetooth that is becoming popular is ZigBee radio (see the official website at www.zigbee.org/en/index.asp). It has similar capabilities to Bluetooth and the communications modules cost about the same.

❑ **If you can, get a robot with a WiFi (wireless Ethernet) interface.** More expensive robots, especially research robots, often use what is called *serial Ethernet* whereby the serial port on the robot is connected to an Ethernet interface. Packets are sent to and from the robot exactly as though the PC were connected via a COM port and a serial cable — the WiFi connection is transparent to the robot.

## Robots Discussed in This Book

In selecting robots for this book, several factors were taken into consideration. The purpose of the book is not to teach robotics, but rather to demonstrate how to use Microsoft Robotics Developer Studio to control robotics hardware. Therefore, the robots were chosen for their educational value. They are not industrial robots. However, they do cover several categories of robots, including mobile robots (both autonomous and tethered) and articulated robotic arms. As stated previously in this chapter, humanoid and legged robots have deliberately been omitted. Much of what you learn from a robotic arm is applicable to legged robots because they both rely on joints.

A good starting point for information about other robots is the MRDS Community Page, which contains the latest list. It can be accessed from the Community tab on the MRDS home page or using the following URL: `http://msdn2.microsoft.com/en-au/robotics/aa731519.aspx`.

You should also check the vendor's website for the latest information and software for your robot. Most robot vendors have discussion forums or support pages, including Frequently Asked Questions (FAQs) or Knowledge Bases.

The last chapter in this book (Chapter 17) explains how to set up new robot hardware, but it is recommended that you first gain some experience with one of the robots listed here. The two robots in Chapter 17, the Integrator from Picblok and the Hemisson from K-Team, are not mentioned below because they don't have the same level of support as the other robots.

Note that the robots are listed in order of complexity for new users, so if you are a novice you might like to start with the first robot and work your way through the list. Over a period of time, you might want to buy more than one of these robots. Each of them has its own advantages and disadvantages, and they will all teach you different lessons about robotics.

## LEGO Mindstorms NXT

The primary advantage of the LEGO Mindstorms NXT kit is that it doesn't require any tools or special skills. Everyone knows how to assemble LEGO, right?

The LEGO Mindstorms NXT kit, which is the successor to the RCX, is a great robot to get started with. It sells for different prices around the world, but in the United States it retails for around $250. It is easy to build and has built-in Bluetooth. You can also design your own robots and even switch from using wheels to legs. You are limited only by your imagination and how many LEGO blocks you own. See `http://mindstorms.lego.com/Products/Default.aspx`.

The basic intelligence is provided by a pre-built "brick," and Bluetooth is built right into the brick. LEGO offers a wide variety of sensors for the NXT, and there are third parties that sell NXT sensors. A disadvantage of the LEGO NXT brick is that it can only handle four sensors and three motors. The motors have internal encoders to measure rotations, but the maximum resolution is 360 ticks per rotation.

The LEGO NXT Tribot (which can be used in the simulator and is supported by MRDS) requires about an hour to assemble. However, the LEGO Mindstorms NXT kit can also be used to build other robots. There is an active online LEGO NXT community. Due to its popularity, the LEGO NXT Tribot is covered in Chapter 14. Figure 13-3 shows a LEGO NXT Tribot loaded up with a range of sensors.

**Figure 13-3**

In October 2007, Microsoft released updated services for the LEGO NXT that provide additional features and support third-party sensors such as the HiTechnic Compass. Make sure that you download and install these services, or preferably install the MRDS V1.5 Refresh.

## Parallax Boe-Bot

Parallax also makes a robot called the Boe-Bot that is available in kit form specifically for MRDS, which includes a Bluetooth module and a USB-to-serial device, for about $210 (U.S.). The Boe-Bot uses a BASIC Stamp for its onboard intelligence — the same processor used in the Scribbler discussed below. See www .parallax.com/ProductInfo/Robotics/BoeBotRobotInformation/tabid/411/Default.aspx.

*If you are considering buying a Boe-Bot, make sure you order the kit specifically for MRDS, which is shown in Figure 13-4. You can see an assembled Boe-Bot in Chapter 14.*

**Figure 13-4**

The Boe-Bot is a slightly more complicated robot to build than the LEGO NXT, but it is cheaper. The instruction book that comes with the Boe-Bot is quite easy to understand. The most difficult task is adjusting the servo motors for the wheels so that they don't turn when you set the speed to zero. If you are an experienced programmer, though, you might find that it is too low-level. This is because the intended audience is students age 13 and older.

A primary advantage of the Boe-Bot, at least in an educational setting, is that it has a small breadboard on the Board of Education (BOE) circuit board. A set of electronic components, included in the kit, can be used to build sensors directly on the robot. These components include LEDs for displaying program status, a piezoelectric speaker for making sounds, and the necessary parts for simple infrared detectors. If you have a background in electronics, you can design and implement your own sensors or other output circuits without soldering.

Additional sensors are also available from Parallax, including sonar and a line-follower kit. These are not supported under MRDS, so you would have to write your own services.

The Boe-Bot uses a BASIC Stamp for its onboard intelligence. The BASIC Stamp Editor is a language-sensitive editor that understands and compiles PBASIC code. It also takes care of downloading to the Boe-Bot, making it a trivial task. Extensive documentation is provided and PBASIC is fairly easy to learn.

In any case, you shouldn't need to use PBASIC because you can simply load the monitor program into the Boe-Bot and then control it remotely using MRDS on your PC via Bluetooth.

Although the Boe-Bot is supported in MRDS V1.5, the support is patchy. Some of the code is supplied with MRDS; the rest you have to download from Parallax. Therefore, the authors provide improved versions of the Parallax services for the Boe-Bot with this book in Chapter 14. The updated services enable you to flash LEDs and sound the buzzer. They also implement the `DriveDistance` and `RotateDegrees` operations, which are missing from the MRDS services. Lastly, they also include Compact Framework (CF) versions of the services. Chapter 16 provides instructions on how to drive your Boe-Bot using a PDA with built-in Bluetooth, such as the Dell Axim X50v.

## Lynx 6 Robotic Arm

If you are interested in articulated arms, the Lynx 6 is a reasonably priced (about US$390) and fully functional arm, which is shown in Figure 13-5. Lynxmotion also has more robust, and expensive, arms available. See `www.lynxmotion.com/Category.aspx?CategoryID=25` for more information.

Figure 13-5

Don't be misled by the picture of a Coke can in Figure 13-5. The arm is nowhere near strong enough to lift a full can. It is just there to show you the relative size of the arm.

The Lynxmotion L6 is a five DOF (Degrees of Freedom) articulated arm with a gripper. The axes include the Base (rotate), Shoulder, Elbow, and Wrist (Rotate and Tilt). There is also an L5 model, which is cheaper because it doesn't include the Wrist Rotate, but the L6 is a better choice.

Because dealing with joints in articulated arms is so different from the wheels on a differential drive robot, it is something that you would do only if you had a particular interest in or reason for using an arm. These differences are one of the reasons why the L6 is covered separately in this book in Chapter 15.

There is some very complicated mathematics involved in moving an arm, called *inverse kinematics*. The software supplied with this book takes care of this, but there is more to it than calculating the necessary joint angles to move the end effector to a specified pose. You also need to consider the speed and torque limitations of the robot. Otherwise, you can damage your robot by inappropriate movements.

The L6 comes as a kit that you have to assemble yourself. It is a more involved task than the Boe-Bot and takes much longer to put together and test. It is not a job for inexperienced users. The parts are made out of lexan (a very tough plastic) and have very tight tolerances, making them hard to fit together.

There is software supplied with the L6 called LynxTerm that can be used to control the servos. This is a good way to test the robot prior to using MRDS. Another much more sophisticated program called RIOS is also included.

> **All of the assembly guides for the L6 are online, so you have to download or print them. You should read the guides before purchasing the L6 because the construction is fairly involved and you need some tools that are not included in the kit. If you have no experience in building kits, then this is probably not a good robot to start with.**

## RoboticsConnection Stinger CE

The final robot, shown in Figure 13-6, is the Stinger CE from RoboticsConnection, which costs around US$525. It is an autonomous robot that uses the eBox-2300 running Windows CE (hence the "CE" in the name of the robot). You can get more information at `www.roboticsconnection.com/pc-78-3-stinger-windowsce-kit.aspx`.

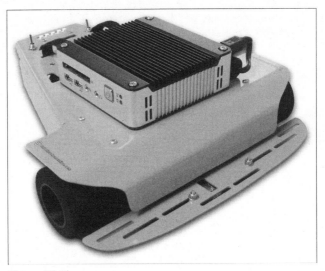

**Figure 13-6**

RoboticsConnection originally released a robot called the Traxster (because it uses tracks or treads) that was supported under MRDS V1.0. It uses the Serializer controller board. The Serializer is the core component of a new robot called the Stinger. See `www.roboticsconnection.com/c-3-robot-kits.aspx` for more information. The Stinger is the successor to the Traxster.

Communication is via a serial port. The eBox attaches to the top of the Stinger and then connects to the Serializer using a serial cable, thereby making the robot autonomous. However, there are alternatives with a variety of communication modules available for the Serializer, including Bluetooth, ZigBee, and WiFi.

For example, you can purchase a standard Stinger robot with a Bluetooth module. It can then be driven in the same way as the LEGO NXT or the Boe-Bot. This enables you to become familiar with the Stinger and the Serializer board without the additional complexity and cost of the eBox-2300.

The eBox-2300 is the same PC104 computer that is used on the Sumo robots based on the iRobot Create. Some of the instructional material on the Stinger is also applicable to the Sumo robots because they use the same onboard PC. However, the Stinger has to be built from a kit, whereas the Create comes already built. This gives you a lot more flexibility regarding how you arrange the sensors, and which sensors you put on the robot, but at the cost of your time and labor.

Unless you have some prior experience with building kits, this should probably not be your first robot. The Stinger itself is easy enough to build. It comes with a great set of instructions in a multimedia PDF file that includes 3D animations and synthesized voice instructions. However, setting up Windows CE and developing code for it is more complex than programming for Windows XP or Vista. This makes the overall task more difficult.

> *The Stinger instructions require Adobe Acrobat version 8.1 or later and a sound card in your PC. The actual PDF file is around 21MB and must be downloaded from the RoboticsConnection website.*

Support for the Stinger under MRDS V1.5 is available from the RoboticsConnection website. Chapter 16 discusses how to use the Stinger CE with MRDS and run services directly on the robot. An optional WiFi card is also available for the eBox, although it is not necessary because the eBox can run on its own.

## Other Robots Supported under MRDS

The following is a very brief summary of some other robots that are available and suitable for use with MRDS without a significant amount of additional programming.

Because bipedal, or humanoid, robots tend to be much more expensive than wheeled robots, no recommendations are made here. The only humanoid robot supported by MRDS V1.5 is the Kondo KHR-1. Humanoid robots rely on joints in a similar way to robotic arms, so using a robotic arm is a good starting point.

Similarly, quadruped robots (usually robotic dogs) have also been ignored. There have been competitions for these types of robots in the past, but the robots are expensive, and until Microsoft ran a competition using the RobuDog simulation from Robosoft (www.robosoft.fr/eng/index.php), they were not supported under MRDS. The RoboCup RoboSoccer Four Legged League (www.robocup.org) has used the Sony AIBO in the past, but since 2008 the league has been renamed the Standard Platform League and will be based on a new humanoid robot called the Aldebaran Nao from Aldebaran Robotics. Refer to www.aldebaran-robotics.com/eng/index.php for more information.

The list of robots supported by MRDS is constantly changing, especially as third parties develop their own services. This type of information becomes obsolete very quickly. The following list is in order of increasing price.

## Parallax Scribbler

If you are simply interested in learning about robotics and have no prior experience, then a prebuilt robot is probably the best starting point. For example, the Scribbler robot from Parallax (`www.parallax .com/ProductInfo/Robotics/ScribblerRobotInformation/tabid/455/Default.aspx`) is relatively cheap (around US$80) and comes preassembled and ready to use out of the box. It is more like a toy — it is very basic and cannot be expanded. You need to buy a Bluetooth device to attach to the robot so that you can operate it without using a serial cable. The Sena Serial-to-Bluetooth device (see Chapter 16) is suitable for this.

MRDS services for the Scribbler were originally developed by Ben Axelrod for the Institute of Personal Robotics Education (IPRE) while he was a graduate student at Georgia Tech. Services for MRDS V1.5 are available from the IPRE resources page: `www.ipre.org/resources.html`. Note that you have to update the firmware inside the Scribbler to use the IPRE services.

IPRE has developed a hardware "dongle," called a Fluke, that has an onboard color camera, IR range sensing, internal voltage sensing, an extra-bright LED, and Bluetooth. These became available for purchase in early 2008 and cost about US$80. It plugs into the Scribbler's serial port and can reprogram the Scribbler's firmware by downloading code over Bluetooth. This is an attractive alternative to buying a serial-to-Bluetooth device. The Fluke is shown attached to a Scribbler in Figure 13-7, courtesy of Tucker Balch at Georgia Tech.

Figure 13-7

This device is interesting for two reasons. First, the inclusion of a camera shows the importance that educators place on vision for robots. Second, it is likely that small robots combining the functionality of the Scribbler and the Fluke will become available, which raises the bar for entry-level robots.

Note that IPRE provides educational materials for teaching robotics, but the Myro (My Robotics) software is based on Python and it does not use MRDS. This will change over time.

## iRobot Create

Another alternative is the iRobot Create (`www.irobot.com/sp.cfm?pageid=305`), which is basically a Roomba vacuum-cleaning robot without the vacuum cleaner, redesigned specifically for educational purposes. It is relatively cheap, at about US$130, and it comes preassembled. You will also need to buy a Bluetooth device. Unfortunately, at the time of writing the Create is not available for sale outside of the U.S.A. and Canada.

The Create is discussed in Chapter 9 in relation to the sumo competition that Microsoft ran in 2007. Figure 13-8 shows one of the robots used in the sumo competition with an eBox-2300 and a Logitech webcam mounted on top. A simulated iCreate is also used in Chapter 12 for line following and sumo.

Figure 13-8

The iRobot Roomba and Create robots have a microcontroller that can be controlled directly via a serial port. You can attach a Serial-to-Bluetooth device to the serial port on the robot and control it. The firmware is already installed and the serial protocol is clearly documented. The RooTooth (www.roombadevtools.com) is a popular device for Bluetooth connections.

## Surveyor SRV-1

If you are interested in computer vision, then you might want to consider the Surveyor SRV-1 for about US$465 (www.surveyor.com).

This robot was undergoing a major revision at the time of writing, with the onboard computer being upgraded to a much faster Blackfin processor. Figure 13-9 shows the original SRV-1 (left) and the upgraded SRV-1b (right). Its main advantages are that it is preassembled; includes a camera; and offers a choice of Bluetooth, ZigBee, or WiFi. There is software for computer vision available from RoboRealm that supports the SRV-1 and works with MRDS (www.roborealm.com/help/MSRS.php).

Figure 13-9

For vision applications, it is important to get the fastest possible connection to the robot. WiFi is at least five times faster than Bluetooth, at 11Mbit/sec for 802.11b, and 25 times faster for 802.11g, at 54Mbit/sec.

The SRV-1 is another one of the robots that is targeted by IPRE for use with Myro.

The authors have updated the MRDS services for the SRV-1 to support the generic webcam contact and reduce flooding effects on the DifferentialDrive service so that the robot can be used with the TeleOperation service in Chapter 4. These updated services are available from the book's website. Services for the SRV-1b that operate using TCP/IP over WiFi will also be made available.

## CoroWare CoroBot

The Corobot from CoroWare (www.corobot.net) is a four-wheeled robot with an onboard PC. Figure 13-10 shows a Corobot with the optional robotic arm attached. It comes with a webcam as standard and there is an optional robotic arm attachment. CoroWare has some sophisticated software that runs under MRDS.

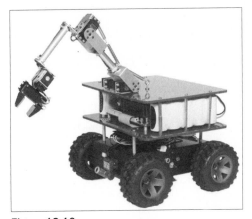

Figure 13-10

Chapter 6 develops a Corobot simulation, which is used again in Chapters 7, 9, and 12. However, there is no coverage of a real Corobot in this part of the book because the authors did not have access to one.

With a price around US$3,000, this robot is probably out of the price range of the average hobbyist and is targeted more at researchers, academics, and the military.

## White Box Robotics 914 PC-BOT

The PC-BOT, as its name implies, includes a full-blown PC that runs Windows XP and has a DVD drive, hard disk, speakers, webcam, wireless Ethernet, and so on. It has been available on the market for some time, but at US$8,000 it is not likely to be the first choice for a beginner. The original version came with MRDS installed, as well as all the necessary services and software to support directly programming the robot in .NET applications. See the PC-BOT home page for more information (www.whiteboxrobotics.com/PCBOTs/index.html).

One exciting announcement in late 2007 was that the 914 PC-BOT is also to be sold by Heathkit under the name HE-RObot. Heathkit has a strong reputation in hardware training and educational systems. You can see the Heathkit version of the robot at www.heathkit.com/herobot.html.

At the time of writing, the PC-BOT does not support MRDS V1.5. Services were released for the original V1.0 but the White Box has not updated them. However, it is their intention to support MRDS in the future, so perhaps this will happen with V2.0.

### MobileRobots Pioneer 3DX

The Pioneer robot family has been used for years in research institutions. However, a 3DX with a SICK Laser Range Finder costs over US$10,000. MobileRobots sell a whole range of robots, some of which are cheaper, such as the AmigoBots. All these robots are available with WiFi as an option and have an onboard processor, wheel encoders, sonar sensors, and everything that you might expect in a high-quality research robot.

The Pioneer 3DX is one of the robots supported in the simulator. The ExplorerSim in Chapter 9 and the VPLExplorer in Chapter 12 use this robot specifically because it has an LRF, which is a very accurate range sensor.

For more information, see www.activrobots.com/ROBOTS/p2dx.html. (The URL is correct — MobileRobots used to be called ActivRobots, and for some reason the P3DX is on a page called p2dx.)

# Fundamental Concepts: Sensors, Actuators, and Controllers

Although you can learn a lot about MRDS simply by working through the Simulation chapters in this book, there is no substitute for using real hardware. Yes, hardware is messy, unpredictable, and often frustrating. It can be complicated to build and sometimes it doesn't do exactly what it is commanded to do. However, nobody will ever believe that you are a true roboticist unless you have experience with real hardware.

This section briefly addresses the various components of a robot. The key points to note are that hardware is often nonlinear and that it has limitations. It should be obvious that interfacing a computer with a robot requires specialized hardware.

- ❑ A robot requires input data, which is collected from its environment. This data comes from *sensors* (also called *transducers*). Without sensory input, the robot would drive around blind and bump into things — not a very useful outcome.

- ❑ Once the robot has received some inputs, it needs to process this data to make decisions. Based on these decisions, it can then take some action. A *controller* does the work of making decisions. Usually the controller is given some task to perform. Depending on the type of controller, it might be a low-level task such as driving in a straight line, or a high-level task such as finding victims after a disaster.

❑    Regardless of the task to be performed, the robot needs to interact with the real world, and for this it needs *actuators*. Actuators take a variety of forms, but two are very common: the differential drive, consisting of two independently driven wheels, and the joints in robotic arms, which are based on servo motors. In addition to actuators, there might be other output or display devices. Status information is commonly displayed using LEDs.

One important distinction between various types of sensors and actuators is whether they are *digital* (binary values — usually represented as a `bool`) or *analog* (a continuous range of values — usually stored as a `double`). Obviously, switches and lamps are digital devices because they can only be on or off. Analog input devices require an analog-to-digital converter (ADC) to convert quantities such as voltages, temperatures, pressures, and so on, from the real world into numbers inside a computer. A digital-to-analog converter (DAC) does the opposite and can be used to convert a number, such as a speed, to an output, such as a voltage to control a motor.

The sensors, controller, and actuators are usually connected in a *feedback loop* or *control loop* (sometimes also referred to as *closed-loop control*). One of the most commonly used types of feedback controllers is the Proportional-Integral-Derivative (PID) controller, which is explained in a little more detail below.

One of the objectives of MRDS is to create abstract versions of the various sensors and actuators and then connect to the real devices at runtime by defining partners in a manifest. For this reason, MRDS defines *generic contracts* for a variety of sensors and actuators. When you build a new MRDS service, you can use these generic contracts so that your service can connect to a variety of devices at runtime. For example, an Analog Sensor service included in MRDS can be used to build a service for any sensor that outputs an analog signal (usually a voltage), as opposed to a digital (or binary) signal.

The following three sections discuss sensors, actuators, and controllers in more detail. There is no attempt to cover all types of sensors and actuators, but rather to discuss the ones that you will commonly encounter when using MRDS. If you have experience with robots, then you might want to skim through these sections because they are background material.

# Sensors

There are two types of sensor data: *proprioceptive* and *exteroceptive*. These terms just mean internal and external, respectively.

Internal information could include things such as the battery voltage, which a robot needs to monitor to prevent it from running out of power and to signal the need for a recharge. MRDS includes a Battery service. Other examples include temperature sensors to ensure that the robot is not overheating, and stall sensors on the motors that detect when the robot is stuck and can be used to shut down the motors to prevent them from burning out.

External information is obtained from the world around the robot, such as the distance to the nearest obstacle, a GPS location, the current compass heading, the ambient temperature, or even images from a video camera. Notice that in most cases the sensor information can be represented by a single number that has a minimum and a maximum value. The Webcam service is the obvious exception because it provides an array of pixel values.

The following sections discuss various types of exteroceptive sensors.

# Contact Sensors

The simplest of all sensors is a *contact sensor*, also called a *bumper*. A *bumper* is just a switch that senses when the robot runs into something, so it only has two states: pressed or not pressed. The LEGO NXT has a touch sensor that is a bumper. Some robots, such as the Boe-Bot, have "whiskers," which are thin pieces of wire that flex when the robot runs into something, causing an electrical contact to close.

Bumpers are called *contact sensors* because the robot actually makes contact with (bumps into) the obstacle. If this obstacle happens to be a human, then it could be harmful to the human! If it is a wall, then it might damage the robot. Either way, this is not a good sensor to use for general navigation. Therefore, bumpers are usually used as a last resort in case the other sensors have failed to detect an obstacle. Even expensive robots such as the Pioneer 3DX have bumpers on the front and back as a fail-safe measure. In the case of the 3DX, there are several bumpers, which MRDS refers to as a *bumper array*.

It is also possible to have *virtual bumpers* that don't involve physical contact. In Chapter 17, for example, the Integrator robot has an infrared sensor that detects obstacles, but the output of the sensor is a binary value — on or off.

# Range Sensors

Obviously, it is much better if the robot can sense obstacles from a distance, rather than bump into them. Several types of sensors, usually referred to as *range sensors,* can provide distance information. Examples include infrared (IR) detectors, sonar sensors, and laser range finders. These are also commonly called *time of flight* sensors because they send out a signal (infrared light, ultrasound pulse, or laser beam) and wait for a reflection to come back. The time required for the echo is a measure of the distance. You have probably seen movies involving submarines and have heard the "pings" sent out as the sonar searches for enemy ships.

## Infrared Sensors

On many hobby robots, infrared sensors are the preferred type of sensor because they're cheap. However, they're noisy and nonlinear, meaning that they do not always produce reliable readings and the values from the sensor are not directly correlated to the robot's distance from an obstacle. (See Chapter 17 for a graph showing the response of IR sensors on a Hemisson robot.) They are also susceptible to interference from strong sunlight (so they might not work at all outdoors), and they work over a very short range (typically 5 to 50cm).

*Passive* IR sensors just look for reflected infrared signals, which are all around us (but we can't see them). *Active* IR sensors work by sending out pulses of infrared light at a fixed frequency and then watching to see whether there is a reflection. Both of these types of sensors are effectively on/off sensors, or what MRDS refers to as "bumpers" although in this case there is no physical contact.

The cheapest IR sensors only indicate that an obstacle is somewhere within a fairly narrow range of distances. The exact range depends on the frequency of the IR pulses. In this case, the input signal is digital, i.e., 0 or 1. Many people do not realize this when they first start working with hobby robots. The absence of range data means that it is not possible to build reliable maps of the surrounding environment. It is also possible to miss obstacles if they are too close to the robot, such as a foot if a person suddenly steps in front of the robot.

The frequency and sensitivity can be adjusted to change the distance at which the sensor is triggered by an obstacle. This is usually done by changing a resistor or adjusting a potentiometer. An approach used in the Boe-Bot documentation is to change the frequency of the IR signal. A sweep of multiple frequencies enables the distance to an obstacle to be determined, but only in very rough bands. (MRDS does not use this method.)

More expensive IR sensors give an actual measurement of the range. There are usually several sensors located around the outside of the robot so that it can see in many directions simultaneously. For example, the Stinger robot in Chapter 16 uses three Sharp sensors across the front. These can measure distances from 10 to 80cm. An example of a Sharp GP2D12 IR sensor is shown on the left in Figure 13-11; shown on the right is its response to obstacles, sometimes referred to as the *measurement model*.

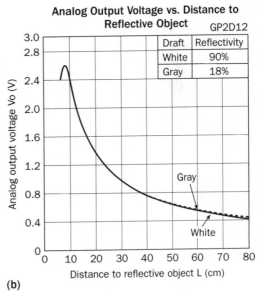

**Analog Output Voltage vs. Distance to Reflective Object** — GP2D12

| Draft | Reflectivity |
|-------|--------------|
| White | 90% |
| Gray | 18% |

(a)                                        (b)

**Figure 13-11**

## Sonar Sensors

Sonar sensors use ultrasonic "pings" to sense obstacles, usually at a frequency around 40 kHz. This is high enough that most people cannot hear the pings. It is unlikely to bother your dog, either, even though dogs can hear higher frequencies than humans. The measurement range for sonar sensors is from a few centimeters up to several meters, and they are usually accurate to within a centimeter or two. More important, modern sonar sensors are linear, so no processing is required to convert the input value to an actual range (this is usually done on the sensor circuit board).

Shown on the left in Figure 13-12 is the PING))) sensor manufactured by Parallax. A typical "beam" footprint (top-down view) from a sonar sensor (on the right) shows concentric circles, which represent distance from the sensor, with the numbers around the outside of the figure representing angles in degrees from the axis of the sensor.

The outline in Figure 13-12 (on the right) shows the point where the reflected signal can no longer be detected — that is, the maximum range of the sensor. Notice that it is not a narrow beam. The range returned by the sensor is the distance to the nearest obstacle (according to the concentric circles) that falls inside the outlined area. If multiple obstacles are inside the outlined area, the distance measurement will be the range to the closest one.

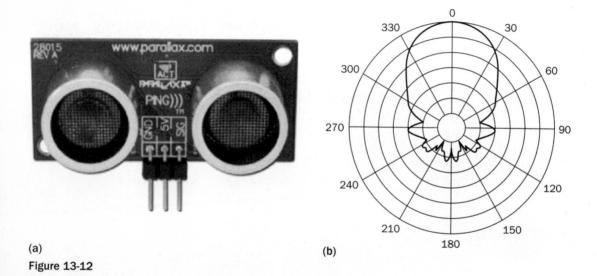

(a)

(b)

**Figure 13-12**

It might seem like sonar is a much better option than IR; but as with all sensors, sonar has several deficiencies and failure modes. Because a sonar beam is based on sound, it spreads out in a cone shape the farther it travels from the sensor. This means that a reading only gives an indication of the closest obstacle within an arc of several degrees (up to 60 degrees in the right image in Figure 13-12), and it is not possible to tell the exact direction to an obstacle. Some materials can also absorb sound, so there might not be sufficient reflection to get a reading. Lastly, if the sensor is at an angle to a wall, then it is possible for the reflection off the wall to bounce away from the sensor, not back toward it. In this case, the sensor records a maximum range value, which is a "miss." Obviously, this is a serious problem if the robot then charges ahead at full speed thinking that nothing is there!

Sonar sensors are usually arranged in a *sonar ring*, equally spaced around the robot, often with 12 or even 24 sensors. They can't be located too close together because of "cross-talk" between the sensors — that is, one sensor sees the reflection of the signal from a different sensor and therefore registers the wrong range.

MRDS does have a generic sonar contract, but the latest version of the LEGO NXT services uses the analog sensor contract for the LEGO sonar sensor.

## Laser Range Finders

Much of the recent research into robotics has used laser range finders (LRFs), which are far more accurate and have a much larger range compared to IR or sonar. However, a LRF costs thousands of dollars, whereas a sonar sensor costs less than $50.

A LRF uses a low-power laser beam, which is often a red beam, but newer devices use infrared. The LRF scans the beam across a wide arc using a rotating mirror and captures a *range scan* consisting of an array of measurements at fixed angles, often one degree apart.

Typical LRFs have maximum ranges from 8m to over 100m and they usually have a *field of view* (the spread from first value to last value on the right in Figure 13-13) between 100 and 180 degrees. The SICK LMS200 LRF, shown on the left in Figure 13-13, is a popular model. Although it is very expensive compared to IR or sonar sensors, it is highly accurate, with a 180° field of view, resolution of 10mm ±15mm, and a maximum range of 10m.

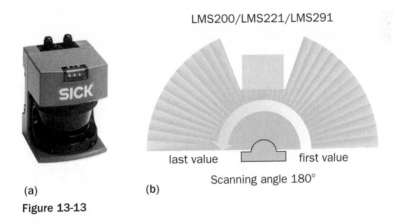

LMS200/LMS221/LMS291

last value      first value

Scanning angle 180°

(a)          (b)

**Figure 13-13**

The simulated Pioneer 3DX robot has a LRF and you can see red dots appearing on obstacles a couple of times per second. With a range of 10m or more, a LRF in an indoor environment can nearly always see obstacles. Only long corridors result in a maximum range reading being returned.

## Light Sensors

The Boe-Bot kit comes with a couple of light-dependent resistors (or photoresistors). These detect the amount of light falling on them, and their resistance varies accordingly. When incorporated into a circuit, they can be used to build a robot that either seeks out light (a photovore) or avoids light (a photophobe). With appropriate programming, the robot can be made to follow a flashlight or hide in the shadows like a cockroach.

The LEGO Mindstorms NXT kit also includes a light sensor, which incorporates an LED that can be turned on to illuminate the target so it can operate in either a passive mode (using ambient light) or an active mode. A color sensor is also available, which uses colored LEDs to illuminate the target and return a number that represents a code for the particular color.

*These types of sensors just detect the presence of light, or the color of the light. They cannot be used to "see" because they don't construct an image.*

# Video/Image Sensors

Cameras have become popular input devices in recent years as the price of digital camera sensors has dropped dramatically and the resolution has continued to increase. MRDS supports both Internet Protocol (IP) cameras and webcams (which usually attach via a USB connection). Image sensors provide an array of pixel values whereby each pixel usually consists of red, green, and blue (RGB) components.

In effect, a camera is the equivalent of thousands or millions of individual light sensors. In Chapter 12, downward-facing simulated webcams are used as light sensors for the line-following example by averaging small areas of the input images and coming up with values for "lightness." In reality, nobody would waste a webcam for this task because it can be done with much cheaper sensors.

Processing video images is a complex task. It is not easy to obtain range measurements from a video image because of factors such as perspective. Cameras have therefore tended to be used to recognize objects, people, and places by performing what is called *feature* (or *blob*) *extraction*. For instance, robot soccer relies on the soccer ball being a bright color, like orange, so that the robot can easily locate it in the image. (See the FollowBall example in Chapter 12.) Using *blob tracking*, the robot can then follow the ball. MRDS V1.5 includes sample calibration and blob-tracking services.

The simulated Pioneer robot has a camera mounted on top of it. You can see the view from this camera by pressing F8 in the simulation window or selecting it from the Camera menu. If you use the Dashboard or TeleOperation services (located in the `ProMRDS\Chapter4` folder), then you can select the camera from the list of services and the video will be displayed in a separate window that automatically maintains the correct aspect ratio. This Dashboard also works with real webcams if you include a Webcam service with your robot.

If you are interested in computer vision, you should download RoboRealm (`www.roborealm.com`). It works with MRDS and is fairly easy to set up.

# Other Sensors

Several other types of sensors are commonly used on robots. This is not a complete list, but it does provide some idea of what is available. Many of these sensors are available for the LEGO NXT, which is why they have been included here.

## Global Positioning System Sensors

A GPS (global positioning system) sensor provides an estimate of the robot's location to within an area of several meters. Unfortunately, the signals from the GPS satellites are extremely weak, so a GPS receiver does not work indoors. Moreover, the accuracy of the position fixes is so poor that the robot might believe that it is in the room next door. MRDS includes services for the Microsoft GPS.

## Compass Sensors

The robot's orientation is one of the most important pieces of information for navigation. If the robot becomes disoriented, either by bumping into obstacles or due to wheel slippage, then it can quickly become lost. A compass is valuable because it provides an absolute reference, i.e., errors do not accumulate over time.

Compass sensors work off the Earth's magnetic field. The HiTechnic compass sensor (www.hitechnic .com) for the LEGO NXT can measure bearings to the nearest degree, but the NXT only uses a single byte to represent analog values, so the readings are divided by two, resulting in a resolution of two degrees. The compass sensor can be read in the same way as an Ultrasonic (sonar) sensor because it just provides an analog value.

Just like a real compass, a compass sensor can be affected by large metallic objects and magnetic interference caused by motors and computers. Therefore, the compass sensor must be placed some distance from the LEGO NXT motors and even the NXT "brick." (The Tribot shown in Figure 13-3 has a compass mounted on a "stalk" at the back of the robot for this reason.) A compass must also be kept level or it produces incorrect readings. The effectiveness of compasses is reduced indoors, but they do work inside most buildings.

## Acceleration/Tilt Sensors

Acceleration/tilt sensors can be used to measure changes in speed and orientation. This type of sensor is available for the LEGO NXT (from LEGO or HiTechnic). Tilt sensors are useful to determine whether the robot is driving uphill, or for a legged robot to realize when it has fallen over so that it can get up again.

## Sound Sensors

Sound sensors vary in their capabilities. Because sounds can contain high frequencies, it is not usually feasible to measure them in real time if the robot is remotely controlled. Therefore, sound sensors tend to produce an output that is proportional to the volume of noise, or they selectively filter (listen for) particular frequencies.

Obviously, a microphone is a form of sound sensor, but it requires a high sampling rate to obtain useful information. Microsoft provides a Speech Recognition service as part of MRDS V1.5. For this to work, the microphone must be attached to the PC running MRDS, not to the robot if it is remotely controlled. Of course, if there is a PC on the robot and it has sufficient processing power, then it might be able to do both speech recognition and speech synthesis — that is, it could actually talk to you!

## Wheel Encoders

A *wheel encoder* measures rotations of a wheel so that you can determine how far a robot has traveled, assuming that you know the radius of the wheels. Wheel encoders are usually rated in terms of the number of "ticks" per revolution of the wheel. An encoder might give 1200 ticks per rotation, or even 3600 ticks. The encoders built into the LEGO NXT motors, however, can only measure 360 ticks per revolution, or, even worse, in low-resolution mode only 9 ticks.

Wheel encoders, also sometimes called *odometers*, are notoriously unreliable. If you rely only on input from wheel encoders to keep track of your position, a process called *dead reckoning*, then you will invariably get lost. This happens because the wheels slip against the ground as the robot moves; it bumps into objects, which makes the wheels jump off the ground momentarily; or a variety of other reasons.

However, wheel encoders are essential if you expect to make accurate turns using the `RotateDegrees` request for a Differential Drive service, or move precise distances using `DriveDistance`. Even then, the small errors that occur on each request will accumulate over time.

# Actuators

Actuators are output devices that affect either the robot or the real world in some way. The obvious examples are motors (and wheels) that enable robots to drive around. Robotic arms have grippers to pick up objects. A bomb-disposal robot might have a rifle to shoot bombs, although you might imagine that this actuator would only ever be used once.

The main point about actuators is that they have an operating range: motors have a maximum speed; joints have minimum and maximum bend, or rotation angles; and so on.

## Motors

You are probably familiar with electric motors that run continuously when power is applied to them. By adjusting the voltage, the speed of the motor can be varied. Reversing direction requires the polarity of the power to be reversed.

Motor control is usually done using a circuit board called an *H-bridge*. This board must be able to handle the large currents, sometimes several amps, that are necessary to drive a motor. In contrast, the typical circuitry inside a computer operates at a low voltage (5V or 3.3V) and low current (a few milliamps or even less). The H-bridge, therefore, translates the small signals from the computer to much larger signals for the motors, and it can also handle polarity reversals.

## Servo Motors

Servo motors are designed to move to a particular position and then hold that position. The position information is encoded in the signal sent to the servo motor. Typical servos operate on a pulse width modulation (PWM) system, with pulses being repeated 50 times per second. As long as the servo is receiving pulses, it resists any change to its position, which effectively locks it in place. This makes servos perfect for use in a robotic arm that has to lift loads of different weights, or for grippers.

> *A servo has an internal encoder to measure the angle of the motor shaft. This is usually a potentiometer. Most servos operate over a range from −90 degrees to +90 degrees and have physical stops at the extremes of their travel. Trying to drive a cheap servo beyond its range can damage it.*

Just to confuse you, some hobby robots use servo motors for the wheels. By deliberately disabling the positional feedback circuit inside the motor, it can be made to rotate continuously (always searching for the desired position). The Boe-Bot uses continuous rotation servos. There are many advantages to using a small servo motor instead of a conventional electric motor, including cost and lower rotational speed.

## Digital Outputs

Some types of actuators take a binary signal, such as light-emitting diodes (LEDs) or buzzers. You can turn these actuators on or off.

## *Sound and Voice Synthesis*

MRDS includes support to play WAV files. However, this only works on devices (such as PCs) that can handle WAV files and have an audio system capable of playing different frequencies of sound. Even if your robot does not have audio capability, you might still want to use sounds in your program to alert the user. Remember that the program is likely to be running on a PC that is remotely controlling the robot.

One unusual form of actuator is the Microsoft Text to Speech service. This can be used to add a voice to your robot, but the output plays on the PC controlling the robot, not the robot itself (unless it has an onboard PC).

# Controllers

As mentioned previously, PID controllers are very popular for process control. These controllers have a *set point*, i.e., a desired value of the system output, and the objective is to hold that value as closely as possible. The controller calculates the difference between the desired value and the current value of the output (which is the error), and adjusts the input accordingly. In the case of a motor, the objective might be to run at a constant speed, so the set point is the specified speed (which can be changed as necessary).

## *Understanding a Closed-Loop Control System*

Figure 13-14 shows a simplified *closed-loop control system* where the controller is not onboard the actual robot. The controller can issue commands to the robot and it receives sensor information. On the basis of this *feedback*, the controller calculates an updated value for the output and the loop continues in this fashion. Eventually the error should reach zero and the output will remain steady.

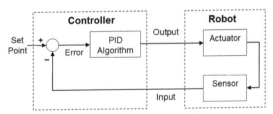

**Figure 13-14**

In Figure 13-14, the output to the actuator could be a power setting for a drive motor. The sensor in this case would be a *wheel encoder*, which measures rotations of the wheel. Wheel encoders are usually rated in terms of the number of "ticks" per revolution of the wheel. By counting ticks, the wheel speed can be determined.

The proportional component of the PID algorithm multiplies the error by a factor, called the proportional *gain*, so that the output is driven toward the desired value as quickly as possible. However, there is one problem with proportional control on its own — it often results in some residual offset from the desired value. (A discussion of the reason for this is beyond the scope of this chapter.)

Integral control involves integrating the error over time. Again there is a multiplying factor for the gain. (In process control, this factor is often expressed as a *time constant*, but this is not important here.) Integration can easily be performed by keeping a running sum of the errors each time through the control loop. Note that the error can be positive or negative. One effect of integral control is to cause the output to oscillate around the set point until the error is eventually reduced to zero.

To try to make the output settle quickly, differential control multiplies the derivative of the error by a gain factor (also sometimes expressed as a time constant). The objective is to have a *damping* effect. Calculating the derivative of a signal is easy if the sensor updates arrive at a regular time intervals — you simply subtract the previous value from the new value and divide by the time interval. It is not even necessary to divide by the time interval if it is constant because it can simply be lumped into the differential gain factor.

To avoid being overly sensitive, most controllers have a small *dead zone* — an acceptable threshold for the error below which no changes are made to the signal sent to the actuator. Without a dead zone, an actuator tends to *jitter* or *hunt* around the correct value.

Even though the diagram in Figure 13-14 shows the PID controller separate from the robot, in practice this is not a good idea. If the communication between the controller and the robot is over a slow connection, there is a delay in obtaining the feedback from the sensor. This delay has a destabilizing effect on the control loop. If it is too long, it causes oscillations, which might increase in magnitude and eventually be catastrophic. This is a classic problem in control theory when the system being controlled has inherent delays. It is quite obvious that if the feedback signal is sufficiently out of phase with the controller output, then the calculated value of the error will be wrong.

Fast feedback is also important for fundamental operations such as `DriveDistance` and `RotateDegrees` that are defined for the DifferentialDrive service. These operations enable the robot to be moved precisely. For this reason, they might be implemented onboard the robot, rather than remotely.

## Behavior-Based Control

PID control is applicable if you are trying to control an analog quantity. However, some control schemes are based on digital inputs, so they use Boolean logic. For example, the Boe-Bot has two IR sensors and two "whiskers." Both of these are the equivalent of contact sensors, even though the IR sensor triggers without physical contact. The logic for avoiding obstacles is therefore very simple.

If you want to see how this is implemented on the Boe-Bot, go to the MRDS directory in Windows Explorer and navigate to `samples\platforms\Parallax` and open `BASICStamp2.sln`. Once Visual Studio has started, open `BoeBotControl.cs` from the Solution Explorer. Locate the `ExecuteMain` routine. (You can select it from the drop-down list of class members.) Inside this routine you should find code similar to the following.

The first part of the code checks the whiskers (`wLeft` and `wRight`) to determine whether they have been triggered. Depending on which combination of whiskers has been depressed, the robot is sent a pre-defined command to execute backup, turn left, or turn right behavior. (The monitor program running on the Boe-Bot understands these commands and executes the necessary moves.)

```
if (_autonomousMode)
{
    if (wFlag)
    {
        if (wLeft && wRight)
        {
            Maneuver('U');
        }
        else if (!wLeft && wRight)
        {
            Maneuver('L');
        }
        else if (wLeft && !wRight)
        {
            Maneuver('R');
        }
    }
}
```

The next section of code checks the IR sensors (`irLeft` and `irRight`) to determine whether either of them has registered an obstacle. The logic is basically the same as for the whiskers, except that if no IR sensors have been triggered, then the robot moves forward, which causes it to keep wandering around. If a sensor is triggered, then the robot turns in an arc rather than turning on the spot as it does when a whisker is triggered:

```
if (irFlag)
{
    if (irLeft && irRight)
    {
        autonDir = AutonomousDirection.BKWD;
    }
    else if (!irLeft && irRight)
    {
        autonDir = AutonomousDirection.LEFT;
    }
    else if (irLeft && !irRight)
    {
        autonDir = AutonomousDirection.RIGHT;
    }
    else
    {
        autonDir = AutonomousDirection.FWD;
    }
}
```

The last part of the code sets the speed of the motors. (In this implementation, 0 means full-speed clockwise, 200 is full-speed counterclockwise, and 100 is stopped. These values are interpreted by code running on the Boe-Bot, as covered earlier in this chapter. Note that the motors are mounted with their shafts facing in opposite directions, so you have to rotate them in opposite directions to drive in a straight line.)

```
switch (autonDir)
{
    case AutonomousDirection.FWD:
        {
            pwmLeft = 200;
            pwmRight = 0;
        } break;

    case AutonomousDirection.LEFT:
        {
            pwmLeft = 0;
            pwmRight = 0;
        } break;

    case AutonomousDirection.RIGHT:
        {
            pwmLeft = 200;
            pwmRight = 200;
        } break;

    case AutonomousDirection.BKWD:
        {
            pwmLeft = 0;
            pwmRight = 200;
        } break;
    }
    SendSpeed(pwmLeft, pwmRight);
}
}
```

This type of control can be described in words, or behaviors, rather than mathematical equations (like a PID controller) — for example, "if you see an obstacle on the left, veer to the right." The exact definition of "veer" is up to the programmer, and choosing different values will make the robot behave differently. For example, if the programmer were paranoid and interpreted "veer" to mean "turn around and run away" then you would have a very shy robot.

Obviously, the variety of possible controllers is very large. A significant part of the effort in programming a robot is creating the control strategies and algorithms to make it perform the desired tasks, and then testing this code. Quite often, unexpected "emergent behaviors" appear when you combine a variety of different algorithms.

# Summary

This chapter covered some of the basics of robotics hardware and outlined some of the robots that are available on the market and supported by MRDS. Depending on your finances, choose a robot or two, or perhaps you have been given a robot to work with. In any case, you should find that it falls into one of the categories listed in this chapter. Support under MRDS might be provided by Microsoft or the hardware vendor.

It is important to understand that Microsoft Robotics Developer Studio is a framework or platform for running robots. Microsoft makes the analogy between the Windows operating system for a PC and MRDS for a robot. In the same way that Windows requires each brand of PC to implement the standard BIOS calls for low-level hardware access, MRDS needs some low-level drivers. Microsoft does not intend to support every robot on the market — that will be left to the manufacturers of the different robots. If the robot is not supported, you need to write the services yourself. This is not a trivial task and it is not advisable for a novice roboticist to embark on this path.

The next few chapters demonstrate how to use several real robots, culminating in the development of new services for a robot.

# 14

# Remotely Controlling a Mobile Robot

Remotely controlling a robot is one of the most common robotics scenarios. It is usually necessary because the robot is too small to carry an onboard PC, or the cost of an onboard PC would be prohibitive. For example, teaching robotics in a classroom requires a lot of robots, so they need to be relatively inexpensive.

This chapter covers how to connect real robots to Microsoft Robotics Developer Studio and make them perform simple tasks. Two different robots are used in the examples: the LEGO NXT Tribot and the Parallax Boe-Bot. There are differences between the examples due to different hardware capabilities, which has the benefit of demonstrating different tasks. Regardless of which type of robot you have, you should read the entire chapter.

Subsequent chapters explain how you can build your own mobile robot with an onboard PC so that it can operate autonomously.

## Remote Control and Teleoperation

The previous chapter explains how remote control of a robot works and points out that the robot does not run MRDS but instead runs a small monitor program. In this case the MRDS service sends commands to the robot and receives sensor information back from the robot. The robot then operates without human intervention. The complexity of the task that the robot can perform depends on the capabilities of the robot and the sophistication of the MRDS service. This chapter shows you how you can write a simple service to remotely control a robot.

*Teleoperation* refers to a human controlling a robot remotely. This is possible using the Simple Dashboard that comes with MRDS or the enhanced Dashboard that is supplied with this book. A joystick or game controller can be used in conjunction with the Dashboard so that you can easily drive your robot around. However, this assumes that you can see your robot. If you cannot see the robot, you need to rely on the sensor readings, or install a camera that can provide a video feed in real time. Small wireless cameras are available that can easily be attached to a robot and run off a 9V battery. They come with a wireless receiver with a composite video output, and might even provide audio. A Swann MicroCam is shown in Figure 14-1. Smaller "spy" cameras are also available from other manufacturers.

Figure 14-1

MRDS supports webcams, and the enhanced Dashboard can display the video from a webcam so you don't have to write any code. All you need to do is buy a suitable camera and a video capture device that is compatible with DirectX. There are many USB video capture devices available on the market. Attach the camera to your robot, plug the capture device to the wireless receiver, and you have a live video feed that you can use to drive your robot when it is out of sight.

# Setting Up Your Robot

The first step is obviously to build your robot! The instructions provided with each of the robots are quite clear. The time required to build the robot depends on your past experience, but it would be wise to allow at least a day to build it and get it working with MRDS.

Remote control requires a communications link. For both the LEGO NXT and the Parallax Boe-Bot, this link is provided via a virtual serial port over Bluetooth. The instructions for setting up Bluetooth are basically the same in both cases. You should read the LEGO NXT section even if you have a Boe-Bot because the Boe-Bot section is abbreviated.

Note that the examples in this chapter assume that you have set up the software for this book, available from the website (www.wrox.com and www.proMRDS.com). The files are copied into a folder called ProMRDS under your MRDS installation. These files are necessary to complete the examples in this chapter.

## *Using a LEGO NXT Tribot with MRDS*

The LEGO NXT Tribot can be built straight out of the box with a LEGO Mindstorms NXT kit using the instructions supplied with the kit. It has a two-wheel differential drive and a third castor, or jockey wheel, for balance. It should only take about an hour to build the robot.

If you have read the section on simulation, then you will already have seen the Tribot in the simulator. The simulated version has only a touch sensor at the front. However, the NXT kit comes with a range of sensors, and you can build onto the Tribot to create much more complex robots, as shown in Figure 14-2, which shows a Tribot loaded with sensors.

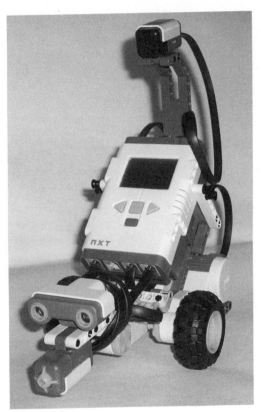

**Figure 14-2**

The NXT "brick" is a large block with an LCD display and buttons that contains an ARM processor and runs its own operating system. (Microsoft uses the term *brick* to refer to the central intelligence onboard any robot.) It takes six AA batteries, for a total of 9V. Fully loaded with sensors, it will chew through these batteries in a couple of hours. However, do *not* use rechargeable batteries because they only provide 1.2V (not 1.5V) even when fully charged.

The brick can accept up to four sensors that plug into the bottom and three motors that plug into the top. The black cables are clearly visible in Figure 14-2.

The robot in Figure 14-2 has an ultrasonic sonar sensor in the front that is used to detect obstacles. (It looks like two eyes and is often used as part of a face in LEGO designs.) A touch sensor pokes out from underneath the ultrasonic sensor and is a last resort in case the ultrasonic sensor fails to see an obstacle.

On top of the robot, mounted on a stalk, is a HiTechnic Compass sensor. It should be well away from the motors and brick to avoid magnetic interference. It works indoors but is a little slow to respond, so it is a good idea to turn slowly or stop and wait after a turn before using the compass direction.

Support for the LEGO NXT was included in the original version of MRDS because of the popularity of LEGO. MRDS is not directly supported by LEGO because they sell their own Mindstorms software, which is effectively a competitor to VPL.

---

### LEGO NXT Services Version 2

In October 2007, Microsoft released an update to the LEGO NXT services referred to as V2. These services are much more sophisticated than the older services, although the emphasis in the new release was on using VPL. They support the HiTechnic and MindSensors sensors and make more use of the wheel encoders.

To use the V2 services, you must make sure that your LEGO NXT brick has firmware version 1.05. You can download and install this using the LEGO Mindstorms software if necessary. The V2 services are significantly different from V1.5 in the way that they operate. This chapter covers using both the old and the new services.

Throughout this chapter you will see references to LEGO NXT services and LEGO NXT V2 services. You need to be careful to use the correct versions because they are incompatible with each other.

---

## Establishing a Bluetooth Connection

Create a Bluetooth connection between your PC and the LEGO NXT Tribot so that you can control the robot. This is a one-time task — once you have set it up you will not need to do it again. This process is called *pairing* and it involves establishing the necessary security credentials — in this case a *passkey* — so that the Bluetooth device can talk to your PC.

You first need to install your Bluetooth device on your PC according to the manufacturer's instructions. If Bluetooth was built into the PC when you bought it, this has probably been done for you. However, if you do not have Bluetooth, then you need to buy a suitable device. LEGO sells a USB-to-Bluetooth device (commonly called a *dongle*), but many other devices are available.

Note that Windows XP and Vista come with Bluetooth drivers from Microsoft, although problems have been reported in the past using these drivers. If your hardware came with a CD, install the software from the CD and don't rely on the Microsoft drivers. In some cases this might result in two Bluetooth items in your Control Panel — one from the hardware manufacturer and the other from Microsoft — and you will probably find that the Microsoft control panel applet won't work.

If you have trouble getting Bluetooth to work, one brand of software that has worked successfully for many people was written by WIDCOMM. This company was purchased by Broadcom in 2004, but the name was retained. You can download the latest version of the BTW software from Broadcom's website (www.broadcom.com/products/bluetooth_update.php). Note that this should be a last resort because there is no guarantee that this software will work with your device.

The instructions here are based on using an Anycom Blue USB-240 Adapter. (For more information see www.anycom.com.) This is a Class-1 device, which means that it should have a range of up to 100m. It comes with a CD and installation of the software is very simple, so it is not covered here.

Once you have a Bluetooth icon in your taskbar, you can set up the connection:

1. Plug in your Bluetooth USB dongle or turn on Bluetooth on your PC or laptop.

2. Turn on your Tribot and set it up as follows:

   a. On the main menu, scroll through the options using the gray arrow buttons on the brick until you find Bluetooth and then select it by pressing the orange Enter button.

   b. In the Bluetooth menu, find the option labeled On/Off and select it.

   c. Select the On option.

   Now that you have done this, Bluetooth will always be enabled when you turn on your Tribot, unless you replace the batteries — there is no need to repeat this step every time you use the Tribot.

3. Either from the icon in the system tray or from the Windows Control Panel, start the Bluetooth Devices applet (see Figure 14-3).

---

### Bluetooth Discovery

Bluetooth devices can broadcast their name and the services that they offer so that other devices can find them. The process of finding other devices is called *discovery*. Note that you do *not* have to turn on discovery in the Options panel as shown on the right in Figure 14-3. *Discovery should be left off on the PC for security reasons.* MRDS will make an outgoing connection to the NXT; it is not necessary for the PC to broadcast its availability.

The NXT, however, does have to have discovery enabled. There is an option in the Bluetooth menu on the NXT called Visibility. You must make sure that this is set to Visible (which is the default setting).

---

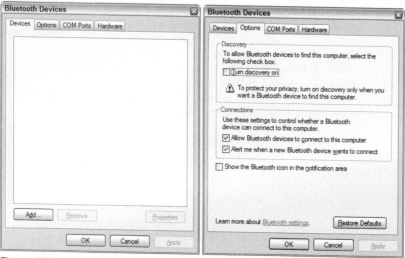

**Figure 14-3**

**4.** Click the Add button in the Devices panel. This starts the Add Bluetooth Device Wizard, shown in Figure 14-4. Enable the checkbox "My device is set up and ready to be found."

**Figure 14-4**

**5.** When you click Next, the wizard will start looking for Bluetooth devices. After a short delay, the LEGO NXT should appear in the list. Select it and click the Next button.

---

**Bluetooth Device Names**

If you have a Bluetooth-enabled mobile phone or PDA and they are within range, you might see them listed also. The LEGO NXT has a default name of NXT, so it is easy to identify. However, if you are working in a classroom environment, you will probably want to change the names on all of the NXT bricks so that they are unique. Refer to the LEGO documentation for how to do this.

---

**6.** The next Wizard screen asks you for a *passkey*. Before Bluetooth devices can communicate with each other, they must be *paired*. The passkey is used as a security measure. Without it, you could sit at a curbside cafe and read information off every Bluetooth-enabled phone or PDA that passed by! The LEGO NXT has a default passkey of 1234. Therefore, you should select the option to enter a passkey as shown in Figure 14-5 and ignore the warning that it should be at least eight digits long.

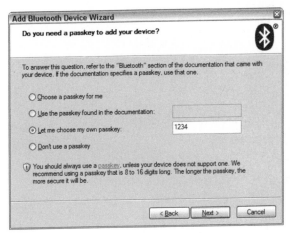

**Figure 14-5**

**7.** When you click the Next button, the LEGO NXT will make a sound and display a message asking you to enter a passkey. You can simply click the Enter (orange) button on the NXT because you are using the default passkey. Note that this pairing process will time out if you are not quick enough. In that case, you can back up in the wizard and try again.

**8.** After the Bluetooth device has been installed (which takes a few seconds), the wizard displays one final screen, shown in Figure 14-6. This lists the COM ports that have been set up for you. Figure 14-6 indicates that there is an outgoing port COM5 and an incoming port COM6. *Make a note of the outgoing port number because you will need it later.* If you forget, you can open the Bluetooth Devices applet and look in the COM Ports panel.

Figure 14-6

9.  Click Finish. You have now paired your LEGO NXT with your PC. This pairing will be remembered in the future unless you delete it in the Bluetooth Devices applet on the PC or in the Bluetooth Connections menu on the LEGO NXT.

Now that you have a working Bluetooth connection, it is time to establish a link using MRDS. You might want to turn your LEGO NXT off and on again just to be sure that you are starting from scratch.

## Communicating with the LEGO NXT Brick

The LEGO NXT brick runs the monitor program that controls the Tribot. MRDS downloads this program automatically when the Brick service connects to the NXT.

The monitor code for V1.5 is available in `samples\Platforms\LEGO\NXT\Resources`. (This code is no longer used by the V2 services.) If you want to change it, you have to use the Mindstorms software from LEGO to modify it and recompile. This process is specific to the LEGO NXT and is outside the scope of this book.

The following instructions apply to using the V1.5 services. The V2 services are covered later.

1.  Start a MRDS command prompt by clicking Start ⇨ All Programs ⇨ Microsoft Robotics Studio (1.5) ⇨ Command Prompt.

2.  At the command prompt, enter the following command to start the LEGO NXT Brick service:

```
dsshost /p:50000 /t:50001 /m:"ProMRDS\Config\LEGO.NXT.Brick.manifest.xml"
```

### Backwards or Forwards Slashes?

You might notice in examples of `dsshost` commands that sometimes the manifest paths are shown using backslashes (\) and sometimes using forward slashes (/). Technically, they should be forward slashes because they are supposed to be URLs. However, Windows uses backslashes for directory paths so either of them is acceptable. This is not an important point, but users are sometimes puzzled by it.

Another possible source of confusion is that the `dsshost` command line can use either slashes (/) or hyphens (-) to introduce parameters. Furthermore, you can abbreviate commands to their first letter — for example, `/manifest` simply becomes `/m`.

In the Command Prompt window you will see the DSS node starting up. A web browser window should also appear and display the LEGO NXT Brick service, as shown in Figure 14-7. Notice that there is an error message saying that the NXT is not connected, and the COM port is shown as 0 (zero).

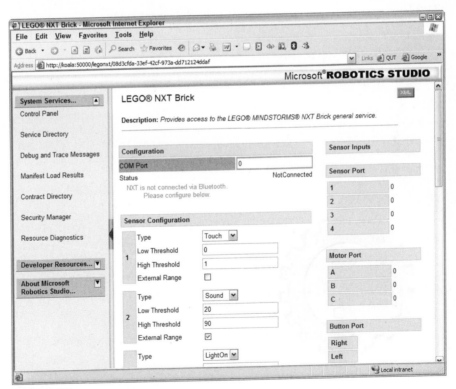

Figure 14-7

**3.** Enter the COM port number for the *outgoing* port that you made a note of when you set up the Bluetooth connection. Scroll down to the bottom of the web page and click the Connect button. The NXT will display "Microsoft Robotics Studio" and an hourglass. After a couple of seconds, the hourglass should disappear and the NXT will beep. It is now ready to be controlled via MRDS.

**4.** You can close down the MRDS command prompt and browser windows and turn off the NXT. (Press the dark gray Cancel button on the brick to exit the MRDS program, and then a couple more times until you get to the "Turn off?" prompt.)

After successfully connecting to the LEGO NXT, MRDS writes the COM port number into the `config` file so that you won't have to enter it in the future. You can look in `ProMRDS\Config\LEGO.NXT .Brick.Config.xml` to find the `ComPort` setting. If you have problems with your NXT and need to start again from the beginning, you can always edit this `config` file and put 0 (zero) in as the `ComPort` setting and run the Brick service again to reconfigure it.

> *If you upgrade the firmware in your NXT, you will lose the configuration information in the brick. In this case, you should set the* `ComPort` *back to 0 and repeat the preceding steps.*

Although you have made a connection between MRDS and the NXT brick, you need a suitable service to be able to control the Tribot. The Dashboard is a good starting point (it is covered later in the chapter). If you can't wait and want to jump ahead, you can run the following command:

```
dsshost /p:50000 /t:50001 /m:"ProMRDS\Config\Tribot.manifest.xml"
/m: "ProMRDS\Config\Dashboard.xml"
```

> *Make sure you type this all on one line, not on two lines as it appears here.*

At this stage you have successfully configured your LEGO NXT using the original LEGO services and you can quite happily continue to use these services. However, you have not configured the sensors on your NXT, which can be done on the Brick web page. This needs to be done before you can use the sensors, but you will come back to that later when you set up the V2 services, which are much better. Please continue on to the next section.

## LEGO NXT V2 Services

When you download the LEGO NXT V2 services (available from the MRDS Downloads web page) and run the installation, the source code will be copied into the following folder:

```
samples\Platforms\LEGO\NXT\V2
```

> *Source code is provided for all of the supported platforms. The original LEGO NXT services will not be overwritten when you install the V2 services.*

There is a solution file in this directory called `nxtbrick.sln` that you can open in Visual Studio. This contains all of the services in four different projects.

---

**MRDS Service Versioning**

MRDS uses strong versioning to ensure that the various services are all compatible with one another. If you make even minor changes to any of the code while you are editing nxtbrick.sln and then recompile, you might have to do a complete rebuild of all the LEGO NXT V2 services.

This has caused some people trouble in the past because the interdependencies between services are quite complex. If you get errors when you are trying to start MRDS services that are related to version mismatches, it might be necessary to recompile all of the affected services *twice*. The second compilation is required to pick up dependencies that could not be resolved during the first compilation.

Note that there is no need to change the V2 services at this stage!

---

To use the V2 services, you simply need to set up the appropriate config files and a manifest. The way that V2 works is quite different from V1.5. In particular, the individual sensors have their own config files, which specify where the sensors are connected on the brick. (In V1.5 the sensor configuration was attached to the brick as shown in Figure 14-7.)

Initially, it is not important to get the sensor configuration correct. You can change it later using the Manifest Editor or simply by editing the config files. However, as with V1.5, you have to specify the correct COM port for the Bluetooth serial connection.

Look in the ProMRDS\Config directory. You will find several config files related to the LEGO NXT V2 configuration:

❑   Lego.Nxt.v2.Brick.Config.xml

❑   Lego.Nxt.v2.Drive.Config.xml

❑   Lego.Nxt.v2.TouchSensor.Config.xml (Sensor1)

❑   Lego.Nxt.v2.UltrasonicSensor.Config.xml (Sensor4)

❑   HiTechnicCompassSensor.Config.xml (Sensor3)

There are also several manifests in the directory, but the two of interest are TriBotV2.manifest.xml and MyTriBotV2.manifest.xml. The first one is a simple manifest that just loads the basic Tribot with a differential drive. The second manifest includes all of the sensors listed above (assuming they are attached to the sensor ports as specified in the list).

Unlike the V1.5 services, if you run the LEGO NXT V2 services with no serial port specified, then the browser window does not allow you to edit the setting.

If you open Lego.Nxt.v2.Brick.Config.xml in Notepad, you will see that it is only a short file:

```xml
<?xml version="1.0" encoding="utf-8"?>
<NxtBrickState xmlns:s="http://www.w3.org/2003/05/soap-envelope"
xmlns:wsa="http://schemas.xmlsoap.org/ws/2004/08/addressing"
xmlns:d="http://schemas.microsoft.com/xw/2004/10/dssp.html"
xmlns="http://schemas.microsoft.com/robotics/2007/07/lego/nxt/brick.html">
  <Configuration>
    <SerialPort>5</SerialPort>
    <BaudRate>115200</BaudRate>
    <ConnectionType>Bluetooth</ConnectionType>
    <ShowInBrowser>true</ShowInBrowser>
  </Configuration>
</NxtBrickState>
```

The `SerialPort` is specified as number 5 in the preceding code. You need to change this to the same outgoing port that you used previously to configure the LEGO NXT Brick service in the web browser. That is the only change required. You can then run your LEGO NXT Tribot using the manifests listed above. For example,

```
dsshost /p:50000 /t:50001 /m:"ProMRDS\Config\TribotV2.manifest.xml"
```

starts the Tribot and launches the Dashboard (because it is in the manifest). The Dashboard is covered later in this chapter. The LEGO NXT should beep, and then you can use the Dashboard to drive it around.

An interesting point about this configuration is that `ShowInBrowser` is set to `true`, so a web browser is displayed showing the details of the brick. An example of the main LEGO NXT V2 Brick service is shown in Figure 14-8 for `MyTribotV2.manifest.xml`, which includes several sensors.

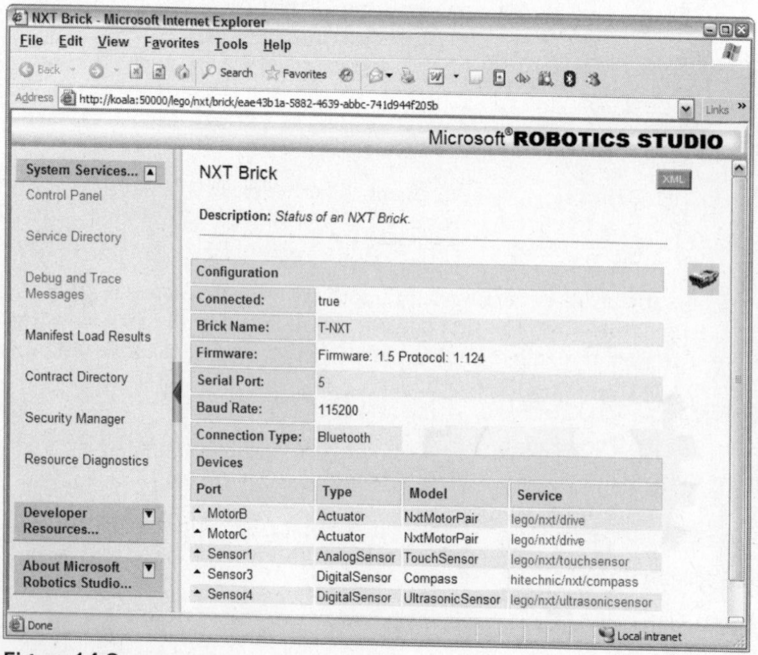

Figure 14-8

If you have not used the web browser interface before, you should poke around and become familiar with it. In particular, look in the Service Directory and click some of the services. Figure 14-9 shows thumbnails of the new V2 Touch Sensor, HiTechnic Compass, Ultrasonic Sensor, and Differential Drive. These interfaces update dynamically and are quite impressive.

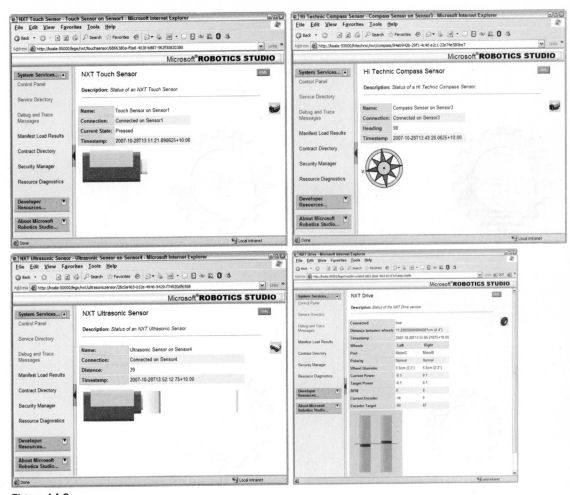

Figure 14-9

## *Using a Parallax Boe-Bot with MRDS*

Building a Parallax Boe-Bot is relatively easy but it is time consuming. It is not necessary to work through the entire book *Robotics with the Boe-Bot* that comes with the robot, so it will not take you the 40 hours indicated on the front of the book. The relevant chapters are as follows:

❑ Chapter 1: Activities 1, 2, 3, and 4

❑ Chapter 2: Activities 3, 4, and 6

❑ Chapter 3: Activities 1, 2, and 3

If you are not familiar with *breadboarding,* you should read Appendix D in the Parallax book as well.

The instructions in the Parallax book will assist you in building the basic robot, and you can start driving it around immediately, but it will be blind because it has no sensors.

In order to use the Boe-Bot with MRDS, you need to download the relevant instructions from the Parallax website and then build the sensor package on the breadboard area. At the time of writing, this document was available from the Boe-Bot Kit for Microsoft Robotics Developer Studio page on the Parallax website, and it was called `MRDS-Bluetooth-Boe-Bot-v1.5.pdf`.

Note that some of the I/O port assignments for MRDS do not match those used in the *Robotics with the Boe-Bot* book, but you might find Chapter 5 (about the whiskers) helpful, as well as Chapter 7 (about the infrared "headlights").

Once you have completed your Boe-Bot, it should look like the image in Figure 14-10.

Figure 14-10

Notice that the Boe-Bot in Figure 14-10 has the eb500 Bluetooth module installed (standing up in the middle of the Board of Education) and a lot of circuits in the breadboard area. The BASIC Stamp 2 has 16 I/O pins, allocated to the various devices as indicated in the following table:

| Pin | Function |
| --- | --- |
| 0 | eb500 Do Not Use (Serial In) |
| 1 | eb500 Do Not Use (Serial Out) |
| 2 | Right IR Detector |
| 3 | Right IR LED |
| 4 | Piezoelectric Speaker |
| 5 | eb500 Do Not Use (Connected Flag) |
| 6 | eb500 Do Not Use |
| 7 | Left Whisker |
| 8 | Right Whisker |
| 9 | Left IR LED |
| 10 | Left IR Detector |
| 11 | (No connection — available for use) |
| 12 | Right Servo Motor |
| 13 | Left Servo Motor |
| 14 | LED |
| 15 | LED |

As well as building the robot, you must download the monitor program into it. You can do this using the BASIC Stamp Editor that is supplied with the kit.

A revised version of the Parallax software is supplied with the code for this book in the Parallax folder under Chapter14. This updated version has some additional features that you might like to use. It can be found in the directory ProMRDS\Chapter14\Parallax\Boe-Bot and the file is BoeBotControlForMRDS.bs2.

When you double-click the .bs2 file in Windows Explorer, it launches the BASIC Stamp Editor. You do not need to make any changes to the code. Just plug your Boe-Bot into a serial port on your PC and download the program by selecting Run from the menu. If you don't have a serial port, then you can buy a USB-to-Serial device from Parallax.

Once you have downloaded the program, the Boe-Bot will reset itself and start running the program. It will beep once and flash the LED on pin 15. From then on, every time you turn on the Boe-Bot, it will beep and flash. This indicates that the monitor program has started running successfully.

## Establishing a Bluetooth Connection

Connecting the Boe-Bot to MRDS via Bluetooth is quite easy and is very similar to the procedure described earlier in the chapter for the LEGO NXT:

1. Plug in your Bluetooth USB dongle or turn on Bluetooth on your PC or laptop.

2. Turn on your Boe-Bot, with the eb500 module plugged into it. No setup is required.

3. Open the Bluetooth applet from the Control Panel or the system tray. It should be called Bluetooth Devices or Bluetooth Configuration.

4. In the Devices tab, click the Add button (refer to Figure 14-3). The Add Bluetooth Device Wizard will appear (refer to Figure 14-4).

5. Click the checkbox that says "My device is set up and ready to be found" and then click Next. Your PC will search for Bluetooth devices in range and should find the eb500. The wizard will display a dialog asking for a *passkey*.

6. Enter the passkey for the eb500, which is **0000** (see Figure 14-5 for an example, but make sure you enter **0000**).

   After entering the passkey, there will be a slight delay while the PC determines what services are offered by the eb500 and creates the necessary virtual devices. Eventually the final dialog of the wizard will be displayed, showing the COM ports that have been allocated. (See Figure 14-6 for an example.)

7. Make a note of the *outgoing* COM port number. You will need this to configure your Boe-Bot. Then click Finish.

If you have installed both a LEGO NXT and a Boe-Bot, then the Devices tab in the Bluetooth applet should look something like the left-hand diagram in Figure 14-11. The eb500 is the new device and the T-NXT is the existing LEGO NXT.

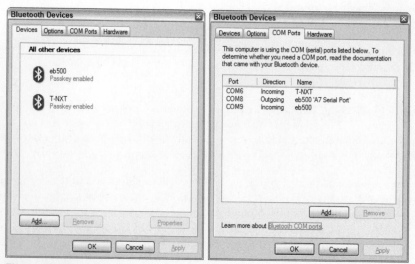

**Figure 14-11**

The right side of Figure 14-11 shows the COM Ports tab. Note that there are two ports for the Boe-Bot: COM8 and COM9. The outgoing port, COM8, is connected to an A7 serial port. So port number 8 is the one that should be used when setting the Boe-Bot configuration. The ports that you see will probably be different from these. It depends on the hardware installed on your PC.

## Communicating with the Boe-Bot

In order to use a Boe-Bot, you must set up the COM port number. Open the configuration file, `Parallax.BoeBot.Config.xml`, which is in the `ProMRDS\Config` directory. It should look like the following:

```xml
<?xml version="1.0" encoding="utf-8"?>
<BasicStampState xmlns:s="http://www.w3.org/2003/05/soap-envelope"
xmlns:wsa="http://schemas.xmlsoap.org/ws/2004/08/addressing"
xmlns:d="http://schemas.microsoft.com/xw/2004/10/dssp.html"
xmlns="http://schemas.microsoft.com/robotics/2007/06/basicstamp2.html">
  <Configuration>
    <Delay>0</Delay>
    <SerialPort>8</SerialPort>
  </Configuration>
  <AutonomousMode>false</AutonomousMode>
  <Connected>false</Connected>
  <FrameCounter>0</FrameCounter>
  <ConnectAttempts>0</ConnectAttempts>
  <MotorSpeed>
    <LeftSpeed>0</LeftSpeed>
    <RightSpeed>0</RightSpeed>
  </MotorSpeed>
  <Sensors>
    <IRLeft>false</IRLeft>
    <IRRight>false</IRRight>
    <WhiskerLeft>false</WhiskerLeft>
    <WhiskerRight>false</WhiskerRight>
  </Sensors>
</BasicStampState>
```

You can change the `SerialPort` (which is shown as 8 in the preceding code) to the correct *outgoing port* that you recorded when you set up Bluetooth.

# Using the Dashboard

MRDS comes with a service called the Simple Dashboard. This is useful for getting started with your robot because it enables you to move it around manually and confirm that the communications link is working.

An enhanced version of the Dashboard is provided in the code for Chapter 4 of this book. It has extra features that you might find useful. This Dashboard is shown in Figure 14-12 controlling a LEGO NXT using an Xbox 360 gamepad. Unnecessary parts of the dialog can be collapsed so that only the driving controls are visible. It also remembers its location on the screen. There are controls (the large arrow buttons) that use the `DriveDistance` and `RotateDegrees` requests (more on these later), which enable you to control your robot more precisely, and it won't run away from you. One of the best features might be that it remembers the name of the remote node. This means that all you have to do when you start up is click Connect and select the services you want to use, without typing a thing.

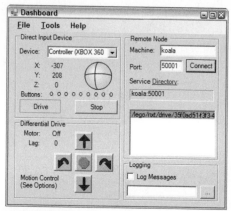

**Figure 14-12**

Figure 14-13 shows the options dialog (in the Tools menu) for the Dashboard. The dead zone has been set quite large to allow for some "slop" on the gamepad thumbstick. (This also prevents one of the bugs from the Simple Dashboard whereby the robot spins in the opposite direction if the vertical crosshair is not exactly centered.) Rotation speed is also scaled down. Notice that the Laser Range Finder and Articulated Arm displays are turned off in Figure 14-13. In addition, the `DriveDistance` and `RotateDegrees` parameters are set here.

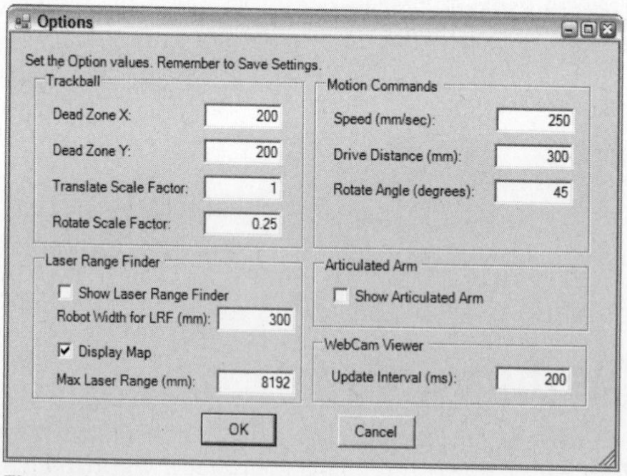

**Figure 14-13**

## Creating a Manifest

Starting the LEGO NXT Tribot service and the Dashboard service at the same time is an example of *orchestration*. This can be done by creating an appropriate manifest file. Manifests were discussed in earlier chapters, but a brief explanation is given here again because it is good to know how things work "under the hood."

A manifest is an XML file that contains a list of services to run and the relationships between the services (partners). All services should have a manifest. If you look in the MRDS directory under samples\ Config you will find dozens of manifests supplied by Microsoft. You will also see a lot of config files, which are discussed later.

Manifests are read by the Manifest Loader service when you start a DSS node. The basic structure of a manifest is as follows:

```
Service List
     Service Record
          Contract
          Partner List
               Partner
               . . .

     . . .
```

For more information, refer to Service Manifests in the MRDS documentation.

## Setting Up a Manifest

To set up a new manifest, follow these steps:

1. Make sure that you have a folder for it. Assuming that you have installed the ProMRDS samples from the book's website, there should be a folder under Microsoft Robotics Studio (1.5) called ProMRDS\Config.

2. Take a copy of the manifest file called ProMRDS\Config\LEGO.NXT.TriBot.manifest.xml and name it MyTribot.manifest.xml.

3. Open the manifest in Notepad. (You can edit a manifest in Visual Studio, but it is a plain-text file so you can just use Notepad). It should look like the following. Note that whitespace in an XML file is ignored, so the layout of the following code has been rearranged slightly to fit on the page:

```
<?xml version="1.0"?>
<!--
//  This file is part of the Microsoft Robotics Studio Code Samples.
//
//  Copyright (C) Microsoft Corporation.  All rights reserved.
//
//  $File: LEGO.NXT.TriBot.manifest.xml $ $Revision: 11 $
-->
<Manifest
xmlns:legonxt="http://schemas.microsoft.com/robotics/2006/05/legonxt.html"
xmlns:bumper="http://schemas.microsoft.com/2006/06/lego.nxt.bumper.html"
xmlns:legonxtdrive="http://schemas.microsoft.com/robotics/2006/10/legonxtdrive.
html"
xmlns:this="urn:uuid:e7401b21-b3c2-4a80-bb72-29204083cf87"
xmlns:dssp="http://schemas.microsoft.com/xw/2004/10/dssp.html"
xmlns="http://schemas.microsoft.com/xw/2004/10/manifest.html">
```

The first part of the manifest specifies the different contracts that will be used and gives them short names, like legonxtdrive, for ease of reference later in the manifest.

Next in the manifest is the service list. The first service in the list is the Brick service, which is the basic LEGO NXT. It partners with a DSSP State service and specifies a configuration file (called `lego.nxt.brick.config.xml`) that is loaded when the service starts. This `config` file is assumed to be in the same folder as the manifest because no path is specified. If it doesn't exist, the service creates it the first time the manifest is run:

```
<CreateServiceList>
  <ServiceRecordType>
    <dssp:Contract>http://schemas.microsoft.com/robotics/2006/05/legonxt.html
    </dssp:Contract>
    <dssp:PartnerList>
      <dssp:Partner>
        <dssp:Contract>http://schemas.microsoft.com/robotics/2006/05/legonxt.html
        </dssp:Contract>
        <dssp:Service>lego.nxt.brick.config.xml</dssp:Service>
        <dssp:PartnerList />
        <dssp:Name>dssp:StateService</dssp:Name>
      </dssp:Partner>
    </dssp:PartnerList>
    <Name>this:Brick</Name>
  </ServiceRecordType>
```

The drive service has to partner with the brick. Note that the LEGO NXT drive implements the generic Differential Drive contract so that it can be controlled by services that know nothing about the specifics of the LEGO NXT, such as the Dashboard:

```
    <ServiceRecordType>
<dssp:Contract>http://schemas.microsoft.com/robotics/2006/10/legonxtdrive.html
</dssp:Contract>
      <dssp:PartnerList>
        <dssp:Partner>
<dssp:Contract>http://schemas.microsoft.com/robotics/2006/10/legonxtdrive.html
</dssp:Contract>
          <dssp:Service>lego.nxt.tribot.drive.config.xml</dssp:Service>
          <dssp:PartnerList />
          <dssp:Name>dssp:StateService</dssp:Name>
        </dssp:Partner>
        <dssp:Partner>
          <dssp:Contract>http://schemas.microsoft.com/robotics/2006/05/legonxt.html
          </dssp:Contract>
          <dssp:PartnerList />
          <dssp:Name>legonxtdrive:Nxt</dssp:Name>
          <dssp:ServiceName>this:Brick</dssp:ServiceName>
        </dssp:Partner>
      </dssp:PartnerList>
      <Name>this:legonxtdrive</Name>
    </ServiceRecordType>
```

Lastly, the bumper service is added to the manifest and the service list, and the manifest tags are closed. Notice that the bumper also partners with the brick:

```
    <ServiceRecordType>
      <dssp:Contract>http://schemas.microsoft.com/2006/06/lego.nxt.bumper.html
```

```
        </dssp:Contract>
        <dssp:PartnerList>
          <dssp:Partner>
            <dssp:Contract>http://schemas.microsoft.com/2006/06/lego.nxt.bumper.html
            </dssp:Contract>
            <dssp:Service>lego.nxt.tribot.bumper.config.xml</dssp:Service>
            <dssp:PartnerList />
            <dssp:Name>dssp:StateService</dssp:Name>
          </dssp:Partner>
          <dssp:Partner>
            <dssp:Contract>http://schemas.microsoft.com/robotics/2006/05/legonxt.html
            </dssp:Contract>
            <dssp:PartnerList />
            <dssp:Name>bumper:LegoNxt</dssp:Name>
            <dssp:ServiceName>this:Brick</dssp:ServiceName>
          </dssp:Partner>
        </dssp:PartnerList>
        <Name>this:bumper</Name>
      </ServiceRecordType>

    </CreateServiceList>
  </Manifest>
```

## Starting the Simple Dashboard

Go to the bottom of the manifest. Just before the closing tag `</CreateServiceList>` insert the following code to start the Simple Dashboard supplied by Microsoft:

```
    <!-- Dashboard -->
    <ServiceRecordType>
        <dssp:Contract>http://www.promrds.com/contracts/2007/01/↲
simpledashboard.html</dssp:Contract>
    </ServiceRecordType>
</CreateServiceList>
```

As you can see, this code adds a new `ServiceRecordType` to the manifest. The first line is a comment, and obviously it is not required for the code to work. The format for comments is the same as for HTML. This can be useful for temporarily removing part of a manifest, or if you want to store different options in a manifest, such as running your service with the simulator or a real robot.

If you want to use the Dashboard from the website for this book (`www.proMRDS.com` or `www.wrox.com`), then the contract is slightly different:

```
<dssp:Contract>http://www.promrds.com/contracts/2007/10/dashboard.html
</dssp:Contract>
```

Notice that the name is `dashboard.html` and that the month and year are different. You must have installed the software from the book's website in order to use this service. It might be necessary to recompile the service as well.

---

### Alternative Approach for Running the Dashboard

This example shows you how to add the Dashboard to a manifest. However, because the Dashboard establishes its partnerships dynamically at runtime, there is no need to put it in the manifest. In fact, it would be wrong to specify partnerships in the manifest.

You can therefore just use a separate Dashboard manifest on the command line in addition to the manifest for the robot that you want to control. Here is an example:

```
dsshost /p:50000 /t:50001↵
/m:"ProMRDS\Config\Tribot.manifest.xml"↵
/m:"ProMRDS\Dashboard.manifest.xml"
```

Note that the preceding command must be typed all on one line.

An example that uses this approach is provided in the MRDS `bin` folder for the Boe-Bot. It is called `BoeBot.cmd`. See the discussion below on creating shortcuts to run your robots.

---

Save the manifest you have been editing and exit from Notepad. You can verify that the manifest is okay by loading it with the Manifest Editor. Select Microsoft DSS Manifest Editor from the `Microsoft Robotics Studio (1.5)` folder in the Start menu. Open the manifest that you have created. It should look like Figure 14-14.

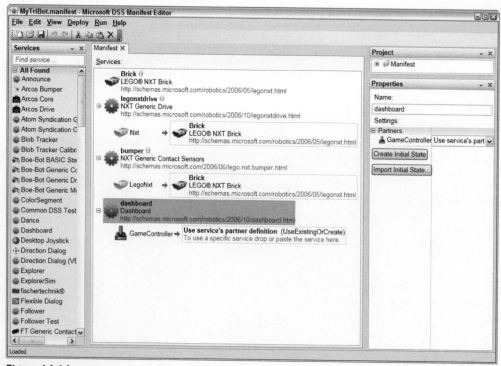

**Figure 14-14**

The Manifest Editor is the preferred way to create and edit manifests because it understands the format and syntax of manifests, and lists all of the available services. If you enter a string, such as LEGO, into the textbox at the top of the Services panel on the left-hand side, you can search for just the related services.

In the future, you might prefer to use the Manifest Editor. For example, you can try creating a manifest for the LEGO NXT V2 by using only V2 services. It should look very similar to Figure 14-12, but you will need to do some configuration. If you are not comfortable building your own manifests yet, there is a manifest already supplied as ProMRDS\Config\MyTribotV2.manifest.xml. This manifest includes all of the sensors for the robot shown in Figure 14-2 and the web browser in Figure 14-9.

If you are using a Boe-Bot, then you can start with the Parallax.BoeBot.manifest.xml and add the Dashboard to it by following the instructions given earlier. Call your new manifest whatever you like. If you are feeling lazy, just use the MyBoeBot.manifest.xml file already provided for you.

## Running the Manifest

Now that you have created an appropriate manifest, you can use it in a dsshost command. However, typing long command lines to start up a DSS node is painful and prone to error, so you will probably want to create a batch command file that does it for you.

The following instructions are for the LEGO NXT. If you have a Boe-Bot, then you can read through the instructions and simply use the Boe-Bot version of the code that is supplied on the book's website, or you can try to figure it out for yourself. There is a BoeBot.cmd batch file in the MRDS bin folder that corresponds to the RunLego.cmd file in the following paragraphs (which is also supplied if you don't like typing). In the example that follows, the batch file is created in the top-level MRDS folder. However, the sample files in the bin folder are designed to run from there and are a little different.

Using Notepad, create a new file called RunLego.cmd and place it into the top-level MRDS folder. It should contain the following commands:

```
@ECHO ON
REM Run a Lego NXT Tribot
REM Type Ctrl-C in this window when you want to stop the program.
dsshost -port:50000 -tcpport:50001 -manifest:"ProMRDS/Config/MyTribot.manifest.xml"
```

Now open an MRDS Command Prompt window and run the command procedure by typing **RunLego**. Figure 14-15 shows what the command window should look like. (This figure shows a complete test run. Note that when you press Ctrl+C to cancel the program, a prompt asks whether you want to terminate the batch job. You should enter **y**, but don't do this right now!)

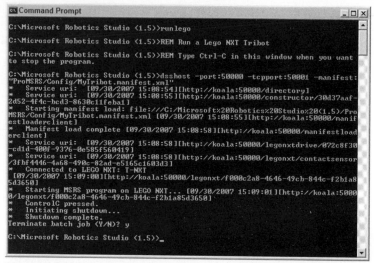

Figure 14-15

*If you want to use the LEGO NXT V2 services, then you can use* TribotV2.manifest.xml, *which is provided for you.*

If the NXT is turned off, or MRDS cannot establish a connection, you will see a red error message in the Command Prompt window. Figure 14-16 shows an example of this error message (without the color, of course). In this case, the LEGO NXT Brick service web page is also displayed. This is a roundabout way to do it, but it's one way to change the configuration of your NXT.

Figure 14-16

*The LEGO NXT V2 services will not generate red error messages if the LEGO NXT is not turned on. Instead, it simply waits for a connection.*

If you made any typing errors when you created the manifest, then red error messages will also appear in the Command Prompt window from the Manifest Loader. You will also see errors appear if you turn off the LEGO NXT before shutting down the service.

Assuming that you did everything correctly, a Dashboard window will appear, as well as a LEGO NXT Brick window. You will have to enter the COM port number again. This happens because you have copied the original manifest into a different folder, but you did not copy the config files. Look in the ProMRDS\Config folder now and you will see a set of config files.

If you continue to run the Tribot from this batch command file, then this will not be a problem. However, if you start your Tribot by running the original manifest, you will have two sets of config files and things can very rapidly get confusing. This is why it is a good idea to place all of your manifests together in a single Config directory.

In the Dashboard, enter **localhost** for the machine name and click Connect. (The machine name will change to the actual computer name.) Double-click the LEGO drive in the service list to select it, and then click the Drive button. You can now drive the robot around. When you drag the mouse on the "trackball" (just above the Stop button), the Tribot should drive around. Try it out.

*The port you use in the Dashboard should be 50001, which is the TCP port. Remember that two ports are specified on the command line. If you mistakenly enter **50000**, you will be waiting a long time because there won't be any response.*

At this stage you will probably want to click File ⇨ Save Settings to remember your remote node, the window position, and so on. When you get tired of playing with the Tribot, press Ctrl+C in the Command Prompt window. Note that closing the Dashboard window does not terminate DssHost.

## Using a Joystick or Gamepad

Driving a robot using the "trackball" in the Dashboard is a little tricky. It is much easier to use a control that is designed specifically for the purpose. If you have a joystick or an Xbox 360 GamePad, then you can use that to drive the robot. The only requirement is that the controller must be supported by DirectX. Simply plug in the joystick or gamepad and run a robot manifest that includes the Dashboard.

The example in Figure 14-17 shows a Logitech Attack 3 USB joystick being used to drive a LEGO NXT Tribot using the Simple Dashboard.

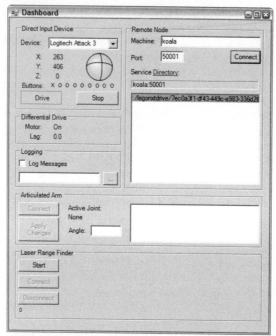

**Figure 14-17**

If you look carefully at the dialog, you will notice that there is an "x" for the leftmost button in the row of "o" symbols. This indicates that the trigger button on the joystick is depressed (and none of the other buttons on the joystick are currently pressed). You can use the trigger so that the robot will only move when the trigger is depressed *and* the joystick is off-center. Some of the other buttons also have functions, but they are not described here — look in the code for the Dashboard.

If you have an Xbox 360 USB GamePad and a joystick, then it will appear in the drop-down list for Direct Input Devices. Simply select it instead of your joystick. You can use the left-hand thumbstick to drive the robot.

## Creating a Shortcut to Run Your Robot

Although the `dsshost` command is now safely embedded in a command procedure, you can't double-click the command procedure in Windows Explorer to run the program. If you watch the screen carefully, you will see a DOS Command Prompt window appear and then quickly disappear again.

The problem is that the environment for running DSS has not been set up. This is easily fixed by a small modification to the command file. Edit `RunLego.cmd` by adding the two extra lines shown here:

```
@ECHO ON
REM Run a LEGO NXT Tribot
REM Type Ctrl-C in this window when you want to stop the program.
setlocal
call "%~dp0sdkenv.cmd"
dsshost -port:50000 -tcpport:50001 -manifest:"ProMRDS/Config/MyTribot.manifest.xml"
```

The purpose of these extra commands is to run `sdkenv.cmd`, which sets up the MRDS environment. This creates the necessary environment variables. When you open an MRDS Command Prompt window from the Start menu, these variables are already defined.

From Windows Explorer, double-click `RunLego.cmd`. This should start the program just as if you had run it from an MRDS command prompt.

Taking this one step further, right-click `RunLego.cmd` in Windows Explorer and click Create Shortcut from the pop-up menu. This will create a shortcut called `Shortcut to RunLego.cmd`. You can rename this to whatever you like. For example, you might call it just `RunLego`. If you right-click this shortcut and then click Properties, you can change the icon. A batch file does not contain any icons, but you can select from the default set of icons in `Shell32.DLL`.

Finally, you can copy or move the shortcut anywhere you like, such as onto your Desktop. This gives you immediate access to the LEGO NXT Tribot right next to all your other important business applications (see Figure 14-18).

Obviously, you can create other command procedures and icons this way for the Boe-Bot or for the LEGO NXT V2 services. Even more complex command procedures can be used to select among a variety of robots.

**Figure 14-18**

# Making a Robot Dance

Now that you have your robot working, it would be good to make it do something on its own without requiring a human to drive it around. The first step is to get it to perform some pre-programmed tasks, which might be referred to as a *behavior*.

The code for the final version of this example is available in the `ProMRDS\Chapter14\Dance` directory once you have installed the sample software from the book's website (`www.proMRDS.com` or `www.wrox.com`). You do not have to type in all of the code.

The initial example aims to get the robot to drive around the sides of a square. This might sound like a very simple task, but it actually introduces some real-world problems. How can you determine how far the robot has driven or how far it has turned? The simplest approach is to use timers. For example, you can run the robot for a short period of time and measure how far it travels. If you do this for several different time periods, you can plot a graph of distance versus time. This is a very scientific approach. (See the example with the Hemisson robot in Chapter 17.)

The steps that follow do not assume that you are experienced in MRDS. If you have followed through the chapters on simulation you will have created a new service before. Therefore, this chapter repeats all of the steps but with a little less explanation.

## Building a New Service

This section contains instructions for building a new service from scratch. The completed service, with several additional features, is available in the `ProMRDS\Chapter14\Dance` folder.

> *An extra service called SynchronizedDance was added late in the development of this book. This service controls multiple robots and makes them all perform the same dance. It is not discussed here.*

Create a new directory called `ProMRDS\Dance` and then create a new service as per the commands below. Once the service has been created, open the solution called `Dance.sln`. To open this in Visual Studio, you can either type the filename at the command prompt as in the following example, or double-click it in Windows Explorer:

```
C:\Microsoft Robotics Studio (1.5)>cd ProMRDS
C:\Microsoft Robotics Studio (1.5)\ProMRDS>mkdir Dance
C:\Microsoft Robotics Studio (1.5)\ProMRDS>cd Dance
C:\Microsoft Robotics Studio (1.5)\ProMRDS\Dance>dssnewservice /service:"Dance"↵
/namespace:"ProMRDS.Robotics.Dance" /year:"2007" /month:"10"
C:\Microsoft Robotics Studio (1.5)\ProMRDS\Dance>cd Dance
C:\Microsoft Robotics Studio (1.5)\ProMRDS\Dance\Dance>Dance.sln
```

If you are using Visual Studio Express Edition and you have multiple languages installed, make sure that you select C# in the Choose Application dialog, as shown in Figure 14-19.

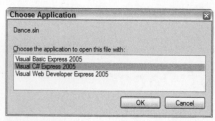

**Figure 14-19**

Open `Dance.cs` and add the following `using` statement at the bottom of the list:

```
using drive = Microsoft.Robotics.Services.Drive.Proxy;
```

In order to actually use the drive proxy, you must also add a reference to the appropriate DLL, which in this case is `RoboticsCommon.Proxy`. You can right-click the Solution Explorer panel and then click Add Reference. Alternatively, click Project ⇨ Add Reference. It might take a little while to display the list of DLLs, so be patient. Set the Local Copy and Specific Version properties on the Reference to `false`.

Scroll down in the code to just below the main port declaration and insert a drive partner as shown here:

```
[ServicePort("/dance", AllowMultipleInstances=false)]
private DanceOperations _mainPort = new DanceOperations();

#region partners
[Partner("Drive", Contract = drive.Contract.Identifier, CreationPolicy =
PartnerCreationPolicy.UseExisting)]
drive.DriveOperations _drivePort = new drive.DriveOperations();
#endregion
```

It is a good idea to use regions in the code so that you can collapse sections of code to make it easier to find your way around. It also helps to make the code self-documenting.

This code sets up a generic differential drive partner, which means that you can connect to a specific type of robot at runtime by specifying the appropriate contract in the manifest. Notice that the CreationPolicy is set to UseExisting. This causes the Dance service to wait until the partner becomes available. It would not make sense to create a new instance of the drive partner because there is no way to tell which of the many different robot drive services should be used. It is the responsibility of the Manifest Loader to associate a real differential drive with the generic one at runtime.

Add some constants at the top of the class definition as shown:

```
public class DanceService : DsspServiceBase
{
    #region constants
    // Constants for motor speeds and timing
    const int repeatCount = 2;          // Number of times to repeat dance
    const float driveSpeed = 0.5f;      // Speed driving forward
    const float rotateSpeed = 0.25f;    // Speed during rotation
    const int driveTime = 1200;         // Time to drive forward (millisec)
    const int rotateTime = 920;         // Time to rotate
    const int settlingTime = 500;       // Time to wait after each move
    #endregion
```

The purpose of these constants should be obvious from their names and the comments. Basically, the robot will drive around the sides of a square by driving forward for a certain amount of time, turning for a different amount of time, and repeating this three times. An overall repeatCount enables the entire "square dance" to be repeated.

Because some robots are quite primitive, using timers to control motions is the only option that you have. More expensive robots have *wheel encoders* that can be used to monitor how much the wheels have turned. The LEGO NXT motors do actually have encoders, but they are not used in this example. This is discussed later.

The numbers listed in the preceding code were determined by trial an error. They are not exact, so the robot will not necessarily traverse a perfect square. They also depend on whether you have fresh batteries in the robot because motor speed can decrease considerably as the battery voltage drops.

The settlingTime is a period of time between moves to enable the robot to settle down. Robots cannot instantaneously change speed and/or direction. Without this settling period, driving forward and turning tend to blend into one another, reducing accuracy.

Go to the bottom of the class now and add the following method called `Behavior`:

```
#region behavior

// Iterator to execute the Behavior
IEnumerator<ITask> Behavior()
{
    // Wait for the Lego NXT to initialize, otherwise it will
    // miss the initial command
    yield return Arbiter.Receive(
        false,
        TimeoutPort(10000),
        delegate(DateTime timeout) { });

    LogInfo(LogGroups.Console, "Starting now ...");

    // Make sure that the drive is enabled first!
    _drivePort.EnableDrive(true);

    for (int times = 0; times < repeatCount; times++)
    {
        // Drive along the four sides of a square
        for (int side = 0; side < 4; side++)
        {
            LogInfo(LogGroups.Console, "Side " + side);

            // Start driving and then wait for a while
            _drivePort.SetDrivePower(driveSpeed, driveSpeed);
            yield return Arbiter.Receive(
                false,
                TimeoutPort(driveTime),
                delegate(DateTime timeout) { });
            // Stop the motors and wait for robot to settle
            _drivePort.SetDrivePower(0, 0);
            yield return Arbiter.Receive(
                false,
                TimeoutPort(settlingTime),
                delegate(DateTime timeout) { });

            // Now turn left and wait for a different amount of time
            _drivePort.SetDrivePower(-rotateSpeed, rotateSpeed);
            yield return Arbiter.Receive(
                false,
                TimeoutPort(rotateTime),
                delegate(DateTime timeout) { });
            // Stop the motors and wait for robot to settle
            _drivePort.SetDrivePower(0, 0);
            yield return Arbiter.Receive(
                false,
                TimeoutPort(settlingTime),
                delegate(DateTime timeout) { });
        }
    }
```

```
        // And finally make sure that the robot is stopped!
        _drivePort.SetDrivePower(0, 0);

        LogInfo(LogGroups.Console, "Finished");

    }

    #endregion
```

The `Behavior` method is an iterator. It periodically yields control of the running thread using `yield return`. When an `Arbiter.Receive` has received a message, the CCR resumes execution of the task at the point where it yielded. This means that the code is executed sequentially, but not necessarily by the same thread.

The first step is to wait for 10 seconds (10,000 milliseconds). This ensures that the partner service has started. It is not a reliable or efficient way to wait for a partner, but it is used here for simplicity.

Notice the calls to `LogInfo(LogGroups.Console, "...")` in the code, which are there for debugging and to monitor progress. Various logging methods are available for use with DSS, including `LogError`, `LogWarning`, `LogInfo`, and `LogVerbose`. In this case, the output will be sent to the console (the MRDS Command Prompt window) just as if you had used `Console.WriteLine("...")`. If you omit the `LogGroups.Console` parameter, then the output is sent to the log, not the console. Either way, the output appears in the Console Output web page.

*The drive must be enabled before it will respond to any other commands.*

Next are two nested loops. The outermost loop simply repeats the "dance" pattern, while the innermost loop executes the four sides of the square that make up the basic behavior. The code consists of a series of calls to `SetDrivePower` followed by a wait.

---

### Speed versus Power

In MRDS, there are two ways to make a robot move — by setting its speed or the power to the motors. However, some service implementations don't distinguish between these two methods.

Power normally ranges from 0 to 1. If the power is negative, then the motor runs backwards.

Speed, conversely, is set in meters/second, so it must be calculated based on the radius of the wheels and knowledge of the relationship between the RPM (revolutions per minute) of the wheels and the applied power. This information is rarely available.

*`SetDriveSpeed` does not work with the LEGO NXT in V1.5, so you must use `SetDrivePower`. You can waste a lot of time wondering why your robot is not moving if you try to use `SetDriveSpeed`. For the Boe-Bot, the `SetDriveSpeed` handler simply passes the values across to `SetDrivePower`, which is not correct but at least the robot moves.*

---

The last piece of code you need to add is to execute the `Behavior` from the `Start` method:

```
protected override void Start()
{
    base.Start();
    // Add service specific initialization here.
    SpawnIterator(Behavior);
}
```

Try building the service to ensure that you have not made any mistakes. There is no point in running it yet because it needs a drive service for a robot as a partner.

## Partnering with a Robot Using a Manifest

You now have a service that can control a robot, but you need to associate it with an actual robot so that it can run.

Open `Dance.manifest.xml`, which `DssNewService` created for you. You should see it in the Solution Explorer panel in Visual Studio and you can open it from there. Visual Studio understands XML and will color-code it for you. Add a new Service Record just above the existing Service Record for the Dance service:

```
<!-- Use a Simulated robot -->
<ServiceRecordType>
    <dssp:Contract>http://schemas.tempuri.org/2006/06/simulationtutorial2.html
    </dssp:Contract>
</ServiceRecordType>
```

Notice that the additional service is Simulation Tutorial 2 that comes with MRDS. This creates two robots: a Pioneer 3DX and a LEGO NXT Tribot. Save the manifest. (A complete copy of the manifest is supplied as `ProMRDS\Config\Dance.SimTut2.manifest.xml`.)

You can now run this new service by simply starting the Dance program in the debugger in Visual Studio.

After the simulator starts, there is a short delay and then the Pioneer robot will start doing the "dance." Figure 14-20 shows the Pioneer partway through its behavior. The LEGO NXT Tribot (in the foreground) will not move.

**Figure 14-20**

When the Pioneer robot has finished, the MRDS window will look like Figure 14-21. Notice the status messages as it progresses through each side of the square; they include a timestamp and the name of the service. Dance, dance, dance!

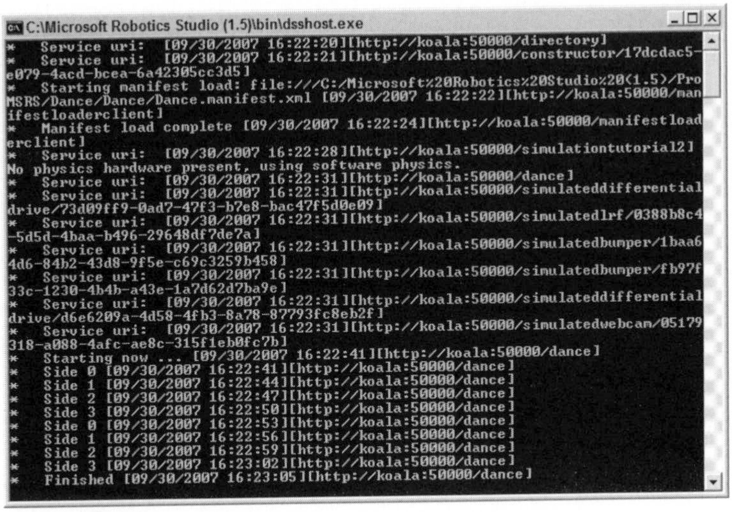

**Figure 14-21**

You might be wondering why the Pioneer robot did the dance when this chapter has been dealing with LEGO NXT robots. The reason is that Simulation Tutorial 2 creates the Pioneer robot first. Because the

Pioneer has a differential drive, the Dance service immediately partners with it. It is possible to enumerate the available services and select the LEGO service, but that's far too complicated for this simple example. Look in the code of the Dashboard for details about how to list the available services.

*If you only want to play with the LEGO NXT simulation, then you can use a saved simulation state and start up the necessary simulation services by running the manifest from the book's website. Located in* ProMRDS\Config\Dance.Simulation.manifest.xml, *this manifest is an example of how to use saved simulator state, a topic that is not covered in this part of the book.*

## Using the LEGO NXT Tribot

Now it is time to switch to a real robot so you can see the power of MRDS. The same code you have just used in the simulator will work on a LEGO NXT Tribot.

Go back to the Dance.manifest.xml in Visual Studio and either remove the service record for Simulation Tutorial 2 or comment it out. In its place, insert the code for the LEGO NXT V2, which is highlighted in the following manifest listing. (The complete manifest is supplied with the name Dance .LegoV2.manifest.xml if you want to check your work.) The simulated robot has been commented out in the following code:

```xml
<?xml version="1.0" ?>
<Manifest
    xmlns="http://schemas.microsoft.com/xw/2004/10/manifest.html"
    xmlns:dssp="http://schemas.microsoft.com/xw/2004/10/dssp.html"
>
  <CreateServiceList>
    <!-- Create a new Lego NXT service -->
    <ServiceRecordType>

        <dssp:Contract>http://schemas.microsoft.com/robotics/2007/07/lego/↵
nxt/brick.html</dssp:Contract>
        <dssp:PartnerList>
            <dssp:Partner>
                <dssp:Contract>http://schemas.microsoft.com/robotics/2007/07/↵
lego/nxt/brick.html</dssp:Contract>
                <dssp:Service>legov2.nxt.brick.config.xml</dssp:Service>
                <dssp:PartnerList />
                <dssp:Name>dssp:StateService</dssp:Name>
            </dssp:Partner>
        </dssp:PartnerList>
        <Name>Brick</Name>
    </ServiceRecordType>
    <ServiceRecordType>

        <dssp:Contract>http://schemas.microsoft.com/robotics/2007/07/lego/↵
nxt/drive.html</dssp:Contract>
        <dssp:PartnerList>
            <dssp:Partner>
                <dssp:Contract>http://schemas.microsoft.com/robotics/2007/07/↵
lego/nxt/drive.html</dssp:Contract>
                <dssp:Service>legov2.nxt.drive.config.xml</dssp:Service>
                <dssp:PartnerList />
                <dssp:Name>dssp:StateService</dssp:Name>
```

```
                </dssp:Partner>
                <dssp:Partner>
                    <dssp:Contract>http://schemas.microsoft.com/robotics/2007/07/↵
lego/nxt/brick.html</dssp:Contract>
                    <dssp:PartnerList />
                    <dssp:Name>Nxt</dssp:Name>
                    <dssp:ServiceName>Brick</dssp:ServiceName>
                </dssp:Partner>
            </dssp:PartnerList>
            <Name>legonxtdrive</Name>
        </ServiceRecordType>

    <!-- Use a Simulated robot -->
<!--
    <ServiceRecordType>

            <dssp:Contract>http://schemas.tempuri.org/2006/06/↵
simulationtutorial2.html</dssp:Contract>
        </ServiceRecordType>
-->
        <ServiceRecordType>
            <dssp:Contract>http://www.promrds.com/contracts/2007/10/↵
dance.html</dssp:Contract>
            <dssp:PartnerList>
                <dssp:Partner>

                    <dssp:Contract>http://www.promrds.com/contracts/2007/10/↵
dance.html</dssp:Contract>
                    <dssp:Service>dance.config.xml</dssp:Service>
                    <dssp:PartnerList />
                    <dssp:Name>dssp:StateService</dssp:Name>
                </dssp:Partner>
            </dssp:PartnerList>
        </ServiceRecordType>
    </CreateServiceList>
</Manifest>
```

Make sure that you save the manifest.

There is one last step before you can try out the program using a real LEGO NXT Tribot: Make sure that you have the appropriate config files. Otherwise, you will have to configure your Tribot yet again. You can get all of the config files from the ProMRDS\Config directory and copy them into your solution folder along with the manifest. Alternatively, you could copy your new manifest to the Config directory and run it from there.

Rebuild the Dance service just to be sure everything is okay, and then run the Dance service in the debugger again. Make sure that your LEGO NXT is turned on first and that you have a couple of meters of clear space around the robot. You don't want it running into anything!

### Using the Boe-Bot

If you have a Boe-Bot, then you need to create a different manifest. You can copy the necessary code from Parallax.BoeBot.manifest.xml. It has a similar pattern to the LEGO NXT, including a brick service (BASICStamp2) and a drive (BSDrive). Copy the service records that are highlighted below and paste them into `Dance.manifest.xml` in place of the LEGO code (if you did the previous section) or the Simulation Tutorial 2 (if you skipped the previous section):

```xml
<?xml version="1.0" encoding="utf-8"?>
<Manifest
    xmlns="http://schemas.microsoft.com/xw/2004/10/manifest.html"
    xmlns:dssp="http://schemas.microsoft.com/xw/2004/10/dssp.html"
    >

  <CreateServiceList>

      <!--Start BasicStamp2 Brick -->
      <ServiceRecordType>
        <dssp:Contract>http://schemas.microsoft.com/robotics/2007/06/↵
basicstamp2.html</dssp:Contract>
        <dssp:PartnerList>
          <!--Initial BasicStamp2 config file -->
          <dssp:Partner>
            <dssp:Service>Parallax.BoeBot.Config.xml</dssp:Service>
            <dssp:Name>dssp:StateService</dssp:Name>
          </dssp:Partner>
        </dssp:PartnerList>

      </ServiceRecordType>

      <!--Start the BoeBot drive service-->
      <ServiceRecordType>
        <dssp:Contract>http://schemas.microsoft.com/robotics/2007/06/↵
bsdrive.html</dssp:Contract>
        <dssp:PartnerList>
          <!--Initial Drive Configuration File -->
          <dssp:Partner>
            <dssp:Service>Parallax.BoeBot.Drive.Config.xml</dssp:Service>
            <dssp:Name>dssp:StateService</dssp:Name>
          </dssp:Partner>
        </dssp:PartnerList>

      </ServiceRecordType>

  </CreateServiceList>

</Manifest>
```

Save the manifest. Make sure that you copy the necessary `config` files from the `ProMRDS\Config` folder into your solution folder so that they are with the manifest. Run the Dance service in the Visual Studio debugger.

## Managing Multiple Manifests

In the previous sections, you were told to copy the `config` files into your solution folder so that they can be found when the manifest is started. However, it is preferable to keep all of your `config` files and manifests together because this makes maintenance easier.

Copy `Dance.manifest.xml` from the solution folder to `ProMRDS\Config`. Then, in Visual Studio, click Project ⇨ Dance Properties. Select the Debug tab in the Properties window. In the command-line arguments, you should see the parameters that will be passed to `DssHost.exe` in order to start the Dance service. You need to change the manifest parameter as shown here:

```
-port:50000 -tcpport:50001 -manifest:"ProMRDS/Config/Dance.manifest.xml"
```

Notice that the path points to the `ProMRDS\Config` folder (and there is no slash at the beginning). Because you have previously run the Tribot and/or Boe-Bot from manifests in this directory, appropriate `config` files are already in the directory.

Having to copy your manifest to `ProMRDS/Config` every time you change it is a little inconvenient if you are likely to change it often (although usually you would set it up and leave it alone). To avoid getting out of sync, you can add a pre-build command in the Project Properties in the Build Events tab. The command line to add is as follows:

```
copy "$(ProjectDir)dance.manifest.xml" "$(ProjectDir)..\..\Config\"
```

An alternative to constantly changing the manifest is to create different manifests for different robots. Five manifests are provided in the `ProMRDS\Config` directory:

❑   `Dance.BoeBot.manifest.xml`

❑   `Dance.Lego.manifest.xml`

❑   `Dance.LegoV2.manifest.xml`

❑   `Dance.SimTut2.manifest.xml`

❑   `Dance.Simulation.manifest.xml`

In addition, there is a batch file in the MRDS `bin` folder called `Dance.cmd` that can be used to run any of these five manifests using a command-line parameter. This might seem a little excessive, but it is intended to illustrate different ways to handle manifests.

## Reviewing the Results

Watch carefully as the robot attempts to inscribe a square on the floor. If you have fresh batteries and you are lucky enough to have a well-matched pair of motors, then the robot might do a reasonably good job. However, most people find that the robot does not drive around a square at all.

Stop the debugger and run the program again. Do this several times. What do you notice? It is quite probable that each time you run the program you get a different result. Even though it only does two iterations of the square, it is unlikely that your robot will end up where it started or even be facing in exactly the same direction. (Of course, you have not tuned the values of timing parameters, but that isn't the point.)

If you look closely, you will notice that when the robot is supposed to be driving forward it often veers slightly to one side — usually the same side — every time it moves forward. This consistent error in the motion of the robot accumulates into larger and larger errors in the *pose* (position and orientation) of the robot the more times it dances around the square. Welcome to the real world.

There are many reasons for these errors — motors that are not exactly matched in terms of their performance; wheel slippage; slightly different wheel diameters; motors or axles that are not aligned squarely; and so on. You cannot avoid these problems. The best you can do is try to design your application so that the errors are reduced, compensated for, or not relevant.

What you have seen here is an *open-loop control system* — there is no feedback to help the robot perform its task correctly. If anything goes wrong, the robot has no way to determine that. Contrast this with the PID controller described in the previous chapter.

Many robots rely on *reactive behaviors*. This means that the robot constantly reacts to what it sees and acts accordingly. For example, following a wall and even wandering around a maze can be done via reactive behavior. In this case, it does not matter much if the robot can drive straight or not because the control program continually adjusts the robot's heading and speed so that it does not bump into walls.

Conversely, if your objective is to build an accurate map of the environment, then you will want the robot to know exactly where it is at all times. Otherwise, the information in the map will be wrong. This task is known as *simultaneous localization and mapping*, or *SLAM*. (In Europe it is also known as *concurrent mapping and localization*, or *CML*). It is an active area of research.

Localization refers to the process of finding where you are on a map. Unfortunately, if your objective is to build a map, then you don't have a map to begin with! This is a "chicken or the egg" situation, i.e., which comes first? This is why SLAM includes the word "simultaneous."

Without putting too fine a point on it, performing SLAM with the limited capabilities of the LEGO NXT or the Boe-Bot is not really possible. One of the requirements for SLAM that is often overlooked is the information content necessary to produce reliable localization. A laser range finder typically has a 180-degree field of view, with measurements taken every degree or even half a degree and distances measured to an accuracy of about 1cm or 2cm out to a range of 20 meters. Contrast this with the LEGO NXT Ultrasonic Sonar, which is one single data point (not 180 measurements) with a maximum range of 255cm and an accuracy of only ±3cm. If you were planning to have your robot map your house, you might want to reconsider.

However, if your robot is operating in a fairly small environment and you already have a map, then you might be able to localize by taking a series of readings. Provided that the robot can turn around accurately, even half a dozen distance readings might be sufficient to localize itself against an existing map.

If you are using a Boe-Bot, however, there is more bad news. The IR sensors on the Boe-Bot are not range sensors — they cannot provide reliable distance measurements. Therefore, it is almost impossible to localize a Boe-Bot or build a reasonable map. Of course, none of this matters if your objective is to learn about robotics and have some fun.

## Improving the Behavior

Part of the problem with the original "square dance" is that it is highly unlikely to be square. Using timers to control motions is not very accurate for several reasons:

❑ The timers under Windows are not very reliable, and the CCR doesn't really improve on them. If the PC suddenly performs garbage collection or flushes the disk cache, then a timer might run for longer than you requested.

❑ Motor speed varies depending on the battery level and it is not linear. For very short time periods, the motor might not have enough time to get up to full speed. Conversely, once you reach a certain power level the motor is rotating at its top speed, so increasing the power will have no effect.

These problems mean that there is no simple relationship between the running time of the motors and the distance traveled.

This is where wheel encoders might seem like a good solution. A wheel encoder measures a wheel's amount of rotation. This is usually specified as a number of "ticks per revolution." In the case of the LEGO NXT motors, for example, the encoder can be set to either 6 or 360 ticks per revolution.

Unfortunately, the Boe-Bot does not have wheel encoders, so it has no option except to use timers. This leaves you with two options: let the PC do the timing or let the robot do the timing. It is more accurate to get the robot to time the moves, but there are disadvantages as well, as discussed in Chapter 16.

There are two methods defined by the generic Differential Drive that are designed to take advantage of wheel encoders: `DriveDistance` and `RotateDegrees`. As its name suggests, `DriveDistance` can be used to drive the robot forwards or backwards a specified distance (in meters, with negative values meaning backwards). `RotateDegrees` rotates the robot by turning the wheels in opposite directions. Counterclockwise rotations are positive, and clockwise is negative.

*The `RotateDegrees` and `DriveDistance` methods are not implemented for the LEGO NXT, and they throw an exception that will crash your program. At least you will know they are not implemented! However, the newer LEGO NXT V2 services do implement `RotateDegrees` and `DriveDistance` using the wheel encoders. You can actually see the encoders in operation if you view the Drive service in a web browser (refer to Figure 14-9).*

*The Parallax BoeBot services that come with MRDS do not implement `RotateDegrees` or `Drive-Distance` either, but two implementations are provided in Chapter 16 and are included in the Parallax BoeBot services from the website for this book (www.proMRDS.com. or www.wrox.com).*

It is easy to modify the Dance service to use these new methods:

**1.** Insert a `using` statement at the top of the code:

```
using drive = Microsoft.Robotics.Services.Drive.Proxy;
// Added for the Fault class
using W3C.Soap;
```

The `using W3C.Soap` is not essential, but it introduces the `Fault` class, which can be returned if there is an error, or if a `DriveDistance` or `RotateDegrees` request is interrupted by another request (and therefore cancelled).

**2.** In the Constants region, add the following values:

```
// Values for "exact" movements using DriveDistance and RotateDegrees
const bool controlledMoves = true;  // Use controlled (not timed) moves
const bool waitForMoves = true;     // Wait for response messages
const float driveDistance = 0.30f;  // Drive 30cm
const float rotateAngle = 90.0f;    // Turn 90 degrees to the left
```

The purpose of these constants is as follows:

❑ controlledMoves determines whether DriveDistance and RotateDegrees are called or the simple approach of using timed moves is used instead.

❑ waitForMoves indicates whether the code should wait for DriveDistance and Rotate-Degrees to send a response message indicating that the motion is complete.

❑ driveDistance is the actual distance to drive forwards expressed in meters.

❑ rotateAngle is the amount to rotate by (with positive values being to the left).

3. Now find the Behavior method and replace the code inside the innermost for loop as shown in the following code. The first section of code uses the DriveDistance and RotateDegrees requests, but calls them inside an Arbiter.Choice in order to wait for the response message that is sent when the motion completes. This means that timers are not used. Note that these requests might return a Fault instead of the default response. The error handling isn't good but it illustrates how a local variable, called success, can be accessed inside a delegate:

```
// Drive along the four sides of a square
for (int side = 0; side < 4; side++)
{
    LogInfo(LogGroups.Console, "Side " + side);

        if (controlledMoves && waitForMoves)
        {
            bool success = true;

            // Drive straight ahead
            yield return Arbiter.Choice(
                _drivePort.DriveDistance(driveDistance, driveSpeed),
                delegate(DefaultUpdateResponseType response) { },
                delegate(Fault f) { success = false; }
            );

            // Wait for settling time
            yield return Arbiter.Receive(
                false,
                TimeoutPort(settlingTime),
                delegate(DateTime timeout) { });

            // Now turn left
            yield return Arbiter.Choice(
                _drivePort.RotateDegrees(rotateAngle, rotateSpeed),
                delegate(DefaultUpdateResponseType response) { },
                delegate(Fault f) { success = false; }
            );

            if (!success)
                LogError("Error occurred while attempting to drive robot");
        }
```

If the world worked perfectly, then the preceding code would be all that you need. However, there is a problem with the LEGO NXT V2 services: Sometimes the robot stalls before reaching its target destination. This happens because the motor speed is ramped down as the target is approached in order to avoid overshooting it. However, below a certain power level, there is not enough torque to keep the robot moving and it stalls. This is why the `waitForMoves` flag was added to the code — to turn off this behavior.

To overcome this problem but still take advantage of the wheel encoders on the LEGO NXT, the following section of code is a hybrid of the timed moves and the exact moves. If you look carefully, you will see that it is the same code that was originally used for the timed moves except that now the `DriveDistance` and `RotateDegrees` requests replace the `SetDrivePower` requests if `controlledMoves` is set to `true`:

```
else
{
    // Start driving and then wait for a while
    if (controlledMoves)
        _drivePort.DriveDistance(driveDistance, driveSpeed);
    else
        _drivePort.SetDrivePower(driveSpeed, driveSpeed);

    yield return Arbiter.Receive(
        false,
        TimeoutPort(driveTime),
        delegate(DateTime timeout) { });

    // Stop the motors and wait for robot to settle
    _drivePort.SetDrivePower(0, 0);
    yield return Arbiter.Receive(
        false,
        TimeoutPort(settlingTime),
        delegate(DateTime timeout) { });

    // Now turn left and wait for a different amount of time
    if (controlledMoves)
        _drivePort.RotateDegrees(rotateAngle, rotateSpeed);
    else
        _drivePort.SetDrivePower(-rotateSpeed, rotateSpeed);

    yield return Arbiter.Receive(
        false,
        TimeoutPort(rotateTime),
        delegate(DateTime timeout) { });

    // Stop the motors and wait for robot to settle
    _drivePort.SetDrivePower(0, 0);
    yield return Arbiter.Receive(
        false,
        TimeoutPort(settlingTime),
        delegate(DateTime timeout) { });
}
}
```

The time delays in this new code are simply to wait for the motions to complete. You need to make sure that they are long enough. For example, change both time delays to 1500. That way, the settling period between driving and turning might be unnecessary. However, stopping the motors after each motion is still a good idea in case the motor was stalled.

Try out this new code and see if it improves the square-drawing ability of your robot. Sometimes it might draw close to a perfect square, but if you run it often enough, you will probably see that it draws some misshapen squares as well, so using wheel encoders is not the perfect solution you might have expected.

## *Flashing and Beeping*

Children love lights and sounds. The Boe-Bot has two LEDs (and you could add a third one using the spare I/O pin) as well as a speaker. The LEGO NXT brick has a speaker but no lights. (Although you can display text messages with the LEGO NXT V2 services, they are hard to read on the LCD display.)

Unfortunately, MRDS does not define a generic service that can be used to manipulate digital I/O ports or play tones on a robot. Therefore, you have to use the Brick services directly.

To connect to a Brick, you need to establish a partnership. The process for this is quite simple:

1. Open the Dance service in Visual Studio. You need to add a reference (see Chapter 3) to the appropriate brick. Depending on the robot that you have, you will need either of the following:

   ```
   nxtbrick.y07.m07.proxy   (for LEGO NXT V2)
   BASICStamp2.Y07.M06.Proxy   (for Boe-Bot)
   ```

   It does no harm to add references to both of these, but it is clearly not necessary if you only have one of these robots. The sample code has both references.

2. Once you have added the references, insert the corresponding `using` statement at the top of `Dance.cs`. The two statements are shown here:

   ```
   using legonxt = Microsoft.Robotics.Services.Sample.Lego.Nxt.Brick.Proxy;

   using stamp = Microsoft.Robotics.Services.BasicStamp2.Proxy;
   ```

   *The LEGO NXT V2 services have departed from the previous pattern and are located under the* Sample *branch of the namespace. Also notice that each of them uses an alias to make it easier to reference methods and properties.*

3. Locate the region called `partners` and expand it. The Dance service needs to partner with the correct Brick service. Because the code will be specific to the particular brick, it doesn't make sense to put the partnership in the manifest. At this point you start to lose the hardware independence.

   ❑ If you are using a LEGO NXT, then insert the following code:

   ```
   // Add a LEGO NXT V2 partner to allow direct access to its methods
   [Partner("LegoNxtV2", Contract = legonxt.Contract.Identifier,
   CreationPolicy = PartnerCreationPolicy.UseExisting)]
   legonxt.NxtBrickOperations _legonxtPort = new legonxt.NxtBrickOperations();
   ```

   ❑ If you are using a Boe-Bot, insert this code instead:

```
        // Add a BASICStamp2 partner to allow direct access to its methods
        [Partner("Stamp", Contract = stamp.Contract.Identifier, CreationPolicy =
PartnerCreationPolicy.UseExisting)]
        stamp.BasicStamp2Operations _stampPort = new stamp.BasicStamp2Operations();
```

*Do not add both of these to the service at the same time! It makes no sense to partner with two different bricks (unless you want to control two robots simultaneously). If you check the code supplied on the book's website (www.proMRDS.com or www.wrox.com), you will find that conditional compilation has been used to enable the same source code to be used for the LEGO NXT, the Boe-Bot, and Simulation. This conditional code is omitted here for clarity.*

**4.** The `PartnerCreationPolicy` for the brick is set to `UseExisting`. This is important because your manifest will be creating the brick anyway, and you don't want the service to create a second instance of the Brick service, which would happen if you used a different policy.

When a service partner is specified this way, dsshost will try to establish the partnership for a short period of time before giving up. This enables all of the services enough time to start up and add themselves to the service directory. The code also creates a new port that can be used for sending requests to the Brick service (and receiving responses).

**5.** With the partnership established, you can call the appropriate brick method to play a tone:

❑ For the LEGO NXT V2, the following code should be added at the top of the innermost `for` loop in the `Behavior` method:

```
for (int side = 0; side < 4; side++)
{
    LogInfo(LogGroups.Console, "Side " + side);

    // Make a beep (no lights to flash on a LEGO brick!)
    _legonxtPort.PlayTone(3500 + side * 250, 50);
```

❑ The `PlayTone` request takes a frequency (in hertz) and a duration (in milliseconds) as its parameters. This causes the LEGO NXT brick to "beep." Notice that the frequency increases for each side of the square just to make it a little more interesting.

❑ For the Boe-Bot, there is a little more code:

```
for (int side = 0; side < 4; side++)
{
    LogInfo(LogGroups.Console, "Side " + side);

    byte op1, op2;

    // Toggle the lights to make it pretty and beep as well :-)
    op1 = ((side & 1) != 0) ? (byte)stamp.PinOperations.OUTPUT_HIGH :
(byte)stamp.PinOperations.OUTPUT_LOW;
    op2 = ((side & 2) != 0) ? (byte)stamp.PinOperations.OUTPUT_HIGH :
(byte)stamp.PinOperations.OUTPUT_LOW;
    // Pins 14 and 15 are LEDs
    _stampPort.SetPins(14, op1, 15, op2);
    // The speaker is on pin 4, but we don't need to know this because
    // the firmware does the tone
    _stampPort.PlayTone(3500 + side * 250, 50);
```

*The code for the Parallax Boe-Bot on the book's website has been enhanced so that it supports some additional functions. The SetPins and PlayTone requests are two of these functions. You must use the Parallax services from the book's website (www.proMRDS.com or www.wrox.com) in order for this code to work. It will not work with the MRDS V1.5 Parallax services.*

The first step in the Boe-Bot code is to figure out the binary pattern for the two LEDs so that they display the side number for the square (as well as look pretty).

The Boe-Bot uses a microcontroller called a BASIC Stamp 2 to control the robot. Microcontrollers typically have several digital I/O ports that are grouped together in bytes or words. The BASIC Stamp 2 on the Boe-Bot has 16 I/O pins. These can be configured as either inputs or outputs. In theory, you should set pins 14 and 15 to output mode before using them, but this is the default mode.

The SetPins request accepts two pin numbers and the operations to perform on those pins. (It is modeled directly on an internal function that already existed in the Parallax service). The operations that can be performed are defined in an enum called PinOperations, as shown in the following table:

| Value | Function |
|---|---|
| NO_OP | No operation |
| SET_OUTPUT | Set pin to output mode |
| SET_INPUT | Set pin to input mode |
| OUTPUT_HIGH | Set pin to logic High (5V) |
| OUTPUT_LOW | Set pin to logic Low (0V) |

The NO_OP value enables the manipulation of a single pin. The others are self-explanatory.

*Be careful if you update the pins directly because all of them except pin 11 have specific purposes. In particular, pins 0, 1, 5, and 6 are connected to the eb500 Bluetooth module and you should leave them alone!*

The last step in the Boe-Bot code is to call PlayTone. This has been deliberately designed to have the same parameters as the PlayTone request for the LEGO NXT, i.e., frequency and duration. Recall that there are no standards yet for bricks (see Chapter 17). Using the same signature, however, enables the code to be the same regardless of which brick is used. However, due to the protocol used by the Boe-Bot, these values are down-sampled by a factor of 50 before being sent to the robot.

You might want to add some code to play a final parting tone and turn the LEDs off when the robot has finished its dance. See if you can figure this out yourself, and then check the sample code.

Now run your new service. It is not exactly musical. The LEGO NXT does a reasonable job of playing tones, but the Boe-Bot is not very good at it. The Boe-Bot speaker volume varies with the frequency, which is not something that you can control. However, you now have a way to warn people to get out of the way or for your robot to express its emotions.

> Although *you* might enjoy it, your office co-workers might get annoyed with the constant beeping of your robot. Be thoughtful in your coding.

# Exploring Using Sensors

Although you now have a working service that can exercise your robot, the robot is still effectively blind. To make it interact with its environment, you need to add some code that uses the sensors.

Because the sensors on the LEGO NXT Tribot and the Parallax Boe-Bot are quite different, this section develops two separate services. You should read through both of them because they illustrate different aspects of MRDS.

You built a service from scratch at the beginning of this chapter, so to save time, this section steps you through code supplied on the book's website. You can create a new service if you want the practice, but it isn't necessary. These services can be found in the `ProMRDS\Chapter14` folder and are called `WanderLegoV2` and `WanderBoeBot`.

## *Reading from the Sensors*

First, you must know what type of sensors you are going to use so that you can include the appropriate reference in your project.

If you have created a new service, you should add a reference to `RoboticsCommon.proxy`. This contains definitions for the differential drive and for generic sensors like the contact sensor array. For other sensors, you will have to reference the appropriate DLL as explained below.

Once you have added a reference, insert a `using` statement at the top of your code with a short alias. This is not essential, but it saves a lot of typing.

Second, you need to know how the sensors are connected to the robot, i.e., to which ports they are attached. You probably won't be rebuilding your robot every day, so the assignment of sensors to ports on the robot will most likely remain fixed. The port assignments have to be entered into `config` files for the LEGO NXT V2, but for the Boe-Bot the pin connections are hard-coded into the services.

To read from a sensor, you must first subscribe to it. You do this by issuing a `Subscribe` request specifying both the name of the handler routine that you want called whenever a new update arrives, and which notification port to use. The handler receives information from the sensor inside the `Body` of an `Update` or `Replace` message (depending on the particular implementation).

When you subscribe to a sensor, it sends updates either on a regular basis or whenever the state of the sensor changes. In the case of a *bump sensor*, for example, you will only receive an update when the sensor is pressed or released. As long as it remains in the same state, there won't be any notification messages. A *sonar sensor*, however, tends to send updates fairly frequently because the distance it measures is constantly changing unless the robot is standing still, in which case you might not see any notifications.

# *Making a LEGO NXT Wander Around*

Open the solution from `ProMRDS\Chapter14\WanderLegoV2`. This solution is for the LEGO NXT using the V2 services.

If you look in the References, you will see `NXTCommon` and `NXTBrick`, as well as `RoboticsCommon`. The corresponding `using` statements appear at the top of `WanderLegoV2.cs`:

```
// References to the required services
using drive = Microsoft.Robotics.Services.Drive.Proxy;
using sonarsensor = Microsoft.Robotics.Services.Sample.Lego.Nxt.SonarSensor.Proxy;
using touchsensor = Microsoft.Robotics.Services.Sample.Lego.Nxt.TouchSensor.Proxy;
```

Notice that the sonar sensor and the touch sensor are referenced explicitly. Because of this, the WanderLegoV2 service cannot be used with other types of robots. However, for most people this isn't a problem because they only have one robot!

The `constants` region contains a variety of different constants used in the code. The names and comments should be self-explanatory:

```
#region constants

const float drivePower = 0.5f;   // Power driving forward
const float rotatePower = 0.25f; // Power during rotation
const float veerPower = 0.6f;    // Power during veering away
const float rotateAngle = 45.0f; // Angle for Left turns
const int driveTime = 1200;      // Time to drive forward (millisec)
const int rotateTime = 920;      // Time to rotate

const float backupPower = 0.5f;  // Power while backing up
const int backupTime = 1000;     // Time to back up on physical contact

#endregion
```

The connections to the service partners are made in the `partners` region. Notice that two ports are defined for each of the sensors — one to send requests to the sensor and the other for receiving notification messages containing updated sensor information. As usual, the partner creation policy is specified as `UseExisting` and the actual connections are established via the manifest and `config` files:

```
#region partners

// Partner: NxtDrive, Contract:
http://schemas.microsoft.com/robotics/2006/05/drive.html
[Partner("NxtDrive", Contract = drive.Contract.Identifier, CreationPolicy =
PartnerCreationPolicy.UseExisting)]
drive.DriveOperations _nxtDrivePort = new drive.DriveOperations();

// Partner: NxtUltrasonicSensor, Contract:
http://schemas.microsoft.com/robotics/2007/07/lego/nxt/sonarsensor.html
[Partner("NxtUltrasonicSensor", Contract = sonarsensor.Contract.Identifier,
CreationPolicy = PartnerCreationPolicy.UseExisting)]
sonarsensor.UltrasonicSensorOperations _nxtUltrasonicSensorPort =
new sonarsensor.UltrasonicSensorOperations();
```

```
sonarsensor.UltrasonicSensorOperations _nxtUltrasonicSensorNotify =
new sonarsensor.UltrasonicSensorOperations();

// Partner: NxtTouchSensor, Contract:
http://schemas.microsoft.com/robotics/2007/07/lego/nxt/touchsensor.html
[Partner("NxtTouchSensor", Contract = touchsensor.Contract.Identifier,
CreationPolicy = PartnerCreationPolicy.UseExisting)]
touchsensor.TouchSensorOperations _nxtTouchSensorPort =
new touchsensor.TouchSensorOperations();
touchsensor.TouchSensorOperations _nxtTouchSensorNotify =
new touchsensor.TouchSensorOperations();

#endregion
```

You need to add some code to the `Start` method to set up the subscriptions to the sensors. Subscribe requests are made to both of the sensors, and the handlers are set up as well:

```
protected override void Start()
{
    base.Start();
    // Add service specific initialization here.

    // Subscribe to partners
    _nxtUltrasonicSensorPort.Subscribe(_nxtUltrasonicSensorNotify);
    _nxtTouchSensorPort.Subscribe(_nxtTouchSensorNotify);

    // Add notifications to the main interleave
    base.MainPortInterleave.CombineWith(
        new Interleave(
            new ExclusiveReceiverGroup(
                Arbiter.ReceiveWithIterator<sonarsensor.SonarSensorUpdate>(true,
_nxtUltrasonicSensorNotify, NxtUltrasonicSensorUpdateHandler),
                Arbiter.ReceiveWithIterator<touchsensor.TouchSensorUpdate>(true,
_nxtTouchSensorNotify, NxtTouchSensorUpdateHandler)
                ),
            new ConcurrentReceiverGroup()
            )
    );

}
```

Note that the handlers are added to the `ExclusiveReceiverGroup`, and this is combined with the existing handlers using `MainPortInterleave.CombineWith`.

The handler for the sonar updates receives a `SonarSensorUpdate` as its parameter. The current distance can be obtained from the message `Body`:

```
IEnumerator<ITask> NxtUltrasonicSensorUpdateHandler(sonarsensor.SonarSensorUpdate
message)
{
    if (message.Body.Distance < 35)
    {
        LogInfo(LogGroups.Console, "Turn Left");
```

*(continued)*

*(continued)*

```
            drive.RotateDegreesRequest request = new drive.RotateDegreesRequest();
            request.Power = rotatePower;
            request.Degrees = (double)rotateAngle;
            _nxtDrivePort.RotateDegrees(request);
        }
        else if (message.Body.Distance < 60)
        {
            LogInfo(LogGroups.Console, "Veer Left");
            drive.SetDrivePowerRequest powerRequest = new drive.SetDrivePowerRequest();
            powerRequest.RightWheelPower = veerPower;
            powerRequest.LeftWheelPower = veerPower/2;
            _nxtDrivePort.SetDrivePower(powerRequest);
        }
        else
        {
            drive.SetDrivePowerRequest powerRequest = new drive.SetDrivePowerRequest();
            powerRequest.RightWheelPower = drivePower;
            powerRequest.LeftWheelPower = drivePower;
            _nxtDrivePort.SetDrivePower(powerRequest);
        }
        yield break;

    }
```

The algorithm is quite simple. When the robot gets too close to an obstacle, it turns to the left. If there is an obstacle farther away, it veers to the left (drives in an arc); otherwise, it drives straight ahead. When you run it, the robot tends to get quite close to obstacles before it reacts. This is dependent on how fast it is moving and how well it can see the obstacle. If the obstacle tends to absorb sound, then it is not as "visible" as a hard object that reflects sound well.

The last piece of code is a safety measure. The touch sensor detects a collision with an obstacle and backs the robot away:

```
IEnumerator<ITask> NxtTouchSensorUpdateHandler(touchsensor.TouchSensorUpdate
message)
{
    if (message.Body.TouchSensorOn)
    {
        LogInfo(LogGroups.Console, "Bump!");

        // Back up to get away from the obstacle
        _nxtDrivePort.SetDrivePower(-backupPower, -backupPower);
        LogInfo(LogGroups.Console, ">>> Backing Up");
        yield return Arbiter.Receive(
            false,
            TimeoutPort(backupTime),
            delegate(DateTime timeout) { });
        _nxtDrivePort.SetDrivePower(0, 0);


    }

    yield break;
}
```

Messages are written to the console in various places in the code just so you can see what is happening. They slow down the running of the program, however, so when you are satisfied that your code is working, you should remove them or comment them out. The logging facility in MRDS enables you to set the trace level so that messages can be selectively filtered out.

Note that the reaction to a bump (when the touch sensor is pressed) involves moving backwards for a short period of time. During this time, the robot is not watching the sensors. However, it doesn't have any sensors facing backwards, so there is not much point.

If you run this program with the manifest and `config` files supplied, then the touch sensor should be plugged into sensor port 1 and the sonar sensor should be in port 4.

The robot will wander around trying to avoid obstacles. You can wave your hands in front of it and shepherd it around. You can also push the touch sensor if you want to see it react.

## Making a Boe-Bot Wander Around

Open the solution located in `ProMRDS\Chapter14\WanderBoeBot`. This solution is for the Parallax Boe-Bot using the updated Parallax services supplied on the book's website.

Have a look in the References. The only additional DLL is `RoboticsCommon` because this example only requires the generic Contact Sensor Array to work. The Parallax services implement a Contact Sensor Array using the two infrared sensors and the two whiskers.

*It is possible to use the Contact Sensor Array with the LEGO NXT as well. However, the NXT services attempt to use every possible sensor as a bumper, including the ultrasonic sonar sensor, sound sensor, and light sensor. Under the V2 services, the sonar sensor is further divided into virtual "near" and "far" bumpers. This makes dealing with the NXT Contact Sensor Array much more complicated. Furthermore, the sonar bumpers do not always issue release messages, which makes the logic very convoluted.*

**1.** The following `using` statements are at the top of the file:

```
// Add a reference to a generic Bumper (in Robotics.Common.Proxy)
using bumper = Microsoft.Robotics.Services.ContactSensor.Proxy;
// Add a reference to the Differential Drive
using drive = Microsoft.Robotics.Services.Drive.Proxy;
// Required for Fault class
using W3C.Soap;
```

**2.** As usual, there is a `constants` region:

```
#region constants

const float drivePower = 0.5f;  // Speed driving forward
const float rotatePower = 0.25f; // Speed during rotation
const int driveTime = 1200;     // Time to drive forward (millisec)
const int rotateTime = 920;     // Time to rotate

const float backupPower = 0.5f; // Power while backing up
const int backupTime = 1000;    // Time to back up on physical contact

#endregion
```

**3.** By now you should be familiar with setting up partners. Notice that there are two bumper ports:

```
#region partners

// Partner with a Bumper (Contact Sensor) service
[Partner("bumper", Contract = bumper.Contract.Identifier,
     CreationPolicy = PartnerCreationPolicy.UseExisting)]
// Create a port to send requests
private bumper.ContactSensorArrayOperations _bumperPort =
new bumper.ContactSensorArrayOperations();
// Create the bumper notification port
private bumper.ContactSensorArrayOperations _bumperNotificationPort =
new bumper.ContactSensorArrayOperations();

// Partner with a Differential Drive service
[Partner("drive", Contract = drive.Contract.Identifier,
     CreationPolicy = PartnerCreationPolicy.UseExisting)]
// Create a port to talk to the diff drive
drive.DriveOperations _drivePort = new drive.DriveOperations();

#endregion
```

**4.** The Start method is modified so that it subscribes to the contact sensor array:

```
protected override void Start()
{
    base.Start();

    // Add service specific initialization here.

    // Allocate the necessary storage for the bumper state
    // The size of this array will depend on the particular robot
    _state.bumperStates = new bool[4];
    _state.bumperNames = new string[4];
    for (int i = 0; i < 4; i++)
    {
        _state.bumperStates[i] = false;
        _state.bumperNames[i] = null;
    }

    // Now subscribe to bumper notifications
    SpawnIterator(SubscribeToBumpers);

    SpawnIterator(Countdown);
}
```

This example takes a different approach to handling the bumpers. Additional fields are added to the service state to remember the latest logic states of the various bumpers, including their names. This makes the values visible in a web browser via the service directory. However, this is not the reason for adding them to the state. When you receive a notification from the Contact Sensor Array, it is for a single sensor. The logic for wandering around needs to keep track of the current state of all of the bumpers.

**5.** The additions to the state are made in `WanderBoeBotTypes.cs`.

```
[DataContract()]
public class WanderState
{
    [DataMember]
    public bool[] bumperStates;
    [DataMember]
    public string[] bumperNames;
}
```

Notice that the new properties are marked with the `[DataMember]` attribute and declared as `public`. This is important for making them visible outside the service.

**6.** During initialization, a subscription is made to the bumpers. This is done in an iterator, although it does not have to be:

```
IEnumerator<ITask> SubscribeToBumpers()
{
    // Subscribe to the bumper service
    // Receive notifications on the bumperNotificationPort
    _bumperPort.Subscribe(_bumperNotificationPort);

    // Start listening for updates from the bumper service
    // Note that we use the Exclusive Receiver Group so that
    // two bumper messages cannot be processed at the same time
    Activate(
        Arbiter.Interleave(
            new TeardownReceiverGroup(),
            new ExclusiveReceiverGroup(
                Arbiter.ReceiveWithIterator<bumper.Update>
                    (true, _bumperNotificationPort, BumperHandler)
            ),
            new ConcurrentReceiverGroup()
        )
    );

}
```

**7.** To finish off the initialization, the `Countdown` method is called. This delays starting the robot for 10 seconds. In this case it does have to be an iterator because it uses `yield return` to wait for timeouts:

```
IEnumerator<ITask> Countdown()
{
    // Wait for the robot to initialize, otherwise it will
    // miss the initial command
    for (int i = 10; i > 0; i--)
    {
        LogInfo(LogGroups.Console, i.ToString());
        yield return Arbiter.Receive(
            false,
            TimeoutPort(1000),
            delegate(DateTime timeout) { });
    }
```

*(continued)*

*(continued)*

```
      LogInfo(LogGroups.Console, "Starting now ...");

      // Make sure that the drive is enabled first!
      _drivePort.EnableDrive(true);

      // Start the robot on its way!
      _drivePort.SetDrivePower(drivePower, drivePower);

  }
```

**8.** The drive must be enabled before sending commands to it, and finally the robot sets off driving forwards.

**9.** The code that handles the bumper notifications needs to know which bumper is which. Although it is possible to look at the names of the contact sensors, they will never change so an enum is used:

```
private enum BumperIDs { IR_LEFT = 0, IR_RIGHT, WHISKER_LEFT, WHISKER_RIGHT };
```

**10.** Finally, the bumper handler is responsible for adjusting the motor power based on the combination of active bumpers. The first part of the code simply displays a message indicating which bumper has changed and its state:

```
IEnumerator<ITask> BumperHandler(bumper.Update notification)
{
    string message;
    string bumperName;

    // Find out which bumper this is
    // BoeBot numbers from 1 to 4
    int num = notification.Body.HardwareIdentifier - 1;

    if (string.IsNullOrEmpty(notification.Body.Name))
        bumperName = "NO NAME";
    else
        bumperName = notification.Body.Name.ToLowerInvariant();

    if (!notification.Body.Pressed)
    {
        message = "Bumper " + num.ToString() + " (" + bumperName +
") was released.";
        LogInfo(LogGroups.Console, message);
    }
    else
    {
        message = "Bumper " + num.ToString() + " (" + bumperName +
") was pressed.";
        LogInfo(LogGroups.Console, message);
    }
```

## Processing Bump Notifications

If the hardware identifier is between 1 and 4, then the new state of the bumper is recorded, including the name if it has not been seen before:

*The LEGO NXT V2 issues release messages for each bumper initially. The Parallax services do not do this.*

```
if (num >= 0 && num < _state.bumperStates.Length)
{
    _state.bumperStates[num] = notification.Body.Pressed;
    if (_state.bumperNames[num] == null)
        _state.bumperNames[num] = bumperName;
}
else
{
    message = "Invalid Hardware ID: " + num.ToString();
    LogInfo(LogGroups.Console, message);
}
```

The whiskers are physical contact sensors. If one of them is triggered, then the robot has either collided with an object or it is going to very soon! Therefore, the whiskers are checked first. If either of them is depressed, then the robot backs up for a little while and then turns to the left or right, depending on which of the whiskers was depressed:

```
// Handle the physical touch sensors first because these are the
// most important
if (_state.bumperStates[(int)BumperIDs.WHISKER_LEFT] ||
_state.bumperStates[(int)BumperIDs.WHISKER_RIGHT])
{
    // Back up to get away from the obstacle
    _drivePort.SetDrivePower(-backupPower, -backupPower);
    LogInfo(LogGroups.Console, ">>> Backing Up");
    yield return Arbiter.Receive(
        false,
        TimeoutPort(backupTime),
        delegate(DateTime timeout) { });
    _drivePort.SetDrivePower(0, 0);

    if (_state.bumperStates[(int)BumperIDs.WHISKER_LEFT])
    {
        // Turn right
        LogInfo(LogGroups.Console, ">>> Turn Right");
        _drivePort.SetDrivePower(rotatePower, -rotatePower);
    }
    else
    {
        // Turn left
        LogInfo(LogGroups.Console, ">>> Turn Left");
        _drivePort.SetDrivePower(-rotatePower, rotatePower);
    }
```

```
            // Wait a while
            yield return Arbiter.Receive(
                false,
                TimeoutPort(rotateTime),
                delegate(DateTime timeout) { });

            // Stop the motors and wait for robot to settle
            _drivePort.SetDrivePower(0, 0);

        }
```

Next, the code checks the infrared sensors. If both of them see an obstacle, then the robot backs up. However, if only one of them can see something, then the robot turns in the appropriate direction. If no obstacles are visible, it simply charges straight ahead:

```
        // Now that everything has settled down
        if (_state.bumperStates[(int)BumperIDs.IR_LEFT] && _state.bumperStates[(int)Bum
    perIDs.IR_RIGHT])
        {
            // Both IRs see something so back up
            // Eventually one of them must lose sight of the obstacle
            LogInfo(LogGroups.Console, ">>> Reverse");
            _drivePort.SetDrivePower(-drivePower, -drivePower);
        }
        else if (_state.bumperStates[(int)BumperIDs.IR_LEFT] &&
    !_state.bumperStates[(int)BumperIDs.IR_RIGHT])
        {
            // Obstacle on the left, so turn right until gone
            LogInfo(LogGroups.Console, ">>> Veer Right");
            _drivePort.SetDrivePower(rotatePower, -rotatePower);
        }
        else if (!_state.bumperStates[(int)BumperIDs.IR_LEFT] && _state.bumperStates[(i
    nt)BumperIDs.IR_RIGHT])
        {
            // Obstacle on the right, so turn left until gone
            LogInfo(LogGroups.Console, ">>> Veer Left");
            _drivePort.SetDrivePower(-rotatePower, rotatePower);
        }
        else
        {
            // No obstacles so drive forwards
            LogInfo(LogGroups.Console, ">>> Forward");
            _drivePort.SetDrivePower(drivePower, drivePower);
        }
    }
```

That completes this section on wandering around. Try out the code if you have a Boe-Bot. It is more fun to play with than the LEGO NXT.

# Summary

Setting up and programming a LEGO NXT Tribot and a Parallax Boe-Bot are covered in detail in this chapter. It should be apparent that a lot of real-world challenges affect the operation of a robot, and even seemingly simple tasks might not be easy.

You now know how to create manifests to orchestrate services, and you should understand how to use `config` files. The last part of the chapter explained how to subscribe to sensors and use the information to make simple steering decisions. Along the way, you played with the Dashboard and created a shortcut so that you can easily run your robot.

This chapter has used the services that come with MRDS. The next couple of chapters explain how to write your own services and develop more complex applications.

# 15

# Using a Robotic Arm

The previous chapter discussed remotely controlling a wheeled robot. This chapter covers using a robotic arm, which is substantially different. Although robotic arms can be attached to a mobile robot, they are usually fixed to the floor or a table, and can therefore be directly connected to the PC that controls them with no need for wireless communication or an onboard PC.

There are many different types of articulated, robots, and arms are only a small subset. For example, bipedal humanoid robots are articulated, and so are robots that emulate snakes. However, the robot arms that build cars are probably one of the first and most successful applications of robotics in industry.

Relatively cheap hobby arms are available from several sources, such as Lynxmotion. This chapter shows you how to use a Lynx 6 robotic arm to do some simple operations. The L6 is a small, lightweight arm that is intended for hobbyists and education. The software on the book's website (www.proMRDS.com or www.wrox.com) includes a Lynx6Arm service for controlling the arm, as well as some examples demonstrating how to use it. The original service software and several of the photos in this chapter are courtesy of Lynxmotion.

A Lynx arm is not suitable for industrial use. For that you need something like a heavy-duty arm from KUKA. You might be interested in trying out the KUKA Educational Framework for MRDS, available for download from www.kuka.com.

The KUKA framework contains several tutorials using a simulation of a KUKA LBR3 arm. However, the tutorials rapidly get into complex mathematics. The objective in this chapter is to provide an overview without going into the great depth of the KUKA tutorials.

> *If you have not read Chapter 8, it would be a good idea to go back and do so now. It introduces a lot of key concepts for articulated arms and does so using a simulation so that you don't have to worry about breaking anything if you make a mistake! This chapter assumes that you are familiar with the material in Chapter 8.*

# Introduction to Articulated Arms

The common factor for all articulated robots is *joints*. Joints can take a variety of forms, with the most common type based on a servo that can rotate around a single axis. In this case, the position of the servo is referred to as the *axis angle* or *twist* around the main axis (also called the *local axis*).

The `Joint` class in MRDS describes the type of joint, how it is connected, and several other properties. Each joint has one or more values associated with it that describe the current pose of the joint. Usually, the application maintains these values itself so that it always knows the configuration of the arm. Some arms enable you to query their current position or they provide continuous feedback. This is important for ensuring that the state information in your service actually matches the real state of the arm.

There are two primary types of joints in MRDS: angular and linear. This chapter deals exclusively with angular joints because that is what the L6 arm uses.

Rotation of a *servo* can be changed into a linear motion using appropriate gears. For example, a *gripper* can be built using a servo to spread the fingers apart or squeeze them together. Figure 15-1 shows a close-up of a L6 arm gripper in which the gears and linkages are clearly visible. The gripper servo is underneath the hand and is not visible. The servo in the middle of the wrist is for rotating the wrist, whereas the one at the end of the arm (on the right-hand side of the photo) is for tilting the wrist up and down.

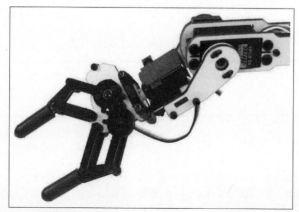

**Figure 15-1**

Pictures of the L6 arm are courtesy of Lynxmotion and are available from its web site (www.lynxmotion.com).

*In Chapter 8, the simulation uses two linear joints to make the gripper. This complexity is hidden from you if you connect to the simulated L6 arm using the generic contract for articulated arms.*

The state of an articulated arm consists of, at a minimum, a list of joints and an *end effector* pose. The purpose of the joint list is obvious, but if you have never used an arm before, the term *end effector* might be new to you.

Most robotic arms have a gripper or some other device at the end of the arm to do useful work. This is called an end effector or sometimes a tool. The location and orientation of the end effector is the most valuable piece of information that you can have. Unfortunately, as explained in a moment, determining the position of the end effector is a complicated process.

The KUKA documentation uses two terms related to the end effector:

❑ **End of Arm (EOA):** The EOA is the very extremity of the arm. In the case of the L6 arm it is the tip of the gripper fingers. As the gripper opens and closes, the linkages cause the tips of the fingers to move slightly closer or farther away from the wrist (as well as side to side). This small variation is not worth worrying about.

❑ **Tool Center Point (TCP):** The TCP is the location inside the tool that is important for operating the tool. For a gripper, this is midway along the fingers, where the gripper makes contact with an object that it is supposed to pick up. If you refer to Figure 15-1 and imagine an object in the center of the gripper, you will understand the difference between EOA and TCP.

## Servo Basics

*Servo motors* are used in remote-controlled (R/C) toys such as cars and airplanes. Due to the large number of motors produced each year, they have become commodity items that are readily available in a range of sizes and types. There are both digital and analog servos, but the difference is not important here.

To move a servo, you send it a series of pulses 50 times per second (20 milliseconds apart), with the width of the pulses controlling the rotation angle of the *servo horn*. Figure 15-2 shows a servo with the corresponding pulses and rotation angles. The white disc on the servo is the horn, although servos often come with a horn shaped like a cross. Notice its orientation.

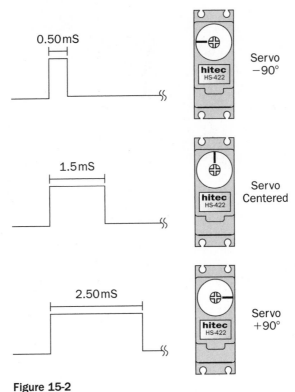

**Figure 15-2**

Reproduced from the SSC-32 Users Manual version 2.0 with permission from Lynxmotion.

The L6 arm has a Lynxmotion SSC-32 controller that is used to send the appropriate pulses to the servos, which means that the MRDS service does not have do to this every 20ms. The SSC-32 can handle up to 32 servos, but the arm only has 6. It interfaces to a PC using a serial port.

A servo motor continues to drive the servo horn until the correct angle is obtained. (Servos have an internal sensor, usually a potentiometer, to measure the angle.) If the servo is prevented from reaching the desired angle, it will keep pushing in the appropriate direction, effectively applying pressure, but this causes high current drain on the power supply, and there is a limited amount of torque that the servo can apply. Once the servo reaches the correct orientation, it resists external attempts to move it and effectively "locks" itself in place. However, as soon as the power is turned off, the servo will "slump" down depending on the weight of the linkages that it is holding.

Most servo motors only need to operate over a range of −45 to +45 degrees, e.g., for the control surfaces for flying a plane. However, it is quite common for servos to have a range of −90 to +90 degrees, and some might operate over −180 to +180 degrees.

Some inexpensive robots, including the Boe-Bot, use modified servo motors in which the potentiometer has been removed or disabled. Because the servo can never find its target location, it keeps driving in the same direction and, in effect, becomes a conventional electric motor.

The last point to note about servos is that when you power them up, the servo controller usually has no idea where the servos are positioned. The interface to a servo consists of only three wires: power, ground, and pulse. There is no way for them to provide feedback. As a result, as soon as you start sending pulses to the servo, it jumps from its initial position to the specified position. For an arm, this can be quite surprising and even dangerous.

# Controlling Joints: Speed and Travel Constraints

Driving an arm to reach a particular position involves sending commands to the various joints. Because the joints cannot instantaneously reach their destinations, you need to employ some strategy to ensure that the servo motors are not overloaded. There are physical limits to how fast a servo can move based on the maximum torque that the motor can apply.

## Speed

In general, the speed of a joint should be kept well below its physical maximum to avoid burning it out. Another speed consideration is that the arm has inertia, and if it moves too fast it will overshoot, causing it to oscillate backwards and forwards as it settles down to the final destination.

The Lynx6Arm service provided with this chapter limits the speed that joints are allowed to move because the L6 arm is a little fragile. The service does this by calculating the difference between the current joint angle and the desired joint angle, calculating the time required based on a maximum angular speed, and then instructing the SSC-32 controller to make a timed move. If a time parameter is not supplied in a command to the SSC-32, the joint jumps to its destination as fast as possible.

The following code in the Lynx6Arm service handles the SetJointTargetPose request that sets the value for a single joint. You don't need to know how it works in order to use the request, but it is interesting to understand how the service works under the hood.

The function starts by looking up the joint details and converting the joint orientation to an angle in degrees. As with many MRDS operations, there is a lot of converting to and from quaternions and axis angles when you are working with joints. You need to become familiar with these — for example, to make a `SetJointTargetPose` request!

```
private IEnumerator<ITask> SetJointPoseHandler(armproxy.SetJointTargetPose update)
{
    // Find out which joint it is from the name
    string name = update.Body.JointName;
    int index = _channelLookup[name];

    // Get the joint
    Joint j = _jointLookup[name];

    // Figure out the desired angle in degrees
    AxisAngle orientation = update.Body.TargetOrientation;
    float orientationAngle = orientation.Angle * Math.Sign(orientation.Axis.X);
    float jointAngle = Conversions.RadiansToDegrees(orientationAngle);
```

The time parameter is calculated based on the change in the joint angle and the `MaxAngularSpeed`, which is a value set in the service state. If you want to make the arm move faster, you can change the value of `MaxAngularSpeed` and recompile the service. (It is not exposed through a configuration file.) However, be aware that making the arm move too fast can make it wobble!

```
    float angleDiff;
    int moveTime;

    // Calculate the difference between the current angle and the
    // requested angle and then figure out the time to take
    angleDiff = Math.Abs(jointAngle - _state.Angles[index]);
    // Calculate the elapsed time for the move
    moveTime = (int)((angleDiff / _state.MaxAngularSpeed) * 1000.0f);
    // If the time is very small, ignore it
    if (moveTime < 20)
        moveTime = -1;
```

The `jointAngle` is saved in a cache in the state for use the next time. It must be converted to a servo angle and then to a pulse width to send to the SSC-32. The conversion function called `JointAngleToServoAngle` allows remapping of the joint angles and is used to make the real arm match the conventions used in the simulated arm. Following the conversion, the information is put into a `SSC32ServoMove` request and sent to the SSC-32 controller. Note that the joint properties also contain a speed parameter that can act to limit the speed of movement.

```
    // Update the joint angle in the state now
    // NOTE: This will also be updated during a Get request
    // with info from the arm itself
    _state.Angles[index] = jointAngle;

    int servoAngle = JointAngleToServoAngle(index, ⏎
(int)(orientationAngle * 180 / Math.PI));
```

*(continued)*

*(continued)*

```
        int pulseWidth = AngleToPulseWidth(servoAngle);

        ssc32.SSC32ServoMove moveCommand = new ssc32.SSC32ServoMove();
        moveCommand.Channels = new int[1] { index };
        moveCommand.PulseWidths = new int[1] { pulseWidth };
        moveCommand.Speeds = new int[1] {(int) physicalmodel.Vector3.Length(
                ProxyConversion.FromProxy(j.State.Angular.DriveTargetVelocity))};
        // Always apply some sort of reasonable time so that the arm does
        // not move suddenly
        moveCommand.Time = moveTime;

        ssc32.SendSSC32Command command = new ssc32.SendSSC32Command(moveCommand);
        _ssc32Port.Post(command);
        yield return Arbiter.Choice(command.ResponsePort,
            delegate(ssc32.SSC32ResponseType response)
            {
                update.ResponsePort.Post(DefaultUpdateResponseType.Instance);
            },
            delegate(Fault fault)
            {
                update.ResponsePort.Post(fault);
            }
        );
    }
```

Notice that the code posts a response back to the caller as soon as the command is sent to the SSC-32. At this stage, the motion will not have completed. There are various options here. In the case of the sample applications in this chapter, the code simply waits for a period of time before issuing a new motion request. Alternatively, the arm service could interrogate the SSC-32 to determine when the motion has completed and only send the response after the command has finished executing. The simulation takes this approach.

## Travel Constraints

There is another architectural issue here. The Lynx6Arm service implements a hard constraint on the speed of arm movements. In the case of the simulation, the limit is applied by the caller specifying the amount of time allowed for the motion to complete. It is up to you whether you want to impose an internal limit and treat the arm as a "black box" or take complete control and suffer the consequences of getting it wrong. However, the Lynx6Arm service tries to protect novice users from themselves.

Ideally, to minimize the stress on the motors and the jerkiness of the motion, each joint should gradually accelerate to some maximum speed, travel at this speed for a while, and then decelerate so that it arrives at its destination. The KUKA documentation defines three different approaches for point-to-point (PTP) motion:

❑ **Asynchronous PTP:** All joints get to their target destinations in the shortest possible time.

❑ **Synchronous PTP:** Joints travel at different speeds so that they all arrive at the same time.

❑ **Fully Synchronous PTP:** The acceleration and deceleration phases are also synchronized to minimize wear and tear on the joints.

As shown in the preceding code, the Lynx6Arm service does not do any accelerating or decelerating. It does apply a maximum speed, however, and most of the time this works fine. The extra sophistication and overhead of a motion planner are not essential.

Although it might be obvious, it is important to realize that an arm cannot move to every possible position within a hemisphere around its base. There are combinations of joint angles that will violate constraints either on the way that the joints and linkages are interconnected or on the servo motors themselves. In addition, of course, there are combinations of angles that will result in the arm crashing into the ground. In fact, an arm has a fairly limited range over which is it safe to operate.

Another problem is that an arm is a cantilever and there can be large stresses on the joints when the arm picks up a heavy object. The L6 arm has a load limit of only about 85g. It can successfully pick up an AA battery, but it is not a good idea to have it pick up anything much heavier than this.

Lastly, if you have never used an arm before, it might not be apparent to you that moving the EOA in a straight line is a nontrivial task. Joints rotate, so changing the angle of a joint invariably means that you have to change the angle of another joint to compensate and keep the EOA moving in a straight line.

This is referred to as *linear motion*, and most industrial applications require linear movements. If you have an industrial robotic arm, the controller itself might be able to do the necessary calculations on-the-fly. In that case, you are off the hook, but if you have a hobby robot arm like the L6, then you need to interpolate a series of points along the motion path and execute them in sequence.

Luckily, if you are only moving a small distance, such as approaching an object to pick it up, the motion of the arm is approximately linear provided that all of the joints move simultaneously in a synchronous PTP fashion. The Lynx6Arm service does not implement linear motions, but as you will see from the sample applications this is not a significant problem.

The KUKA tutorials discuss at length the various approaches to motion planning and control for moving robotic arms. The calculations are quite complex, and are not implemented in the Lynx6Arm service.

# Forward Kinematics

An articulated arm consists of a set of linkages with servos at the joints. The more joints there are, the more complex the calculations are to determine the location of the end effector. Given the values of all of the joint angles, the process of calculating the end effector pose is called a *forward*, or *direct*, *kinematic transformation*.

## Arm Linkage Configuration

Forward kinematics can be done using simple geometry if you know the configuration and measurements of the arm linkages. Using the L6 arm as an example, the forward kinematics calculations can be done as explained here.

*This chapter does not cover the general case of kinematics for all robotic arms, but concentrates on the specific case of the Lynx 6 arm. Unless you are building your own robotic arm, you shouldn't need to write your own code to perform these calculations; the manufacturer should provide you with the necessary software. In the case of the Lynx 6, the RIOS (Robotic arm Interactive Operating System) software that comes with the arm does the necessary calculations and can be used to control the arm. However, this book is about MRDS, so the calculations are embedded in the Lynx6Arm service.*

The physical dimensions of the arm are provided in the following table. All measurements are between joints' centers, except for the last one, which is measured to the tip of the gripper. The values in the table are in inches because these are the units that the manufacturer uses. However, MRDS works in meters so the values need to be converted. (The conversion factor of 1 inch equals 2.54cm is exact because the two systems were aligned a long time ago to allow precise conversions.)

| Link | Length |
|------|--------|
| Base to Shoulder (H also called L0) | 2.5" |
| Upper Arm (L1) | 4.75" |
| Lower Arm (L2) | 4.75" |
| Hand (L3) | 5.75" (L3 = L4 + L5) |
| Wrist to Wrist Rotator (L4) | 2.25" |
| Gripper (L5) | 3.5" |

Forget about the wrist rotation (which happens between L4 and L5). It is not really relevant, as most of the time the wrist will be horizontal. The "tip" of the gripper is always taken to be the point midway between the tips of the two fingers, regardless of how far open the gripper is.

## Finally, a Use for Grade-School Geometry

First, note that the shoulder, elbow, and wrist joints all rotate in the same plane. This is an important point because it allows simple geometry to be used, rather than 3D matrix algebra.

If you look at the arm side-on, the linkages for a typical pose of the robot might look like the simplified diagram in Figure 15-3. The wrist is rotated in the diagram so you can see the two fingers. The hand consists of the wrist rotation mechanism and the gripper, but for the purposes of the example it is only necessary to know the total length of the hand (L3).

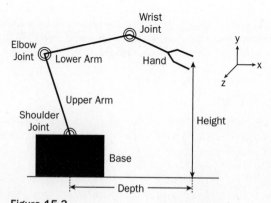

**Figure 15-3**

The height and depth specify the location of the tip of the gripper relative to the center of the base, which is the origin. A right-hand coordinate system is shown in Figure 15-3. It is intended to show the respective axis directions, although the origin should be located at the center of the base. Ignore the coordinate system for now and assume that the robot is in the X-Y plane. Note that rotations are measured counterclockwise around a joint.

As you traverse the arm from the base to the tip, the angles of the joints add up to determine the angle of each link measured with respect to the horizontal. In Figure 15-3, the shoulder is about 110 degrees with respect to the base, the elbow is at −95 degrees with respect to the upper arm, and the wrist is at −30 degrees with respect to the lower arm. (The diagram is not drawn to scale.)

Consider each link as a simple straight line that is the hypotenuse of a right-angled triangle. The contribution of each link to the total depth and height are the horizontal and vertical edges of this triangle and can be calculated based on the angle of the link, α, and its length, L. Figure 15-4 shows the side-on view of a single link, which might be the upper arm, lower arm, or hand.

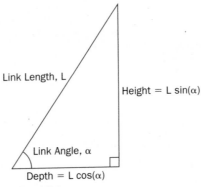

**Figure 15-4**

For the L6 arm in the position in Figure 15-3, the calculations are shown in the following table:

| Element | Length | Angle (Total) | Height | Depth |
|---|---|---|---|---|
| Base to Shoulder (H) | 2.25" | | 2.25 | 0.0 |
| Shoulder | | 110° (110°) | | |
| Upper Arm (L1) | 4.75" | | 4.46 | −1.62 |
| Elbow | | −95° (15°) | | |
| Lower Arm (L2) | 4.75" | | 1.23 | 4.59 |
| Wrist | | −30° (−15°) | | |
| Hand (L3) | 5.75" | | −1.49 | 5.55 |
| Totals | | | 6.45 | 8.52 |

To determine the angles for each of the links, you need to take into account the way that the servo is fixed to the respective links and translate the servo angles to link angles. For example, when the shoulder servo has a setting of zero degrees, it is standing upright, i.e., the upper arm (L1) is at 90 degrees. A summary of the servo orientations and directions of travel is given in the following table:

| Servo | Zero Position | Positive Direction (Increasing Angle) |
|---|---|---|
| Base | Facing away from SSC-32 | Clockwise (right) |
| Shoulder | Vertical | Lean backwards |
| Elbow | Right angle to upper arm* | Bend down |
| Wrist | In line with lower arm | Bend up |
| Wrist Rotate | Horizontal | Counterclockwise (left) |

*If you look at the physical mounting of the elbow, you can see that the screws are offset so that it actually makes a 70-degree angle between the upper arm and lower arm when it is in the zero position. However, the software adjusts for this difference, so if you request an angle of zero for the elbow servo, it moves to make a 90-degree angle between the upper and lower arms. This offset is necessary because if the upper arm moved all the way to a servo setting of +90, it would hit the cross bracing in the lower arm. It is free to move to −90 in the opposite direction, although this is not advisable. In addition, the simulation uses a direction of rotation for the elbow that is opposite to the real arm, but this is hidden from you by the arm service.

The depth and height show the position of the tip relative to the base. The height immediately translates to the Y coordinate in 3D space. However, the base can also rotate, so to obtain the X and Z coordinates you use the depth and the angle of base rotation. The depth corresponds to the radius of rotation measured on the ground (looking down at the arm from above).

To complicate matters, the base servo starts with the arm facing away from the SSC-32 controller board and it moves clockwise for increasing servo angles. If you are sitting beside the robot with the SSC-32 on your right, which is the view that first appears in the simulation, the X coordinate is always negative. Unlike the simulation, the real L6 cannot rotate by 180 degrees in either direction.

Combining all of these calculations results in the (X, Y, Z) position of the tip, given all of the servo angles. The code to do this is in a routine called `ForwardKinematics`, which is in `Lynx6Arm.cs` in the folder `ProMRDS\Chapter15\Lynxmotion\Lynx6Arm`. There is no point in reproducing the code here. If you are interested you can read through it.

# Inverse Kinematics

Unless you have a very expensive arm, it is not possible to instruct it to go to a particular location in 3D space. Instead, you must calculate the values of all of the joint angles that correspond to the desired pose. This is called *inverse kinematics*, and you can find books with entire chapters dedicated to this topic.

> As noted in Chapter 8, there are spreadsheets available on the Lynxmotion website for performing the forward and inverse kinematics calculations.

The Lynx6Arm service developed for this chapter was based on the original MRDS V1.0 code from the Lynxmotion website. In order to simplify the code, the Lynx Inverse Kinematics service was dropped and replaced by direct calculations inside the Lynx6Arm service itself. The algorithm was taken from Hoon Hong's ik2.xls spreadsheet for inverse kinematics posted on the Lynxmotion website. This is much simpler and easier to understand (and debug).

Providing a general service to solve the kinematics equations is a good idea, but every articulated robot has a different design, and it is a difficult problem to generalize. For example, a humanoid robot is a very different configuration from a robotic arm, but it still needs to solve inverse kinematics equations to make it walk.

If you followed the forward kinematics discussion, then you can imagine the inverse kinematics as the opposite set of calculations. Unfortunately, in general, there can be more than one possible solution or no solution at all.

Multiple solutions to reaching a particular 3D point in space occur when you have a wrist. You can tilt your wrist up or down, but the tips of your fingers can still be at the same place. To resolve this issue, you must also specify the tilt angle of the wrist. This is also important when using a gripper because you often want to have the gripper horizontal when it picks up an object. That means four pieces of information are required: X, Y, Z, and P, where P is the wrist/hand angle.

There might be no solution if you have specified a point that is outside the reach of the robot. Even close to the robot, there can be places that it cannot reach. For example, the Lynx 6 would have a hard time trying to scratch itself — for the same reason that you cannot touch your elbow with your fingers (the elbow on the same arm as your hand — not the other arm, that's cheating!).

Even if you do find a solution to the equations, the arm might be unstable if placed in that position. The Lynx6Arm service places some constraints on the servo angles to try to avoid overtaxing the servos. However, it is still possible to damage the arm if you are not careful.

# Setting Up a Lynxmotion L6 Arm

The Lynx 6 Robotic Arm Combo Kit for PC available from Lynxmotion is a relatively inexpensive kit that the authors use to demonstrate many of the principles of robotic arms. For full details see www .lynxmotion.com/Category.aspx?CategoryID=25.

The L6 kit has six degrees of freedom: Base Rotate, Shoulder, Elbow, Wrist, Wrist Rotate, and Gripper. The somewhat cheaper L5 kit does not include the Wrist Rotate, but we recommend that you save your pennies and buy the L6 because the extra degree of freedom is useful. Figure 15-5 shows the L6 arm. The SSC-32 controller board is at the right-hand side and you can see the serial port socket.

Figure 15-5

# Building the Lynxmotion L6 Arm

Building the arm will take you several hours. The assembly instructions are available online at the Lynxmotion website, and it might be wise to read them before you decide to buy an arm. As you build, consider the following:

❑ The pieces are made out of lexan, which is very tough. They are laser cut, but the cuts don't go all the way through in some places. Pushing out the smaller holes can be difficult, and the pieces have very tight tolerances. You should carefully drill out the holes with the correct size drill bit, and you might want to use a small file to trim a little off the various tongues before fitting them into their mating slots.

❑ An SSC-32 servo controller is supplied as part of the kit. It is already assembled so you don't need to do any soldering. It can handle up to 32 servos via a serial port connection with a PC. You can see the circuit board on the right-hand side in Figure 15-5. A very good manual is available for the SSC-32, and you should read it in conjunction with this chapter. It explains the format of the serial commands that can be sent to the arm. This information isn't necessary, but it gives you a better idea of the capabilities of the SSC-32.

❑ Although the L6 arm comes with some sophisticated control software called RIOS (robotic arm interactive operating system), the objective of this book is to show you how to control it using MRDS. The Lynx6Arm service on the book's website (www.proMRDS.com or www.wrox.com) offers a simple interface for the arm with the advantage that you have the source code.

❑ LynxTerm is basic software that enables you to manipulate the robot's joints individually. The Motion Recorder service included in this chapter provides similar functionality for moving the joints. You can use either of these to test your arm once you have built it, although the assembly instructions refer to LynxTerm.

# Starting the Lynxmotion L6 Arm

Of course, once you've built the arm you'll want to try it out. Before you do, you might want to consider the following:

❑ Make sure that you have plenty of clear space around the robot. It might do unexpected things if you enter a bad value for a parameter or make a coding mistake.

❑ Starting up the arm frequently results in it jerking. This happens because the L6 has no feedback from the servos. Consequently, when the SSC-32 powers up, it assumes that all of the servos are in their centered positions. In general, this is a bad assumption. The first time you try to move a servo, it might jump rapidly from its current position to the new position that you request. Therefore, try to hold up the arm when you start it up so that it does not have as far to travel. You can move the linkages carefully while it is powered off without damaging the arm.

❑ The arm has a tendency to oscillate if you move it too fast or try to pick up very heavy loads. There are springs that can be attached between the elbow and the shoulder to try to compensate for the load (look carefully at Figure 15-5 just behind the upper arm), but these add extra load and might even exacerbate the problem for some arm positions.

However, the springs do stop the arm from slumping to the ground when the power is turned off. This helps to protect the arm, and minimizes jerking when the arm is initialized because it is more likely to be near its starting position. It's a good idea to reset the arm to the default position every time you use it before turning it off.

❑ To make the robot more stable, it is highly advisable to screw it to a board or table. Unfortunately, this makes it difficult to move around if you want to take it somewhere to do a demo. Based on the length of the arm, a board 60cm by 90cm is more than big enough to cover the whole area that the arm can reach.

# Controlling the L6 Arm

The code to control the L6 arm is not included with MRDS. A version of the code can be downloaded from the Lynxmotion website, but the companion websites for this book (www.wrox.com and www.ProMRDS.com) contain a heavily modified version that is used in the examples. This new version is used with the permission of Lynxmotion. All of the code is contained in the folder ProMRDS\Chapter15\Lynxmotion\Lynx6Arm.

There are two parts to the code: the Lynx6Arm service and the SSC32 service. The simulated Lynx 6 robotic arm is in the Chapter 8 folder.

Before you attempt to use the software, you need to connect your arm to a serial port on your PC and make the appropriate change to the code. The SSC32 service has COM1 hard-coded into the SerialPortConfig class in SSC32State.cs:

```
public class SerialPortConfig
{
    //Serial port configuration parameters

    //Baud rate is 115.2k with both jumpers
    private int _baudRate = 115200;
    private Parity _parity = Parity.None;
    private int _dataBits = 8;
    private StopBits _stopBits = StopBits.One;
    // IMPORTANT: This is the COM port that will be used by the arm.
    // It is hard-coded here. There should be a config file associated
    // with this service, but this has not been implemented yet.
    private string _portName = "COM1";
    private int _timeout = 1000;
```

Note that there is a timeout value of one second. Don't touch the other serial port settings because they are set in the firmware on the SSC-32 controller board.

# Using the Generic Arm Contract

The Lynx6Arm service implements the MRDS generic Articulated Arm contract. This contract is used by the examples to communicate with the arm. It is also implemented in the simulator, so the same code works on both the simulator and the real arm in most cases.

The generic Articulated Arm contract included with MRDS has only a small number of requests. You can find the code under MRDS in `samples\Common\ArticulatedArmState.cs` and `ArticulatedArmTypes.cs`.

The following requests are available for the generic contract in addition to the usual requests such as `Get`, `Drop`, and `Subscribe`:

- **SetJointTargetPose:** Sets the pose for a single joint
- **SetJointTargetVelocity:** Sets the velocity for a single joint
- **GetEndEffectorPose:** Gets the pose of the end effector (using forward kinematics)
- **SetEndEffectorPose:** Sets the pose of the end effector (changes all joints as necessary using inverse kinematics)

The standard set of requests might not be suitable for some applications. For example, there are no explicit requests to control the opening/closing of the gripper, although it can just be considered another joint. Humanoid robots might not have grippers, but a gripper is a fundamental feature of an arm — without a gripper an arm would be useless!

## Setting Joint Angles

For basic operations, these requests are sufficient. In fact, you can control the arm using only `SetJointTargetPose` if you know all of the relevant joint positions for the trajectory of the arm. If the arm will be performing some repetitive task, you can just record the joint positions using the Motion Recorder service developed for this chapter and then replay them.

The Lynx6Arm service treats the gripper as the last joint in the set, but limits the angle to $\pm35$ degrees because the gripper servo is a very lightweight servo, and it could become overloaded if the gripper mechanism were driven too far. In addition, be *very* careful when closing the gripper on an object so that you do not try to drive it too far.

There are no timing parameters for `SetJointTargetPose` so you are restricted to using whatever value is implemented in the service itself. `SetJointTargetVelocity` can be used to specify the maximum speed for joint motions. Its name might suggest that it starts a joint moving at a specified speed, but this is not the case.

The firmware on the SSC-32 has been designed so that a sequence of joint motion commands can be sent in quick succession and it will execute all of them. However, this is not an ideal approach. Unless the

Time parameter is the same for all of the joints in a sequence, they will move to their destinations at different speeds. This is equivalent to Asynchronous PTP in KUKA terminology.

## Using GetEndEffectorPose

You can use `GetEndEffectorPose` to obtain the 3D coordinates of the EOA and the tilt angle of the wrist. Note that the wrist tilt angle is *not* the same as the angle of the wrist servo angle. The wrist tilt is the accumulation of all of the angles for all the joints and specifies how the wrist is tilted with respect to a horizontal line.

When you call `GetEndEffectorPose`, the Lynx6Arm service calls the SSC-32 service to query the arm. This means that the values returned should be current. This operation is done synchronously. In other words, the SSC-32 service sends a request to the SSC-32 and then waits for a reply. However, it does *not* guarantee that the arm is physically located at the corresponding location. If the arm's motion is restricted for any reason, the servo settings might not be reached. The SSC-32 is only telling you how it is pulsing the servos, not whether the servos are at the desired positions, which it has no way to know.

The EOA is at the tip of the gripper, but to pick up an object you need to position the gripper so that the object is at the TCP. There is a difference of about an inch (a couple of centimeters). It is a simple matter to change the kinematics code to calculate the TCP rather than the EOA. Open `Lynx6Arm.cs`, locate the following code, and reduce the value of L3:

```
// Physical attributes of the arm
float L1 = Conversions.InchesToMeters(4.75f);    // Length of upper arm
float L2 = Conversions.InchesToMeters(4.75f);    // Length of lower arm


// Not required
//float Grip = Conversions.InchesToMeters(3.5f);   // Length of gripper

// NOTE: The gripper is basically in two parts because it contains
// the wrist rotate joint which is NOT at the same place as the
// wrist joint.
// Measurements are:
// Wrist joint to rotate plate = 2.25"
// Rotate place to tip of fingers = 3.5"
// The actual length of the "hand" varies slightly depending on
// whether the gripper is open or closed.
// L3 measures the distance to the tip of the fingers or End Of Arm (EOA).
// A Tool Center Point (TCP) would be shorter than this.
// The rubber fingers are just over 1" long, so to keep the grip action
// happening on the fingers, then L3 needs to be shorter by about 0.5".
float L3 = Conversions.InchesToMeters(5.75f);    // Length of whole hand

float H = Conversions.InchesToMeters(2.5f);      // Height of the base
```

If you have an older Lynx 6 arm, you will have a yellow base instead of a black one. This base is a different height so you also need to modify the value of H. Note that the base height is measured up to the center of the shoulder servo.

The `SetEndEffectorPose` request performs the inverse kinematics calculations. You specify both the position and orientation (wrist tilt) of the end effector.

The SSC-32 has a command that sets multiple joints in motion at the same time but with an overall `Time` parameter, and `SetEndEffectorPose` uses this command. (Speeds can still be specified individually, and the motion will take the longer of the times specified or the time based on the speed.) This is basically the equivalent of the KUKA Synchronous PTP mode of operation.

One consequence of this is that motions specified using `SetEndEffectorPose` are usually much smoother than setting all of the joint angles (via multiple calls to `SetJointTargetPose`).

## Using the Dashboard to Control the Arm

By now you should be familiar with the Dashboard. It includes a panel for controlling an arm and it will work with the Lynx6Arm service.

*The generic arm contract requests do not allow for setting multiple joints simultaneously. KUKA created a new generic contract with a request called `SetMultipleJointAngles` for just this purpose. However, the KUKA implementation is not backwardly compatible with the existing generic arm contract, so standard programs such as the Dashboard do not work with the KUKA simulator.*

To try using the Dashboard, run the following command from a MRDS Command Prompt window (remembering that it has to be on one line):

```
dsshost -port:50000 -tcpport:50001 ↵
-manifest:"ProMRDS/Config/Lynxmotion.Lynx6Arm.manifest.xml" ↵
-manifest:"ProMRDS/Config/Dashboard.manifest.xml"
```

For your convenience, a batch command file called `LynxL6Arm.cmd` is also provided.

When the Dashboard starts up, connect to localhost. You should see the lynx6arm in the services list. Make sure that you have enabled the articulated arm display under Tools ⇨ Options and connect to the arm. This will enumerate the joints and list them for you. To make it easier to understand, the joint names include the joint number. This is the servo number on the SSC-32. The display should look like Figure 15-6.

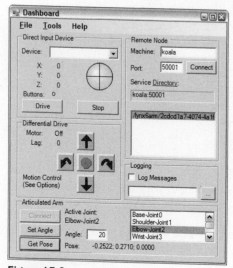

**Figure 15-6**

You can select a joint by clicking it. Enter a value for the angle and then click the Set Angle button. The joint will move to the specified angle. When you are first experimenting, use fairly small angles until you know how the arm will behave. You can use both positive and negative angles. If you click the Get Pose button, the current X, Y, Z coordinates of the EOA will display.

Clearly, moving the arm around in this fashion is tedious. The example is simply included here to illustrate that the generic arm contract works. An alternative way to control the arm is using one of the VPL programs in Chapter 12 and an Xbox game controller. The two VPL programs for robotic arms in Chapter 12 use the thumbsticks and buttons in different ways. This is deliberate, and highlights the difference between adjusting servo angles and specifying the position of the end effector. Some people will find one way more intuitive than the other, but in some situations both approaches are useful. Playing with an Xbox controller is a very good way to get a practical understanding of the different approaches to moving an arm.

## Recording Motions

The next step beyond the Dashboard is to use a more sophisticated application to control the arm. This chapter includes a service called the Motion Recorder. The service can be used with either the simulated arm or a real arm, so even if you don't have a Lynxmotion L6 arm you can still experiment with the simulator. This section explains how the Motion Recorder works and how to use it.

In the MRDS `bin` folder you will find a batch file called `MotionRecorder.cmd`. Run this batch file to start the Motion Recorder.

The Motion Recorder service enables you to move the arm around either by moving joints individually or by specifying a pose. You can save the arm positions to a file and then later replay the sequence of commands. The service GUI looks like Figure 15-7.

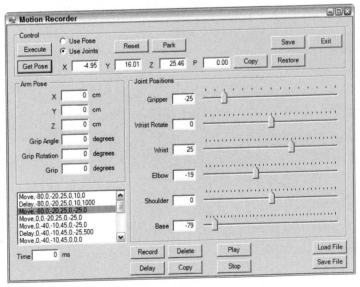

Figure 15-7

The purpose of the Motion Recorder is to help you develop applications. It could be improved in a lot of ways, but it is only intended as a simple tool.

## Using the Simulated Lynx Arm

You will notice some similarities between the Motion Recorder and the GUI for the Lynx arm simulation in Chapter 8. In fact, it can be used with the simulator if you want. To use the simulator, edit the manifest in the ProMRDS\Config directory called MotionRecorder.manifest.xml. At the bottom of the manifest you will see that the simulated arm is commented out (this is highlighted in the following code). Remove the commenting and instead comment out the real arm service, which is just above the simulated one.

```xml
<?xml version="1.0" ?>
<Manifest
  xmlns="http://schemas.microsoft.com/xw/2004/10/manifest.html"
  xmlns:dssp="http://schemas.microsoft.com/xw/2004/10/dssp.html"
  >
  <CreateServiceList>
    <ServiceRecordType>
        <dssp:Contract>http://schemas.tempuri.org/2007/11/↵
motionrecorder.html</dssp:Contract>
    </ServiceRecordType>
<!-- Use the real arm -->
    <ServiceRecordType>
        <dssp:Contract>http://schemas.coroware.com/2007/11/↵
lynx6arm.html</dssp:Contract>
    </ServiceRecordType>

<!-- Use the simulated arm
    <ServiceRecordType>
        <dssp:Contract>http://schemas.tempuri.org/2007/10/↵
SimulatedLynxL6Arm.html</dssp:Contract>
    </ServiceRecordType>
-->
  </CreateServiceList>
</Manifest>
```

Now when you run the Motion Recorder, the simulator will start. You can play around as much as you like without potentially damaging your Lynx 6 arm.

## Inside the Motion Recorder Service

The following is a brief description of how the Motion Recorder works. The user interface should be fairly intuitive. Three main files make up the application:

❑ **MotionRecorder.cs:** The main service where most of the action happens

❑ **MotionRecorderTypes.cs:** State, operations port, classes for communication, etc.

❑ **MotionRecorderUI.cs:** Windows Form for the user interface

The MotionRecorderTypes.cs file should not require any explanation, as it is quite short.

`MotionRecorder.cs` does all the work of controlling the arm and is explained further in this section. It specifies the Lynx6Arm service as a partner. Notice that it uses the generic arm contract and creates a port called `_lynxPort`:

```
[Partner("Lynx6Arm", Contract = arm.Contract.Identifier,
    CreationPolicy = PartnerCreationPolicy.UseExisting, Optional = false)]
private arm.ArticulatedArmOperations _lynxPort =
    new arm.ArticulatedArmOperations();
```

A lot of the code in `MotionRecorderUI.cs` is for validating input data, reading and writing files, and so on. This has nothing to do with MRDS, but unfortunately it has to be included.

`MotionRecorderUI.cs` uses the standard process for a WinForm to communicate with the main service. (If you are unsure about how this works, refer back to Chapter 4. Only a very brief explanation is provided here.) It uses a set of classes that are defined in a similar way to operations for a service, except that it is not a full-service interface. Instead, it posts all of its requests using a single operation class and passes across an object that represents the function to be performed. This helps to reduce the number of operations in the main service's port set so that the limit of 20 operations is not exceeded.

For example, when the Form is first created it sends back a handle to itself to the main service. This lets the main service know that the Form is running so that it can enumerate the joints in the arm. In the Form, the code is as follows:

```
// Send a message to say that the init is now complete
// This also passes back a handle to the Form
_fromWinformPort.Post(new FromWinformMsg(FromWinformMsg.MsgEnum.Loaded, null,
this));
```

Back in the main service, the `FromWinformMsg` handler processes the request and executes the appropriate code based on the command type, which in this case is `Loaded`:

```
// Process messages from the UI Form
IEnumerator<ITask> OnWinformMessageHandler(FromWinformMsg msg)
{
    switch (msg.Command)
    {
        case FromWinformMsg.MsgEnum.Loaded:
            // Windows Form is ready to go
            _motionRecorderUI = (MotionRecorderUI)msg.Object;
            SpawnIterator(EnumerateJoints);
            break;
```

`EnumerateJoints` issues a `Get` request to the Lynx6Arm service to get its state. Notice that it does this in a loop because the arm service might not be up and running properly, in which case there will be a `Fault`. To ensure that it does not loop forever, there is a counter:

```
IEnumerator<ITask> EnumerateJoints()
{
    int counter = 0;
    arm.ArticulatedArmState armState = null;

    // Get the arm state because it contains the list of joints
```

*(continued)*

*(continued)*

```
        // Wait until we get a response from the arm service
        // If it is not ready, then null will be returned
        // However, we don't want to do this forever!
        while (armState == null)
        {
            // Get the arm state which contains a list of all the joints
            yield return Arbiter.Choice(
                _lynxPort.Get(new GetRequestType()),
                delegate(arm.ArticulatedArmState state) { armState = state; },
                delegate(Fault f) { LogError(f); }
            );
            // This is not good error handling!!!
            if (armState == null)
            {
                counter++;
                if (counter > 20)
                {
                    Console.WriteLine("***** Could not get arm state. Is the arm ⏎
turned on? *****");
                    //throw (new Exception("Could not get arm state!"));
                    yield break;
                }

                // Wait a little while before trying again ...
                yield return Arbiter.Receive(false, TimeoutPort(500), ⏎
delegate(DateTime now) { });

            }
        }
```

The `armState` contains a list of the joints. A list of the joint names is created and then the Form is updated using `FormInvoke`, which executes a delegate in the context of the Form. The Form contains the function `ReplaceArticulatedArmJointList`, which updates the list of joint names inside the Form so that it can also refer to them by name:

```
        // Create the new list of names
        _jointNames = new string[armState.Joints.Count];
        for (int i = 0; i < armState.Joints.Count; i++)
        {
            _jointNames[i] = armState.Joints[i].State.Name;
        }

        // Now update the names in the Form so that it can do
        // whatever it needs to with them as well
        WinFormsServicePort.FormInvoke(delegate()
        {
            _motionRecorderUI.ReplaceArticulatedArmJointList(_jointNames);
        });
```

In addition to setting the joint names, the initial joint angles are also set, but the code is not shown here. That completes the startup process.

## Using the Motion Recorder

The Motion Recorder does not automatically record anything. You must click the Record button (near the bottom of the window) to record the current command. You can play around using the slider controls, and as you move them the arm will follow (and the angle values in the textboxes will update). When you are happy with the position of the arm, you can record it. By doing this repeatedly, you can build up a complete sequence of motions to perform a particular task.

> **Warning: Do not move the sliders too fast!**

The UI contains four basic areas. At the top is the Control Panel, which enables you to execute motion commands. Next are the Arm Pose and Joint Positions panels, which provide values for motions. Finally, there is an area for recording, saving, and replaying commands.

The most important button is the Execute button, which calls the main service to perform either a SetEndEffectorPose if the Use Pose radio button is selected, or a sequence of separate SetJointTargetPose operations (one for each joint) if the Use Joints radio button is selected. These two operations are the crux of the application, so the code for them is discussed here.

As you move the sliders around, the trackbar controls continually fire off requests to the main service to move the corresponding joints. Once the main service has decoded the FromWinformMsg it executes the MoveJoint function using the values passed from the Form:

```
void MoveJoint(string name, float angle)
{
    arm.SetJointTargetPoseRequest req = new arm.SetJointTargetPoseRequest();
    req.JointName = name;
    AxisAngle aa = new AxisAngle(
        new Vector3(1, 0, 0),
        (float)(Math.PI * ((double)angle / 180)));

    req.TargetOrientation = aa;

    _lynxPort.SetJointTargetPose(req);
}
```

Notice that the SetJointTargetPoseRequest requires the name of the joint and the target angle, expressed as an AxisAngle. By convention, the axis angles are all with respect to the X axis. In addition, the angle from the Form is in degrees, so it has to be converted to radians.

When Use Joints is selected, the Execute button invokes a different method in the main service called MoveMultipleJoints, but this simply calls MoveJoint for each of the joints because the generic arm contract does not have a request to set multiple joints at the same time. The disadvantage of this approach is that the different joint movements are not coordinated.

Conversely, if Use Pose is selected when the Execute button is pressed, it sends a request to run the MoveToPosition function in the main service. This function sends a SetEndEffectorPose request to the Lynx6Arm service, but first it has to create a quaternion for the pose. This is a somewhat painful process that is not helped by the fact that the proxy versions of the classes don't have the necessary static methods. After several lines of code, the request is finally sent. Then the wrist rotate angle and the gripper are set independently because SetEndEffectorPose does not do these:

```
public SuccessFailurePort MoveToPosition(
      float mx, // x position
      float my, // y position
      float mz, // z position
      float p, // angle of the grip
      float w, // rotation of the grip
      float grip, // distance the grip is open
      float time) // time to complete the movement
{
      // Create a request
      arm.SetEndEffectorPoseRequest req = new arm.SetEndEffectorPoseRequest();
      Pose rp = new Pose();
      Vector3 v = new Vector3(mx, my, mz);
      rp.Position = v;
      physicalmodel.Quaternion q = new physicalmodel.Quaternion();
      q = physicalmodel.Quaternion.FromAxisAngle(1, 0, 0, ↵
Conversions.DegreesToRadians(p));
      Quaternion qp = new Quaternion(q.X, q.Y, q.Z, q.W);
      rp.Orientation = qp;
      req.EndEffectorPose = rp;
      _lynxPort.SetEndEffectorPose(req);

      // Set the positions of the wrist rotate and the gripper as well
      MoveJoint(_jointNames[(int)lynx.JointNumbers.WristRotate], w);
      MoveJoint(_jointNames[(int)lynx.JointNumbers.Gripper], grip);

      // Now return success
      SuccessFailurePort responsePort = new SuccessFailurePort();
      responsePort.Post(new SuccessResult());
      return responsePort;

}
```

Note a couple of advantages in using SetEndEffectorPose. You can set precisely where the arm will go by entering values into the Arm Pose panel. You can also specify the tilt angle of the gripper, which is important if you want the robot to pick something up. More importantly, the motion is better coordinated because, as explained earlier, it sends a single command to the SSC-32 to move all the joints.

The disadvantage of using the Arm Pose panel is that it isn't always easy to pick valid values. Moreover, sometimes the results might be a little surprising. If the arm doesn't move, look in the MRDS DOS command window for an error message. It is possible that you selected an invalid pose, which causes the arm service to generate a Fault.

When you click the Get Pose button, the code sends a FromWinformMsg request to the main service containing the command GetPose. In the main service, the FromWinformMsg handler decodes the command and calls GetArmHandler. This handler makes a GetEndEffectorPose request to the Lynx6Arm service and then uses FormInvoke to return the results to the Form, where they appear in the textboxes across the top. The slider controls are also updated to match the current positions of the joints. A Copy button beside the pose values enables you to transfer the pose information to the Arm Pose textboxes.

Alternatively, you can enter values into the textboxes beside the joint sliders. Then when you click Execute (with Use Joints selected), the joint angles will be set.

The Save button provides a quick way to save all the values in the Arm Pose and Joint Position textboxes. You can get them back later using Restore.

The Reset button sets all of the angles to zero, which should be the initial setting when the arm starts up. Park sends the arm to a "parked" position, but it is quite arbitrary. You can change the joint angles for each of these at the top of the main service:

```
// Joint angles for the Reset and Park commands
float[] _resetAngles = { 0, 0, 0, 0, 0, 0 };
float[] _parkAngles = { 0, 20, -50, 30, 0, 0 };
```

The remaining buttons across the bottom of the window perform the following functions:

- ❑ **Record:** Saves the values of either the Arm Pose or Joint Positions, depending on whether Use Pose or Use Joints is selected, into the command history list at the bottom left

- ❑ **Delay:** Adds a delay command to the command history list using the value in the Time field

- ❑ **Delete:** Removes the currently selected item from the command history list

- ❑ **Copy:** Copies the values from the currently selected command to either the Arm Pose or Joint Positions textboxes depending on the type of command that is selected

- ❑ **Play:** Replays all of the commands in the list starting from the top

- ❑ **Stop:** Stops replaying commands

- ❑ **Load File:** Loads a previously saved command history list file

- ❑ **Save File:** Saves the command history list to a CSV file

Saved command history files are comma-separated value files. These are plain text and can be edited with Notepad, or loaded into Excel. An example is given in the next section. You might find it quicker to edit a file rather than use the Motion Recorder once you have saved a few key commands. Just be careful not to mess up the format.

# Pick and Place

"Pick and place" is a very common operation for robotic arms. They can spend the whole day picking things off a conveyor belt and dropping them into bins—a mindless task that you would probably find exceedingly boring.

The basic principle is to program the arm so that it knows where to go to get an object (pick) and where to go to put it down (place). Even today, many robotic arms are manually programmed by moving them through a sequence of motions and recording the path.

A saved command history file is included in the software package on the book's website (www.proMRDS .com and www.wrox.com). It should be in the folder with the Motion Recorder source files and is called take_object.csv. The contents of this file are as follows:

```
Move,-80,0,-20,25,0,10,0
Delay,-80,0,-20,25,0,10,1000
Move,-80,0,-20,25,0,-25,0
Move,0,0,-20,25,0,-25,0
Move,0,-40,-10,45,0,-25,0
Delay,0,-40,-10,45,0,-25,500
Move,0,-40,-10,45,0,0,0
Move,0,-20,-10,45,0,0,0
Delay,0,-20,-10,45,0,0,1000
Move,0,-40,-10,45,0,0,0
Delay,0,-40,-10,45,0,0,500
Move,0,-40,-10,45,0,-25,0
Move,0,-20,-10,45,0,-25,0
Move,0,0,-20,25,0,-25,0
Move,-80,0,-20,25,0,-25,0
Delay,-80,0,-20,25,0,-25,500
Move,-80,0,-20,25,0,10,0
```

Notice that the file only contains Move and Delay commands. There is no particular reason for this other than that it was easier to move the sliders than to figure out coordinates. However, it could use Pose commands.

This file executes the following basic steps:

1. Swing the arm to the left (which should be to where you are sitting) and open the gripper in expectation of being handed something.

2. Close the gripper. The gripper setting is for a AA battery. If you want to use a different object, you should first practice with the Motion Recorder and set the gripper value accordingly. You don't want to force the gripper too far past the point where it contacts the object.

3. Turn the arm back so it faces straight ahead and put the object down.

4. Raise the arm and pause for effect. (This might be an appropriate point for applause.)

5. Pick the object up again.

6. Take it back and drop it where it came from. If you are not ready to catch it, the object — whatever it is — will fall on the ground.

To begin, load the file into the Motion Recorder and play it without using an object. Just sit and watch. Then you can try handing an object to the arm when it extends its hand out to you.

This sequence can definitely be improved, but that isn't the point. What you have is a program that enables you to easily determine the necessary joint parameters for a sequence of operations. With this knowledge you can write services that replay the commands, or perhaps generate commands automatically based on some formula.

You need to play and replay your motion sequence repeatedly, looking for flaws. One of the main problems is that the arm might take a shortcut to its destination that takes it too close to an obstacle. Similarly, when picking up or putting down an object, you need to make clear the up and down movements. Don't simply drag the object off to its new location straight away because that is exactly what might happen — it drags along the ground.

## Logjam

When moving objects around, you need to remember where you put them. The arm cannot sense anything in its environment. Another problem is that the arm needs a certain amount of clearance to put down and pick up an object. Placing objects too close together is almost certain to result in some of them getting knocked over. The further the arm is extended, the less leeway there is to "reach over" an object to get to one behind it.

You will find that the Lynx 6 arm tends to wave around (oscillate) when it is extended a long way. If it waves too much, it can knock over other objects. The examples using dominoes in the simulation in Chapter 8 are probably not possible with a real L6.

In addition, remember that the gripper on the arm can only approach an object along a radial line from the base of the arm out to the tip. The wrist rotate servo is no help in turning the end of the arm to face a different direction — it just tips the gripper on its side. You might therefore find that it is easier to place objects in semi-circular patterns, rather than straight lines.

## Using an Overhead Camera

MRDS includes support for a webcam. If you start a webcam service and run the Dashboard, you will see your webcam listed. Selecting it brings up a separate window to show the video stream. The source code for this is included in the Dashboard service in Chapter 4.

If you placed a camera above your arm looking down, it should be a simple matter to locate objects in the image for the arm to pick up. (The simulated arm in Chapter 8 includes a camera overlooking the arm, as well as a camera mounted just behind the gripper.) As long as the objects are a significantly different color from any of the colors on the ground, you can use image subtraction to separate objects out from the background. You first take an image of the empty area as a reference. Then, as objects are placed in front of the robot, you subtract the original image from the current image. Anything that is not ground will stand out.

A blob tracker example included with MRDS is another alternative. It can be trained to follow objects of different colors. Regardless of how you locate the objects in the image, you have to calibrate the arm to the image. The simplest way to do this is probably to position the arm roughly near the four corners of the image and use GetEndEffectorPose to determine the X and Z coordinates (the Y coordinate is the height, which you can't tell from an overhead image). By interpolating between these points in both directions, you can determine the coordinates of any point in the image.

Note that your camera needs to be mounted directly overhead and looking straight down. If it is off to one side, then perspective will make the interpolation nonlinear. This is left for you as an exercise.

# Summary

After completing this chapter successfully, you should have a good grounding in using robotic arms. If you are a student of mechatronics, then you might feel that something is missing, and there is — a heap of complex equations. If you want more details, visit the KUKA Educational Framework website (www.kuka.com).

Robotic arm applications are limited only by your imagination (and, of course, the physical constraints on your arm).

The next chapter shows you how to write services to run on an onboard PC running Windows CE or a PDA using the .NET Compact Framework (CF).

# 16

# Autonomous Robots

And now for something completely different . . .

So far you have seen robots controlled directly from your desktop PC. However, it is much more fun to let them roam free. To do this you need to make them autonomous, i.e., give the robots onboard brains.

In writing a book like this it is hard to predict the skills of the audience. Some of you might have worked with MS-DOS in what now seems like a previous life, and others might be young enough to have grown up without having to deal with the horrors of command prompts, `Autoexec.bat`, Windows for Workgroups, and so on. Unfortunately, in some ways working with onboard computers is taking a step backwards from the comfortable environment of Windows XP or Vista that you have on your desktop PC to the bygone era of DOS.

Some of this chapter, therefore, is about introducing you to what might be new concepts and ways of programming and debugging. The robotics content is not very high, but unless you understand how to work in these environments you will not be able to create autonomous robots.

Regardless of whether you want to build an autonomous robot or not, you might find parts of this chapter useful. For example, it discusses using the .NET Compact Framework (CF) on a personal digital assistant (PDA) to control a robot via Bluetooth.

You should read the entire chapter even if you are using Windows CE because some of the material in the section on PDAs is relevant to Windows CE as well.

## PC-Based Robots

It has already been pointed out several times in this book that MRDS runs on Windows. It does not generate native code for a BASIC Stamp (Boe-Bot or Scribbler), a PIC16F877 (Hemisson), an ARM (LEGO NXT), or any other microcontroller. MRDS services can talk to a monitor program running on one of these small microcomputers, but you still need Windows. Therefore, somehow you have to work a Windows-based computer into the scenario.

## Windows XP-Based PCs

In late 2007, White Box Robotics announced a new robot called the 914 PC-BOT (www.whiteboxrobotics .com/PCBOTs/index.html). It's also sold by Heathkit as the HE-RObot (www.heathkit.com/herobot .html). The PC-BOT, which looks a little like R2D2, has an internal PC with a hard disk, DVD drive, speakers, webcam, and just about everything that you would expect on a PC except for a monitor, mouse, and keyboard (which you can plug in if the robot will just stand still).

The 914 PC-BOT runs Windows XP and comes with its own control software. Unfortunately, it costs many thousands of dollars and is not the type of robot that you are likely to buy if you are a beginner. It certainly qualifies as an autonomous robot, but because it has a full-blown PC embedded in it, there is no real challenge for software development apart from trying to remotely debug your application.

Even remote debugging is relatively easy with the PC-BOT because you can use the WiFi interface to connect to the robot and run a remote desktop connection, which is as good as being logged onto the robot itself. You can install Visual Studio and the full MRDS directly onto the robot's PC and compile and debug your code onboard — no messy downloads or issues with synchronizing source code.

If you have enough money, you might want to buy a PC-BOT and skip the rest of this chapter.

*Despite having demonstrated the PC-BOT with MRDS V1.0 at the launch, at the time of writing White Box was redeveloping their MRDS services, so MRDS support was not available. Check the White Box website for updates (www.whiteboxrobotics.com).*

## Onboard Laptops

In terms of ease of use, putting a laptop on your robot has to be the simplest solution. Ignoring the obvious question of how you connect to the sensors and actuators, the laptop runs exactly the same software that you would run on your desktop and has a monitor, keyboard, touchpad, and so on. It is no different from using a desktop. (The PC-BOT uses a USB interface called an *M3 module* to connect to the motors and sensors).

Evolution Robotics released a robot a few years ago called the ER1 that was basically a set of wheels onto which you could strap your laptop. The key component was a USB interface for controlling the wheels. Unfortunately, the ER1 is now only sold in a bundle with some expensive software that has pushed it out of the price range of the hobbyist or beginner roboticist.

The problem with a laptop is that it would squash most of the robots covered in this book — the Boe-Bot and LEGO NXT, for example, simply cannot carry a laptop. Even the Stinger robot used in this chapter might have trouble with a laptop, so this option is ignored.

## PDAs

The next step down the food chain as far as computers are concerned is handheld devices, commonly called PDAs (personal digital assistants) or Pocket PCs.

PDAs run an abridged version of Windows that has been known by various names over the years, including Pocket PC, Windows Mobile, or Windows CE (in different flavors and versions). The main thing that these operating systems have in common is that they all use a subset of the .NET runtime, called the .NET Compact Framework (CF), for running MRDS services. This has implications for how you write your code.

The other issues you have to face with PDAs are limited memory and slower CPU speeds. These restrictions mean that a PDA is probably not suitable for a very sophisticated application due to the overhead of MRDS. However, it is possible to run a simple application on a PDA just to illustrate the principles involved.

In the folder for Chapter 4 you can find a sample application, DriveByWire, that enables you to remotely control a Boe-Bot using a PDA. This chapter provides a service for driving a Stinger robot from a PDA. These are really toys, but they can be cool examples to use to impress your friends.

## Embedded PCs

The term "embedded PC" can refer to a variety of different physical formats, ranging from a complete "PC in a box" to bare circuit boards. For our purposes, an embedded PC is any computer that runs Windows CE. (Embedded XP could also be used, but it is not covered in this chapter).

PC motherboard designs have undergone a series of changes since the original IBM PC was introduced. Motherboards have become progressively more compact and less power hungry, ranging from the IBM AT to the ATX and the microATX form factors.

The ITX series of boards promoted by VIA Technologies are available in Mini-ITX, Nano-ITX, and Pico-ITX varieties. Pico-ITX is a tiny 10cm × 7.2cm (3.9in × 2.8in). The advantage of the VIA boards is that they are convection cooled — because they don't require a fan, they are quiet and low power.

At the same time that smaller PC motherboards were being refined, *single-board computers* (SBCs) were also introduced for use in embedded systems. One of the early standards was PC104, so called because there were 104 pins on the bus connector (which corresponded to the original ISA bus from the IBM PC). PC104 was different from standard PC motherboards in that it did not have slots for plugging in extra cards. Instead, boards were designed to stack on top of one another. Over the years there have been successive revisions of PC104 — for example, to include the PCI bus.

To avoid the heat and weight of hard drives, embedded PCs often use *solid-state storage devices* or *flash memory* such as CompactFlash cards. With no rotating parts, they are less susceptible to damage from vibration and shocks.

The eBox-2300, shown in Figure 16-1, is used in the latter part of this chapter. It comes in a metal case with standard connectors for monitor, keyboard, USB, Ethernet, and so on, so that it is effectively a tiny PC. It runs off a single 5V supply and does not have a fan (which is why there are heat sinks on the case). The eBox-2300 is manufactured by ICOP Technology (www.icoptech.com).

**Figure 16-1**
*Photo courtesy of EmbeddedPC.NET (www.embeddedpc.net).

A "Jump Start Kit" based on the eBox-2300 is available specifically for use with MRDS (www.embeddedpc .net/eBox2300MSJK/tabid/111/Default.aspx). However, if you buy a Stinger CE robot from RoboticsConnection, then the Jump Start Kit is included, so you don't have to buy it separately. Furthermore, the Stinger CE kit includes a 5V regulator that you need in order to run the eBox off a rechargeable battery.

The eBox runs Windows Embedded CE 6.0. The CE stands for either Compact Edition or Consumer Electronics, but it doesn't matter and most people just refer to it as CE or WinCE. See the Windows CE home page for documentation and downloads: http://msdn2.microsoft.com/en-us/embedded/ aa731407.aspx.

As with PDAs, Windows CE uses the .NET Compact Framework (CF), so the services for embedded PCs are generally referred to as CF services in MRDS terminology.

## The Development Environment

Developing MRDS services for PDAs is similar, but not identical, to developing for embedded PCs running Windows CE. In both cases you use Visual Studio with its built-in support for mobile devices.

Visual Studio has a cool emulator for PDAs that can be used for debugging conventional applications. However, in the case of MRDS this environment has several shortcomings. In particular, it does not include support for USB Bluetooth dongles, so it is not used in this chapter.

> In order to develop code for Windows Mobile or Windows CE, you must have the correct versions of the software. Visual Studio Express Editions do not support development for mobile devices, i.e., the Compact Framework. You need the full version of Visual Studio 2005 with Service Pack 1 installed. Also make sure that you have the .NET Compact Framework V2.0 Service Pack 2 installed on your PDA or embedded PC. The .NET runtime and the service packs are available from the Microsoft Downloads page but you might need to search for them because they have been superseded by Visual Studio 2008 and .NET 3.5. This book was written using Visual Studio 2005 and the code was not tested with later versions. Please refer to the book's website for the most up-to-date information.

# Setting Up a Stinger Robot

The examples in this chapter are based on the Stinger CE robot from RoboticsConnection (www.roboticsconnection.com) shown in Figure 16-2.

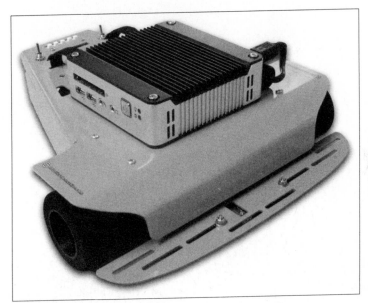

**Figure 16-2**
*Photo courtesy of RoboticsConnection

In Figure 16-2 the eBox-2300 is shown mounted on top of the robot. However, an alternative using a PDA is developed before discussing how to use the eBox.

You can buy a Stinger robot on its own and build your own configuration using the optional accessories. Even if you are not interested in autonomous robots, you can still use the Stinger (but without the 'CE' bit).

It is recommended that you buy a Bluetooth module and the infrared sensors as well as the standard Stinger CE kit, which has the line-following sensors (not visible in Figure 16-2 because the sensors are mounted under the front bumper).

The Bluetooth module is not used in the Stinger CE kit because the eBox plugs into the RS-232 module using a cable connected to COM1 on the eBox. However, Bluetooth is good for getting started because it enables remote control of the robot, meaning you can develop code on your desktop PC initially.

Figure 16-3 shows a Stinger configured with Bluetooth (the antenna is visible sticking up in the middle of the photo) and three IR sensors across the front bumper. The serial connector poking out of the middle of the robot is for the eBox and is not used in this configuration.

**Figure 16-3**

Note that the IR sensors should be mounted on standoffs so that they do not see the ground immediately in front of the robot. The GP2D12 sensors have a range of 10cm to 80cm, whereas GP2D120 sensors have a range of 4cm to 30cm. Which ones you choose depends on how cluttered your environment is and how fast you plan to drive the robot. The Stinger moves quite fast at full power!

There are very good multimedia assembly instructions for the Stinger available on the website, including 3D diagrams, which you can manipulate to change the viewpoint, and spoken instructions. However, some of the optional extras such as the IR sensors do not have assembly instructions.

The necessary MRDS services for the Serializer board are available from the RoboticsConnection website to enable you to control the Stinger. Anywhere that you see references to the "Traxster" on the website, you can also take this to mean the Stinger because both robots use the same control board, called the Serializer. These services include both the standard desktop versions and the CF versions.

You should download the Serializer services from the RoboticsConnection website and install them first. Included with the services is a sample program that is very useful for testing your robot once you have assembled it.

# Setting Up Your PDA

To run the first example in this chapter you must have a PDA with built-in Bluetooth, such as the Dell Axim X50v or X51v. Most PDAs in recent years include Bluetooth.

Although the Axim also has built-in WiFi, this is not required to use a PDA for MRDS. Note that these Axim models run at 624MHz, which is slow compared to a desktop PC but fairly fast for a PDA.

Figure 16-4 shows an Axim running the Stinger Drive-By-Wire program. You can find this program in the `ProMRDS\Chapter16` folder. (Astute readers might be able to tell that this is a composite picture — a screen capture from an X50v PDA has been pasted over an image of an X51v from the Dell website. It looks better than an actual photo.)

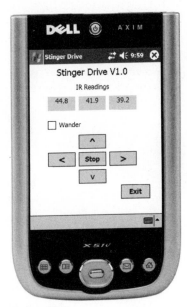

Figure 16-4

You should install the latest version of ActiveSync, available free from the Downloads link on the Microsoft website — select Windows Mobile from the list of Product Families. At the time of writing, the latest version is 4.5. If you are an avid PDA user, you probably have this software already. From here on it is assumed that you have ActiveSync working.

Recent versions of Windows Mobile include the .NET Compact Framework, but if you have an older PDA, running Windows Mobile 2003 (also called Pocket PC 2003) for example, then you will have to download and install the .NET Compact Framework Version 2.0 Service Pack 2. You install the software on your desktop PC first, and then on your PDA using ActiveSync, but this is not covered here because you should be able to do this.

When it comes to running and debugging services on your PDA, there are some tools that you will find invaluable. First, download the Windows Mobile Developer Power Toys from `www.microsoft.com/downloads/details.aspx?FamilyID=74473fd6-1dcc-47aa-ab28-6a2b006edfe9&DisplayLang=en`.

This package includes the Pocket PC (PPC) Command Shell, which enables you to enter commands in the same way as you do with the command prompt for MRDS on your desktop PC. The real advantage is that you can see the output from `DssHost` as it starts up and find out about service failures or mistakes in the manifest right away.

After downloading and installing the Power Toys, you need to copy `cmd.exe`, `shell.exe` and `console.dll` to the `\Windows` folder on your PDA using ActiveSync. You can also add a shortcut for `cmd.exe` to the `\Windows\Start Menu\Programs` folder so that you can easily start a command prompt whenever you like.

---

### A Note about Shortcuts

Shortcuts on Windows Mobile and Windows CE are plaintext files with a file type extension of `.lnk`. You can create them using Notepad on your desktop PC and then transfer them to your PDA with ActiveSync. A shortcut is handy for starting MRDS on your PDA.

The format of a shortcut is

```
nn#path\filename.exe /parameters
```

where nn is the number of bytes in the command line following the hash sign (#). In the case of the PPC Command Shell, the executable `cmd.exe` is on the search path so the shortcut is very simple:

```
3#cmd
```

Interestingly, you can enter a number that is larger than the length of the command and it still works. For example, you could put 10 in the previous shortcut instead of 3 and it would still work.

When you want to run an MRDS service, the shortcut becomes more complicated. Later in the chapter you will see the Stinger Drive-By-Wire service. Its shortcut looks like the following:

```
95#"\Program Files\mrds\bin\cf.DssHost.exe" /p:50000↵
/t:50001 /m:StingerCFDriveByWire.manifest.xml
```

Note that this command must all be entered on a single line in the shortcut file. A copy of this shortcut is included with the Stinger Drive-By-Wire service source code in the file called `StartStinger.lnk` in the `CF` subdirectory; it is deployed to your PDA when you install the service.

---

The Power Toys also contain several other useful programs. One that you might find handy for producing documentation is the ActiveSync Remote Display, `ASRDisp.exe`. This enables you to view on your desktop PC a replica of the screen on your PDA so you can make screenshots. For example, Figure 16-5 shows the memory utilization on the PDA while the Stinger Drive-By-Wire service is running. (The title bar and menu across the top are from `ASRDisp` and are not part of the PDA screen.)

Figure 16-5

Note carefully in Figure 16-5 that a large portion of the available memory has been dedicated to program usage, rather than storage. This is necessary so that the MRDS services don't run out of memory. On the Dell Axim used for writing this book, the Bluetooth Manager exits if memory gets too low and it is necessary to perform a warm restart on the PDA. This is one of the problems that you might encounter when using a PDA. If you have too many files on your PDA you might need to clean up so that you can get at least 40MB for Program memory.

For the Bluetooth connection, you have to pair your PDA with the robot. This process is similar to what was covered in Chapter 14. The exact details vary slightly depending on the brand of the PDA and the version of the operating system.

Basically, you need to search for new devices in the vicinity of the PDA while the robot is turned on. The Stinger robot will show up as an EB100 device. You can select this and enter the passkey, which is **0000**. This only needs to be done once. Make sure that you record the outgoing serial port number when you complete the pairing because you will need this later to configure the MRDS service.

If you have more than one Bluetooth device paired with your PDA, then you will be prompted to select a device when you run the Stinger Drive-By-Wire service. This might look like the screenshot in Figure 16-6 where eb500 is a Boe-Bot, eb100 is a Stinger, and TTMobile is a mobile phone, all of which have been paired with the PDA.

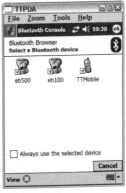

Figure 16-6

Another piece of software that you might need is something like the Process Explorer (PE.exe) from MadeBits (http://madebits.com/tools/pocketpc-process-explorer.php). This program enables you to view and kill processes running on your PDA, which comes in handy sometimes because services do not always shut down cleanly. A sample screenshot is shown in Figure 16-7.

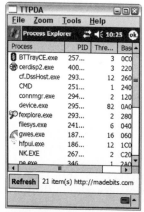

**Figure 16-7**

Notice in Figure 16-7 that the client side of the ActiveSync Remote Display is running (cerdisp2.exe) as well as the CF version of DssHost and a command prompt (CMD).

# Driving a Stinger Remotely

To begin the example for this chapter, a service is discussed that connects to the Stinger using Bluetooth and runs on a desktop PC. Then the service is migrated to a PDA and finally to the eBox-2300.

There is no requirement to use a PDA, but it provides an attractive alternative to using an eBox. For a start, you can either attach the PDA directly to the Stinger or you can simply walk around behind the Stinger. The PDA has a screen that makes it easier to see what is going on, whereas the eBox operates in a "headless" mode when it is attached to the Stinger. In addition, you can use your PDA for other things besides just driving a robot!

This example is similar to the Robotics Tutorial 4 (Drive-By-Wire) with the addition of a mode that enables the robot to wander around on its own. This example also enables you to use the keyboard to drive the robot, whereas the Tutorial does not. The Stinger services do not include a generic Differential Drive service, so the code for this service has to partner directly with the Stinger motors and it is not directly portable to other robots. A different version of the code included in Chapter 4 will work with other robots such as the Boe-Bot.

The Stinger Drive-By-Wire service displays a Windows Form when you run it. Figure 16-8 shows what the GUI looks like running under Windows XP on a desktop PC. (It looks very similar on a PDA). This is the first step in getting the service to work on a PDA or Windows CE device.

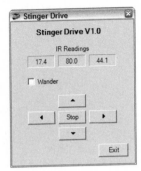

**Figure 16-8**

# Code Overview

The following discussion highlights the main features of the code. It is not a step-by-step explanation. By now, you should be familiar with creating services.

The StingerDriveByWire service in the `ProMRDS\Chapter16` folder uses the Serializer services from RoboticsConnection as well as a Windows Form. Therefore, it must have references to

❏ `RoboticsCommon.Proxy`

❏ `SerializerServices.Y2006.M08.Proxy`

❏ `Ccr.Adapters.WinForms`

❏ `System.Windows.Forms`

and, of course, the corresponding `using` statements that you can find at the top of the main service source file, `StingerDriveByWire.cs`.

Several topics are covered in the following sections:

❏ Using a Windows Form interface: This can be disabled later using a setting in the configuration file for the service. There are some differences in how a Form is handled under CF.

❏ Handling keystrokes in a Form: This is handy when using a PDA because you can use the "rocker switch" to control the robot.

❏ Differences in declaring service operations under CF

❏ Using the Serializer services is discussed briefly.

## Windows Form

In Chapter 4 you saw how to use a Windows Form in conjunction with a service. It was pointed out there that the streamlined CF version of `DssHost` does not have a `WinFormsServicePort`. To get around this minor problem you have to create your own port for Windows operations. (Remember that this is necessary because WinForms use the STA model for multi-threading).

The initialization code in the Start method is as follows:

```
// If we are NOT Headless, then create the WinForm UI
if (!_state.Headless)
{
    // Create the Dispatcher that will execute our delegates that we add to the
    // DispatcherQueue
    Dispatcher dispatcher = new Dispatcher(1, "WinForms Dispatcher");
    _DispatcherQueue = new DispatcherQueue("WinForms DispatcherQueue", dispatcher);

    // Bind the DispatcherQueue to the WinFormsAdapter and
    // create the WinFormServicePort
    _WinFormsPort = WinFormsAdapter.Create(_DispatcherQueue);

    // Use the WinFormsServicePort to create a MainForm
    _WinFormsPort.Post(new RunForm(StartForm));
}
```

Notice here that the code is conditional on the service *not* running "headless." This flag can be set in the config file, which is read using an initial state partner. All the code that refers to the Form uses the same check, which enables the service to run without a GUI. Although this is not really necessary, there is no point in displaying a GUI on the eBox-2300 when the robot is running autonomously because no screen is attached to the eBox. Removing the GUI reduces the overhead on the eBox, which is a good idea because it is a very slow CPU.

The routine to create the new form is trivial:

```
// Create the new WinForm
private System.Windows.Forms.Form StartForm()
{
    StingerDriveByWireForm form = new StingerDriveByWireForm(_mainPort);
    // Keep a handle to the Form
    _driveForm = form;
    return form;
}
```

Notice that the main service port is passed to the constructor for the Form to enable the Form to send messages back to the main service.

## Keyboard Handling

Chapter 4 covers the steps involved in using the keyboard directly. It might seem a little out of place in a chapter on autonomous robots, which do not have keyboards, but it fits in with using the rocker switch on a PDA to drive a robot.

If you open the StingerDriveByWireForm in Design view in Visual Studio, you can go to the Properties panel and click the lightning bolt icon. This displays the event handlers for the Form. Figure 16-9 shows that there are KeyDown and KeyUp event handlers (as well as Form Load).

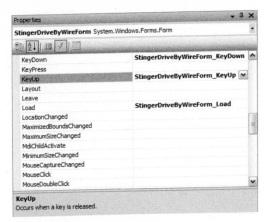

**Figure 16-9**

The keyboard handlers are in a region at the bottom of the Form code:

```
#region Key Handlers

// NOTE: The arrow keys will not normally appear in a KeyDown event.
// This is because they will be pre-processed. However, the operation
// under CF seems to be different and it does no harm to leave them
// in here anyway.
private void StingerDriveByWireForm_KeyDown(object sender, KeyEventArgs e)
{
    switch ((Keys) e.KeyValue)
    {
        case Keys.Up:
        case Keys.W:
            Forward();
            e.Handled = true;
            break;

... (code omitted for brevity) ...

        case Keys.Escape:
            btnExit_Click(sender, e);
            e.Handled = true;
            break;

        default:
            break;
    }
}

// Stop the motors any time that a key is released
private void StingerDriveByWireForm_KeyUp(object sender, KeyEventArgs e)
```

*(continued)*

**693**

*(continued)*

```
    {
        Stop();
    }
    // Overriding the ProcessDialogKey handler allows us to trap the arrow keys.
    protected override bool ProcessDialogKey(Keys keyData)
    {
        switch (keyData)
        {
            case Keys.Up:
                Forward();
                return true;

... (code omitted for brevity) ...

            case Keys.Right:
                TurnRight();
                return true;

            default:
                break;
        }

        return base.ProcessDialogKey(keyData);
    }

    #endregion
```

First, notice that the KeyDown event is used to send a motion command to the robot. (For the convenience of gamers, the A, S, D, and W keys can be used instead of the arrow keys, but obviously this only applies to the desktop version of the service). This event occurs as soon as a key is pressed.

Conversely, a KeyUp event occurs when a key is released. The handler for KeyUp immediately sends a Stop command without even looking to see which key was released. The combined effect of these two handlers is that the robot will continue to move as long as you hold down a key. Note that if you "tap" a key, the robot might jerk a little, but it will not keep driving.

For the PDA version, these two event handlers are all that is required, provided that you change the KeyPreview property for the Form to true. The arrow keycodes (Up, Down, Left, and Right) are generated by the "rocker switch" on the PDA and passed to the keyboard event handlers.

The Windows XP version is a little more complicated. In addition to setting the Form's KeyPreview property to true, you also need to override the ProcessDialogKey handler for the Form. This handler must return true if it handles a particular keystroke. If it does not want to handle the key, it can simply pass the keystroke back to the regular handler. See Chapter 4 for more details.

## Service Operations

The messages that the Form can send are declared in StingerDriveByWireTypes.cs and included in the PortSet for the main service operations. Obviously, the Form has to send commands to the main service whenever one of the buttons is pressed:

```
[ServicePort]
public class StingerDriveByWireOperations : PortSet
```

```
{
    // These changes from the previous way of declaring a PortSet
    // are necessary to work with CF when you have more than 8
    // ports in the set
    public StingerDriveByWireOperations()
        : base(
    typeof(DsspDefaultLookup),
    typeof(DsspDefaultDrop),
    typeof(Get),
    typeof(Replace),
    typeof(Load),
    typeof(MotionCommand),
    typeof(Wander),
    typeof(Quit)
        )
    {}
```

This code illustrates another difference between the desktop version of MRDS and the CF version. Under the CF version, only a maximum of eight ports can be declared in a `PortSet` using the usual syntax:

```
public class xxxOperations : PortSet<DsspDefaultLookup, DsspDefaultDrop, Get>
{
}
```

The alternative form of declaration using `typeof` gets around that problem, but at a cost — the strong compile-time type-checking that used to exist on the `Post` method is lost and the types are not checked until runtime. Even with this alternative method, there is still a limit of 20 operations in total.

You can compensate for this loss of type-checking by overloading the `Post` method as follows, but you have to do this for each of the operations individually:

```
public void Post(MotionCommand m)
{
    PostUnknownType(m);
}
```

The `MotionCommand` operation warrants a little more explanation. This particular operation handles all of the possible motions: Forward, Backward, Turn Left, Turn Right, and Stop. It is much better to condense several operations into one message type than to have several separate operations. Because there is a limit of 20 operations in the CF environment, combining them wherever possible makes sense.

The code for the `MotionCommand` class in `StingerDriveByWireTypes.cs` includes some additional constructors:

```
    public class MotionCommand : Update<MotionRequest,↵
PortSet<DefaultUpdateResponseType, Fault>>
    {
        public MotionCommand()
            : base(new MotionRequest())
        {
        }
```

(continued)

*(continued)*

```csharp
        public MotionCommand(MotionRequest body)
            : base()
        {
            this.Body = body;
        }
        public MotionCommand(int left, int right)
            : base()
        {
            this.Body.LeftPower = left;
            this.Body.RightPower = right;
        }
    }


    [DataContract]
    [DataMemberConstructor]
    public class MotionRequest
    {
        private int _leftPower;
        private int _rightPower;

        [DataMember, DataMemberConstructor(Order=1)]
        [Description("Left Motor Power")]
        public int LeftPower
        {
            get { return this._leftPower; }
            set { this._leftPower = value; }
        }


        [DataMember, DataMemberConstructor(Order=2)]
        [Description("Right Motor Power")]
        public int RightPower
        {
            get { return this._rightPower; }
            set { this._rightPower = value; }
        }

        public MotionRequest()
        {
        }
        public MotionRequest(int left, int right)
        {
            this._leftPower = left;
            this._rightPower = right;
        }
    }
```

Notice the use of the attribute [DataMemberConstructor] under [DataContract] and on class members, which causes DssProxy to generate helpers (also called *constructors* in the text) in the Proxy DLL. These helper methods are the same as the ones you see declared in MotionRequest. Only two constructors can be created this way — one with no parameters and another with the parameters that are specified using the [DataMemberConstructor] attributes (or all of the public fields if no attributes are used on individual data members).

The `Order` parameter enables you to specify the order in which the fields will appear in the constructor, or not at all if `Order = -1`.

The reason for declaring the constructors in the class declarations is so that they can be used in the Windows Form code. This service cannot use its own Proxy, so the automatically generated constructors are not available. Unfortunately, this is a "chicken or the egg" situation because `DssProxy` generates the constructors when you compile, but you need them defined so that the service will compile ...

The last step in this puzzle is back in `StingerDriveByWireOperations` where a shorthand method is declared for sending `MotionCommand` messages:

```
public virtual
PortSet<Microsoft.Dss.ServiceModel.Dssp.DefaultUpdateResponseType,Fault>
MotionCommand(int leftPower, int rightPower)
{
    MotionRequest body = new MotionRequest( leftPower, rightPower);
    MotionCommand op = new MotionCommand(body);
    this.Post(op);
    return op.ResponsePort;
}
```

With all of these declarations in place, the Windows Form can make motion requests simply by using the shorthand version, such as in the `Forward` method, which drives the robot forwards. This method is called by the event handler for the up arrow button on the Form or the keyboard handler when the up arrow key or W is pressed:

```
private void Forward()
{
    // This sample sets the power to 75%.
    // Depending on your robot hardware,
    // you may wish to change these values.

    // Use the short-hand message sender as an example.
    _mainPort.MotionCommand(75, 75);
}
```

This slightly long-winded explanation shows how it is possible to create operations in the Proxy DLL that other services can use. It considerably simplifies sending messages.

Lastly, note that the buttons on the Form are "sticky," meaning that when you click a button, the robot will continue to move in that direction — you don't have to hold down a button as you do with a key. In order to stop the robot, you must click the Stop button.

## Interfacing to the Serializer Board

The Serializer services do not implement a generic Differential Drive so you have to partner directly with the motors. (The manifest specifies Motor1 and Motor2 partners). You do not need to partner with the Serializer itself, although the service does this so that it can send a `Shutdown` message to the Serializer:

*An additional service, Stinger PWM Drive, is included with the code for this chapter that implements the generic Differential Drive contract. It is not discussed here.*

```
// Notice that we only partner with one motor because you can send
// a request to a motor service with either ID and it will work
[Partner("Motor1",
    Contract = PwmMotorProxy.Contract.Identifier,
    CreationPolicy = PartnerCreationPolicy.UsePartnerListEntry)]
PwmMotorProxy.MotorOperations _motorPort = new PwmMotorProxy.MotorOperations();
PwmMotorProxy.MotorOperations _motorNotifyPort = new
        PwmMotorProxy.MotorOperations();
// For unsubscribing
Port<Shutdown> _motorShutdownPort = new Port<Shutdown>();
```

The Stinger Drive-By-Wire service subscribes to the PWM (Pulse Width Modulation) Motor Service, but this is not strictly necessary to use the motors. The notification messages are only used to display the current motor power.

When a motion request arrives from the Form, it is processed by the `MotionHandler`:

```
[ServiceHandler(ServiceHandlerBehavior.Exclusive)]
public virtual IEnumerator<ITask> MotionHandler(MotionCommand motion)
{
    SetMotors(motion.Body.LeftPower, motion.Body.RightPower);
    yield break;
}
```

This calls the `SetMotors` method, which sends `SetMotorPower` messages to each of the motors.

Note that if the Stinger supported the generic Differential Drive contact, then the preceding code would send a `SetDrivePower` request to the Stinger drive instead of calling `SetMotors`.

```
// Common routine to handle all motor commands
// NOTE: This should not be called directly. Post a MotionCommand
// to the main port so that the requests will be queued (it is an
// Exclusive handler). There are a couple of places where this rule
// is violated because it is safe to do so.
void SetMotors(int motor1Power, int motor2Power)
{
    if (_state.MotorEnabled)
    {
        // Add a coordination header to the motor requests
        // so that advanced motor implementations can
        // coordinate the individual motor requests.
        // Note that there are TWO items in the coordination.
        coord.ActuatorCoordination coordination = new
                coord.ActuatorCoordination();
        coordination.Count = 2;

        // Set up the requests
        PwmMotorProxy.SetMotorPowerRequest req1 = new
                PwmMotorProxy.SetMotorPowerRequest();
        req1.Id = 1;
        req1.TargetPower = motor1Power;
        PwmMotorProxy.SetMotorPower leftSmp = new
                PwmMotorProxy.SetMotorPower(req1);
```

```
            leftSmp.AddHeader(coordination);
            PwmMotorProxy.SetMotorPowerRequest req2 = new
                    PwmMotorProxy.SetMotorPowerRequest();
            req2.Id = 2;
            req2.TargetPower = motor2Power;
            PwmMotorProxy.SetMotorPower rightSmp = new
                    PwmMotorProxy.SetMotorPower(req2);
            rightSmp.AddHeader(coordination);

            // Now post to both motors as quickly as possible just in case
            // the coordination does not work
            _motorPort.Post(leftSmp);
            _motorPort.Post(rightSmp);
        }
        else
        {
            // The motors are disabled (for debugging) but show the user
            // what would have happened
            Console.WriteLine("Dummy Motors: {0},{1}", motor1Power, motor2Power);
        }
    }
```

The `MotionHandler` uses an `ActuatorCoordination` to try to synchronize the changes to the motor power settings. Advanced services can use *Coordinations* to specify that a set of messages must all be received before taking an action. In this case there are two motors, so messages must be received in pairs. (The PWM Motor service does not seem to properly support Coordinations).

`SetMotors` uses the `MotorEnabled` state variable to control whether the motor power is actually changed or not. You can edit the `config` file to change this flag to `false` so that the motors on the Stinger will not be activated. This is handy during testing because the Stinger won't run away from you; and it also conserves battery life. However, a message is still displayed on the console so that you know `SetMotors` was called.

## Reading the Infrared Sensors

Three GP2D12 services run independently — one for each IR sensor. (The manifest specifies three partners: IRFront, IRLeft, and IRRight.) This makes partnering with them a little more complicated, but it illustrates an interesting way to obtain information about partners and establish the connections at runtime.

*Each of the IR sensors has its own* config *file specified in the manifest. This enables each sensor to be associated with a different analog input pin. The* Units *settings in these* config *files must be metric (not English) because the values used by the* Wander *behavior are all in centimeters.*

**1.** The necessary ports are declared at the top of the main service:

```
// Ports for the InfraRed sensors
// Notice that there is NO Partner attribute on this port
// See below for subscription
Gp2d12Proxy.Gp2d12Operations _gp2d12Port;
Gp2d12Proxy.Gp2d12Operations _gp2d12NotifyPort;
// For unsubscribing
```

*(continued)*

*(continued)*

```
Port<Shutdown> _gp2d12ShutdownPort = new Port<Shutdown>();
// Port to receive Directory service notifications over for subscribing...
ds.DirectoryPort _directoryPort;
ds.DirectoryPort _directoryNotifyPort = new ds.DirectoryPort();
```

There is no partner declaration on the `Gp2d12Operations` port. Additional ports are declared for communicating with the DSS Directory service.

**2.** The `Start` method calls `SubscribeToIRSensors`, which in turn sends a subscription request to the Directory service and sets up a handler to receive notifications:

```
// Subscribe to the IR sensors
// NOTE: This is a round-about process because there are three
// IR sensors and we want to handle them all together
void SubscribeToIRSensors()
{
    // Create a Directory Service port to receive notifications
    _directoryPort = 
ServiceForwarder<ds.DirectoryPort>(ServicePaths.InstanceDirectory);

    // Send a subscription to the Directory service for the type of
    // contract that we are interested in
    SendSubscriptionToDirectory(Gp2d12Proxy.Contract.Identifier);

    Console.WriteLine("Listening for subscriptions from Directory Service");
    // Listen for Directory insertions
    Activate(
        Arbiter.Receive<ds.Insert>(true, _directoryNotifyPort,
            InsertNotificationHandler));
}
```

**3.** Subscribing to the Directory is a simple process:

```
// Send a subscription to the Directory Service for a type of Contract
private void SendSubscriptionToDirectory(string identifier)
{
    ds.SubscribeRequestType subBody =
        new ds.SubscribeRequestType(null,
        new ServiceInfoType(identifier));

    // Create a subscribe message, and fill the body in...
    ds.Subscribe sub = new ds.Subscribe(subBody);
    sub.NotificationPort = _directoryNotifyPort;

    // Send the message to the Directory
    _directoryPort.Post(sub);

    // Set up a one-time receiver
    Activate(
        Arbiter.Choice(
            Arbiter.Receive<SubscribeResponseType>(false, sub.ResponsePort,
                delegate(SubscribeResponseType response)
                {

                }),
```

```
                Arbiter.Receive<Fault>(false, sub.ResponsePort,
                    delegate(Fault f)
                    {
                            LogError("Subscribe to directory service failed for " + ↵
identifier);
                    })
            )
        );
    }
```

When a new notification arrives from the Directory service, it is processed to determine whether the contract identifier is one of the desired partners; if it is, a subscription is made to the service to receive updates:

```
// Process Directory insertions
void InsertNotificationHandler(ds.Insert insert)
{
    // See if the new service in the Directory is one that we are
    // interested in
    if (insert.Body.Record.Contract == Gp2d12Proxy.Contract.Identifier)
    {
        // Create a new notification port for IR messages
        _gp2d12NotifyPort = new Gp2d12Proxy.Gp2d12Operations();
        // Get a service forwarder so we can send messages to the IR service
        _gp2d12Port = ↵
ServiceForwarder<Gp2d12Proxy.Gp2d12Operations>(insert.Body.Record.Service);
        Gp2d12Proxy.Subscribe subscribe = new Gp2d12Proxy.Subscribe(↵
new SubscribeRequestType());
        subscribe.NotificationPort = _gp2d12NotifyPort;
        // Set the Shutdown port too so we can unsubscribe later
        subscribe.NotificationShutdownPort = _gp2d12ShutdownPort;
        // Now subscribe to the IR service
        _gp2d12Port.PostUnknownType(subscribe);

        // Set up a persistent receiver
        MainPortInterleave.CombineWith(
            new Interleave(
                new TeardownReceiverGroup(),
                new ExclusiveReceiverGroup(
                    Arbiter.Receive<Gp2d12Proxy.UpdateReading>
                        (true, _gp2d12NotifyPort, ReplaceIRHandler)
                ),
                new ConcurrentReceiverGroup()
            )
        );

        Console.WriteLine("Subscribed to Gp2d12 Service");
    }
}
```

Note two other points about this code:

❑   It adds a shutdown port to the subscription message. By sending a Shutdown message to this port later, it is possible to unsubscribe from the service.

❑　The handler uses `CombineWith` to add an `Exclusive` receiver to the existing set of receivers. If the handler had simply created a new receiver, then it would run independently and there could be multiple simultaneous instances of the receiver activated. The receiver should be part of the Exclusive group because it updates the state.

**4.**　The last part of the code is the actual IR handler:

```
// Handle IR Sensor updates
// This handler is essential to the operation of the Wander behavior.
// However, it does not execute the behavior directly. Instead, the
// Wander behavior runs on a timer. This gives a far greater chance
// that all three of the IR sensors will have updated their values.
// Otherwise, we would be reevaluating the status after every new
// individual value, and this is not a good idea.
public void ReplaceIRHandler(Gp2d12Proxy.UpdateReading ur)
{
    // Save the value
    switch (ur.Body.Pin)
    {
        case SerializerIoPin.Pin0:
            _state.IRLeft = ur.Body.Distance;
            break;

        case SerializerIoPin.Pin1:
            _state.IRFront = ur.Body.Distance;
            break;

        case SerializerIoPin.Pin2:
            _state.IRRight = ur.Body.Distance;
            break;

        default:
            break;
    }

    // Display the value if we have a GUI
    if (!_state.Headless)
    {
        _WinFormsPort.FormInvoke(
            delegate()
            {
                _driveForm.SetIRReadings((int)ur.Body.Pin, ur.Body.Distance);
            }
        );
    }
}
...
```

This handler simply updates the state with the latest IR range value and then, if not operating headless, also updates the Form to display the current value. Although this handler is a key component of the Wander behavior, it does not directly call the Wander method. Instead, the Wander method runs on an independent timer, as explained in the next section.

# Wandering Using Sensors

You have seen a Wander behavior in Chapter 14. The behavior in this chapter is somewhat more complex and the robot can wander autonomously because the computer can be onboard.

To initiate wandering you click the Wander checkbox on the Form. Of course, if your robot is operating autonomously then there might not be a Form! To get around this problem, the setting of the Wander checkbox is saved in the config file when the service exits (or you can edit the config file manually). If the flag is already set when the service starts, it immediately goes into wander mode.

## Handling the IR Sensors

In the sample code, the Stinger uses three IR sensors. These all operate independently, and because they are all separate services they can take some time to start up. In fact, the first update will not arrive until the range value changes, so you might have to wave your hand in front of the robot.

To avoid possibly making decisions before the data is available, the Wander behavior is started by the IR sensor handler only *after* data has been received from all three sensors. The code that does this is at the bottom of the ReplaceIRHandler (from the end of the previous section):

```
// Finally, if the timer is not going then start it now.
// This ensures that the timer does not start until IR data
// is arriving. Otherwise the Wander behavior might try to
// process IR values before they are valid and this would
// result in some strange behavior initially.
if (_state.WanderEnabled && !_timerRunning)
{
    if (_state.IRLeft != 0 && _state.IRFront != 0 && _state.IRRight != 0)
    {
        // Indicate that the timer is active and call the Wander
        // routine to kick it off immediately with the new IR
        // data that has just arrived
        _timerRunning = true;
        Wander(System.DateTime.Now);
    }
}
```

The Wander method implements the Wander behavior. It relies on some constants defined at the top of the main service:

```
// Thresholds for the IR sensors
// NOTE: These values are in CENTIMETERS and depend on the config
// settings for the IR sensors being set to Metric. Also, the
// Stinger does not turn around its center -- the back end hangs out!
// This means that it really needs about 30cm clearance to avoid
// banging its backside on a wall as it turns. A value of 20 for
// the Front Threshold is probably too low, but in limited space
// you don't want to set it too high. Bear in mind that the GP2D12
// can only measure down to 10cm and then it starts to go UP again!
private const double FRONT_THRESHOLD = 20;
private const double LEFT_THRESHOLD = 30;
```

*(continued)*

*(continued)*

```
        private const double RIGHT_THRESHOLD = 30;
        // Timer interval for the Wander behavior
        private const int WANDER_INTERVAL = 100;

        // Number of timer "ticks" to execute respective behaviors
        // Multiply these counters times the Wander Interval to get
        // the amount of time spent in each behavior
        private const int TURN_COUNT = 10;
        private const int VEER_COUNT = 10;
        private const int BACKUP_COUNT = 10;

        // Motor power to use for various modes
        // The Stinger power settings range from -100 to +100,
        // but don't include a sign here
        // NOTE: In limited space, don't set these higher than
        // about 50. This little sucker is fast!
        private const int DRIVE_POWER = 50;
        private const int TURN_POWER = 50;
        private const int VEER_POWER = 25;
        private const int BACKUP_POWER = 50;
```

The Wander behavior runs on a timer controlled by the WANDER_INTERVAL. It is a state machine that moves between states based on the values of the IR sensors. The COUNT values are used to ensure that the robot remains in a particular mode for a certain number of iterations. If you take a COUNT and multiply it by the WANDER_INTERVAL, you will get the elapsed time.

## Defining Wander Modes

The various wander modes (or states) are defined in StingerDriveByWireTypes.cs as a public enum and the current mode is stored in the WanderMode field in the service state. Placing the wander mode in the service state enables it to be examined using a web browser. Using an enum declared with [DataContract] makes the values available outside the service and also stored symbolically in the saved state:

```
        // Modes that the robot can be in while wandering
        [DataContract]
        public enum WanderModes
        {
            None,
            DriveStraight,
            VeerLeft,
            VeerRight,
            TurnLeft,
            TurnRight,
            BackUp
        }
```

The modes should not require much explanation (refer to Chapter 14 for more details). Initially, the robot is in the None mode and then immediately transitions to DriveStraight when wandering is enabled. VeerLeft and VeerRight are triggered if the right- or left-side IR sensor, respectively, sees an obstacle. BackUp is triggered if all of the IR sensors are blocked, but it does not occur very often because of the effects of the other modes. In fact, the robot performs a TurnLeft or TurnRight in preference to BackUp.

The whole of the `Wander` method is shown in the following code. It is fairly long, but if you look at each of the Wander modes separately it is relatively easy to understand. There are many other ways that the behavior could be implemented, and this is not necessarily the best way:

```
// The Wander Behavior
// This is a State Machine that runs on a timer.
// It works reasonably well except for the Stinger's fat
// backside which scrapes on the walls from time to time.
void Wander(DateTime t)
{
    // See if we should execute the Wander behavior
    if (_state.WanderEnabled)
    {
        switch (_state.WanderMode)
        {
            case WanderModes.DriveStraight:
                // This code does not check for combinations of Left and Right
                // which should be unnecessary as long as the Front is free
                if (_state.IRFront < FRONT_THRESHOLD)
                {
                    // This presents a problem because the front threshold has
                    // been breached, so we need to take evasive action
                    if (_state.IRLeft < LEFT_THRESHOLD &&
                        _state.IRRight < RIGHT_THRESHOLD)
                    {
                        // If both sides are blocked, back up
                        BackUp();
                    }
                    else
                    {
                        // Otherwise, turn towards the clearest direction
                        if (_state.IRLeft < _state.IRRight)
                            TurnRight();
                        else
                            TurnLeft();
                    }
                }
                else if (_state.IRLeft < LEFT_THRESHOLD)
                {
                    // Left side obstacle seen
                    VeerRight();
                }
                else if (_state.IRRight < RIGHT_THRESHOLD)
                {
                    // Right side obstacle
                    VeerLeft();
                }
                break;

            case WanderModes.None:
                // No mode selected, so start driving
                DriveStraight();
                break;
```

*(continued)*

*(continued)*

```
case WanderModes.BackUp:
    // Decrement the counter
    _state.WanderCounter--;
    // If finished, try something else
    // Otherwise just fall through and keep backing up
    if (_state.WanderCounter <= 0)
    {
        // Turn in the clearest direction.
        // DON'T start driving ahead because that is
        // where we just came from!
        if (_state.IRLeft < _state.IRRight)
            TurnRight();
        else
            TurnLeft();
    }
    break;

case WanderModes.VeerLeft:
    // If veering has cleared the corresponding side,
    // then drive straight again.
    // It makes sense to check here if the way
    // is blocked first, and if so then continue
    // turning but this time harder.
    if (_state.IRFront < FRONT_THRESHOLD)
        TurnLeft();
    else
    {
        if (_state.IRRight > RIGHT_THRESHOLD)
            DriveStraight();
    }
    break;

case WanderModes.VeerRight:
    if (_state.IRFront < FRONT_THRESHOLD)
        TurnRight();
    else
    {
        if (_state.IRLeft > LEFT_THRESHOLD)
            DriveStraight();
    }
    break;

case WanderModes.TurnLeft:
    _state.WanderCounter--;
    if (_state.WanderCounter <= 0)
    {
        // If front is clear now, then drive ahead
        // Otherwise keep turning left
        if (_state.IRFront > FRONT_THRESHOLD)
            DriveStraight();
        else
            TurnLeft();
    }
    break;
```

```
          case WanderModes.TurnRight:
              _state.WanderCounter--;
              if (_state.WanderCounter <= 0)
              {
                  // If front is clear now, then drive ahead
                  // Otherwise keep turning right
                  if (_state.IRFront > FRONT_THRESHOLD)
                      DriveStraight();
                  else
                      TurnRight();
              }
              break;

          default:
              Console.WriteLine("Invalid Wander Mode!");
              break;

      }
  }

  // If the timer is still running, then kick it off again
  if (_timerRunning)
  {
      // Restart the timer
      Activate(Arbiter.Receive(
          false,
          TimeoutPort(WANDER_INTERVAL),
          Wander)
      );
  }

}
```

Notice that the last thing the Wander method does is start the timer again. The global flag _ timerRunning can be used to stop wandering by preventing the timer from being started again.

## *Running the Wander Code*

To run this service on your desktop PC, you need to edit the config file to set the correct COM port for your Bluetooth connection. The config file you need to change is the one for the Serializer, not the one for the Stinger Drive-By-Wire service. The filename is ProMRDS\Config\Stinger.Serializer.config.xml.

When you have changed the value of the ComPort setting, you can run the service either by starting it in the debugger in Visual Studio or by running RunStinger.cmd, which is in the MRDS bin folder.

Eventually, the Form in Figure 16-8 should appear and you can manually drive the robot by using the arrow keys on the keyboard or clicking the direction buttons. Note that you have to click Stop if you use the buttons on the Form, but if you use the keyboard you can just release the key.

When you are satisfied that the robot is working and you can see that the IR sensor values are updating, you can click the Wander checkbox and let it go on its own. Make sure that you turn off wandering before you click Exit because the setting of the Wander flag is saved to the config file.

# Creating a CF Version of the Service

After you have the desktop version of the service working, you can progress to the CF version. (That's the beauty of using Bluetooth on both the desktop and the PDA). To do this, you use DssNewService to make a CF service. This has already been done for you in the sample code, but the instructions are provided here for your reference.

*You must have the full version of Visual Studio 2005 in order to create CF projects. The Express Edition of Visual C# will not work.*

The steps to make a CF service from an existing desktop service are quite simple:

❑ Open a MRDS Command Prompt window and change directory to where the service source code is located.

❑ Enter the command dssnewservice/generatecfproject:existingservicename.csproj.

Some new files will be created with the same names as the old service files but with a prefix of "cf.". For the preceding example, there will be a new project file called cf.existingservicename.csproj. Open your existing solution and add the new project to the solution so that you can easily edit both projects.

An alternative to converting the project is to create the CF project at the outset when you first create the new service. You can do this on the command line to DssNewService by using the /createcfproject command-line qualifier (abbreviated to /ccfp). However, it is more likely that you have existing projects you want to migrate.

Unfortunately, you are not finished yet. Several more steps are required:

1. Windows Forms under Windows XP or Vista are quite different from Forms under Windows Mobile or CE. Therefore, you have to delete the Windows Form from the new CF project and then create a new one from within the CF project.

2. Give the new Form the same name as the old one but with "cf." in front of the name so that it will have a unique filename. Once the Form has been created, go into the code for the Form and remove the "cf." prefix from the Form class name. This little trick is necessary so that the existing code in the service that references the Form will work for either the desktop or the CF versions.

3. Now you need to recreate the Form with exactly the same controls as the desktop version. Add the buttons, labels, and so on, as required and make sure that they have the same names as those in the desktop Form. Do this through the Design view on the CF Form.

4. Once you have added all of the UI controls, add the event handlers. For example, double-click a button to create the corresponding Click event. Do this for all the event handlers. Remember that the keyboard event handlers can be added through the Properties panel for the Form.

5. Copy the code from each of the event handlers in the desktop Form across to the corresponding handler in the CF Form. This is a tedious process, but it is the best way to guarantee that the handlers are registered properly. Simply copying all of the code from the desktop Form over to the CF Form does not create all of the necessary "connections." (The event handlers are specified in the Form designer file, which is automatically generated. Although you can edit this if you know what you are doing, it is not advisable.)

6. Build your new CF project.

Assuming that everything compiles OK, you should find DLLs in the MRDS bin\CF folder with the name of your service prefixed by "cf."

# Deploying a Service to a PDA

*Before you can use your PDA to run a MRDS service, you must ensure that it has the latest version of the V2.0 .NET Compact Framework installed, which was Service Pack 2 at the time of writing. You can download this runtime environment from the Microsoft website. Then you have to run the installer on your desktop PC and install the software to your PDA using ActiveSync.*

Now that you have a CF service ready to run, you need to get it onto your PDA. The first step in deploying your service is to create a package. The DssDeploy tool is designed for this purpose.

Open a MRDS Command Prompt window and change directory to your service source folder. Assuming the manifest for your CF service is in the same directory and is called cf.service.manifest.xml, you can enter the following command:

```
dssdeploy /p /cf /m:cf.service.manifest.xml cfdeploy.exe
```

This command creates a CF package called cfdeploy.exe using the specified manifest. The /cf qualifier indicates that the package should be built for the CF environment.

## Creating a CF Package

Continuing with the Stinger example, perform the following steps:

1. For the example service, a batch file called BuildStingerCFPackage.cmd is supplied in the source folder. This runs DssDeploy and creates an output file called StingerCF.exe. However, before running this, edit the config file ProMRDS\Config\StingerCF.Serializer.Config.xml to change the ComPort setting to the outgoing port number that you found when you paired your PDA with the Stinger robot.

2. Note carefully that this config file is called StingerCF and not Stinger. There are two versions. There is also a separate manifest called, appropriately, StingerCFDriveByWire.manifext.xml.

3. DssDeploy generates a self-expanding executable. You must copy this to your PDA and then execute it from the location where you want it to unpack itself. The package contains all of the files necessary to run the CF version of DssHost so that you can start a DSS node and run your service.

4. Microsoft suggests creating a folder called \Program Files\MRDS, so open File Explorer on your PDA and make a folder called MRDS under Program Files. Then copy your package into this folder using ActiveSync and run it from File Explorer on the PDA by tapping on it.
   After a little while, you should see a bin directory. If you are using the example code from this chapter, there will also be a ProMRDS folder. In the process of deploying the package, all of the necessary executable files are copied from bin\CF on your desktop to the bin folder on your PDA. The manifest and config files, however, are left in ProMRDS\config. Note that none of the source code files are copied because these are not required to run the service.

5. The `cf.dsshost.exe` program operates a little differently than the desktop version. It assumes that all manifest paths are relative to its own folder, not relative to the current directory when you execute the program. Therefore, it is easiest to place all of the manifests and `config` files directly into the bin folder alongside `cf.dsshost.exe`. You need to copy all of the files from `ProMRDS\config` to bin. These should include the following:

- ❑ `Stinger.Gp2d12Front.Config.xml`

- ❑ `Stinger.Gp2d12Left.Config.xml`

- ❑ `Stinger.Gp2d12Right.Config.xml`

- ❑ `Stinger.Motor1.Config.xml`

- ❑ `Stinger.Motor2.Config.xml`

- ❑ `StingerCF.Serializer.Config.xml`

- ❑ `StingerCFDriveByWire.config.xml`

- ❑ `StingerCFDriveByWire.manifest.xml`

## Running the Service

Finally, you need to run your new service. You can open a command prompt on your PDA (if you installed the Power Toys) and then enter the appropriate commands. Alternatively, you can start the DSS node using a shortcut.

Because typing on the SIP (Software Input Panel) on a PDA is painful, you will find that it is best to create command procedures to do most tasks for you. Some useful procedures are supplied in the folder `Chapter16\StingerDriveByWire\CF`, which you will see on your PDA. Other general procedures are in `ProMRDS\CF`.

For example, you might want to create a command procedure called `MRDS.cmd` in the root directory on your PDA containing the following command:

```
cd "\Program Files\MRDS\bin"
```

Then you can simply type `MRDS` after opening a Command Prompt window to go to the correct directory. A copy of this file is included in the `ProMRDS\CF` folder.

In the bin folder, you can create a command procedure called `RunStinger.cmd` containing this command:

```
cf.DssHost.exe /p:50000 /t:50001 /m:StingerCFDriveByWire.manifest.xml
```

Notice that the manifest is in the current directory (which must be the bin folder). This batch file is also in the `ProMRDS\CF` folder for your convenience. Just copy it over to the bin folder on your PDA.

This will start a DSS node and run the service. Be patient. PDAs are not fast, and it will take a little while for `DssHost` to rebuild the contract directory cache. After a short delay, a new console window will appear and you will begin to see the output from `DssHost` as it loads the manifest.

Eventually, the Stinger Drive UI should appear. If you have properly paired your PDA with the Stinger robot as well as some other device(s), a window might appear asking you to select the correct Bluetooth device. Once you have done this, the robot will beep once to indicate that a connection has been established. The IR sensor values on the screen should update. If you notice that one of them is blank, wave your hand in front of the sensor. The sensors don't send updates unless something happens. Now you can drive your robot around.

When you have finished, turn off the robot and click the Exit button. If DssHost does not shut down, try turning off Bluetooth as well. This is an issue that has not been resolved and it is why you might need to use Process Explorer to kill the process. (You can use the Remote Process Viewer if you have ActiveSync running on the PDA). Sometimes Process Explorer is unable to kill DssHost and you have no option but to do a warm restart.

Now you can mount your PDA on your Stinger and let it loose! A universal PDA mounting kit for a car dashboard can be used with appropriate modifications and a little bit of Blutac. Because the PDA has a user interface, there is no need to set it up to run your service automatically — you can start it manually.

## Debugging a Service on a PDA

Before you can run a CF service in the debugger, you must deploy it to the PDA. The deployment step in the previous section copies over a lot of files that will not be copied by Visual Studio when you run the service in the debugger, so make sure that you have followed the instructions in the previous section first.

> You must have Visual Studio 2005 Service Pack 1 or you might not be able to make a connection to your PDA. You must also have the latest version of ActiveSync on your desktop PC.

Debugging requires an active connection to the PDA. A little bit of "magic" happens in the background to enable you to view and control processes running on the PDA. The first step, however, is to verify that you can successfully connect to the PDA using Visual Studio. Then you can set the appropriate options for debugging on a PDA, and proceed to debug in the same way as you would with a desktop application. Apart from being a little slower, the process is really very simple when you have done it at least once before.

### Testing the PDA Connection

You need to confirm that you can establish a connection to the PDA from inside Visual Studio. To test the connection, follow these steps:

1. Place your PDA is in its docking station and plug it into your PC. Start ActiveSync if it does not launch automatically.

2. Click the cf.StingerDriveByWire project in the Visual Studio Solution Explorer panel to ensure that it is the currently selected project.

3. If you do not already have the Device toolbar displayed, then click View ⇨ Toolbars ⇨ Device. (Refer to Figure 16-10, which shows the Device toolbar at the left-hand side.).

4. In the Target Device drop-down list in the Device toolbar, select "Pocket PC 2003 Device."

5. Click the "connect to device" icon just to the right of the device list. (It looks like a PDA with a power plug attached). A dialog box appears, and after a short period of time it should indicate that the connection was successful.

6.   Once the connection is established, you can click Target ⇨ Remote Tools to run the various remote tools that are included with Visual Studio. (Refer to Figure 16-10 for a list of the available Remote Tools.) Help for the Remote Tools is included in Visual Studio help under Smart Devices.

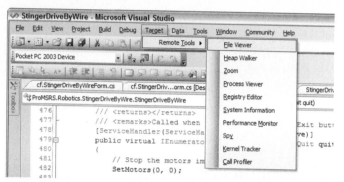

**Figure 16-10**

7.   You can configure the default device using the Device Options. Either click Tools ⇨ Options and navigate to Devices or click the Device Options icon, which is in the Device toolbar just to the right of the "connect to device" icon. The Devices window of the Options dialog is shown in Figure 16-11.

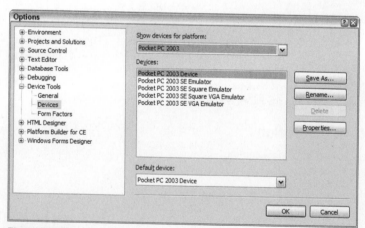

**Figure 16-11**

## Setting Up for Debugging

Once you have confirmed the connection to your PDA and finished playing around with the remote tools, you can set up for debugging:

1.   In the Properties for the `cf.StingerDriveByWire` project, go to the Debug tab. Make sure that it is set as follows:

Start External Program:

```
\Program Files\MRDS\bin\cf.dsshost.exe
```

Command-line arguments:

```
/p:50000 /t:50001
/m:StingerCFDriveByWire.manifest.xml
```

2.  On the Devices tab in the Project Properties, select Pocket PC 2003 Device as the target device. The output file folder should be `ProgramFiles\MRDS\bin`.

3.  Also on the Devices tab, turn *off* (uncheck) the checkbox beside "Deploy the latest version of the .NET Compact Framework." This will make startup faster and avoid possibly overwriting the .NET executables on the PDA with an older version from your desktop PC. Remember that you must install the latest service pack of V2.0 on your PDA before you try to run an MRDS CF service.

4.  Make sure that Bluetooth is started on your PDA.

5.  To start the debugger, right-click the `cf.StingerDriveByWire` project and click Debug ⇨ Start new instance from the pop-up menu. You should almost immediately see the Windows Mobile equivalent of the hourglass appear on the PDA screen. It takes a little while before a console window appears, and then even longer before the Windows Form appears.

6.  You can set breakpoints in your code, and step through the code, just as if you were running the service locally.

When you get tired of debugging, stop the debugger in Visual Studio. This closes down the service on the PDA and cleans up. The PDA should return to the main screen. When you are satisfied that the program is working correctly, there is nothing further to do.

# Setting Up the eBox

When you buy a Stinger CE it comes with an eBox-2300 Jump Start Kit. The eBox already has Windows CE installed on it. There is a large amount of documentation in the Jump Start Kit and on the RoboticsConnection website, which you should read before you begin. This chapter provides only an abbreviated version of the instructions.

Before you mount the eBox on the Stinger and deploy an MRDS service to it, there are many steps involved in setting it up properly. Therefore, this is a long section. It explains how to set up your eBox; configure the serial port; set up Visual Studio; and transfer an OS (operating system) image to the eBox.

> You can use a serial-to-Bluetooth device plugged into COM1 on the eBox instead of mounting it on the Stinger. For initial testing, this might be a lot easier because you can leave the eBox on your desk with a keyboard, mouse, and monitor attached, but still test the code running directly on the eBox.

The Parani-SD100 Bluetooth Serial Adapter from Sena Technologies (www.sena.com), shown in Figure 16-12, is an excellent device. It can run off a 9V battery and it has a 100m range.

**Figure 16-12**

You can set it up as follows:

1.  Configure the SD100 serial interface for 19,2000 baud using the DIP switches.

2.  Plug the SD100 into a COM port on your desktop PC and turn on the Stinger robot. Then run the ParaniWin software to search for the Stinger robot.

3.  Lock the SD100 into Mode 1. Always connect to the same Bluetooth device.

4.  Unplug the SD100 and plug it into your eBox on COM1. (Note that there are two COM ports on the eBox, so make sure you select the correct one.)

Once you have done this, the eBox can communicate with the Stinger via Bluetooth, but no software is required on the eBox because it is unaware of the wireless communication, thanks to the wonders of modern technology! The SD100 acts like a "hardwired" connection.

# Booting the eBox for the First Time

This section explains how the boot process works under Windows CE. To some extent this is a throwback to the old days of DOS. If you have experience with DOS, then you can skim through this section. However, many younger readers might not know how the DOS boot process works.

> *MS-DOS is only used as a vehicle to get Windows CE up and running. If you want a turnkey system, you can set Windows CE to boot directly and avoid all of the complications outlined below.*

## The Boot Menu

First set up the eBox with a keyboard, mouse, monitor, and UTP Ethernet connection. It comes with a universal 5V power pack that you can plug into the mains power. (You need a suitable adapter plug outside the United States, but the voltage is not an issue).

Turn on your eBox and you should see a menu with white text on a blue background, similar to the one that follows. (This is not the infamous "blue screen of death.")

```
MS-DOS 6.22 Startup Menu
========================

     1. Boot CE/PC (local nk.bin with /L:1024x768x32)
     2. Boot CE/PC (ether via eboot.bin with /L:1024x768x32)
     3. Boot CE/PC (ether via eboot.bin with /L:800x600x32)
     4. Boot CE/PC (ether via eboot.bin with /L:640x480x32)
```

```
   5. Boot CE/PC (ether via eboot.bin without display settings)
   6. Boot CE/PC (Static IP: 192.168.2.232, with /L:1024x768x32)
   7. Boot CE/PC (Static IP: 192.168.2.232, with /L:640x480x32)
   8. Boot CE/PC (Static IP: 192.168.2.232, without display settings)
   9. Clean Boot (no commands)

Enter a choice: 1



F5=Bypass startup files F8=Confirm each line of CONFIG.SYS and AUTOEXEC.BAT [N]
```

You can select a menu option by pressing the corresponding key number or highlighting it using the up and down arrow keys and pressing Enter. Note that there is a maximum of nine options because a single keystroke is used to make a selection.

If you don't make a selection from the menu, it will time out and run the default option after a few seconds. To stop the countdown, press the down arrow key to highlight another menu option.

---

### Booting Windows CE

Once you have made a selection from the menu, a program called LOADCEPC.EXE is run to boot the selected Windows CE system image. This image is usually called NK.BIN but the name is not important. You can have several system images on the embedded hard drive.

Windows CE can be set up to boot directly; it does not require MS-DOS. This is how a production system would be configured, but for development work it is much more convenient to use MS-DOS so that you can boot different versions of the operating system and change configuration settings such as screen resolution.

The Startup menu also gives you the option to boot over the Ethernet from a desktop PC that has Platform Builder installed on it. The eBox can obtain an IP address via DHCP or use a static one. Remote booting is not covered here.

The last option in the menu (Clean Boot) simply drops out to MS-DOS. Used for emergency repairs, this option is not covered in the discussion that follows.

---

## MS-DOS Refresher

If you have a background in MS-DOS, then the boot process will be familiar to you. In case you don't remember how it works or you are unfamiliar with MS-DOS, here is a quick refresher.

When MS-DOS boots up (from the hidden files MSDOS.SYS and IO.SYS), it reads a file called CONFIG.SYS that contains the menu, which looks like the following:

```
[menu]
menuitem=LOCAL_1024, Boot CE/PC (local nk.bin with /L:1024x768x32)
menuitem=CEPC_1024, Boot CE/PC (ether via eboot.bin with /L:1024x768x32)
```

*(continued)*

*(continued)*

```
menuitem=CEPC_800, Boot CE/PC (ether via eboot.bin with /L:800x600x32)
menuitem=CEPC_640, Boot CE/PC (ether via eboot.bin with /L:640x480x32)
menuitem=CEPC_xxx, Boot CE/PC (ether via eboot.bin without display settings)
menuitem=IP_1024, Boot CE/PC (Static IP: 192.168.2.232, with /L:1024x768x32)
REM menuitem=IP_800, Boot CE/PC (Static IP: 192.168.2.232, with /L:800x600x32)
menuitem=IP_640, Boot CE/PC (Static IP: 192.168.2.232, with /L:640x480x32)
menuitem=IP_xxx, Boot CE/PC (Static IP: 192.168.2.232, without display settings)
menuitem=CLEAN, Clean Boot (no commands)
Rem menudefault=CEPC_1024,15
menudefault=LOCAL_1024,5
Rem menudefault=IP_1024, 5
menucolor=7,1

[LOCAL_1024]

[CEPC_1024]

[CEPC_800]

... (code deleted for brevity) ...

[IP_xxx]

[CLEAN]

[COMMON]
buffers=10,0
files=30
break=on
lastdrive=Z
dos=high,umb
device=himem.sys /testmem:OFF
```

Note the following about the preceding code:

❑   The file is divided into several sections by headings in square brackets. You can comment out lines in the file using REM.

❑   Each menuitem has a name and a description. In this case, the item-specific sections of the file do not contain any code. The COMMON section at the bottom is used by all menu items. It contains MS-DOS configuration settings that are not relevant here.

❑   The menudefault setting specifies the name of the default menu item and the timeout, which are LOCAL_1024 and 5 seconds, respectively, in the example.

When you make a selection, AUTOEXEC.BAT is executed with the name of the selected menu item passed across as the CONFIG environment variable. The AUTOEXEC.BAT looks like the following:

```
@echo off
verify off
PROMPT $p$g

REM !!!!!!!!!!!!!!!!!!!!!!!!!!!!!!!!!!!!!!!!!!!!!!!!!!!!!!
REM !! NET_IP can be set to specify a static IP      !!
```

```
REM !! address or left blank to use DHCP to obtain   !!
REM !! an IP address.  Format of set should be:      !!
REM !!      set NET_IP=192.168.2.232                 !!
REM !!                                               !!
REM !!!!!!!!!!!!!!!!!!!!!!!!!!!!!!!!!!!!!!!!!!!!!!!!!!!!

set NET_IRQ=0
set NET_IOBASE=0
set NET_IP=

if "%CONFIG%" == "LOCAL_1024" goto LOCAL_1024
if "%CONFIG%" == "CEPC_1024" goto CEPC_1024
if "%CONFIG%" == "CEPC_800" goto CEPC_800
if "%CONFIG%" == "CEPC_640" goto CEPC_640
if "%CONFIG%" == "CEPC_xxx" goto CEPC_xxx
if "%CONFIG%" == "IP_1024" goto IP_1024
if "%CONFIG%" == "IP_640" goto IP_640
if "%CONFIG%" == "IP_xxx" goto IP_xxx
if "%CONFIG%" == "CLEAN" goto CLEAN

:LOCAL_1024
REM ###############################################################
REM    Launch LOADCEPC using a local NK.BIN image.
REM    with 1024x768x32 display

loadcepc /L:1024x768x32 nk.bin
goto END

:CEPC_1024
REM ###############################################################
REM     Set RES=/L:1024x768x32 for use with display driver.
REM

set RES=/L:1024x768x32
goto WITHRES

... (code deleted for brevity) ...

:IP_xxx
REM ###############################################################
REM    Load image without display setting
REM    Assign Static IP Address:  192.168.2.232
REM

Set NET_IP=192.168.2.232
goto NORES

:WITHRES
REM ###############################################################
REM    Here we actually Launch LOADCEPC using the RES, NET_IOBASE,
REM    and NET_IRQ env vars we just set above based on menu
```

(continued)

*(continued)*

```
REM      selections.
loadcepc /e:%NET_IOBASE%:%NET_IRQ%:%NET_IP% %RES% eboot.bin
goto END

:NORES
REM ##################################################################
REM      Load image without display settings
REM

loadcepc /e:%NET_IOBASE%:%NET_IRQ%:%NET_IP% eboot.bin
goto END

:CLEAN

:END
```

The code starts out with a series of IF statements to jump to the correct label based on the CONFIG variable. For example, LOCAL_1024 runs NK.BIN as the system image with a screen resolution of 1024 × 768. Apart from setting some parameters for screen resolution and the network, eventually LOADCEPC.EXE is executed, and from that point Windows CE takes over.

Windows CE displays the desktop, as shown in Figure 16-13. You do not need to log in.

**Figure 16-13**

Windows CE includes a File Explorer, a Command Prompt, and Pocket WordPad (which you can use to edit text files). Although these applications run full-screen, they all have icons on the taskbar and you can switch between them. There is also an icon called Show Desktop at the far right of the taskbar.

## Setting Up the Serial Port

The eBox uses a serial port to communicate with the Stinger robot. You have to either mount the eBox on the Stinger and connect the serial cable that is supplied or use a Bluetooth dongle such as the Sena SD100, as explained earlier. Either way, you need to set up the serial port first.

To set up the serial port, follow these steps:

1. Click Start ⇨ Settings ⇨ Network and Dial-up Connections.

2. Click File ⇨ New.

3. In the text box in the Make New Connection dialog box, enter **COM1**.

4. Select Direct Connection and click Next.

5. In the drop-down list Select a device, select Serial Cable on COM1. You can configure the port as shown in Figure 16-14 if you want, but this is not necessary.

6. Click Finish. An icon labeled COM1 should appear in the folder.

7. Click Start ⇨ Suspend. Be careful not to move the mouse as you click. The screen will go blank and the eBox will save the current settings.

8. Move the mouse or press a key to wake up the eBox.

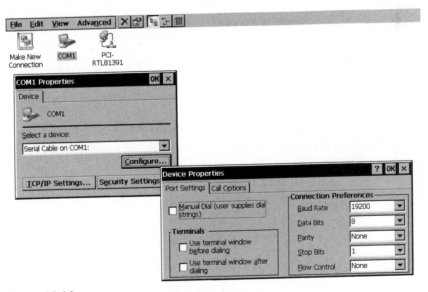

**Figure 16-14**

719

Once you have your eBox set up correctly, you can mount it on the Stinger as per the instructions. You need to remove the Bluetooth module from the Stinger and replace it with the standard serial (RS-232) module that came with the Stinger CE. Plug the serial cable into the serial module and the other end into COM1 on the eBox.

Alternatively, if you have a Bluetooth Serial Adapter for your eBox, then you don't need to swap out the Bluetooth module on the Stinger.

# Setting Up Visual Studio

If you want to develop services for the eBox, then you need to install a lot of software on your development PC. This software comes with the eBox on CD or it can be downloaded from the RoboticsConnection website. After this, you need to establish a connection to the eBox. The following sections outline software requirements and then provide the steps for establishing a connection.

## Software Requirements

The following is a list of software that you need to install to build your services:

❑ **Platform Builder:** This is a plug-in for Visual Studio. An evaluation version is supplied on CD, but you must first go to the Microsoft Windows Embedded website (`www.microsoft.com/windows/embedded/default.mspx`) to obtain a license key. Platform Builder can be used to build Windows CE system images. It can also compile C++ code to produce "native" executables and device drivers (using Embedded Visual C++). However, MRDS uses .NET so the Platform Builder is not needed for compiling code.

*You must have Visual Studio 2005 Service Pack 1 installed before you install Platform Builder.*

When you run the Platform Builder installation, you must *make sure that you select X86 support* because the eBox is an X86 platform but only ARM processor support is selected by default.

❑ **Board Support Package (BSP):** Once you have Platform Builder installed, install the Board Support Package (BSP) for the eBox. A BSP includes the necessary definitions and device drivers to support a particular hardware implementation of an embedded PC. This is available as `ICOP_eBox2300_60B_BSP.msi`.

❑ **Windows CE SDK:** This defines the development environment and a device type that you can use to connect to the eBox over the Ethernet. The installation package is called `eBox2300_WinCE600_SDK.msi`.

❑ **CoreCon component:** This is used for communicating with the eBox. It effectively replaces ActiveSync, which you use with a PDA. The package you need to install is called `CoreCon_x86_VS2005.msi`.

Check the `Windows` folder on your eBox to confirm that the CoreCon files are there. (They should have been included in the system image that is already on the eBox). You should find two programs: `ConmanClient2.exe` and `CMAccept.exe`. If you can't find these files, you should be able to find them on your development PC using the following path: `C:\Program Files\Common Files\Microsoft Shared\CoreCon\1.0\Target\wce400\x86`. The necessary files are as follows:

- `Clientshutdown.exe`
- `ConmanClient2.exe`
- `CMaccept.exe`
- `eDbgTL.dll`
- `TcpConnectionA.dll`

Unless you have these files, you will not be able to connect to your eBox remotely. They perform the equivalent function that ActiveSync provides for a PDA. You have to transfer them to the eBox using a flash drive or CF card (see "Transferring Files," below).

*When you are debugging a service on a PDA, if you look in the Remote Process Viewer you will see the equivalent Connection Manager client running on the PDA, but this is not relevant here.*

## Establishing a Connection to eBox

When you have all of this software installed, open the StingerDriveByWire service in Visual Studio so that you can establish a connection to the eBox. The steps are similar to those for a PDA:

1. Make sure your eBox is connected to the Ethernet and boot it up. It has a UTP connection on the back.

2. On the eBox, open a Command Prompt window and use the `ipconfig` command to determine the IP address of the eBox. You might want to double-check that the network connection is working and that you have the necessary access by pinging the eBox from your desktop PC. It might be necessary to disable the firewall on your desktop PC to enable communication with the eBox.

3. Click the `cf.StingerDriveByWire` project in the Visual Studio Solution Explorer panel to ensure that it is the currently selected project.

4. If you don't already have the Device toolbar displayed, then click View ⇨ Toolbars ⇨ Device. (Refer to Figure 16-15, which shows the Device toolbar at the top left side.)

5. In the Target Device drop-down list in the Device toolbar, select "eBox2300_WinCE600_SDK x86 Device." (This device will not appear unless you installed the SDK.)

6. Click the Device Options icon, which is called out in Figure 16-15 (the second icon to the right of the target device).

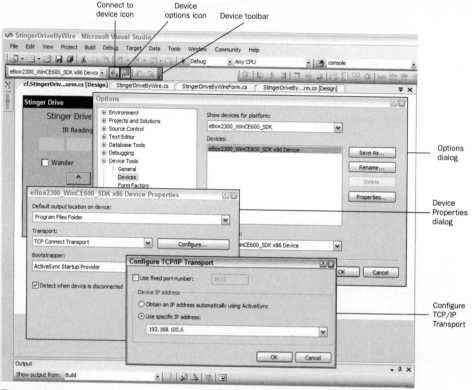

**Figure 16-15**

**7.** As shown in Figure 16-15, you must select the eBox-2300 in the Options dialog. Then click Properties and click Configure in the Device Properties dialog to finally get to the Configure TCP/IP Transport dialog, where you can enter the IP address of the eBox. Close all the dialogs by clicking OK on each of them.

**8.** To connect to the eBox, run the CoreCon client on the eBox. Open File Explorer on the eBox and change to the Windows directory. Double-click ConmanClient2 and then CMAccept to run them.

Alternatively, there is a batch file called CoreCon.cmd in the ProMRDS\CF folder. This contains the following commands:

```
cd \Windows
start ConmanClient2
start CMAccept
```

Note the use of the start command, which starts the programs in the background.

**9.** Back in Visual Studio, click the "connect to device" icon, which is just to the right of the device list. (It looks like a PDA with a power plug attached). A dialog box will appear and after a short delay it should indicate that the connection was successful, as shown in Figure 16-16.

**10.** You can now click Target ➪ Remote Tools to run the various remote tools that are included with Visual Studio. (Refer to back Figure 16-10 for a list of the available Remote Tools).

**Figure 16-16**

Now that you have configured the connection, you should not need to do it again in the future, provided that the IP address of your eBox does not change. However, you will need to run CoreCon on the eBox every time you want to connect. Note also that you must connect within three minutes or the CoreCon client will time out.

## Transferring Files to the eBox

You have several options for transferring files to and from the eBox. The easiest method is using a flash drive. The eBox has two USB ports on the front into which you can plug USB flash drives. When you plug in a flash drive it appears as a top-level folder in File Explorer called USB Storage. There is also a CompactFlash slot. (This is referred to as a CF slot but should not be confused with Compact Framework). A good investment might be a CF card and a universal USB flash card reader for your desktop. This will enable you to transfer large amounts of data.

> *The CF card slot in the eBox does not support hot swap, so if you want to use a CF card you should insert it before turning the power on. Then it appears as a folder called* Hard Disk *in the File Explorer. Hot swap is supported for USB flash drives, however.*

An alternative method is via the Ethernet. You can set up the Visual Studio Remote Tools so that you can use them to access the eBox while you are developing your code. (When the Stinger is roaming free, you will not usually have a network connection unless you buy and install the optional WiFi card for the eBox).

Assuming that you have set up Visual Studio and established a connection to the eBox as in the previous section, you can click Target ➪ Remote Tools ➪ File Viewer. A File Viewer window will appears, similar to the one shown in Figure 16-17.

Figure 16-17 shows the file structure *after* the Stinger Drive-By-Wire service has been deployed to the eBox, which is apparent from the \Program Files\MRDS\bin folder. Also visible is the \Hard Disk folder, which is a CF card, and the \Windows\.NET CF 2.0 folder, which shows that the .NET Compact Framework is installed. You can copy files to or from the eBox using the arrow buttons in the toolbar.

Avoid the temptation to rush off and copy the entire contents of the bin\CF folder from your desktop installation of MRDS to the bin folder on your eBox. This might seem like a good idea in principle, but the next time you start DssHost on the eBox it will rebuild the service directory cache, which takes about 20 minutes on an eBox!

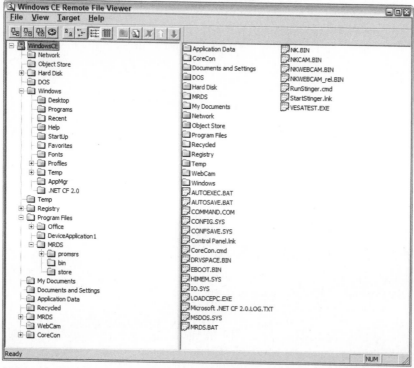

Figure 16-17

It is advisable to only transfer the services that you need to your eBox. When you build a DssDeploy package, DssDeploy analyzes the dependencies between services and only includes the ones you need.

## Building a Windows CE System Image

If you want to make changes to the Windows CE configuration, you have to build a new system image. This can be done using Platform Builder, which is a plug-in for Visual Studio. However, the eBox already has an OS image on it, so there should be no need to replace this unless you want different device driver support.

*It is outside the scope of this book to explain in detail how to create a new Windows CE system image. Instructions are included in the Jump Start Kit. However, a few brief instructions are given below just to give you some idea of what is involved.*

The basic procedure is as follows:

1. Open Visual Studio and create a new Project.

2. For the project type, select Platform Builder for CE 6.0 and then the OS Design template. Platform Builder will take you through the OS Design Wizard.

3. You need to select the BSP for the eBox (which you installed earlier), and then the Industrial Device design template and Internet Appliance. Various options can be selected for inclusion,

such as Media Player, WordPad, Internet Explorer, etc. You must select the .NET Compact Framework 2.0 to support MRDS.

4. When you have completed the wizard, the operating system image will be built. Assuming that you called your OS Design project eBox2300, then the system image will be located in the following folder:

```
C:\WINCE600\OSDesigns\eBox2300\eBox2300\RelDir\↵
ICOP_eBox2300_60B_x86_Release.
```

This can take quite a long time. The resulting file, NK.BIN, is around 20MB in size.

Note that you can build either Debug or Release versions. The Debug version is intended for debugging device drivers and it expects to make a KITL (Kernel Independent Transport Layer) connection when it starts up. It can take an extremely long time before it gives up on this connection, so make sure that you build a Release version.

As shown earlier, you can edit the CONFIG.SYS and AUTOEXEC.BAT files to load any system image that you wish. You need to copy the new system image to the root directory on the eBox and then make the necessary changes. Be careful not to overwrite the existing NK.BIN because you might need this as a fallback in case your new OS image does not boot!

---

### MS-DOS Filenames

When you edit or replace files on the eBox that are involved in booting, you must only use filenames in the old MS-DOS 8.3 format with uppercase letters. This is necessary because MS-DOS does not understand the new filename formats introduced in Windows 95 to support long names and mixed case.

If you make even such a minor change as renaming AUTOEXEC.BAT to Autoexec.bat, your eBox will not boot properly! In this case, you can use Clean Boot to go to MS-DOS and then rename the file. If you run a DIR command you will see a file called AUTOEX~1.BAT, which is the 8.3 version of the name containing lowercase characters. Rename this to AUTOEXEC.BAT and then reboot.

Likewise, be careful creating new system images because the filenames must also be in 8.3 format. For example, NKnew.bin will not work! Use NKNEW.BIN.

---

You can re-open your OS design again in Platform Builder at any time, change some of the options, and rebuild the OS image.

# Running a Service on WinCE

Up to this point you have set up your eBox so that it can boot up, and it is configured to talk to your Stinger. Now it is time to try out your code, and even let the Stinger wander about on its own.

If you have followed through the chapter so far then you have already tested the code on your desktop and a PDA. That means you already have the necessary DssDeploy package for use on the eBox-2300 because it is the same package you used on the PDA. If you didn't read through the previous section "Creating a PDA Version of the Service," you need to go back and read it now because there is a significant overlap.

## Setting Up the Service

To begin, connect up a monitor, keyboard, mouse, and Ethernet to your eBox. (You can do this even if the eBox is attached to your Stinger.) Then follow these steps:

**1.** Make a folder called MRDS under Program Files.

**2.** Transfer the DssDeploy package to the MRDS folder by whatever method you prefer and then run it from there. As with the PDA, this will create a bin folder and a promrds folder.

*Files on the embedded hard drive are not volatile, but each time you reboot the eBox it expands the system image and overwrites any files already on the hard drive that have the same names as files in the image. This means you can actually add files to the Windows directory and they will stay there. However, if you change an existing file, it will be overwritten the next time you reboot.*

**3.** Similar to the instructions for the PDA, copy all of the manifest and config files from promrds\config to the bin folder.

**4.** Edit the StingerCF.Serializer.config.xml file. You can use WordPad for this. (Alternatively, you can edit it on your development PC and then transfer it over to the eBox.) You need to change the ComPort to 1.

> **If you want to test the service while the eBox is mounted on the Stinger and attached to a monitor and keyboard, you must also edit** StingerCFDriveByWire.config.xml **and set** MotorEnable **to** false**. This enables you to run the service while you have the monitor and keyboard plugged into the eBox without worrying about the Stinger trying to run away!**

An alternative is to put the Stinger up on blocks so that the wheels cannot touch the ground. Assuming that you disabled the motors, later you can set MotorEnable back to true. While the motors are disabled, messages are displayed on the console in the following format:

```
Dummy Motors: xx, yy
```

These messages show you what motor power would have been set and therefore give you an indication of what is happening.

Conversely, if you have something like the Parani SD100, then you can use Bluetooth and it isn't necessary to disable the motors because the eBox doesn't have to be attached to the Stinger for testing. You can leave the eBox on your desk and drive the Stinger wirelessly.

**5.** In the ProMRDS\CF folder is a batch file called RunStinger.cmd. Copy this into the root directory on the eBox. (You can put a copy in the bin folder as well if you want to.) It contains the following command (on one line):

```
"\Program Files\MRDS\bin\cf.DssHost.exe" /p:50000 /t:50001 ↵
/m:StingerCFDriveByWire.manifest.xml
```

**6.** You can either execute this batch file or manually open a Command Prompt window, change directory to `\Program Files\MRDS\bin`, and then execute `cf.dsshost.exe` yourself.

## Testing Operation of the Service

If you have set everything up properly, you will see messages from `DssHost` and eventually the GUI will appear. You can drive the robot around using the arrow keys on the keyboard or the buttons on the Form, or click the Wander checkbox to let it "wander" by itself, as shown in Figure 16-18.

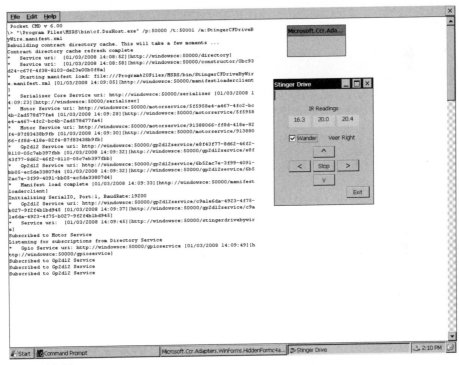

**Figure 16-18**

In Figure 16-18, the Wander checkbox is ticked. In this mode the service displays the current Wander mode — in this case, Veer Right — as the robot moves around. The IR readings are also updated dynamically. These readings will change if you by place your hand in front of one or more of the sensors. This enables you to test various combinations of IR readings to ensure that the Wander mode is updated correctly.

Notice in Figure 16-18 the small window titled `Microsoft.Ccr.Adapters.WinForms` `.HiddenFormxxxxx`. This is the Windows Forms Adapter. When you click Exit, the code sends a `Shutdown` message to the Windows Forms Adapter and this window disappears, along with the GUI.

However, the DSS node might not shut down. If so, close the Command Prompt window and it will take `DssHost` with it.

## Autonomous Operation

When you run the robot on its own, there is no need for the Windows Form. In fact, the robot does not have a monitor when it is running autonomously so there is no point in displaying a Form. However, there is one more task to perform before you let the Stinger run free — you need to set up an auto-start command because you will not have a keyboard or monitor to initiate the service.

To make the robot run autonomously, follow these steps:

1. Edit the `StingerCFDriveByWire.config.xml` file and set `Headless` to `true` and `WanderEnabled` to `true`. For testing purposes, leave `MotorEnabled` set to `false` for now.

2. You also need to set up `DssHost` so that it runs automatically when the eBox boots up, and ensure that you have set the correct menu option as the default in `CONFIG.SYS`.

3. There is a program called `MsrsAutoStart.exe` that is provided by Microsoft in the `bin\CF` folder of the MRDS distribution. The purpose of this program is to launch a startup command after a short delay. A shortcut is included in the `ProMRDS\CF` folder called `MrdsAutoStart.lnk`. You must copy this into the `\Windows\Startup` folder so that it will be executed automatically when Windows CE boots up. The shortcut contains the following:

```
79#"\Program Files\MRDS\bin\MsrsAutoStart.exe" \Windows\cmd.exe ↵
/c \RunStinger.cmd
```

4. If you right-click the shortcut in File Explorer and select Properties, there is a tab in the Properties dialog labeled Shortcut. You can easily edit the command there. However, be aware that there is a limit to the length of the command (somewhere around 110 characters). This is why the shortcut executes the batch file `RunStinger.cmd` instead of the full `cf.dsshost.exe` command.

Look carefully at the command. It executes `MsrsAutoStart.exe`, which in turn executes the following command:

```
\Windows\cmd.exe /c \RunStinger.cmd
```

Notice that you must specify full paths for everything and that the batch file must be located in the root directory. The `/c` qualifier for `cmd.exe` means create a new command prompt.

This is a somewhat complicated procedure, but you can test it out easily — just reboot your eBox. Don't select anything from the boot menu. It will time out and automatically select the default option. After a short delay Windows CE will boot and then the MsrsAutoStart dialog will appear (see Figure 16-19). You can cancel execution at this stage, although you should leave it to run to ensure that it works.

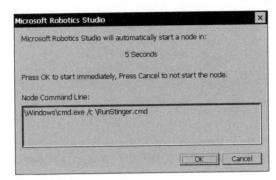

**Figure 16-19**

When MsrsAutoStart has finished counting down, a Command Prompt window should appear and `DssHost` will start. Because you have enabled wandering and disabled the GUI, the robot should spontaneously start to display Dummy Motor messages on the screen. If it does not do this after it beeps, then wave your hand in front of each of the IR sensors to make sure that they register. Then it should start.

When you have reached this point, it is now time to let your creation run loose:

1. Edit the `config` file again and set `MotorEnabled` to `true`.

2. Disconnect all of the cables from the eBox. If you were using Bluetooth, mount the eBox on the robot. The Bluetooth dongle might stick out the side a little bit, so watch the robot carefully to make sure it does not hit any obstacles.

3. Put the robot in a clear area on the floor.

4. Turn on the robot and then turn on the eBox.

5. Sit back and watch.

As the robot wanders around, you can "shepherd" it with your hands or feet by putting them in front of the sensors. If you have a test environment with walls, i.e., a "playpen," then the robot should just wander around avoiding the walls. When you eventually get tired of playing with the robot, you have to catch it and turn it off.

# Where to Go from Here

Rather than just being the end of the chapter, this is really the beginning of a new adventure. There are many, many changes that you can make to the services in this chapter to add more features, make them more robust, and so on. Here are a few suggestions:

❑ You might want to make changes to the Wander behavior in the Stinger Drive-By-Wire service. It is not a very smart behavior. You can also fiddle with the parameters to see whether you can improve it.

❑   The Stinger CE has a Line Following kit that mounts under the front bumper. You might want to try your hand at writing a service that follows lines on the ground. The Stinger Drive-By-Wire service should be a good starting point.

❑   You might also like to consider using a webcam for wandering. However, be warned that webcam support under WinCE is severely limited. About the only camera that will work is the Logitech Pro 5000 (and it must be exactly that model). In addition, the frame rate is not very good — you will only get about 3 or 4 frames per second even at low resolution.

Whenever you rebuild the service, you can run the batch file `BuildStingerCFPackage.cmd` to create a new `DssDeploy` package. If you do this, remember that you might have edited the `config` files on the eBox so you should transfer them back to your desktop PC and place them in the `ProMRDS\Config` folder. This ensures that the updated files are included in the `DssDeploy` package and you don't accidentally revert them to the original versions.

You should regularly visit the RoboticsConnection website to see what new software they might have to offer. From time to time they might also run competitions.

# Summary

After completing this chapter successfully you now have experience in developing MRDS services for mobile devices. Because it is quite a complex area, it is possible that you had a few little hiccups along the way. Unfortunately, it is more than possible, it is quite likely. As usual, Google is your friend. If you get stuck on a particular problem, use Google or search the MRDS Discussion Forums.

Now that you have a robot that is capable of exploring on its own, you can start to design more complex algorithms and behaviors. The Stinger has wheel encoders to help you measure distances, but these were not used in this chapter. It also has a PID (proportional-integral-derivative) mode of operation for the motors that is far more accurate than the PWM method used in this chapter.

The last chapter shows you how to write new services for devices that are not yet supported by MRDS. This might be your chance for a claim to fame.

# 17

# Writing New Hardware Services

For the final chapter in this book, there is a lot of material to draw on. Building services, debugging, and deploying services are all covered in previous chapters as well as Chapter 16, which discusses building services for Windows CE and Windows Mobile.

In this chapter, you'll start with a blank slate and build services for a new robot. You'll begin with a new Generic Brick contract, and then build services based on this for the Integrator and Hemisson robots. To simplify the process of testing, a test service is also available that can execute operations on any robot based on the Generic Brick contract.

The Integrator uses a PICAXE microcontroller that is similar to the BASIC Stamp in that it has an onboard BASIC interpreter. The "monitor program" that is necessary for the robot to communicate with a PC is developed in BASIC.

The Integrator robot was selected because of the popularity of small microcontrollers such as the PIC16F88, on which the PICAXE is based. Although the Integrator is not a widely known robot, the instructions in this chapter should assist you in programming your own robot based on a range of different robots that are available in kit form or preassembled.

The authors also provide services for the Hemisson robot from K-Team in the `ProMRDS\Chapter17` folder. These are based on the Generic Brick contract as a second example, and show you how to approach the task when the robot already has its own firmware and the communications "protocol" is already defined. In this case, you do not have to write a monitor program.

## Integrating New Hardware

Realistically, as a programmer you will rarely start a new programming assignment from scratch. You usually have some existing code to work from. Even though the code in this chapter starts from "nothing," you still have the benefit of existing generic contracts and the code samples provided with MRDS as guides.

If you buy a small robot kit or one that is already built, there will most likely be some software available for it. Many of the hobby and educational robots on the market have active discussion forums, and the source code for software that runs on the robot is readily available. This should assist you if you need to write a monitor program.

The majority of small hobby and educational robots use what is called a PIC (programmable interface controller) or MCU (microcontroller unit) as the processor in their brick. A PIC is a complete system-on-a-chip (SOC) that includes CPU, clock, memory (volatile RAM for data and nonvolatile Flash ROM for programs), and I/O (digital, analog, or both). They have evolved from 8-bit to 16-bit and now 32-bit chips and come in a large variety of sizes — from as small as eight pins in a dual inline package (DIP) to surface mount chips with over 100 pins.

There is an active market for microcontroller chips on circuit boards for embedded systems, especially for use as robot controllers. The circuit board also needs an H-bridge or some power transistors to control motors or servos due to the high current requirements. If you are not familiar with hardware and you are thinking about building your own robot, this is an important point that you should understand: The outputs from a microcontroller cannot drive motors directly.

Microcontroller chips typically include RISC CPUs, which have small instruction sets. In a recent innovation, Parallax introduced the Propeller chip, which has eight CPUs, called *cogs*. This allows for parallel processing, or true multi-tasking, on a single chip. These chips are programmed in a language called Spin. The Spin Stamp is a direct replacement for the BASIC Stamp 2 in the Boe-Bot. Parallax provides Spin code that is a direct replacement for the PBASIC monitor program, but it does not use all of the cogs. Therefore, if you are adventurous, you can install a Propeller chip in your Boe-Bot and rewrite the monitor program to take full advantage of all eight cogs.

Note that there is a trend toward more powerful onboard processors, rather than PICs. For example, the Gumstix boards (www.gumstix.com) have Intel XScale processors and run Linux. Versions are available with built-in Bluetooth. Ethernet and USB connections are also available. To work with a robot, a plug-in board called the Robostix provides the necessary hardware interfaces. This is another whole market segment, however, and Linux-based robots are not covered here. Instead, we target PIC-based robots.

For the sake of completeness, it is worth mentioning another possible brick, the Sun SPOT (www.sunspotworld.com) from Sun Microsystems. Sun SPOTs have a 180 mHz ARM processor, built-in ZigBee radio, a USB interface, and run a Java Virtual Machine. They are intended for use in embedded systems. The key point about Linux or Java-based bricks is that, as the programmer, you have to write the monitor program to run on the brick. This topic is also beyond the scope of this book, but if you are a top-gun Java programmer, you might like to write a monitor program for a Sun SPOT and then create MRDS services for it. If you do, be sure to publicize it on the MRDS Discussion Forum.

With this background in mind, the following sections outline several options for your brick's processor.

## *Microcontrollers*

Microchip Technology (www.microchip.com) named their highly successful series of chips using the "PIC" prefix. The Hemisson robot, for example, uses a PIC16F877 chip. Although the Microchip name has become synonymous with microcontrollers, they are not the only manufacturer. Atmel (www.atmel.com)

makes a competing line of chips under the AVR brand name that is also popular. The ATmega8, for example, is used in several small robots.

Hardware development boards and in-circuit emulators are available for these chips, but generally they are expensive. Luckily, these are not required for most small development projects, which don't involve developing new hardware.

Assemblers and C compilers for these chips are readily available. Although some of them are free, many are commercial products. Several C compilers are based on the GNU GCC cross-compiler tool chain. A variety of other languages are also supported by different vendors, including Java, Forth, Pascal, and others. However, C is by far the most popular language.

Note that C# is not available for these chips because Visual Studio generates code for the .NET environment. It does not generate native code for microcontrollers (a fact that some people clearly do not understand judging by the postings on the MRDS Discussion Forum).

There are also GUI-based programming packages similar in concept to VPL, i.e., you create your program by dragging various "blocks" to the workspace and drawing a flow chart. The capabilities of these software packages vary enormously, and they are not the focus of this chapter.

## Chips with Built-in BASIC Interpreters

The advantages of BASIC are that it is easy for beginners to learn and the development software is typically free. The compiled code can easily be downloaded via a serial port. (If your PC does not have a serial port, you can use a USB-to-serial device.) The disadvantage is that it might be slower than compiled code written in C and the onboard "operating system" does not provide multi-tasking capabilities, so everything has to be done in a large loop, which means that timing can be critical.

Examples of this kind of chip include the following:

❑ The PICAXE chip (www.picaxe.co.uk) sold by Revolution Education is a good example of an application of a PIC. A PICAXE chip is a PIC chip with a boot loader and BASIC interpreter supplied in firmware on the chip. The PICAXE is used in the Picblok Integrator robot discussed in this chapter.

❑ Parallax (www.parallax.com), the manufacturers of the Boe-Bot and Scribbler robots, uses a BASIC Stamp as the onboard controller. This is a small circuit board that holds a 5V regulator, EEPROM, and a PIC that contains a PBASIC interpreter in firmware. Many of the models of the BASIC Stamp use the SX chip that is made by Parallax, while the early BS1 and BS2 models used a PIC chip from Microchip.

❑ An interesting approach has been taken by Savage Innovations with their OOPic hardware (www.oopic.com). They adopted an object-oriented approach in a multi-tasking environment from the outset, and they support multiple language syntaxes (BASIC, C, or Java).

## MRDS Support from Other Vendors

One problem with some companies that make robots is that they have been around for many years and have an existing investment in software. Therefore, some of them are not particularly interested in supporting MRDS, or they don't have people on their staff with MRDS skills so their support is limited. However, there are companies out there that directly support MRDS, including the following:

❑ Devantech Ltd (www.robot-electronics.com) offers MRDS services written by Chris Kilner. They are well written and are available as open source at www.codeplex.com/DevantechMSRS. Chris has since moved on to bigger and better things with Nao humanoid robots at Aldebaran (www.aldebaran-robotics.com/eng/index.php).

❑ The Vex Robotics Design System (www.vexlabs.com) from Innovation First, Inc., also has an open-source project at www.codeplex.com/VexMsrs.

❑ Stinger robots use a board called the Serializer that has an onboard processor. However, you control the Serializer via a serial port (hence its name) and do not program the board directly. Services for MRDS are available from the RoboticsConnection website (www.roboticsconnection.com).

This book provides updated services for Parallax, Lynxmotion, and Surveyor robots to correct bugs or deficiencies, as well as a new Drive service for the Stinger that conforms to the generic Differential Drive contract. There are also services for the Hemisson robot in this chapter because the manufacturer, K-Team, does not provide them.

# Creating a Generic Brick Service

One thing missing from MRDS is a generic "brick" contract, a much-debated topic on the Discussion Forum. The basic problem is that robots are so variable in their hardware components — e.g., sensors, actuators, even the onboard CPU — that it is difficult to generalize.

Consequently, there is no such thing as a "universal remote controller" for a robot. (Even infrared universal remote controls (URCs) for audio-visual equipment have a hard time due to deliberate incompatibilities introduced by the vested interests of competing manufacturers, but that is another story.)

Another issue is the proliferation of services. There needs to be a happy medium between a single monolithic service and "service bloat" caused by too many small services. This is more of a problem when you move services to mobile devices with limited memory and CPU power.

For this chapter, the authors decided to build a Generic Brick service first. This service should be suitable for most small hobby and educational robots. Many elements are common across these types of robots.

The procedure for creating the Generic Brick service contract is quite easy:

1.  Create a new service called GenericBrick and open it in Visual Studio. You can use DssNewService or just make a new DSS service project in Visual Studio. This is covered in Chapters 2 and 3.

2. Edit `AssemblyInfo.cs` and change the Service Declaration to `DataContract` instead of `ServiceBehavior`, as shown here:

```
//For a generic contract, the service is a DataContract NOT ServiceBehavior
[assembly: ServiceDeclaration(DssServiceDeclaration.DataContract)]
```

3. Delete the service implementation file, `GenericBrick.cs`. A generic contract contains only data definitions — no code.

4. Update the service state (`GenericBrickState`) with all the required properties. You might prefer to divide this out from the `GenericBrickTypes.cs` file into a file called `GenericBrickState.cs`. Define additional classes, if necessary, and mark them with the `[DataContract]` attribute.

5. Add the `Request` and `Response` data types for the operations.

6. Define the main operations port using the operations from step 5.

7. Compile!

You should leave the contract identifier alone. If you want to, you can delete the manifest that is automatically created because a generic contract cannot be run as a service, so the manifest is useless.

The final version of the Generic Brick service is supplied in `ProMRDS\Chapter17\GenericBrick`. It was used to create the Integrator and Hemisson services that are also included in this chapter.

## Terminology

Before moving on to discuss the Generic Brick contract, it is important to explain some of the terminology used in this chapter:

❑ **Brick:** A *brick* is the brains of a robot and typically contains a microcontroller chip. It may not physically look like a brick at all, and it certainly won't be the size, shape, or weight of a real brick! However, all communication between MRDS services and the robot passes through the brick. The following are examples of bricks:

    ❑ The brick for the LEGO NXT is something that you can easily point to, and it even looks like a brick. It has I/O ports for motors and sensors, but you cannot open it up and reconfigure the hardware inside. (In fact, it contains two CPUs, but this chapter doesn't discuss the NXT because it is already well supported.)

    ❑ In some cases, such as the Boe-Bot, the brick consists of a small circuit board (the BASIC Stamp 2) and a larger board that includes a breadboard area for building interface circuits. There is also a plug-in Bluetooth module for communication. The entire assembly is considered to be the brick, including all of the I/O devices.

    ❑ The brick in the Stinger robot is a circuit board called a Serializer. It contains a microcontroller and the necessary hardware to interface with sensors and actuators.

    ❑ The Hemisson and Scribbler robots are "sealed" units, so the brick is internal to the robots.

❑ **Device:** Bricks can have a wide range of different hardware attached to them. These pieces of hardware are all referred to as *devices*. For example, a speaker that can play a tone is a device. So is a chip that measures temperature and returns an analog value.

❑ **Pin:** A *pin* quite often refers to a physical pin on the microcontroller. However, there might be logical pins in the implementation that don't exist physically or which set internal configuration parameters in the microcontroller. In the discussion in this chapter, a pin is generally a digital input or output. This is the lowest level of I/O that can be performed.

*Sometimes the terms "pin" and "device" are used interchangeably.*

❑ **Port:** Microcontrollers usually have one or more I/O ports. (These have nothing to do with MRDS ports.) The naming scheme varies from one manufacturer to another, and the number of ports varies from one chip to another. For example, there might be three ports called PortA, PortB, and PortC. The size of each port can also vary from one chip to the next; basically, it is limited by the packaging of the chip. Some ports are 8 bits, and others are 16 or even 32 bits. Certain pins may have dedicated functions so that the effective number of pins in a port is less than 8.

❑ **Hardware Identifier:** To avoid all of this confusion over port/pin naming, the brick implementation must assign a unique hardware identifier to each input and output. The simplest approach is probably to number all the pins sequentially. For example, the BASIC Stamp 2 has 16 pins numbered 0 to 15. The PICAXE, however, has input and output pins, both of which use the same numbering scheme, so you must make up some different numbers.

## Design

There is no doubt that this design is a compromise. A lot of design decisions had to be made to keep the Brick service small and yet flexible. These are discussed below to explain the rationale behind the decisions. You might not agree with them all, in which case you can modify the service and create your own version — that is the advantage of an "open" system. Just remember to change the contract identifier to make your contract unique.

The following design assumptions were made:

❑ One of the authors' ultimate objectives is to write an enhanced version of the Dashboard that can make intelligent decisions about laying out the user interface based on the types of devices connected to the brick. For example, all the digital inputs could be grouped together as an array of checkboxes in numeric or alphabetical order. This requires a clearly defined approach to naming I/O devices and specifying their functions.

❑ It is assumed that the robot uses a two-wheel differential drive. This is by far the most common configuration for cheap educational robots.

❑ It is implicitly assumed that communications will be via a serial port. Whether it uses a wireless connection (Bluetooth, ZigBee, etc.) is immaterial. However, there is no requirement in the design to use a serial port, and it could easily be extended to use wireless Ethernet, for example.

The authors would like to acknowledge that some of the ideas for this design came from Ben Axelrod and others who have contributed to the MRDS Discussion Forums. The following sections address some of the design considerations.

## No Reconfiguration of the Brick (in this First Version)

Hobby robots are either built from a kit or constructed by the hobbyist using whatever components are at hand. Assembly instructions for kits have to be fairly prescriptive, so the resulting robot has a fixed set of sensors and actuators at known hardware "addresses" (pin numbers). For a home-brewed robot, there are no rules, and the exact combination of hardware components might never be duplicated by anyone else. Although it still makes sense to implement the basic MRDS services so that you can take advantage of the cool services available under MRDS, it is highly unlikely that anyone other than the original robot builder will ever need to modify these basic services.

The obvious exception to this rule is the LEGO NXT, which has ports into which you plug sensors. However, the NXT is very well supported by the MRDS team, so there is no need to develop a Brick service for it. Therefore, reconfiguration is *not* possible from a partner service, and only very limited reconfiguration can be done via the service's `config` file. All the necessary configuration is done by the Brick service itself during initialization.

## All Devices Must Have a Name

Although services will invariably use hardware identifiers to manipulate devices, a user interface needs to present information using meaningful names. To avoid situations where different people refer to the same device using different names, or not having a name at all, the implementation must specify names for all devices.

You should try to be consistent in naming devices. For example, you should not have "Left IR Sensor" and "IR Sensor - Rear". Note that these names include a location or position, a device type, and, implicitly, a direction (Sensor = Input).

No guidelines are given here for naming because that is likely to start a heated debate.

## Hardware Identifiers

Hardware identifiers are left to the implementation, and partner services should *not* assume anything except that they are unique (within a single brick implementation, but not across different bricks) and that they are immutable.

There should be no overlap in identifiers between the different types of devices — you cannot have a digital input with id = 3 and an analog output that also has id = 3.

## Timestamps

All I/O values should have timestamps indicating the last time that they were updated. (A value that has never been updated will have the base date.) There are no particular requirements on the values of the date/time except that it must be possible to tell whether a value has been updated since the last time it was retrieved, and the chronological order of updates must be preserved. Several devices can have the same timestamp if the data values all arrived in a single packet.

## Automatic Connection

Given that you use most robots by turning them on and then starting a service, it makes sense for the Brick service to automatically try to connect. This means that the configuration information must be either compiled in or, preferably, in the `config` file.

Explicit `Connect` and `Disconnect` operations are not included in the contract, but if you want to add these you can extend the contract.

## Inputs and Outputs

All devices attached to the brick are classified as one of the following types:

- ❑ Digital Input (Boolean)
- ❑ Digital Output (Boolean)
- ❑ Analog Input (double)
- ❑ Analog Output (double)

There might be some argument for using an integer value (0 or 1) for digital devices, but the decision was made to use a Boolean because this seemed to be more common in the MRDS sample implementations.

Analog devices use doubles, so they can have an enormous range. The minimum and maximum values for each device are specified in the device information.

## Unit Conversion and Calibration

Analog sensors and output devices must translate from binary values to real-world units at some stage. It is highly desirable that any conversions are performed inside the Brick service. These are what are referred to as *normalized measurements* in the MRDS generic Analog Sensor service.

For this reason, the "raw" values of analog sensors are not included in the state as separate properties. This enables the data to be used by young people without the complication of extra math. If you want to expose the raw values of sensors — for example, because you want users to do the mathematical calculations — you can either define pseudo-sensors with different hardware identifiers to hold the raw values or perform no processing of the incoming data.

Wherever applicable, appropriate metric units should be used. For example, when working with small robots, distance measurements should be in centimeters. The exception to this rule is angles that should be in degrees, rather than radians. The objective is to make all values easily understandable.

Calibration of sensors is completely ignored in this specification. If the sensors themselves are not well behaved, then you might want to just return raw values. Conversely, if you can linearize or "condition" the values, then the brick should do so.

Be careful about hiding too much information. For example, an infrared sensor might simply be used as a bump sensor, i.e., a digital input. However, you can expose it as two devices: one that provides an analog value and one that provides a digital value. If the threshold is set correctly in the brick code, the digital input can be used and the analog input can be ignored. If the sensor is unreliable, however, a partner service can still read the analog value and set its own threshold.

One situation in which the raw values might be irrelevant is line following. Manufacturers often adjust the hardware design so that the output from the line-following sensors is very reproducible and there is no need to look at the analog values because the sensors clearly see a line or no line. In fact, the simulated photo sensors in Chapter 12 use webcams to detect light or dark areas on the floor. There is no reason to expose the webcam images as sensor data — only the binary value is required.

## Locations of Sensors and Actuators

The Microsoft samples commonly use a Pose to represent the location of a sensor. Although this is essential for positioning objects in the simulator, it is not intuitively obvious what a Pose means. In particular, the orientation is expressed as a *quaternion*. Very few people can look at a quaternion and tell you which way it is pointing.

Moreover, for simple robots there are only a handful of positions on the robot where sensors are likely to be located. Therefore, an enumeration is used for the location of sensors and actuators based on the 16 principal compass points, with north corresponding to straight ahead in front of the robot. Whether this approach is successful or not is still an open question — perhaps it will revert to a Pose in a future version of the Generic Brick contract.

## Access to Switches and LEDs is Included as a Convenience

The service should internally remap the digital input devices (switches) and output devices (LEDs) to the correct pins (hardware identifiers) and present the information in the form of a binary bitmask. The remapping is an implementation detail that external services should not need to worry about.

Note that setting a single LED this way requires the partner service to remember the current state of all of the LEDs, or to request it beforehand. Even LEDs are not generic. The Roomba has LEDs that can be red, green, or yellow! They will not fit into a binary bitmask unless two bits are used for each LED.

## Motors and Wheel Encoders

The drive system of the robot is assumed to be a two-wheeled differential drive. It makes sense if you are trying to make the code as compact as possible to have operations on the Brick service to control the motors. Ultimately, all communication is channeled via the brick, so it should know the state of the motors anyway.

There is no requirement in the generic Differential Drive contract to actually have motor services. You do have to define some motor objects in the state, but nothing else depends on there being actual motor services. In fact, you could extend the Generic Brick contract to implement the generic Differential Drive contract and eliminate a separate service altogether, but this is not done here.

You can reduce overhead by not implementing a motor service at all. This is how the StingerPWMDrive in Chapter 16 works. It sends commands directly to the Serializer, which is effectively the brick for the Stinger robot.

Given that the target devices are at the low end of the hardware spectrum, wheel encoders are ignored in this definition. You can, of course, add wheel encoders if you want to, but there are no operations defined on the brick to explicitly read or set them. (You can just treat them as analog sensors.) For example, the DriveDistance and RotateDegrees operations are expected to use timers, not encoders (and consequently they are not very accurate).

## I2C Bus is Not Supported

Many microcontrollers have the capability to talk to other devices over an I2C bus. This is explicitly ignored to simplify the design. Otherwise, it raises the issue of how to specify the I2C addresses and what types of devices might be attached.

The implementation is free to assign a hardware identifier to an I2C device, enabling it to be accessed via the normal I/O operations. However, devices that return multiple bytes of data, especially cameras, cannot be accommodated by this mechanism. There is nothing to stop you from extending the brick contract to add more state information and operations if you need to support I2C devices.

# Brick State

The brick state includes the following generic items:

- ❏  Firmware information
- ❏  Configuration information
- ❏  Communications statistics
- ❏  Sensors
- ❏  Actuators
- ❏  LEDs
- ❏  Switches
- ❏  DrivePower

## Definition of the State

The code for the state is in GenericBrickState.cs and is set out in the following example. Note that there is a [DataContract] attribute on the state, which causes it to be copied into the Proxy DLL. In addition, all of the public properties have the [DataMember] attribute. XML comments and a Description attribute are used to provide documentation:

```
/// <summary>
/// The GenericBrick State
/// </summary>
[DataContract]
public partial class GenericBrickState
{
    private string _name;
    private bool _connected;

    /// <summary>
    /// Brick Name
    /// </summary>
    /// <remarks>This is set by the service and not read from the 
Brick</remarks>
    [DataMember, Description("Friendly name for the Brick")]
    public string Name
```

```
        {
            get { return _name; }
            set { _name = value; }
        }

        /// <summary>
        /// Firmware Details (optional)
        /// </summary>
        /// <remarks>This information is read from the Brick</remarks>
        [DataMember, Description("Firmware information")]
        [Browsable(false)]
        public Firmware Firmware;

        /// <summary>
        /// Communications Configuration
        /// </summary>
        /// <remarks>This information should be read from the config file</remarks>
        [DataMember, Description("Communications Configuration")]
        public CommsConfig Configuration;

        /// <summary>
        /// Connection Flag
        /// </summary>
        [DataMember, Description("Indicates if a connection to the Brick is ➡
established")]
        [Browsable(false)]
        public bool Connected
        {
            get { return _connected; }
            set { _connected = value; }
        }

        /// <summary>
        /// Communications Statistics (optional)
        /// </summary>
        [DataMember, Description("Holds the statistics for the communications ➡
link (optional)")]
        [Browsable(false)]
        public CommsStats CommsStats;

        /// <summary>
        /// Summary of the Capabilities of the Brick
        /// </summary>
        /// <remarks>This should NOT be determined from the config file</remarks>
        [DataMember, Description("Brick Capabilities Bitmask")]
        public BrickCapabilities Capabilities;

        /// <summary>
        /// Brick Sensors (Inputs)
        /// </summary>
        [DataMember, Description("All Inputs (Digital and Analog)")]
        [Browsable(false)]
        public List<Device> Sensors;
```

*(continued)*

*(continued)*

```
/// <summary>
/// Brick Actuators (Outputs)
/// </summary>
[DataMember, Description("All Outputs (Digital and Analog)")]
[Browsable(false)]
public List<Device> Actuators;

/// <summary>
/// Status of LEDs (optional -- can use Actuators instead)
/// </summary>
/// <remarks>NOTE: LEDs are also listed as Actuators</remarks>
[DataMember, Description("LED settings as a binary Bitmask")]
[Browsable(false)]
public int LEDs;

/// <summary>
/// Status of Switches (optional -- can use Sensors instead)
/// </summary>
/// <remarks>NOTE: Switches are also listed as Sensors</remarks>
[DataMember, Description("Switch Status as a binary Bitmask")]
[Browsable(false)]
public int Switches;

/// <summary>
/// Drive Power (optional)
/// </summary>
[DataMember, Description("Specifies the current Drive Power settings")]
[Browsable(false)]
public DrivePower DrivePower;

}
```

*Do not add static members to your state (or any other class in a generic contract).* DssProxy *will exit with error code 20 when you try to compile. In addition, all members that you want exposed (and serialized to the* config *file) must have a* [DataMember] *attribute, as well as being marked* public.

## Overview of State Members

As stated in the design, every device has a "friendly" Name, and the purpose of the Connected field is obvious. Several of the members are instances of other classes defined in GenericBrickState.cs. Rather than list all of the code here, these additional classes are discussed briefly:

❑ The information in Firmware is optional. Some robots have version information in their firmware, whereas others do not. If nothing else, requesting the firmware version is a good test to see whether the robot is alive, and it gives the service something to display to the user.

❑ The Configuration field contains information that is basically about communications. It includes the serial port and timing parameters. To set up a robot, you need to run its service once so that it writes out an "empty" config file (saved state). Then you can edit the config and insert the appropriate serial port. It does no harm to run a service and let it fall over. In fact, you can still examine the state using a web browser even when a connection to the robot cannot be established.

*An XSLT file is included with the Generic Brick service and you can use it to format the state information for display on a web page. However, it is not possible to incorporate an XSLT file into a Proxy DLL, so you have to copy the file into your own solution when you implement a brick service. This is discussed later in the section "Adding Embedded Resources."*

❑ Communications statistics, CommsStats, are also optional. It is a good idea to think about instrumenting your services early in the design. For example, you can verify that your service is polling by watching the counters in a web browser. You can also check the error rate.

❑ BrickCapabilities is a bit field that uses an enum in GenericBrickTypes.cs. This enum has the [Flags] attribute so that its values can be combined using a logical OR and it is also marked as a [DataContract] to ensure that it is copied to the Proxy.

❑ LEDs and Switches in the state provide redundant information (because the devices are also in the actuators and sensors), but they are easier to understand. They are both binary bitmasks and are declared as integers. The LEDs and Switches can therefore handle up to 32 devices each — far more I/O pins than are available on any hobby robot currently on the market. Note that this does not imply a maximum of 64 devices because there can be other types of devices as well.

❑ The mapping from I/O pins to LEDs and/or Switches is under the control of the Brick service. For example, there could be two LEDs attached to PortA pins 3 and 5 (out of 8) and two more on PortB pins 0 and 1 (out of 8). These might be presented as four LEDs numbered from 0 to 3. The corresponding hardware identifiers might be 3, 5, 8, and 9 due to the concatenation of the ports to form a sequence of 16 devices.

❑ Lastly, the DrivePower class is a convenient way to pass around the left and right motor power settings in messages.

## Devices

The Sensors and Actuators are deliberately kept as separate lists in the state. This makes it easier to send sensor updates and it reduces the time that would be wasted searching through the entire list for a particular device when you know that the device is a sensor or an actuator.

Sensors and Actuators are represented as lists of type Device. A Device instance contains a complete description of the device, including its name, type, function, and location on the robot. This class is a key component of the Generic Brick contract:

```
/// <summary>
/// Class for all Devices (Analog/Digital, Input/Output)
/// </summary>
/// NOTE: A Device can be Analog or Digital. Therefore, there are properties
/// for both types of values. This is wasteful, but it makes it easier if the
/// device can be reconfigured.
[DataContract]
[DataMemberConstructor]
[Description("Basic details about a Device a.k.a. Pin")]
public class Device
{
    // Basic device information
    private string _name;
    private int _hardwareIdentifier;
    private DeviceTypes _type;
```

*(continued)*

*(continued)*

```csharp
        private DeviceFunctions _function;
        private Location _location;

        // Additional fields to help manage the device
        // These should NOT be changed by external services!
        private int _port;
        private int _pin;

        // Current value properties
        private DateTime _timeStamp;
        private bool _state;
        private double _value;
        private double _minValue;
        private double _maxValue;

        /// <summary>
        /// Device Name
        /// </summary>
        [DataMember]
        [Description("User-friendly Name for the Device")]
        public string Name
        {
            get { return _name; }
            set { _name = value; }
        }

        /// <summary>
        /// Unique Hardware Identifier
        /// </summary>
        [DataMember]
        [Description("A unique, immutable identifier for this particular device")]
        public int HardwareIdentifer
        {
            get { return _hardwareIdentifier; }
            set { _hardwareIdentifier = value; }
        }

        /// <summary>
        /// Device Type
        /// </summary>
        [DataMember]
        [Description("Type of the Device (See DeviceTypes enum)")]
        public DeviceTypes Type
        {
            get { return _type; }
            set { _type = (DeviceTypes)value; }
        }

        /// <summary>
        /// Device Function
        /// </summary>
```

```csharp
[DataMember]
[Description("Function of the Device (See DeviceFunctions enum)")]
public DeviceFunctions Function
{
    get { return _function; }
    set { _function = (DeviceFunctions)value; }
}


/// <summary>
/// Location of the Device on the Robot
/// </summary>
[DataMember]
[Description("Location of the Device (See Location enum)")]
public Location Location
{
    get { return _location; }
    set { _location = (Location)value; }
}


/// <summary>
/// Port on the PIC or MCU
/// </summary>
[DataMember]
[Description("The physical Port the Device is attached to")]
public int Port
{
    get { return _port; }
    set { _port = value; }
}


/// <summary>
/// Pin on the PIC or MCU
/// </summary>
[DataMember]
[Description("The Pin the Device is attached to (within a Port)")]
public int Pin
{
    get { return _pin; }
    set { _pin = value; }
}


/// <summary>
/// Time Stamp for last update
/// </summary>
[DataMember]
[Description("Time Stamp for last update that took place")]
public DateTime TimeStamp
{
    get { return _timeStamp; }
    set { _timeStamp = value; }
}
```

*(continued)*

*(continued)*

```csharp
/// <summary>
/// Binary State of the Device
/// </summary>
[DataMember]
[Description("Indicates if the Device state is currently high/low,
on/off, etc.")]
public bool State
{
    get { return _state; }
    set { _state = value; }
}

/// <summary>
/// Current Value
/// </summary>
[DataMember]
[Description("Current Device value")]
public double Value
{
    get { return _value; }
    set { _value = value; }
}

/// <summary>
/// Minimum Value
/// </summary>
[DataMember]
[Description("Minimum Value")]
public double MinValue
{
    get { return _minValue; }
    set { _minValue = value; }
}

/// <summary>
/// Maximum Value
/// </summary>
[DataMember]
[Description("Maximum Value")]
public double MaxValue
{
    get { return _maxValue; }
    set { _maxValue = value; }
}

}
```

Notice that there is a [DataMemberConstructor] attribute on the Device class. This causes DssProxy to create a constructor for this class, which is important in terms of defining the devices that make up the robot.

Note the following points about device properties:

- ❏ Every device must have a unique HardwareIdentifier (for a given brick implementation). Although the Port and Pin fields also identify the device, they are intended for internal use within the brick code.

- ❏ The current state of a device is stored in either the State field (a bool) or the Value field (a double), depending on whether it is a digital or analog device. Purists might be offended by this mixing of data types, but it is convenient to be able to pass lists of devices around without regard to what types they are.

- ❏ There are fields for MinValue and MaxValue that apply only to analog devices. For consistency, implementations should store 0 and 1 in these fields for a digital device, and set the Value to 1 if the State is true or 0 if the State is false. The reverse also applies.

## Device Enums

The DeviceTypes enum is used to define whether the device is an input (sensor) or output (actuator):

```
[DataContract]
[Flags]
public enum DeviceTypes : short
{
    Other           = 0x0001,    // None of the ones below, or unspecified
    Reserved        = 0x0002,    // Reserved = DO NOT USE THIS DEVICE
    Reconfigurable  = 0x0004,    // Can be one of several types
    DigitalIn       = 0x0008,    // Only Digital Input
    DigitalOut      = 0x0010,    // Only Digital Output
    DigitalInOut    = 0x0020,    // Can be EITHER Input or Output ↲
                                 (but no other type)
                                 // NOTE: This is NOT a logical OR of ↲
                                 Digital In/Out
    AnalogIn        = 0x0040,    // ADC
    AnalogOut       = 0x0080,    // DAC
    CommsIn         = 0x0100,    // RS-232 Rx Data, or Rx Data for ↲
                                 Bluetooth, etc.
    CommsOut        = 0x0200,    // RS-232 Tx Data, or Tx Data for ↲
                                 Bluetooth, etc.
}
```

There might be other types of devices defined in the future, but this is enough for now. Notice that types can be logically combined (because of the [Flags] attribute), so you can mark a device as Reserved if you don't want other services to access it. Of course, you could also leave it out of the device list so that it is invisible outside the Brick service.

Many microcontrollers enable you to reconfigure I/O pins to be either an input or an output. Some can switch between digital and analog as well. This is not easily accommodated, so in the state a device can only be an input *or* an output, and only digital *or* analog. Besides, you don't want naïve programmers inadvertently changing an analog-to-digital converter (ADC) input to a digital output. The consequences for the hardware might be disastrous.

There are enums for several other device properties as well, including `DeviceFunction` and `Location`, but they are not all listed here. Look at them in Visual Studio to see what values they have.

## Brick Operations

Having defined the `GenericBrickState`, you can now move on to the operations that you want to use to manipulate the state. Most operations require either a `Request` or a `Response` data type, or in some cases both.

All of the message data types and the operations themselves (which are `PortSet`s) are defined in the `GenericBrickTypes.cs` file. Rather than list them all here, the main operations port is provided and the operations are discussed briefly:

```
///  <summary>
///  GenericBrick Main Operations Port
///  </summary>
///  <remarks>Keep the number of operations to 20 or less</remarks>
[ServicePort()]
public class GenericBrickOperations : PortSet
{
    public GenericBrickOperations() : base(
        typeof(DsspDefaultLookup),
        typeof(DsspDefaultDrop),
        typeof(Get),
        typeof(HttpGet),
        typeof(HttpPost),
        typeof(Replace),                    // Replaces entire state
        typeof(Subscribe),                  // Subscribe to changes
        typeof(ConfigureBrick),             // Changes Brick Configuration
        typeof(ConfigureDevices),           // Changes Devices (if possible)
        typeof(UpdateSensors),              // Selective update / notification
        typeof(PlayTone),                   // Play a Tone
        typeof(SetLEDs),                    // Turn LEDs on/off
        typeof(GetSwitches),                // Read all switches
        typeof(GetSensors),                 // Get list of Digital and Analog Inputs
        typeof(SetActuators),               // Set list of Digital and Analog Outputs
        typeof(QueryDrivePower),            // Query current Drive Power
        typeof(drive.SetDrivePower),//      Update power to motors
        typeof(drive.DriveDistance),//      Drive a specified distance
        typeof(drive.RotateDegrees) //      Rotate a specified number of degrees
    )
    {
    }
}
```

The first half dozen operations you should recognize as standard DSSP operations. There is some redundancy in the rest of the operations. However, this provides flexibility in how they are used:

❑ `ConfigureBrick` enables some limited changes to the brick such as the `PollingInterval`. With the current design, there is no way to set the `SerialPort` before the service attempts to connect, so you must edit the `config` file.

❏ The `ConfigureDevices` operation is a "placeholder" for a future version, when it will be possible to dynamically reconfigure devices (only those marked as `Reconfigurable`).

❏ `UpdateSensors` is intended for notifications to subscribers, but it might also be used internally for different parts of the brick code to update the central state. Note that because the design is based around lists of devices, it is possible to do selective updates and even update just a single device by supplying a device list in the update message.

❏ Several robot services that already exist have a `PlayTone` operation, or something similar. Lights and sounds are a crucial part of making robotics entertaining for young people. Even the simplest robot should have a buzzer or beeper so that it can indicate its displeasure or that it is in distress.

❏ `SetLEDs` and `GetSwitches` use binary bitmasks, and these operations flow directly from the definition of the state as discussed previously.

❏ `GetSensors` and `SetActuators` both use device lists, so they can be selective.

❏ The last four operations are related to the robot's drive system. Notice that the last three, `SetDrivePower`, `DriveDistance`, and `RotateDegrees`, are actually based on the generic Two-Wheel Differential Drive contract. This is an interesting use of one generic contract inside another.

By defining these operations here and exposing the state of the motors, it is possible to build a drive service that is quite compact. In fact, with some more work, it might even be possible to incorporate the generic differential drive into the brick contract and still remain compatible.

In the code, you can also see some additional methods defined on the main operations port. The declaration of the `PortSet` using `typeof` instead of generics is important for making the contract compatible with the CF environment. However, you lose some of the strong type-checking that generics provide. To overcome this, several overloads of the `Post` method are defined. In addition, implicit operators for extracting a particular port from the `PortSet` are defined. Examples of both of these are shown here for the `Get` operation:

```
/// <summary>
/// Post(Get)
/// </summary>
/// <param name="item"></param>
/// <returns></returns>
public void Post(Get item) { base.PostUnknownType(item); }

/// <summary>
/// Implicit Operator for Port of Get
/// </summary>
/// <param name="portSet"></param>
/// <returns></returns>
public static implicit operator Port<Get>(GenericBrickOperations portSet)
{
    if (portSet == null) return null;
    return (Port<Get>)portSet[typeof(Get)];
}
```

These particular methods are created automatically by DssProxy and do not need to be added to the code explicitly. However, you might still see remnants of this code in services written for earlier versions of MRDS.

It is also common practice to define helper methods to enable you to post messages by calling a method with appropriate parameters that would construct the message, populate it, and then post it. However, DssProxy also does this for you now in most cases.

# Additional Components

In case you have not realized yet, a Proxy DLL only contains data contracts — no code or embedded resources. (It does have some helper methods, but they are created by DssProxy and are not under your direct control.)

As you have seen, the Device class is a core component of the generic brick. It would be handy to have some helper or utility methods for performing various tasks on Devices and Lists of Devices. However, DssProxy does not allow methods to be transferred to the Proxy DLL, only data type definitions.

To finish off the service, an XSLT file enables the state to be displayed in a user-friendly format.

## Creating a Helper DLL

To create a helper DLL, you make a new DLL (Class Library) in Visual Studio. Don't select one of the Robotics templates because you don't want to generate a Proxy for this new DLL — it is not a service.

The best way to keep everything together is to add a new project to the existing solution. The generic brick solution included with this chapter has a second project called GenericBrickUtil. (It is common practice to use the term "utility" functions, hence the abbreviation "util," but it is also common to call them "helper" functions.)

Once you have created a new project, you need to modify the properties as follows:

❑ On the Build tab, change the output directory to be the bin folder under MRDS.

❑ On the Reference Paths tab, add the bin folder under MRDS. You might also want to add the .NET 2.0 Framework and maybe bin\cf. Look in an existing service solution to see what paths are required.

❑ On the Signing tab, click the box that says "Sign the assembly" and then browse to the key file mrisamples.snk, which should be in the samples folder under MRDS.

If you do not sign your DLL, then it won't work with MRDS. When you compile, you will get the following error:

```
Assembly generation failed -- Referenced assembly 'GenericBrickUtil' does not have
a strong name
```

---

**Signing DLLs**

Signing DLLs serves two purposes:

❑  It ensures security of the DLL so it cannot be modified or compromised.

❑  It enables developers to uniquely identify their code, i.e., it determines ownership — a signed assembly can be traced back to the owner/developer.

To sign a DLL, you need a strong key file. You can create your own strong key file if you want. On the Signing tab, select <New . . .> from the drop-down list and enter a filename. You can optionally provide a password on the file for additional security.

However, all of the MRDS samples use `mrisamples.snk`, which is supplied with MRDS. Unless you have a good reason not to do so, you should use this file.

If you create your own key file, you have to distribute it with your source code if you expect other people to modify your services. If you don't supply the key file, there can be problems with version dependencies when users recompile services that have Specific Version set to `True` on references.

Note that if you want to distribute your services commercially, you need to control who can recompile them. In this case, you create your own strong key file but do *not* distribute it outside your organization. By keeping the key file a "trade secret," nobody else will be able to modify the services that you ship or impersonate them.

---

You need to add references to the standard MRDS DLLs:

❑  `Ccr.Core`

❑  `DssBase`

❑  `DssRuntime`

As usual, set Copy Local and Specific Version to `False` in the properties of the references.

Which of these DLLs you need depends on what you intend to do in your helper functions. You might be able to get away without some of them. Conversely, you might need to add more DLLs, such as `RoboticsCommon.Proxy`.

Add a reference to the generic brick proxy, as well as a `using` statement:

```
using brick = ProMRDS.Robotics.GenericBrick.Proxy;
```

Now you can add a class to hold the helper functions. Most functions can be declared `static` because they will be operating on data supplied in their parameters and have no need for private storage. For example, here are a couple of methods from `GenericBrickUtil.cs`:

```
/// <summary>
/// Utility Functions for Generic Brick
/// </summary>
public static class Util
{
    /// <summary>
    /// Find a Device by Hardware Id
    /// </summary>
    /// <param name="list"></param>
    /// <param name="id"></param>
    /// <returns>Index if found or -1 if not</returns>
    public static int FindDeviceById(List<brick.Device> list, int id)
    {
        for (int i = 0; i < list.Count; i++)
        {
            if (list[i].HardwareIdentifer == id)
                return i;
        }
        // Not found!
        return (-1);
    }

    /// <summary>
    /// Is this an Analog Device?
    /// </summary>
    /// <param name="dev"></param>
    /// <returns>True if Analog, False if not</returns>
    public static bool IsAnalogDevice(brick.Device dev)
    {
        if (((dev.Type & brick.DeviceTypes.AnalogIn) != 0) ||
            ((dev.Type & brick.DeviceTypes.AnalogOut) != 0))
            return true;
        else
            return false;
    }
    ...
```

These methods are fairly self-explanatory. There are several more methods in the helper DLL, but they don't need to be covered here. You can look at them in Visual Studio.

To use these methods in a service implementation, you just add a reference to the DLL, just as you would for any other DLL. A `using` statement helps to make the methods easy to use:

```
using util = ProMRDS.Robotics.GenericBrickUtil.Util;
```

Then you can use the methods in your code, as in the following example for the Integrator robot, which handles output devices:

```
/// <summary>
/// SetActuators Handler
/// </summary>
/// <param name="update"></param>
/// <returns></returns>
/// NOTE: Do NOT send commands to Motors or Reserved Pins this way!
[ServiceHandler(ServiceHandlerBehavior.Exclusive)]
public virtual IEnumerator<ITask> SetActuatorsHandler( ⏎
brick.SetActuators update)
    {
        foreach (brick.Device d in update.Body.Outputs)
        {
            // Only Digital devices are handled here
            if (util.IsDigitalDevice(d))
            {
                if (d.Function == brick.DeviceFunctions.LED)
                {
                    _state.LEDs = util.UpdateBitmask(_state.LEDs, LEDIds,
                                            d.HardwareIdentifer, d.State);
                }
                // Now set the pin
                util.UpdateDeviceStateInList(_state.Actuators, ⏎
d.HardwareIdentifer, d.State);
                _control.SetPin(d.Pin, d.State);
            }
        }
        update.ResponsePort.Post(DefaultUpdateResponseType.Instance);
        yield break;
    }
```

This code sets output pins using SetPin (which is discussed later), but along the way it checks for LEDs so that it can maintain the state of the LEDs property. It also has to update the device list in the state to reflect the changes. These tasks are made easier by the utility functions.

## Adding Embedded Resources

It is not possible to include embedded resources, i.e., XSLT files, in a Proxy DLL. An XSLT file for displaying the generic brick state is included with the solution, but it does not appear in the Proxy. This file has to be manually copied over to any new implementation of the generic brick and added as an embedded resource in the new service.

Note that all service implementations based on the generic brick use the same data type for their state — GenericBrickState. Therefore, the XSLT file can be identical for all robots based on the generic brick. Even the name of the robot can be read from the state.

The XSLT file supplied with this chapter includes code to collapse the Drive Power, Sensors, and Actuators sections of the page. This JavaScript was shamelessly copied from the MRDS Debug and Trace Messages page. The resulting output looks like Figure 17-1 for the Integrator robot discussed in the next section.

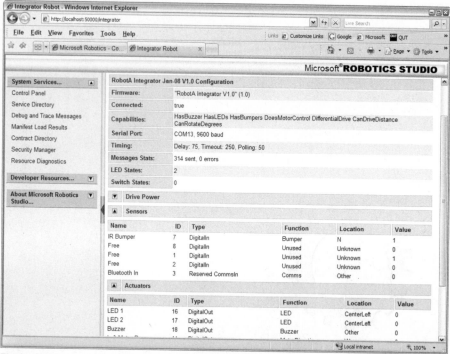

**Figure 17-1**

In Figure 17-1, the Drive Power section has already been collapsed, which you can tell from the small down-arrow next to the heading. The Sensors and Actuators sections show all of the devices that are defined for the robot, including their current values. Note that for simplicity, the state of digital devices is shown as 1 or 0 instead of `true` or `false`. The `Value` field is always maintained in sync with the `State` field for digital devices, and this is one reason why.

# The Integrator Robot

The Integrator robot from Picblok (www.picblokcorporation.com) is designed for use in teaching basic robotics. It uses a PICAXE chip as its onboard controller, and in elementary courses on robotics it is programmed directly in BASIC.

By plugging in a ZX-Bluetooth module from Innovative Experiment Co. Ltd (www.inexglobal.com), the robot can be controlled remotely. Other Bluetooth modules are available that would work just as well. This one is quite cheap, although it is fixed at 9600 baud.

An Integrator robot is shown in Figure 17-2 (on the left). They come in a variety of colors with a clear plastic lid so you can see the circuit board (and in particular the LEDs) inside. The "face" has infrared LEDs and an IR sensor to detect obstacles, as well as two LEDs that flash menacingly when an obstacle is detected. The robot on the right in Figure 17-2 is shown without the lid, and the Bluetooth module has been removed (it fits inside the lid); the PICAXE-18X microcontroller is on a daughterboard in the center of this picture. It is an 18-pin chip based on the Microchip PIC16F88 and runs at 8 mHz.

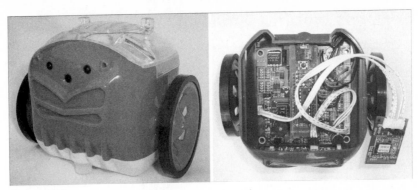

**Figure 17-2**

The PICAXE-18X has a total of eight outputs and five input pins. The pinout for the chip is shown in Figure 17-3. Note that although most of the I/O pins are reconfigurable on the chip, the PICAXE firmware assigns fixed functions to the majority of them.

**PICAXE-18X**

| | | | |
|---|---|---|---|
| ADC 2 / Input 2 | 1 | 18 | Input 1 / ADC 1 |
| Serial Out | 2 | 17 | Input 0 / ADC 0 / Infrain |
| Serial In | 3 | 16 | Input 7 / Keyboard data |
| Reset | 4 | 15 | Input 6 / Keyboard clock |
| 0V | 5 | 14 | +V |
| Output 0 | 6 | 13 | Output 7 |
| Output 1 / i2c sda | 7 | 12 | Output 6 |
| Output 2 | 8 | 11 | Output 5 |
| Output 3 / pwm 3 | 9 | 10 | Output 4 / i2c scl |

**Figure 17-3**

The Inex ZX-Bluetooth module enables the robot to operate wirelessly. This connects to the Input 2 pin and Output 4 pin on the PICAXE — not the pins marked Serial In and Serial Out in Figure 17-3, which are used to program the chip. (There is no particular reason for this choice of pins. They were just available pins.)

The ZX module runs at a fixed baud rate of 9600. Serial communication uses the SerIn and SerOut commands in BASIC. These commands do *bit banging* — that is, they input and output individual bits on the specified pins. Consequently, they are not 100% reliable, and the chip cannot perform other tasks while it is doing serial I/O.

Because it is a very simple robot, the motors can only be on or off; there is no speed control. Enough hardware is built into the robot to enable it to do simple tasks, and programs can easily be downloaded to its flash memory.

# Integrator Robot Monitor Program

All robots controlled using MRDS must have an onboard "monitor" program to communicate with the MRDS services running on your PC. (Or they must have their own onboard PC running MRDS.) This is explained in Chapter 13.

The first step in setting up new robot hardware, therefore, is to develop the monitor program, unless one already exists. (The Hemisson robot discussed later in the chapter already has a built-in monitor program so this step is not required.)

## PICAXE Programming Editor

The Integrator uses a PICAXE chip, and a free integrated development environment (IDE) for this chip can be downloaded from the website (`www.picaxe.co.uk`). The PICAXE Programming Editor uses a simplified version of BASIC. A sample screenshot is shown in Figure 17-4. It also has a Flowchart mode in which you can create programs by using block diagrams, and a Logic mode for creating circuit diagrams, but these are not relevant so they are not shown.

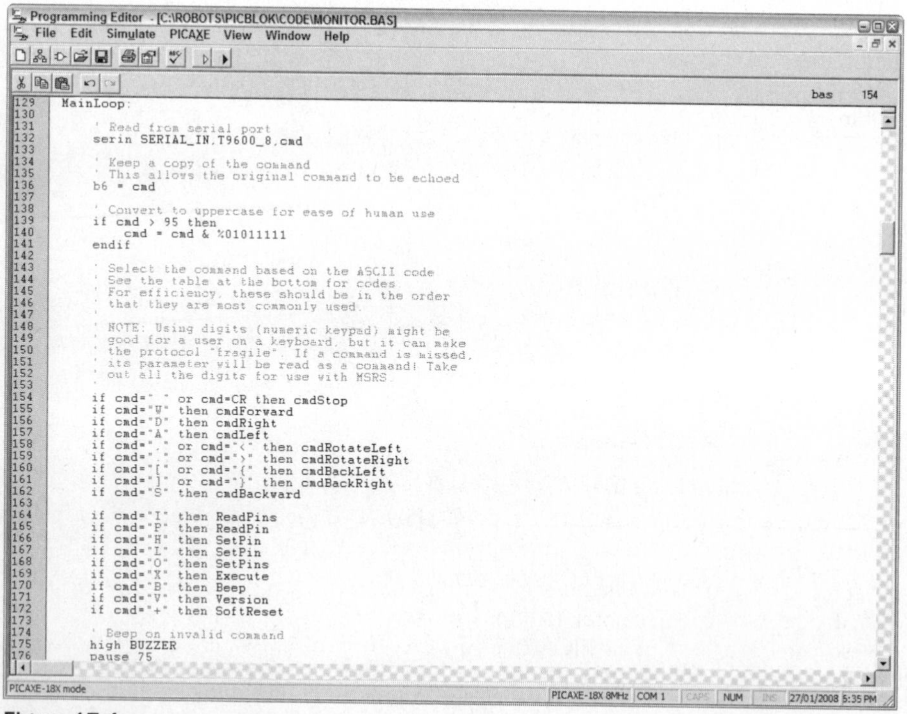
Programming Editor - [C:\ROBOTS\PICBLOK\CODE\MONITOR.BAS]
File Edit Simulate PICAXE View Window Help
bas 154
MainLoop:
' Read from serial port
serin SERIAL_IN,T9600_8,cmd
' Keep a copy of the command
' This allows the original command to be echoed
b6 = cmd
' Convert to uppercase for ease of human use
if cmd > 95 then
cmd = cmd & %01011111
endif
' Select the command based on the ASCII code
' See the table at the bottom for codes.
' For efficiency, these should be in the order
' that they are most commonly used.
' NOTE: Using digits (numeric keypad) might be
' good for a user on a keyboard, but it can make
' the protocol "fragile". If a command is missed,
' its parameter will be read as a command! Take
' out all the digits for use with MSRS.
if cmd=" " or cmd=CR then cmdStop
if cmd="W" then cmdForward
if cmd="D" then cmdRight
if cmd="A" then cmdLeft
if cmd="," or cmd="<" then cmdRotateLeft
if cmd="." or cmd=">" then cmdRotateRight
if cmd="[" or cmd="{" then cmdBackLeft
if cmd="]" or cmd="}" then cmdBackRight
if cmd="S" then cmdBackward
if cmd="I" then ReadPins
if cmd="P" then ReadPin
if cmd="H" then SetPin
if cmd="L" then SetPin
if cmd="O" then SetPins
if cmd="X" then Execute
if cmd="B" then Beep
if cmd="V" then Version
if cmd="+" then SoftReset
' Beep on invalid command
high BUZZER
pause 75
PICAXE-18X mode
PICAXE-18X 8MHz
COM 1
CAPS
NUM
INS
27/01/2008 5:35 PM

Figure 17-4

A simulator is built into the PICAXE Programming Editor. This is useful for teaching students about microcontrollers, but it is also very handy for debugging because you can step through your code one line at a time; view all of the registers and output pins; change the values on input pins; and simulate serial I/O. You can also adjust the execution speed to run your program in "slow motion."

Downloading programs to the chip is very simple. The programming cable has a standard 9-pin serial connector on one end that you plug into a serial port on your PC, and a 3.5mm stereo plug on the other end that plugs into the back of the robot. From the editor menu, you click PICAXE ⇨ Run, or click the Run icon, or press F5 to start the download. As soon as the download is complete, the program begins executing on the robot.

This is not a tutorial on programming PICAXE chips. You should find the commented code provided with this chapter easy enough to read. Full documentation for the PICAXE family, including the BASIC dialect that it uses, is available from the PICAXE website.

## Pin Assignments

To write the code, you need to know the pin assignments for the various I/O devices. The PICAXE refers to pins by number. However, the numbers of the pins used by the output functions (High, Low, Outpins, Outpinx, SerOut) overlap with the numbers used for the input functions (Pins, Pinx, SerIn).

Because there are only 18 pins on the chip, and some of these are required for power, ground, reset, and the serial port for programming, there are only five inputs instead of eight. The Integrator robot pin assignments are shown in the following table. Please note that the term *pin* here is a logical pin number, not a physical pin number on the chip.

| Pin | Input | Output |
| --- | --- | --- |
| 0 | Infrared Remote Control | Left Motor Backward |
| 1 | Microphone | Left Motor Forward |
| 2 | Bluetooth Serial In (Reserved) | Right Motor Backward |
| 3 | (Unavailable) | Right Motor Forward |
| 4 | (Unavailable) | Bluetooth Serial Out (Reserved) |
| 5 | (Unavailable) | LED 1 (Available for use) |
| 6 | IR Obstacle Sensor | LED 2 (Available for use) |
| 7 | (Available for use) | Buzzer |

Notice that the motors are controlled using four pins. There is no speed control — only forward, backward, and stop — for each motor. (Although the PIC16F88 chip has PWM commands and onboard timers, these cannot be used on the PICAXE due to the way that the serial I/O is implemented, so it is not possible to modulate the speed of the motors.)

## The Main Loop

You can look at the source code for the monitor program even if you do not have the PICAXE editor because it is an ASCII text file: ProMRDS\Chapter17\Integrator\Monitor\Monitor.bas.

The code begins by setting the clock frequency on the microcontroller to 8 mHz. This is necessary in order to use the ZX-Bluetooth module, which runs at 9600 baud, because the maximum baud rate in the

PICAXE firmware is 4800 at the default clock speed of 4 mHz. By doubling the clock speed, the baud rate also doubles. This is an important "trick":

```
Start:

' IMPORTANT NOTE:
' The clock frequency is 4 mHz by default.
' Changing it to 8 mHz is necessary to double
' the baud rate from the maximum on a PICAXE
' of 4800 to 9600.
' The ZX-Bluetooth module has a fixed baud
' rate of 9600.
' If this is the first time you have loaded the
' program, you might have to change your config
' in the Programming Editor to 8 mHz or you will
' not be able to download.

SetFreq m8

' Perform a soft reset initially
b0 = "#"
pause 20
goto SoftReset
```

*In the PICAXE editor, you can set the clock speed for downloading. It defaults to 4 mHz. Once you have downloaded a program that sets the speed to 8 mHz, you will have to change the configuration to match or any subsequent attempt to download will fail.*

The code now enters the MainLoop, which it executes forever (or at least until you turn the power off). It spends most of its time waiting for a new character to arrive at the serial input via Bluetooth. It then executes the equivalent of a large switch statement based on the command (character) that it received. The code for the MainLoop is as follows:

```
MainLoop:

  ' Read from serial port
  serin SERIAL_IN,T9600_8,cmd

  ' Keep a copy of the command
  ' This allows the original command to be echoed
  b6 = cmd

  ' Convert to uppercase for ease of human use
  if cmd > 95 then
    cmd = cmd & %01011111
  endif

  ' Select the command based on the ASCII code.
  ' For efficiency, these should be in the order
  ' that they are most commonly used.
  '
  ' NOTE: Using digits (numeric keypad) might be
  ' good for a user on a keyboard, but it can make
```

```
' the protocol "fragile". If a command is missed,
' its parameter will be read as a command!

if cmd=" " or cmd=CR then cmdStop
if cmd="W" then cmdForward
if cmd="D" then cmdRight
if cmd="A" then cmdLeft
if cmd="," or cmd="<" then cmdRotateLeft
if cmd="." or cmd=">" then cmdRotateRight
if cmd="[" or cmd="{" then cmdBackLeft
if cmd="]" or cmd="}" then cmdBackRight
if cmd="S" then cmdBackward

if cmd="I" then ReadPins
if cmd="P" then ReadPin
if cmd="H" then SetPin
if cmd="L" then SetPin
if cmd="O" then SetPins
if cmd="X" then Execute
if cmd="B" then Beep
if cmd="V" then Version
if cmd="+" then SoftReset

' Beep on invalid command
high BUZZER
pause 75
low BUZZER
pause 10
' Send back an error message
serout SERIAL_OUT,T9600_8,("? Invalid",CR,LF)

goto MainLoop
```

If the character does not match any of the known commands, then the program beeps the buzzer and returns an error message. Error messages all begin with a question mark. This is the last section of code before returning to the top of the MainLoop. Note that BUZZER and SERIAL_OUT are symbolic values that are defined at the top of the program to enhance maintainability.

The monitor program was initially designed and tested by using HyperTerminal. This enabled the code to be tested prior to writing the MRDS services. One consequence of this approach is that the robot can actually be driven using the keyboard. In particular, the A, S, D, and W keys are used to control the motors. These keys are commonly used in first-person shooter games, so most children are familiar with them. The spacebar and Enter keys both cause the robot to stop.

The commands are not case-sensitive. (The code in MainLoop takes care of converting to uppercase.) To acknowledge a command, the Integrator echoes back the command followed by a carriage return (CR) and linefeed (LF). When you are using a terminal emulator, this is very handy because you can see the commands that you typed and each one appears on a new line in the emulator window.

## *Protocol Commands*

The "protocol" is extremely simple, and consequently it is not robust if communication is unreliable. Most commands consist of a single character. A few commands have a parameter, which is one or two hexadecimal digits. The full list of Integrator Protocol commands is provided in the following table:

| Code | Parameter | Meaning |
|------|-----------|---------|
| Space or CR | | Stop |
| A | | Turn Left Forwards[1] |
| B | | Beep (Takes 200 ms to execute) |
| D | | Turn Right Forwards[1] |
| H | Pin (0-7) | High (Set pin to 1) |
| I | | Input (Read all pins) |
| L | Pin (0-7) | Low (Set pin to 0) |
| O | xx (Two hex digits) | Output (Write to all pins)<br>**Note:** This command is dangerous because it sets the motors and also interferes with the operation of the Bluetooth module. |
| P | Pin (0-7) | Pin Input |
| S | | Drive Backward |
| V | | Version (Displays version string) |
| W | | Drive Forward |
| X | Behavior<br>L<br>R<br>U<br><br>Behavior xx<br>Axx<br>Bxx<br>Cxx<br>Fxx | Execute Behavior<br>Back up, Rotate Left (90 degrees)<br>Back up, Rotate Right (90 degrees)<br>Back up, Do U-Turn (180 degrees)<br>**Note:** The turn angles are only *approximate*<br>Execute Behavior with Parameter (in hex)<br>Anti-Clockwise (Rotate Left[2]) xx units<br>Backward xx units<br>Clockwise (Rotate Right[2]) xx units<br>Forward xx units<br>**Note:** The units in all cases are *arbitrary* |
| , or < | | Rotate Left[2] |
| . or > | | Rotate Right[2] |

| Code | Parameter | Meaning |
|------|-----------|---------|
| [ or { | | Turn Left Backwards[1] |
| ] or } | | Turn Right Backwards[1] |
| + | | Soft Reset |

[1] Turn operations are performed by leaving one motor *off* and turning on the other motor in the appropriate direction so the robot pivots around the stationary wheel. There are four ways this can be done: Two methods move the robot forward and the other two move the robot backward.

[2] Rotate (spin) operations are performed by running the motors in opposite directions. There are only two possibilities: clockwise (right) or counterclockwise (left).

There is no point including all of the source code for command processing here. A couple of the command routines illustrate how the monitor program works. The version command, for example, returns a string that describes the robot and includes a version number:

```
' Display the Monitor version
'
' NOTE: Make sure to update the version number if this
' code is changed.
Version:
    serout SERIAL_OUT,T9600_8,("RobotA Integrator V1.0",CR,LF)
    goto MainLoop
```

The Brick service can use this information (if necessary) to adjust the way that it communicates with the brick and accommodate different revisions of the firmware.

Obviously, there must be commands to make the robot move. Because the Integrator doesn't support speed control of the wheels, only a small set of motion operations are available. The code to move forward is as follows:

```
cmdForward:
    Gosub SendAck
    Gosub ForwardMotors
    goto MainLoop
```

*PICAXE BASIC commands are not case-sensitive.*

The subroutine SendAck sends an acknowledgment back to the PC. It consists of the character that was received, followed by a carriage return and linefeed. When you are using HyperTerminal, this displays each command that you type on a new line. Then the ForwardMotors subroutine is called, and finally the code jumps back to the top of MainLoop.

*You must configure the editor for the number of subroutines that you want to use in your code. For Monitor.bas, it must be set to 256. In addition, you cannot nest subroutines more than four levels deep because there is a limited amount of stack space. It is therefore common practice to use GoTo, rather than GoSub, and simply jump around in the code. This is not good coding practice, but it's a matter of necessity.*

The `ForwardMotors` subroutine is quite short. It is implemented as a subroutine because it is used from more than one place in the code:

```
' Drive forwards
ForwardMotors:
  low 0
  high 1
  low 2
  high 3
  Return
```

The following table shows how the motors can be controlled using the four direction control pins: 0, 1, 2, and 3. There are nine possible combinations of motor directions that make sense. Another seven combinations make no sense because they would require a motor to go both forward and backward at the same time.

| Function | Left Motor | | Right Motor | | Code |
|---|---|---|---|---|---|
| | Reverse Pin 0 | Forward Pin 1 | Reverse Pin 2 | Forward Pin 3 | |
| Stop | Low | Low | Low | Low | CR/SP |
| Forward | Low | High | Low | High | W |
| Backward | High | Low | High | Low | S |
| Turn Left | Low | Low | Low | High | A |
| Turn Right | Low | High | Low | Low | D |
| Rotate Left | High | Low | Low | High | < |
| Rotate Right | Low | High | High | Low | > |
| Back Left | High | Low | Low | Low | [ |
| Back Right | Low | Low | High | Low | ] |

Other functions commonly used on a microcontroller are setting and reading pins (digital outputs and inputs). These two operations both require a parameter to be supplied in the command. This is read using the `GetHex` routine. The value is returned in a register, which has a symbolic name of `param` set at the top of the program to make the code more readable. The `SetPin` routine is as follows:

```
' Set a single Pin
' Set to high or low based on command (H or L)
' Requires a pin number as parameter
'
SetPin:
  gosub GetHex
  if param > 7 then
      serout SERIAL_OUT,T9600_8,("? Bad Pin",CR,LF)
      goto MainLoop
  else
      serout SERIAL_OUT,T9600_8,(cmd,CR,LF)
```

```
      if cmd = "H" then
         goto SetHigh
      else
         goto SetLow
      endif
   endif

SetHigh:
  high param
  goto MainLoop

SetLow:
  low param
  goto MainLoop
```

To read the value of a pin, the `ReadPin` routine is called:

```
' Read a single Pin
' Requires pin number as a parameter
'
ReadPin:
  gosub GetHex
  ' Only pins 0-7 are supported
  if param > 7 then
     serout SERIAL_OUT,T9600_8,("? Bad Pin",CR,LF)
  else
     b2 = pins
     ' Get bit mask and extract pin status
     lookup param,(1,2,4,8,16,32,64,128),b3
     b2 = b2 & b3
     ' Send back result
     if b2 <> 0 then
        serout SERIAL_OUT,T9600_8,(cmd,"1",CR,LF)
     else
        serout SERIAL_OUT,T9600_8,(cmd,"0",CR,LF)
     endif
  endif
  goto MainLoop
```

It is much quicker to read all of the input pins at once instead of one at a time, so there is a function for this too: `ReadPins`. It sends the information back as two hex digits. Although there is a function for it, setting all of the output pins is not feasible because one of them is Bluetooth Serial Out, and if you write to this pin, you will interfere with the serial communications.

If you want to know more about how the monitor program works, open `Monitor.bas` and read the code. It is fairly well commented.

## *Creating the Integrator Robot Services*

The first step to creating the new Integrator robot service is to make a new service for the brick. Now that you have a Generic Brick service, and a defined communication protocol for the brick, this is fairly easy. To create a new service based on the Generic Brick contract, you use `DssNewService` with the `/alt` (alternate contract) and /i (implement) parameters to specify which service you want and the assembly to look in for the contract.

At the MRDS command prompt, enter the following command (shown in bold text) to create the Integrator service (all on one line):

```
C:\Microsoft Robotics Studio (1.5)\ProMRDS\Chapter17>dssnewservice ↵
/service:"Integrator" /namespace:"ProMRDS.Robotics.Integrator" ↵
/year:"2008" /month:"01" ↵
/alt:"http://www.promrds.com/contracts/2008/01/genericbrick.html" ↵
/i:"..\..\bin\GenericBrick.Y2008.M01.dll"
```

*Don't execute this command in your* ProMRDS *directory! The code already exists. If you want to try it out, make a temporary directory somewhere under the MRDS root and then run the command.*

In the new Integrator solution, the service implementation file, Integrator.cs, contains stubs for all of the Generic Brick operation handlers. Don't worry about them yet. Before you can implement the operations, you need to write code to communicate with the robot.

The new service has a reference to the Generic Brick DLL and a using statement already in the code:

```
using pxgenericbrick = ProMRDS.Robotics.GenericBrick.Proxy;
```

However, this is a bit of a mouthful, so the alias is shortened to brick. The new service does not have a reference to GenericBrickUtil.dll, containing the helper functions, so you need to add this. You also need to add a reference to RoboticsCommon.Proxy.

If you look in IntegratorTypes.cs, initially it contains nothing but the contract identifier. There is no need to define a main operations port or a set of message types — they all come from the generic brick.

## Initialization

Because this first version of the Generic Brick contract cannot be reconfigured, all of the sensors and actuators should be redefined every time the service is run. Otherwise, it would be possible for a user to edit the config file and change device details, perhaps with surprising consequences.

To assist in initializing the Sensors and Actuators in the state, a couple of enums are defined in the IntegratorTypes.cs file: HwIds and Pins. Note that a hardware ID and a pin are *not* the same thing. (Fortunately, no port numbers are involved on the Integrator.) This is not essential, but it helps to keep all of these "magic" numbers together in one place so that you can see if there are duplications or missing values.

When you create the lists of devices during service initialization, you can do it using the Device constructor, as shown here:

```
_state.Sensors.Add(new brick.Device(
    "IR Bumper",
    (int)HwIds.IrBumper,
    brick.DeviceTypes.DigitalIn,
    brick.DeviceFunctions.Bumper,
    brick.Location.FrontCenter,
```

```
                    0,
                    (int)Pins.IrBumper,     // IN6 = IR obstacle sensor
                    DateTime.Now,
                    false,
                    0,
                    0,
                    1
                    )
            );
```

Note that the initial state of some outputs might not be zero (or off). This must be taken into account in the initialization code.

## Serial Communications

Most of the samples provided with MRDS communicate with the robot via a serial port (although it is usually a virtual serial port over Bluetooth). However, if you look at the code for each of the implementations, you will see a range of different approaches to implementing the communications.

Part of the problem is that every robot is subtly different. The Integrator is not very fast and it does "bit banging" instead of using a hardware *universal asynchronous receiver/transmitter (UART)*, so it loses characters if you send them too quickly. Therefore, a small delay is necessary between the characters that make up a command. For the Integrator, this is set to 10 milliseconds because anything less results in lost characters. This is one of the reasons why most commands are a single character — it minimizes the overhead.

You also have to pause between commands to give the robot time to "digest" the command and execute it. The delay between commands has to be determined by trial and error for every new type of robot. You can do this by sending repeated commands and increasing the delay until the robot stops missing commands. You can usually tell if it is missing commands because there will be timeouts on receiving the responses, i.e., the robot never sees a command so it does not respond. For the Integrator, this inter-command delay has to be about 75 milliseconds, which is quite long. (This is one of the properties in the state in the `Configuration`.)

Note that polling (which is covered later) is always occurring. Therefore, commands are constantly being sent to the robot. You cannot have a command interrupted by a polling request — commands sent to the robot must be queued. In addition, a command and its corresponding response must be treated as an atomic operation. As a consequence of these problems, the communications code needs to operate synchronously, i.e., send a command and then wait for a response (or a timeout) before proceeding with the next command. In between commands you must ensure that there is a reasonable delay.

Although the `SerialPort` class in .NET (defined in `System.IO.Ports`) supports asynchronous operations, it is not designed to operate in a CCR environment. The code in `IntegratorControl` therefore uses blocking reads and writes, and throws in a few calls to `Thread.Sleep` for good measure. This is all very antisocial as far as the CCR is concerned, so the main Integrator service class is flagged with an `ActivationSettings` attribute:

```
[ActivationSettings(ShareDispatcher = false, ExecutionUnitsPerDispatcher = 3)]
```

This causes DSS to create a separate dispatcher and allocate three threads. This way, the Integrator service will not interfere with other services running on the DSS node — the serial communications code can block a thread without causing problems.

To isolate the serial communications functions from the rest of the service, there is a separate module called `IntegratorControl.cs`. This is not a separate service, just a separate class in its own source file. The `IntegratorControl` class is responsible for opening and closing the serial port and sending low-level commands to the robot. The main service should not need to know the details of the protocol for talking to the robot, although it is difficult to keep it entirely separate. In particular, there are routines called `SetPin`, `SetPower`, and `SetLEDs` (their names are self-explanatory). Consider the code for `SetPin`:

```
/// <summary>
/// Set Pin
/// </summary>
/// <param name="pin">Pin on the Brick</param>
/// <param name="state">True/False = On/Off or High/Low</param>
/// <returns>Port for the response from the robot</returns>
public Port<byte[]> SetPin(int pin, bool state)
{
    byte[] on = { (byte)'H', (byte)'0' };
    byte[] off = { (byte)'L', (byte)'0' };

    Command cmd = new Command();
    if (state)
    {
        on[1] = (byte)((byte)'0' + pin);
        cmd.CommandString = on;
    }
    else
    {
        off[1] = (byte)((byte)'0' + pin);
        cmd.CommandString = off;
    }

    CommandPort.Post(cmd);

    return cmd.ResponsePort;
}
```

This routine sends either a High (H) or Low (L) command and appends the pin number to the command. Notice that it sends the command by posting a message; it returns a response port. Routines inside `IntegratorControl` send serial commands to an internal port called `CommandPort`. The format of these messages is as follows:

```
/// <summary>
/// Serial Command
/// </summary>
/// <remarks>Used internally to send packets to the serial port</remarks>
public class Command
{
    public byte[] CommandString;
    public Port<byte[]> ResponsePort;
```

```
            public Command()
            {
                CommandString = null;
                // Make sure that there is always a response port
                ResponsePort = new Port<byte[]>();
            }
        }
```

The CommandString is a byte array and it is sent verbatim. All characters in the response up to, but excluding, the following carriage return and linefeed are sent back to the ResponsePort as an array of bytes.

The handler for these Command messages is as follows:

```
            /// <summary>
            /// Serial Port Command Handler
            /// </summary>
            /// <param name="cmd"></param>
            /// <returns></returns>
            /// NOTE: This handler executes serial commands which block the thread
            private IEnumerator<ITask> CommandHandler(Command cmd)
            {
                byte[] response = new byte[0];

                if (cmd.CommandString == null || cmd.CommandString.Length == 0)
                {
                    // Pathological case -- no command!
                    // Somebody sent us an empty packet, so send back an empty response
                    cmd.ResponsePort.Post(response);
                }
                else
                {
                    // Send the packet and wait for a response
                    SendPacket(cmd.CommandString);

                    // Got a response?
                    if (_responseCount > 0)
                    {
                        byte[] resp = new byte[_responseCount];
                        for (int i = 0; i < _responseCount; i++)
                            resp[i] = _response[i];
                        cmd.ResponsePort.Post(resp);
                    }
                    else
                        cmd.ResponsePort.Post(response);
                }

                // Wait one time for a new Command
                Arbiter.Activate(_taskQueue, ←
        Arbiter.ReceiveWithIterator<Command>(false, CommandPort, CommandHandler));

                yield break;
            }
```

The handler sends the command to the robot using `SendPacket`, which does not return until it has received a response or the command has timed out. (The timeout is set in the state `Configuration`.) An empty response indicates that some sort of error occurred, which in most cases is a timeout. The protocol requires the robot to echo all commands back to the PC, so there should always be a response. The response is then posted to the original sender of the command. This enables senders to wait for the response (if they want), but in the usual CCR fashion — by waiting on a port, which frees up threads. Only the thread running the command handler will block and become unusable.

Before finishing, the command handler sets up another receiver to get the next command from the `CommandPort`. (The process is kicked off in the constructor for the `IntegratorControl` class.) By using a nonpersistent handler, it is guaranteed that only one instance of the command handler can ever execute at a time. In effect, this implements an Exclusive behavior.

Notice that a separate task queue (and dispatcher) is used for the command handler. Do you get a sense of paranoia about blocking CCR threads? This is not necessary given the `ActivationSettings` on the service, but it illustrates another approach.

Lastly, the `SendPacket` routine is as follows:

```
// Carriage Return
static char[] CR = { '\r' };

private DateTime _lastPacketTime;

// SendPacket()
// Baud Rate is 9600 which is approximately 960 chrs/sec
// which is about a millisecond per character.
// NOTE: SendPacket() operates SYNCHRONOUSLY, although it does
// have timeouts to prevent it from hanging forever.
private bool SendPacket(byte[] buf)
{
    TimeSpan timeDiff;
    int msDiff;
    bool ok = true;

    // Always check! There could be outstanding messages when the
    // connection is closed.
    if (!_serialPort.IsOpen)
        return false;
```

Notice that the code first checks whether the serial port is open. During shutdown, the port might be closed while commands are still waiting in the queue, and this avoids an exception.

Next it checks to see how much time has elapsed since the last command. The `_lastPacketTime` is set after every command has been processed, so you can figure out how long ago it was. If the new command is too soon after the previous one, then the code waits for the difference in time. This is a better approach than always waiting for a fixed amount of time after a command:

```
// Check that this packet is not too close to the last one.
// It is ESSENTIAL that there is a delay after sending
// a command because the Integrator cannot process the
// commands very quickly and will start to lose data!
```

```
    timeDiff = (DateTime.Now - _lastPacketTime);
    msDiff = timeDiff.Milliseconds;
    if (msDiff < Delay)
    {
        // Wait for the difference in time
        Thread.Sleep(Delay - msDiff);
    }
```

Now the code writes out the serial command and then reads the response. The read and write timeout values are set when the port is opened, so if there is no response within the specified period (currently 250 milliseconds, which is a very long time), an exception occurs. In this case, the code sends a carriage return to the robot to try to illicit a response. This causes the robot to stop (if the command is received properly), which is probably the safest approach. It would be possible to resend the original command, but command retries are not implemented.

```
    _responseCount = 0;

    try
    {
        int i;
        // Throw away any left-over data in the input buffer
        _serialPort.DiscardInBuffer();

        // Output the packet one byte at a time to give the
        // Integrator time to digest the characters.
        // Otherwise it gets a data overrun and locks up!
        for (i = 0; i < buf.Length; i++)
        {
            _serialPort.Write(buf, i, 1);
            Thread.Sleep(10);
        }
    }
    catch (Exception ex)
    {
        errorCounter++;
        Console.WriteLine("Comms Write Error on command " +
            (char)buf[0] + " (Error count " + errorCounter + ")");
        Console.WriteLine("Exception: " + ex.Message);
        // Try to re-synch
        _serialPort.Write(CR, 0, 1);
        ok = false;
    }
    finally
    {
        _lastPacketTime = DateTime.Now;
    }
    if (!ok)
        return false;

    ok = true;
    try
```

*(continued)*

*(continued)*

```
    {
            int i, val;

            i = 0;
            while ((val = _serialPort.ReadByte()) != '\r')
            {
                _response[i] = (byte)val;
                i++;
            }
            _responseCount = i;

            // Throw away the rest of the response (only a LF)
            _serialPort.DiscardInBuffer();
    }
    catch (Exception ex)
    {
            errorCounter++;
            Console.WriteLine("Comms Read Error on command " +
                (char)buf[0] + " (Error count " + errorCounter + ")");
            Console.WriteLine("Exception: " + ex.Message);
            // Send a CR to try to re-synch
            _serialPort.Write(CR, 0, 1);
            ok = false;
    }
    finally
    {
            _lastPacketTime = DateTime.Now;
    }
    if (!ok)
        return false;

    messagesSentCounter++;
    return true;
}
```

Note the use of `DiscardInBuffer`. This prevents unsolicited characters, noise, or leftover characters from a previous command from interfering with the current command.

The last point is that communications statistics are updated as well.

## Handling Sensors Using Polling

Sensor information has to somehow get transferred from the robot to the PC. There are only two options:

❑ The PC polls the robot for sensor data.

❑ The robot periodically sends sensor data to the PC.

Polling is the preferred option in this case because the robot cannot simultaneously send and receive. In other words, once the monitor program on the PICAXE requests a character from the serial port, it is forced to wait (block) until a character arrives. This is a deficiency in the firmware, although it actually arises from the use of "bit banging."

The main Brick service must therefore send commands to the robot at regular intervals. This is easy to set up using a timer. Toward the bottom of `Integrator.cs` is a region called Polling. To kick off the polling process, `SetPollTimer` is called from the `Start` method once a connection to the robot has been successfully established:

```
/// <summary>
/// Set a Timer for Polling
/// </summary>
// NOTE: This does NOT guarantee an exact polling interval.
// Instead it adds the polling interval to the current time.
// There is only ever one receiver active at a time.
private void SetPollTimer()
{
    // Stop the timer if we are shutting down
    if (_shutdown)
        return;

    int wait = -1;
    // If the PollingInterval is negative, then polling is off
    if (_state.Configuration.PollingInterval >= 0 &&
        _state.Configuration.PollingInterval < _minimumPollingInterval)
        wait = _minimumPollingInterval;
    else
        wait = _state.Configuration.PollingInterval;

    if (wait >= 0)
        Activate(Arbiter.Receive(false, TimeoutPort(wait), ←
PollingTimerHandler));
}
```

The `PollingInterval` is one of the few configuration parameters that can be changed. If it is negative, no polling occurs. If it is zero or less than a minimum value, it is set to the minimum. For Bluetooth with a slow serial port (9600 baud in this case), a minimum of 50 milliseconds is optimistic. In fact, you cannot poll the Integrator faster than about 10 Hz, which is a polling interval of 100ms.

The handler for the polling timer is as follows:

```
/// <summary>
/// Polling Timer Handler
/// </summary>
/// <param name="time"></param>
private void PollingTimerHandler(DateTime time)
{
    // Stop the timer if we are shutting down
    if (_shutdown)
        return;

    // Ignore timer if not connected, but keep it ticking over
    if (!_state.Connected)
    {
        SetPollTimer();
        return;
    }
```

*(continued)*

*(continued)*

```
                        // Is Polling turned off?
                        if (_state.Configuration.PollingInterval < 0)
                        {
                            // Set the timer, but don't do anything
                            SetPollTimer();
                            return;
                        }

                        // Get Sensors
                        Port<byte[]> response;
                        response = _control.Poll();

                        Activate(Arbiter.Choice(
                            Arbiter.Receive<byte[]>(false, response, ProcessPollResult),
                            // Make sure that we don't hang here forever!
                            Arbiter.Receive<DateTime>(false, TimeoutPort(1000), ↵
                    SetPollTimer)));
                        }
```

The top of the routine contains a couple of safety checks. Then it calls the `Poll` method in the `IntegratorControl` class.

Notice that the code waits on a response from the poll request, or a timeout of one second. This shows how to avoid getting hung if a response never arrives. In this case, it is unnecessary because the serial port code has a 250ms timeout for the write and read, but the principle is important. As an aside, there is an overloaded version of `SetPollTimer` that accepts a `DateTime` that can be used for the timeout.

The last step is to process the results of the poll:

```
                        /// <summary>
                        /// Handle results from a Poll and reset Timer
                        /// </summary>
                        /// <param name="result"></param>
                        private void ProcessPollResult(byte[] result)
                        {
                            // NOTE: The result should always be 3 bytes long or there is
                            // something wrong
                            if (result.Length == 3)
                            {
                                bool changed = false;

                                // Process the response
                                // Format should be Ixx
                                if (result[0] == (byte)'I')
                                {
                                    // Convert the hex result to binary
                                    int n = HexByteToInt(result, 1);
                                    for (int i=0; i<_state.Sensors.Count; i++)
                                    {
                                        // Only process Digital sensors this way
                                        if (util.IsDigitalDevice(_state.Sensors[i]))
```

```
            {
                // Construct a mask for this Pin
                int mask = 1 << _state.Sensors[i].Pin;
                bool oldState = _state.Sensors[i].State;
                // Check if it is High or Low
                if ((n & mask) != 0)
                {
                    _state.Sensors[i].State = true;
                    _state.Sensors[i].Value = 1;
                }
                else
                {
                    _state.Sensors[i].State = false;
                    _state.Sensors[i].Value = 0;
                }
                // Check for a state change
                if (_state.Sensors[i].State != oldState)
                {
                    // If it is the IR bumper, toggle an LED
                    // This is for debugging ...
                    if (_state.Sensors[i].Pin == (int)Pins.IrBumper)
                    {
                        _state.LEDs = util.UpdateBitmask( ↵
_state.LEDs, LEDIds, (int)HwIds.LED2, _state.Sensors[i].State);
                        _control.SetPin((int)Pins.LED2, ↵
_state.Sensors[i].State);
                    }
                    changed = true;
                }
            }
        }

        // Only send notifications on changes
        if (changed)
            SendNotification<brick.UpdateSensors>(_subMgrPort, ↵
new brick.UpdateSensorsRequest(_state.Sensors));

    }
}
// Finally, set the timer again
SetPollTimer();
}
```

This routine updates the sensor state. It makes use of some of the helper functions that are defined in the GenericBrickUtil DLL prefixed with "util." For debugging purposes, LED 2 reflects the current status of the IR bumper. This is not necessary, but it is an easy way to verify that polling is working. The code also checks to see if any of the values have changed; if so, it sends a notification to all subscribers with the new sensor information. Notice that it does not send the entire state. Before it finishes, the routine kicks off a new timer so that it will be called again.

## Completing the Brick Operations

Now that the infrastructure is in place, you can go back to the operation handler stubs that were automatically created as part of the new service. Filling these in should be straightforward, so they are not covered here.

As an example, consider the `PlayTone` operation. It has to make the buzzer sound on the robot. The Integrator cannot control the frequency of the sound, although some robots can do this. The duration is controlled by the service turning the buzzer on and then off again after the appropriate time delay:

```
/// <summary>
/// PlayTone Handler
/// </summary>
/// <param name="update"></param>
/// <returns></returns>
[ServiceHandler(ServiceHandlerBehavior.Exclusive)]
public virtual IEnumerator<ITask> PlayToneHandler(brick.PlayTone update)
{
    // Turn on the buzzer and don't bother to wait for a response
    util.UpdateDeviceStateInList(_state.Actuators, (int)HwIds.Buzzer, ↵
true);
    _control.SetPin((int)Pins.Buzzer, true);

    // Set up a timer to turn the buzzer off
    Activate(Arbiter.Receive(false, TimeoutPort(update.Body.Duration),
        delegate(DateTime dt)
        {
            // Turn off the buzzer
            util.UpdateDeviceStateInList(_state.Actuators, ↵
(int)HwIds.Buzzer, false);
            _control.SetPin((int)Pins.Buzzer, false);
        }
    ));

    // Send back the default response
    update.ResponsePort.Post(DefaultUpdateResponseType.Instance);

    yield break;
}
```

This function uses the `SetPin` routine mentioned earlier to turn on the appropriate pin. Notice that the current state is also updated to reflect this change. However, actuator changes do not result in notifications.

Then a timeout is set with a delegate that turns the buzzer off (and updates the state again). A response is sent back before the routine finishes. Note that this response is sent back before the timeout has completed because although a receiver is activated, the code does not wait for it. There is no point to keeping the caller waiting.

As a somewhat different example, look at the `GetSensors` handler. Due to the design, it is possible to request any combination of sensors that you like. However, this makes the processing a little more complicated because you have to look up each device. If an invalid hardware identifier is encountered in the requested device list, then the operation throws an exception:

```
/// <summary>
/// GetSensors Handler
/// </summary>
/// <param name="query"></param>
/// <returns></returns>
[ServiceHandler(ServiceHandlerBehavior.Concurrent)]
public virtual IEnumerator<ITask> GetSensorsHandler(brick.GetSensors query)
{
    int n;

    // Create a new result list (which might end up empty!)
    List<brick.Device> list = new List<brick.Device>();

    foreach (brick.Device d in query.Body.Inputs)
    {
        // Look in the State for a matching Hardware Id
        n = util.FindDeviceById(_state.Sensors, d.HardwareIdentifer);
        if (n < 0)
        {
            // No such device! Throw an exception and give up!
            throw new ArgumentOutOfRangeException("No such hardware ↵
identifier: " + d.HardwareIdentifer);
        }
        else
        {
            // Add the device to the result list
            list.Add((brick.Device)_state.Sensors[n].Clone());
        }
    }
    // Return the list
    query.ResponsePort.Post(new brick.GetSensorsResponse(list));

    yield break;
}
```

The code builds a new device list to send back. Notice that it clones the requested devices from the state.

## Creating a New Drive Service

For the Integrator you should implement the generic Differential Drive service so that it can be controlled by existing applications such as the Dashboard and TeleOperation. However, you might want to keep everything together in a single DLL, rather than have separate DLLs for each service.

First create a new Drive service:

```
C:\Microsoft Robotics Studio (1.5)\ProMRDS\Chapter17>dssnewservice ↵
/service:"IntegratorDrive" /namespace:"ProMRDS.Robotics.Integrator.Drive" ↵
/year:"2008" /month:"01" ↵
/alt:"http://schemas.microsoft.com/robotics/2006/05/drive.html" ↵
/i:"..\..\bin\RoboticsCommon.dll"
```

This creates an empty service based on the differential drive contract in a separate folder. Now you can take the `IntegratorDrive.cs` and `IntegratorDriveTypes.cs` files and move them into the existing solution folder for the Integrator. Then, in Visual Studio, use Project ⇨ Add Existing Item to add the files to the solution. This places the Drive service inside the same DLL as the Brick service.

## Command Overload!

Implementing the various operations in `IntegratorDrive.cs` seems easy at first glance, especially as the brick implements `SetDrivePower`. However, taking the simple approach of sending all drive requests to the brick causes a subtle problem that only surfaces once you start to use a joystick. When you "wiggle" a joystick, it can generate dozens of position updates per second. As you have already seen, the Integrator has a slow CPU and commands are sent synchronously. As a result, commands to the motors back up in the internal communications port inside the Brick service. Therefore, long after you have stopped playing with the joystick, the robot is still faithfully executing your instructions. It might be fun to watch, but it makes the robot difficult to control.

In a TeleOperation environment, it is known that delays of more than half a second between action and response make it difficult for a human to control a robot. A delay of more than a second makes it almost impossible because most people are used to virtually instantaneous responses when they move their hands. A backlog of commands acts like a delay. Even worse, how can you stop a runaway robot that has hundreds of queued-up commands?

This "flooding" problem is a common flaw in drive services, and it results from the implicit assumption that requests are executed almost instantaneously. This might be true in simulation, but in real-world robotics, nothing is instantaneous.

Consider a serial port running at 9600 baud, 8 bits, no parity. It takes 10 "bit times" per character (start bit, 8 data bits, and a stop bit), so you can send roughly 960 characters per second. That's over a millisecond per character, without allowing for time between characters or processing delays on the robot. In that time, the CCR might process 50 or 100 messages. It's easy to see how a backlog can develop.

The answer to this dilemma is to always clear out the queue first, and only ever execute the most recent command. The robot might not execute exactly the sequence of commands that correspond to how you move the joystick around, but if you tell it to stop by releasing the joystick, it will stop immediately.

Your first thought might be to look at the length of the queue on the `SetDrivePower` port and clear it out. Unfortunately, the queue length is always zero. The `SetDrivePower` operation is one of many that are handled by an `Interleave`, and it is Exclusive. Therefore, incoming messages are held in the interleave's pending queue and only passed to the handlers one at a time. This queue contains a mixture of different message types and you can't easily extract all of the `SetDrivePower` requests.

The solution to this problem is to funnel all of the `SetDrivePower` requests from the handler through to another internal port. To ensure that only one request is executed at a time, you use a nonpersistent receiver on the internal port and set up a new receiver after each request has been processed. This port is declared at the top of `IntegratorDrive.cs`:

```
// Used for driving the motors. Always execute the newest pending drive request.
private Port<drive.SetDrivePower> _internalDrivePowerPort =
    new Port<drive.SetDrivePower>();
```

The SetDrivePower handler posts a message to the internal port, rather than send the command directly to the robot:

```
/// <summary>
/// SetDrivePower Handler
/// </summary>
/// <param name="update"></param>
/// <returns></returns>
[ServiceHandler(ServiceHandlerBehavior.Exclusive)]
public virtual IEnumerator<ITask> SetDrivePowerHandler( ↵
drive.SetDrivePower update)
    {
        // Return a Fault if the drive is not enabled
        if (!_state.IsEnabled)
        {
            update.ResponsePort.Post(Fault.FromException( ↵
new InvalidOperationException("Drive is not enabled")));
            yield break;
        }

        // All motion requests must do this first!
        ClearPendingStop();

        if (update.Body.LeftWheelPower == 0 && update.Body.RightWheelPower ↵
== 0)
            _state.DriveState = drive.DriveState.Stopped;
        else
            _state.DriveState = drive.DriveState.DrivePower;

        // Pass the request to the internal handler
        _internalDrivePowerPort.Post(update);


    }
```

The internal handler, at the bottom of the source file, clears out any extra requests waiting in the port and then processes the most recent one:

```
/// <summary>
/// Process the most recent Drive Power command
/// When complete, self activate for the next internal command
/// </summary>
/// <param name="driveDistance"></param>
/// <returns></returns>
public virtual IEnumerator<ITask> ↵
InternalDrivePowerHandler(drive.SetDrivePower power)
    {
        try
        {
            // Take a snapshot of the number of pending commands at the time
            // we entered this routine.
            // This will prevent a livelock which can occur if we try to
            // process the queue until it is empty, but the inbound queue
```

*(continued)*

*(continued)*

```
                // is growing at the same rate as we are pulling off the queue.
                int pendingCommands = _internalDrivePowerPort.ItemCount;

                // If newer commands have been issued, send success responses
                // to the older commands and move to the latest one
                drive.SetDrivePower newerUpdate;
                while (pendingCommands > 0)
                {
                    if (_internalDrivePowerPort.Test(out newerUpdate))
                    {
                        // Timed motion requests do not have a response initially,
                        // that comes later when the motors are stopped
                        if (power.ResponsePort != null)
                            power.ResponsePort.Post( ↵
DefaultUpdateResponseType.Instance);
                        power = newerUpdate;
                    }
                    pendingCommands--;
                }

                yield return Arbiter.Choice(_brickPort.SetDrivePower( ↵
new drive.SetDrivePowerRequest(power.Body.LeftWheelPower, ↵
power.Body.RightWheelPower)),
                    delegate(DefaultUpdateResponseType ok)
                    {
                        _state.LeftWheel.MotorState.CurrentPower = ↵
power.Body.LeftWheelPower;
                        _state.RightWheel.MotorState.CurrentPower = ↵
power.Body.RightWheelPower;
                        if (power.ResponsePort != null)
                            power.ResponsePort.Post(ok);
                    },
                    delegate(Fault fault)
                    {
                        if (power.ResponsePort != null)
                            power.ResponsePort.Post(fault);
                    }
                );

            }
            finally
            {
                // Wait one time for the next InternalDrivePower command
                Activate(Arbiter.ReceiveWithIterator(false, ↵
_internalDrivePowerPort, InternalDrivePowerHandler));
            }
            yield break;
        }
```

Note a couple of key points here:

- ❑ The routine gets the current queue length and then removes only that number of requests. If it kept looping and removing requests until the queue length was zero, it is possible that it could loop forever.

- ❑ When requests are taken out of the queue to be discarded, a response is sent back to the sender. This is essential because the sender might be waiting on completion of the request, and it would hang forever if the request were simply thrown away.

Notice that there is a check to see whether the `ResponsePort` is null. This is discussed in the next section. Requests with a `null` `ResponsePort` are the initial requests that are generated by timed motions (`DriveDistance` and `RotateDegrees`). It is more useful to notify the sender once the motion has completed, rather than when it starts, so these operations set the `ResponsePort` to `null` before posting a `SetDrivePower` request to the internal port. When the motion completes, a response message is sent when the stop request is processed. The mechanism for this is explained later in the section "Timed Operations."

Having found a request to execute, the code sends it to the brick. The `_brickPort` is set up at the top of the code where the brick is declared as a partner service:

```
/// <summary>
/// Integrator Brick Partner
/// </summary>
[Partner("Integrator", Contract = ↵
"http://www.promrds.com/contracts/2008/01/integrator.html", CreationPolicy = ↵
PartnerCreationPolicy.UseExistingOrCreate, Optional = false)]
    private brick.GenericBrickOperations _brickPort =
        new brick.GenericBrickOperations();
```

Lastly, the handler sets up a new nonpersistent receiver to get the next message from the internal port. The `Start` method kicks off the whole process by setting up the initial receiver.

There is another benefit of this approach: The timed operations (`DriveDistance` and `RotateDegrees`) now mesh nicely with the `SetDrivePower` operations. They are all Exclusive operations, but the timed operations cannot afford to wait until their operations complete or they would block all other requests. There has to be some way to cancel a timed operation that is in progress. Splitting the timed operations into two requests — start the motors running and stop the motors some time later — solves this problem.

## Timed Operations

In the Dashboard (in Chapter 4), the arrow buttons in the user interface use the `DriveDistance` and `RotateDegrees` operations. Because these are very handy functions to have, it is a good idea to implement them even on robots that don't have wheel encoders and therefore cannot make accurate moves.

To emulate these operations, the code calculates a time delay that corresponds as closely as possible to the amount of time required to complete the requested move. Then it starts the motors, waits for the delay period, and finally stops the motors. If these operations did not return until *after* the delay, then they could never be interrupted because they are Exclusive operations. This is not a good design, especially if the robot is about to crash and the Stop button has no effect!

Therefore, these operations set a timer to fire off a Stop request after the appropriate time interval has elapsed. The only problem with this approach is, what if another command arrives during the delay period? You don't want the Stop to be executed if a new drive command arrives before the delay has expired. Therefore, you have to cancel any pending Stop command (there will only ever be one of them) whenever another command arrives.

To make a long story short, the requirement to be able to interrupt the DriveDistance or RotateDegrees operations adds a lot of extra logic and complexity. However, as is the case with many parts of MRDS, once you have some working code, you can reuse it in similar services. The same code is therefore used for the Hemisson robot.

Consider the RotateDegrees handler. After checking whether the drive is enabled, the first thing it does is to cancel any pending Stop request. All the drive handlers do this, but it was not pointed out on the SetDrivePower handler earlier. Then it checks the rotation angle, because zero is pointless, and sets the wheel power based on the direction of rotation:

```
/// <summary>
/// RotateDegrees Handler
/// </summary>
/// <param name="update"></param>
/// <returns></returns>
[ServiceHandler(ServiceHandlerBehavior.Exclusive)]
public virtual IEnumerator<ITask> RotateDegreesHandler( ⏎
drive.RotateDegrees update)
{
    // This handler is basically the same as DriveDistance except for the
    // calculation of the distance. It could use DriveDistance, but that
    // is an unnecessary overhead.
    // See DriveDistance for more comments.

    // Time delay for the motion
    int delay;
    // Speeds for wheels to move at
    double leftSpeed;
    double rightSpeed;
    // Distance wheels will travel during rotation
    double arcDistance;

    if (!_state.IsEnabled)
    {
        update.ResponsePort.Post(Fault.FromException( ⏎
new InvalidOperationException("Drive is not enabled")));
        yield break;
    }

    // All motion requests must do this first!
    ClearPendingStop();

    // Check for zero degrees first because we don't want the wheels ⏎
to jerk
    // if the robot is supposed to go nowhere!
    if (update.Body.Degrees == 0.0)
```

```
    {
        update.ResponsePort.Post(new DefaultUpdateResponseType());
        yield break;
    }
    else if (update.Body.Degrees < 0.0)
    {
        // Note that the speeds are opposite to make the robot
        // rotate on the spot
        leftSpeed = update.Body.Power;
        rightSpeed = -update.Body.Power;
    }
    else
    {
        leftSpeed = -update.Body.Power;
        rightSpeed = update.Body.Power;
    }
```

Next, it calculates the time to move based on the distance the wheels will travel around the circumference of a circle with the diameter equal to the wheel base (the distance between the wheels). The property DistanceBetweenWheels is one of the standard properties in the generic Differential Drive state. It must be filled in during service initialization. Then it posts a SetDrivePower request to the internal port:

```
            // Calculate the distance that the wheels must travel for the given
            // turn angle. They will move around the circumference of a circle
            // with a diameter equal to the wheel base. Each 360 degrees of  ↩
    rotation
            // is one full circumference.
            arcDistance = Math.Abs(update.Body.Degrees) *  ↩
    _state.DistanceBetweenWheels * Math.PI / 360;

            // Calculate the delay time
            delay = CalculateTimeToMove(arcDistance, update.Body.Power);

            // Start the motors running (in opposite directions)
            drive.SetDrivePower power = new drive.SetDrivePower();
            // Set the power
            power.Body.LeftWheelPower = leftSpeed;
            power.Body.RightWheelPower = rightSpeed;
            // Do NOT respond to this message (Happens later on Stop)
            power.ResponsePort = null;

            // Use the internal handler for throttling
            _internalDrivePowerPort.Post(power);
            _state.DriveState = drive.DriveState.RotateDegrees;

            // Set the timer and forget about it
            SetStopTimer(delay, update.ResponsePort);

    }
```

The fun is not over yet. The timed operations are not supposed to send back a response until they have completed. Therefore, the handler cannot post back a reply straight away — the response must be posted when the Stop command is executed. The internal port handler has to be able to distinguish between normal SetDrivePower requests, which are acknowledged immediately, and DriveDistance and

RotateDegrees, which are not acknowledged until the Stop request. Therefore, the message sent on the internal port has its ResponsePort set to null, and the original ResponsePort is passed to the SetStopTimer routine:

```
// Set a timer to post a Stop later
void SetStopTimer(int delay, PortSet<DefaultUpdateResponseType, Fault> p)
{
    int num;

    // Classic case of "should never happen"
    ClearPendingStop();

    // Increment the ACK counter to make it unique and
    // keep a copy for later
    _ackCounter++;
    num = _ackCounter;
    // Remember the current response port in case it has
    // to be cancelled
    _ackPort = p;
    // Enable ACK
    _ackPending = true;

    // Setup a delegate for the specified time period
    // It will post a StopMotion message when it fires
    Activate(Arbiter.Receive(
        false,
        TimeoutPort(delay),
        delegate(DateTime timeout)
        {
            // Send ourselves a Stop request using the ACK number
            _internalStopPort.PostUnknownType(new StopMotion(num));
        }
    ));
}
```

SetStopTimer uses some global variables and has its own internal port to which Stop messages are posted. It keeps a counter that uniquely identifies the Stop request and sets a flag indicating that a stop is pending. This enables ClearPendingStop to cancel the request and not worry about the timer that has been set on the _internalStopPort. Remember that other requests might arrive while the timer is running, and they all call ClearPendingStop as their first step:

```
// Issue a fault if a pending Stop was terminated and
// clear the flag
void ClearPendingStop()
{
    if (_ackPending)
    {
        _ackPending = false;
        Fault f;
        f = Fault.FromException(new Exception("Stop after timed ↵
motion aborted"));
        _ackPort.Post(f);
        _ackPort = null;
    }
}
```

Notice that ClearPendingStop sends a Fault to the original caller to indicate that the motion was interrupted. The response port is in a global variable. You cannot put it in the timer message because then it would be inaccessible.

If there are no interruptions, then the timer eventually fires and the StopMotionHandler is executed:

```
// Stop Motion Handler
// Called in response to a Stop message when a timed motion finishes
public virtual IEnumerator<ITask> StopMotionHandler(StopMotion stop)
{
    // Somebody beat us to it!
    if (!_ackPending)
        yield break;

    // Another request has arrived and overridden this one
    // so there is still one pending, but it is not us
    if (stop.Body.AckNumber != _ackCounter)
        yield break;

    // Now stop the robot and post a response back to say the move
    // is complete
    drive.SetDrivePower power = new drive.SetDrivePower();
    // Set the power
    power.Body.LeftWheelPower = 0;
    power.Body.RightWheelPower = 0;
    // Set the response port
    power.ResponsePort = _ackPort;

    // Use the internal handler for throttling
    _internalDrivePowerPort.Post(power);
    _state.DriveState = drive.DriveState.Stopped;

    // Clear the ACK
    _ackPending = false;
    _ackPort = null;

}
```

This handler posts a Stop request to the internal drive port and sets the ResponsePort in this message to the response port that was provided in the original message. When the internal handler executes, it posts a response back to the original caller and everyone is happy.

This is a complicated piece of code. You might want to read over it again, or step through it in the debugger. However, once you have figured out a mechanism like this and verified that it works, you can use it repeatedly.

## Using XSLT

Although Microsoft supplies XSLT files with the source code for many different services, there are no generic XSLT files included with the generic contracts. The generic Differential Drive has a standard state data type, so an XSLT file can be used with any implementation of a drive service. You can find the `IntegratorDrive.xslt` file in the `Resources` folder in the Integrator solution.

The first few lines of the file define the data types (contracts) that are used to interpret the XML data from the `HttpGet` operation. Notice that these are all generic contacts:

```
<?xml version="1.0" encoding="utf-8"?>
<xsl:stylesheet version="1.0"
    xmlns:xsl="http://www.w3.org/1999/XSL/Transform"
    xmlns:s="http://www.w3.org/2003/05/soap-envelope"
xmlns:drive="http://schemas.microsoft.com/robotics/2006/05/drive.html"
xmlns:motor="http://schemas.microsoft.com/robotics/2006/05/motor.html"
xmlns:physical="http://schemas.microsoft.com/robotics/2006/07/physicalmodel.html">
```

*If you open this XSLT file in Visual Studio you might see warning messages that it could not find the schemas. Ignore these warnings. They do not affect compilation and will go away when you close the file.*

This file uses the standard Microsoft `MasterPage`:

```
<xsl:import href="/resources/dss/Microsoft.Dss.Runtime.Home.MasterPage.xslt" />
```

It also includes a small image in the page heading to make it look more professional. This PNG file is an embedded resource that is also in the `Resources` folder:

```
<xsl:template match="/">
  <xsl:call-template name="MasterPage">
    <xsl:with-param name="serviceName">
      <img ↵
src="/resources/Integrator.Y2008.M01/ProMRDS.Robotics.Integrator.Resources.↵
Integrator.Image.png" align="middle"/> Integrator Generic Drive
    </xsl:with-param>
    <xsl:with-param name="description">
      Provides access to the Integrator Drive
      (Uses the Generic Drive contract)
    </xsl:with-param>
  </xsl:call-template>
</xsl:template>
```

Everything in the remainder of the file refers to fields in the Differential Drive state using generic data types. It is just a large table, so there is no point in reproducing the code here. The resulting output is shown in Figure 17-5.

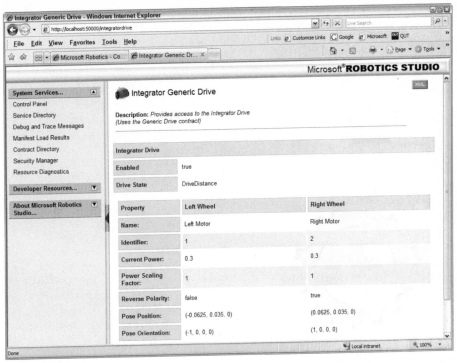

**Figure 17-5**

Notice that the current power is 0.3 for both wheels. Anything above 0.25 turns the motors on. Below this they do not move. In addition, the Drive State is `DriveDistance`, which is the command that is currently being executed.

# Testing a Generic Brick Robot

Once you have built a robot based on the Generic Brick contract, you need to test it. A lot of operations are defined (19 in fact) for the generic brick. This is a lot to test, although many operations can be tested simply by driving the robot around using TeleOperation or the Dashboard and viewing its state in a web browser.

To make life easier for you, there is a test service for the generic brick. It is in the folder `ProMSRS\Chapter17\GenericBrickTest`. To use this service with a robot, you must edit the manifest and insert the necessary code to start your robot. The sample manifest that is provided has the code for the Integrator and the Hemisson. Just comment or delete the one you don't want.

Because the robot is based on the Generic Brick contract, the test service can partner with a generic brick and it does not have to know what type of robot it is. The service is not very sophisticated, but it can exercise all of the operations, including `Drop`! It uses console output to display a list of operations by number (see Figure 17-8 later in the chapter). In some cases an operation requires further information, which you are prompted to enter.

If you select an operation that is not implemented (such as `DriveDistance` or `RotateDegrees`), the result will depend on whether you are running the test service with the debugger or not. The default behavior for operations that are not implemented is to throw an exception.

When the debugger is running, it sees an unhandled exception and breaks execution at that point in the code. Normally you would not continue from an exception, but in the DSS environment it is safe to do so. Just continue running the program and DSS will convert the exception to a `Fault` and send a response back. If you are running the test service without the debugger, then the `Fault` is generated automatically.

The test service sets a timeout on all operations so that it does not hang if the implementation has a problem or you forgot to post back a response in your handler code. One possible problem is that you can easily create a deadlock inside an operation. For example, if you call another operation inside a handler *and* wait for a response, this is a recipe for deadlock. If the handler is Exclusive, no other handlers can run until it completes, so you cannot call any other operations. Even if it is Concurrent, if you make a request to an Exclusive handler, then it cannot run until the Concurrent one completes. Calling a Concurrent operation from inside a Concurrent handler is okay though.

Although the test service has some limited "knowledge" of what the operations do, most of this knowledge is involved in displaying the results. It should be possible to create a more sophisticated service that displays a graphic depiction of the robot (based on the locations of the sensors and actuators), but this is left as another exercise for you.

This XSLT file can be used for any Differential Drive service with only minor changes, i.e., the title at the top of the page and the icon.

# The Hemisson Robot

Hemisson robots are manufactured by K-Team (`www.k-team.com`), which is better known in the research field for its Khepera robots. The Hemisson is designed to be a much cheaper alternative to the Khepera for the educational market.

Figure 17-6 shows a Hemisson loaded up with a Bluetooth dongle and a wireless camera, plus their batteries. This is a heavy load for the poor little fellow, which does not have very powerful motors and only tiny plastic wheels.

**Figure 17-6**

The Sena Parani-SD100 Bluetooth-to-serial converter runs off a 9V battery. The DIP switches are configured for 115200 baud and no hardware flow control. Using the ParaniWin software, the SD100 is placed into Mode 3 (accept connections from any other Bluetooth device). It must be connected to the Hemisson using a null modem connector, which also changes the gender of the serial port. This isn't visible in Figure 17-6 because it is behind the batteries. It is no bigger than two back-to-back 9-pin serial connectors and actually comes in the full Sena SD100 kit.

The Swann wireless MicroCam in Figure 17-6 is stuck in a lump of BluTac. The camera is unplugged from its 9V battery because it does not have a power switch, which is why there is a plug dangling at the front of the Hemisson. Bluetooth interferes with the video signal to some extent, but the reverse does not seem to be the case in this configuration, and the Hemisson can drive around quite reliably with the camera running.

## Hemisson Brick Service

Because it has LEDs, switches, and analog range sensors, the Hemisson is a better example of the generic brick than the Integrator. The code for the Hemisson services is substantially the same as for the Integrator, but there are some points worth mentioning:

❏ The Hemisson comes with an onboard monitor program already loaded. For the purpose of the exercise, the authors chose not to modify this code and instead work with the existing command set. You can download replacement firmware if you want to, but using the existing firmware enables other people to use the services without the hassle of downloading different firmware.

❑ The Hemisson requires a pause of at least 1ms between characters, and up to 100ms between commands. It has a particular problem that the monitor program will hang if a data overrun occurs, i.e., if you send a command before it has finished processing the previous one, then the robot locks up and you have to turn it off and on again to reset it. This appears to be a bug in the firmware, but K-Team support could not shed any light on the matter.

❑ The Hemisson has four LEDs. One of these (the On/Off LED) flashes constantly when the Hemisson is running in program mode, so setting it has no effect because it is immediately overridden. (If you watch carefully, it does turn on solidly for a very brief period and then it starts flashing again.) However, the service allows you to set all four. That leaves three LEDs that can be used to indicate status — two yellow ones at the front and one red one at the back right.

❑ There is a set of four switches on the Hemisson. However, these must all be in the "up" position when the robot is turned on so that it enters "program" mode. The other positions initiate different behaviors such as autonomous wandering and line following. Once the power is on, you can change the switches and then start the MRDS service. Therefore, you can use the switches to select different functions within your own service.

The switch values are obtained during polling as part of the fast binary read (explained under "IR Sensors"), but notifications are not sent for changes in the switch states.

❑ The switches are not polled in the current service implementation — they are only read when the service starts. This was a deliberate design decision because the switches are rarely used, especially while the robot is moving around, and the extra overhead of sending notifications is not warranted. Reading them would slow down the polling of the IR sensors, which are much more important. You can, however, issue an explicit GetSwitches request if you want to get the settings.

## IR Sensors

There are eight infrared range sensors on the Hemisson, although two of them are at the front facing downward and are intended for line following. (The generic brick DeviceFunctions enum has a function explicitly for line following.) To read these using the standard command involves quite a long message because the protocol is designed to be human-readable.

There are new, undocumented commands in the latest version of the Hemisson firmware (V1.51) called "fast binary read" and "fast binary write" that transmit the data in binary. These can save a considerable amount of time because the response packet is shorter than it would be for the ASCII version. If you want more information, download the source code for the HemiOS and read through it to see how the commands work.

The IR sensors have a very limited range and are highly nonlinear. The Hemisson documentation includes a graph showing the response of the IR sensors. Very rough testing showed that they do not register anything until the distance is about 3.5cm or less. Then the values climb rapidly.

The response curves for some of the IR sensors are shown in Figure 17-7. (The Excel spreadsheet is in the solution folder and is called Calibration-IRRange.xls.) Note that measuring distances from the sensors is very difficult, and when you get close to the sensors, even a difference of 1 millimeter can make a substantial difference to the reading. Therefore, this graph is provided for illustration purposes only — it is not intended to be definitive.

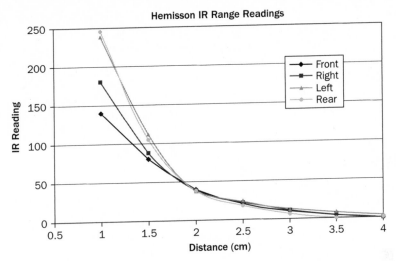

Figure 17-7

The main point to note about Figure 17-7 is that if you drive the robot too fast, it might not sense obstacles in time to avoid them.

Because there are so many range sensors, a *contact sensors array* (also called *bumpers*) has been implemented for the Hemisson. (This wasn't done with the Integrator because it only has a single digital input for an obstacle sensor.) This uses the MRDS generic contract for contact sensors. The service was created in a similar way to the Drive service for the Integrator, i.e., DssNewService was used to create a new service and then the source files were moved to the existing Hemisson solution.

The bumper service subscribes to the brick to receive updates from the IR sensors. The code to implement the bumper service is almost trivial. The most important operation is the one that receives the sensor update notifications from the brick:

```
/// <summary>
/// Receive infrared notifications from the Hemisson service
/// </summary>
/// <param name="notification"></param>
private void UpdateSensorsHandler(brick.UpdateSensors notification)
{
    int i;
    bool changed;
    bool found;
    bool newState;

    foreach (bumper.ContactSensor b in _state.Sensors)
    {
        changed = false;
        found = false;
        foreach (brick.Device dev in notification.Body.Devices)
```

*(continued)*

*(continued)*

```
            {
                if (dev.HardwareIdentifer == b.HardwareIdentifier)
                {
                    found = true;
                    // Check on the purpose of the device
                    // NOTE: IR sensors are NOT marked as Bumpers!
                    // A Bumper is a physical switch, or an IR sensor that only
                    // has a Digital output. The Hemisson IR sensors are range
                    // sensors, except for the Line Followers.
                    if (dev.Function == brick.DeviceFunctions.LineFollower)
                        newState = LineDetected((int)dev.Value);
                    else
                        newState = BumperPressed((int)dev.Value);
                    if (newState != b.Pressed)
                    {
                        changed = true;
                        b.Pressed = newState;
                        b.TimeStamp = DateTime.Now;
                        break;
                    }
                }
            }

            if (!found)
                LogError(LogGroups.Console, "IR sensor not found in update");

            if (changed)
                this.SendNotification<bumper.Update>(_subMgrPort, new
    bumper.Update(b));
        }
    }
```

Notice that bumper notifications are sent only when a bumper state changes, so here you have a notification message triggering yet another notification message, possibly to different subscribers.

The `BumperPressed` function applies a threshold to determine whether the bumpers have been triggered. Looking at Figure 17-7, a value of 10 corresponds to about 3cm, so this is used to trigger the obstacle sensors. (The noise in the IR measurements is less than 10 units, so this is not a problem.)

For the line-following sensors, the behavior of the two sensors is not consistent. This means that applying a single threshold (in the routine `LineDetected`) to both of them does not give optimum results. Unfortunately, the Contact Sensors Array contract does not take into account "virtual" bumpers, so there is nowhere to store a threshold value on a per-sensor basis. If you have a Hemisson, you will want to test it yourself and perhaps change the threshold settings, and maybe even provide an array of thresholds based on the hardware IDs.

IR sensor updates occur at about 10Hz. Occasionally you will see an error message in the console window because a poll times out. The inter-command delay is squeezed as low as possible (currently 50ms), but perhaps it needs to be increased.

## Testing the Hemisson Brick

Because the Hemisson has several devices, it is more important to have a test program than it was for the Integrator because there are more operations to test. For example, Figure 17-8 shows the GenericBrickTest service being used to test the `SetLEDs` operation on a Hemisson.

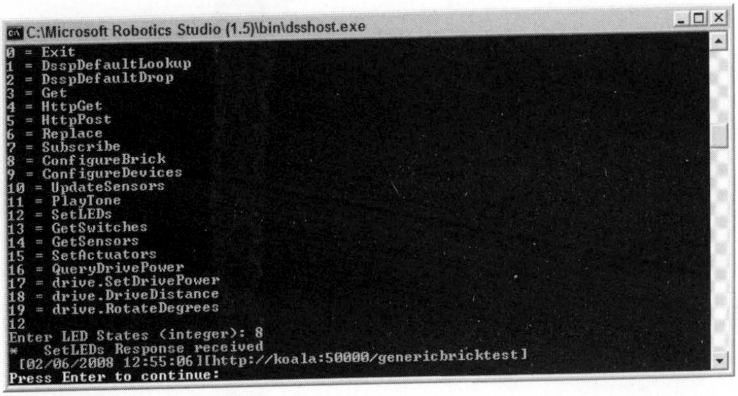

**Figure 17-8**

You can also test subscriptions using the test service, although it only displays the contents of the first notification message. To use the `ConfigureBrick`, `SetActuators`, and `UpdateSensors` operations, you need to first do a `Get`. The service then uses the data from the `Get` in these other operations. This is necessary because the service has no understanding of what the data means, and if it simply generated arbitrary outputs, there is no telling what the robot might do!

While you are using the test service, you should also have a web browser open and use it to view the state information for the various services — Robot Brick, Drive, and Bumpers.

# Hemisson Drive Service

The Hemisson has nine different drive power settings. At the lowest settings, 1 and 2, the robot cannot move with all of the additional equipment loaded on it because the motors are not very strong.

To implement timed motions, it is necessary to map each motor power setting to a time factor for calculating the elapsed time for a given distance or rotation angle. This is a straightforward, but time-consuming, process. The steps are as follows:

1. An arbitrary value is chosen initially as a time factor. In this example it was 5550. The distance specified in a `DriveDistance` request is multiplied by this factor to determine the delay between starting the motors and stopping them. This deliberately ignores the power setting.

2. The Hemisson is run under the control of the TeleOperation service. It is told to drive forward 30cm several times at different speeds and the actual distances are measured and averaged. Using a web browser, the TeleOperation "Motion Speed" (actually power times 1000) can be changed. Values of 100, 200, 300 . . . 1000 are used and the results are recorded for each power level.

**3.** The recorded distances can then be plotted using Excel. Figure 17-9 shows the graph of distance traveled versus power setting. (The spreadsheet is included in the Hemisson folder and is called `Calibration-Drive.xls`.)

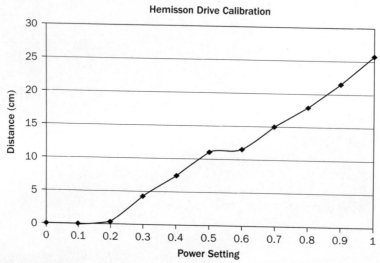

**Figure 17-9**

You might wonder why there is a flat spot between 0.5 and 0.6. This is because the range for drive power is a double from 0 to 1.0, which TeleOperation expresses as 0 to 1000, but the Hemisson only has power settings of 0 to 9. When remapping the power to Hemisson settings, 0.5 and 0.6 both map to the same power level. It might be more sensible to change the mapping and throw away anything lower than 100. You can investigate this if you're interested.

Apart from these problems, the curve is fairly straight. In the real world, of course, there is no such thing as a linear system. Therefore, Excel is used to calculate the required time factor for each of the power settings so that the robot will move the correct distance, which is 30cm in this experiment. These values are then placed in an array, rather than trying to obtain an equation for a straight line.

When a `DriveDistance` request is processed, the time factor is looked up and multiplied by the given distance to obtain the delay time in milliseconds. The code to do this is as follows:

```
// Experimentally determined time factors to travel a specified distance
// Then adjusted slightly after retesting
private static int[] TimeFactors =
    {
        50000,      // 0
        50000,      // 1
        50000,      // 2
        39200,      // 3
        22100,      // 4
        14500,      // 5
        14500,      // 6
```

```
            11100,      // 7
            9200,       // 8
            7740,       // 9
            6500        // 10
    };

    /// <summary>
    /// CalculateTimeToMove
    /// </summary>
    /// <param name="distance">Distance to travel</param>
    /// <param name="power">Drive power to use</param>
    /// <returns>Number of milliseconds</returns>
    public int CalculateTimeToMove(double distance, double power)
    {
        // Drive distance is in meters, but we have to convert it to a
        // time delay in milliseconds. The "magic" numbers here were found
        // by trial and error using a 30cm drive distance. It will change
        // as the batteries run down and due to a variety of other factors.
        int pwr = (int)(Math.Abs(power) * 10 + 0.5);
        double delay = Math.Abs(distance) * TimeFactors[pwr];
        return (int)(delay);
    }
```

The results of timed motions will never be exactly correct, nor reproducible, due to variations in the timing on the `TimeoutPort`. Because it uses Windows timers, the `TimeoutPort` is only accurate to the nearest 15ms. In addition, the service is running on a multi-tasking operating system that was not designed for real-time use.

*Other programs running on your PC might have control of the CPU when a timer fires, forcing your service to wait its turn. Therefore, it is advisable to close down all non-essential programs before you run an MRDS application.*

Even if the timer were perfectly accurate, there is still a chance that a sensor poll might be in progress when it is time to stop the motors. As explained earlier, there must be a short pause between commands, so in general there can be a variation of up to 100ms in the total elapsed time due to serial port contention. At top speed, 100ms equates to several centimeters of travel.

The errors tend to average out, so over a series of motions the result will be close to correct. For example, if you make four 90-degree rotations, the robot will turn close to 360 degrees even though the individual rotations might not be 90 degrees.

Note that the Hemisson does not drive straight. The motor speeds are neither regulated nor matched, so your Hemisson might drift to the left or the right. The problem is worst at lower speeds.

Lastly, the Hemisson also implements the `HttpPost` operation for the Drive service. If you use a web browser, you can see that there are additional buttons at the top of the web page that are not present on the Integrator Drive page. The values of the left and right motor power are also in textboxes so that you can change them (see Figure 17-10).

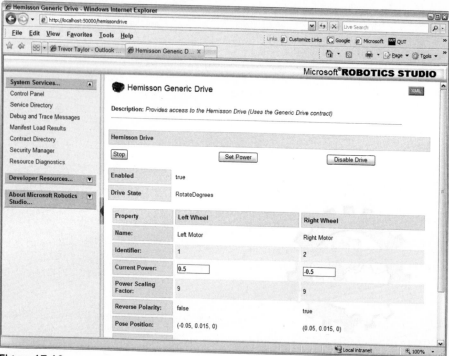

**Figure 17-10**

In Figure 17-10, the robot is executing a `RotateDegrees` operation. This is not implemented in the main Hemisson (Brick) service, but in the Drive service. You can move the code from the Drive service to the main service if you want. In that case, the Drive service would be almost trivial because the main service already implements the `SetDrivePower` operation. In effect, the Drive service would just be a shell that offered an operations port that complies with the generic Differential Drive contract. All the real work would be done in the main service.

If you view the code in the `HttpPostHandler`, you can see how it handles the different buttons at the top of the web form:

```
if (parameters["Stop"] != null)
{
    drive.AllStop req = new drive.AllStop();
    _mainPort.Post(req);
    success.Body.WaitPort = req.ResponsePort;
}
else if (parameters["EnableDrive"] != null)
{
    drive.EnableDrive req = new drive.EnableDrive();
    req.Body.Enable = true;
    _mainPort.Post(req);
    success.Body.WaitPort = req.ResponsePort;
}
else if (parameters["DisableDrive"] != null)
```

```
{
    drive.EnableDrive req = new drive.EnableDrive();
    req.Body.Enable = false;
    _mainPort.Post(req);
    success.Body.WaitPort = req.ResponsePort;
}
else if (parameters["SetPower"] != null)
{
    if (_state.IsEnabled)
    {
        if (!string.IsNullOrEmpty(parameters["LeftPower"]))
        {
```

Each of the submit buttons has a name. When you click one of them, a value is sent along with the other fields on the page. The other buttons are ignored, so you only need to check for the existence of a parameter with the name of a button to determine which button was pressed.

There is another important "trick" here — a deferred response. Notice that in the preceding code, the response port from the original message is placed into a success message as the WaitPort property. As the form is processed, error messages are appended to the ErrorMessage string. The last step in the handler is to check this string:

```
if (ErrorMessage == string.Empty)
{
    // Post a message to finally issue the HTTP response
    // IMPORTANT NOTE:
    // If we simply redisplay the state now, it will be WRONG!
    // It is quite likely that we have issued a request to an
    // Exclusive handler in the code above. If we wait for a
    // response, then we create a Deadlock!
    // The solution is to post a message, which gives the other
    // handler a chance to run, then wait for the response, and
    // finally send the HTTP page back to the user. Got that?
    success.Body.Submit = submit;
    // This is the ONLY place where these messages are sent
    // Do NOT wait for a response!
    _httpPostPort.Post(success);
}
else
    HttpPostFailure(submit, ErrorMessage);
```

If there is an error, then a response can be posted back to the caller immediately. This is not a problem. However, if you want the refreshed page to contain the correct information, you must wait until another operation has completed. To avoid deadlocks, the current operation must finish and the other operation has to be given a chance to execute. This is done by posting a message to an internal port called _httpPostPort.

In the body of the HttpPost handler, a request has already been posted to one of the other drive operations: AllStop, EnableDrive, or SetDrivePower. This other request is already waiting in the queue when the `HttpPostSuccess` message is posted, so as soon as the `HttpPost` handler completes, the other operation will execute first. Then the `HttpPostSuccessHandler` is called:

```
/// <summary>
/// Send Http Post Success Handler
/// </summary>
/// Processes successful HttpPost messages by first waiting for the
/// requested task to be performed, then sending back the web page
/// NOTE: This is a complicated process, but it is necessary to avoid
/// creating a Deadlock!
private IEnumerator<ITask> HttpPostSuccessHandler(HttpPostSuccess success)
{
    Fault fail = null;

    // Wait for the request to be processed
    // Otherwise, the settings displayed on the web page are the
    // OLD state settings, not the new ones!
    if (success.Body.WaitPort != null)
    {
        yield return Arbiter.Choice(
            success.Body.WaitPort,
            delegate(DefaultUpdateResponseType resp) { },
            delegate(Fault f) { fail = f; }
        );
    }

    // Make sure that the requested task completed successfully
    if (fail == null)
    {
        // Post a HTTP response (finally!)
        HttpResponseType rsp =
            new HttpResponseType(HttpStatusCode.OK, _state, _transform);
        success.Body.Submit.ResponsePort.Post(rsp);
    }
    else
    {
        // There is still a chance for this to fail!
        HttpPostFailure(success.Body.Submit, fail);
    }
}
```

The success handler does the waiting, *not* the `HttpPost` handler, which has already completed execution. It does this using the `WaitPort`. Note that there is still a chance that the other operation might fail, in which case the "success" handler actually sends back a "fail" message (as a `Fault`). However, if all goes well, the web page is redisplayed using the XSLT file, just as it is for a `HttpGet` — but now it has the most up-to-date state information.

Note that in order for this approach to work, the success messages must participate in the main interleave. Therefore, the receiver is added to the main interleave in the `Start` method:

```
// Set up the Stop Motion handler which must be in the Exclusive group
// along with the other drive requests
MainPortInterleave.CombineWith(
```

```
                new Interleave(
                    new ExclusiveReceiverGroup(
                        Arbiter.ReceiveWithIterator<StopMotion>(true, ↵
_internalStopPort, StopMotionHandler)
                    ),
                    new ConcurrentReceiverGroup(
                        Arbiter.ReceiveWithIterator<HttpPostSuccess>(true, ↵
_httpPostPort, HttpPostSuccessHandler)
                    )
                )
            );
```

It doesn't matter that the `HttpPostSuccessHandler` is Concurrent because it will never execute at the same time as an Exclusive handler.

Overall, the implementation of the Hemisson services demonstrates more functionality than the Integrator. The Hemisson can be driven around using the TeleOperation program while you watch the live video. The range of the Sena Bluetooth dongle is quite good. The only known problem is that sometimes the Hemisson does not connect when the service starts, but you can fix this by powering it off and on again and restarting the service.

# Where to Go from Here

That's the end of the book, but the fun does not stop here. New applications will continue to be posted on the book's website. You can develop your own robotics applications and, it is hoped, share them with the MRDS community.

If we can believe the pundits, robotics is at a similar stage to the early development of the PC. Over the next few years we can expect dramatic improvements in the capabilities of robots in terms of physical agility, manual dexterity, object recognition, intelligence, and so on. The hobby robots and robotic toys of today will pale in comparison to the commercially available robots on the market in ten years.

One rapidly developing area is *telepresence* — similar to the TeleOperation service in Chapter 4. In late 2007, a whole raft of robots was announced that can be remotely controlled over the Internet with full video and audio capability — Spykee from Meccano, ConnectR from iRobot, and Rovio from WowWee Robotics. These should become available in 2008. This approach overcomes the artificial intelligence "roadblock" that researchers hit some 20 years ago, and it extends the concept of robotic toys to something useful in the home. Grandma will be able to "tele" in on the family robot and check whether you are keeping your room tidy.

With an aging population in many countries, such as Spain and Japan, placing "companion robots" into the homes of the elderly is one way to care for them. Research has shown that elderly people living in nursing homes become more active and responsive when they are given small, furry robotics pets, such as seals, pandas, and so on, to play with. Meanwhile, children's toys are getting more intelligent and lifelike, such as Pleo the dinosaur. These robots, however, are just as much about human interaction as they are about robotics. More effort is required on the user interface than on the robotics.

It's a great time to be on the ground floor of a new industry, but don't expect to be writing services to control differential drives in 2020. Will we see domestic robots powered by MRDS roaming around homes? Microsoft would like to think so. Only time will tell.

# Summary

This chapter has shown you how to build a set of services for a completely new robot, albeit a simple one. From hardware and design to timed operations and polling, you have seen what is required to build and test a new robot.

Of course, despite the fact that it is over 800 pages long, this book still does not cover some areas of MRDS. You need to read the online documentation to fill in the gaps, and when V2.0 comes out you will probably have to start all over again. Welcome to the constantly evolving world of robotics.

Now go out and program your robots!

# Index

## Symbols

## A

# W

# Take your library wherever you go

Now you can access more than 200 complete Wrox books online, wherever you happen to be! Every diagram, description, screen capture, and code sample is available with your subscription to the **Wrox Reference Library**. For answers when and where you need them, go to wrox.books24x7.com and subscribe today!

## Find books on

- ASP.NET
- C#/C++
- Database
- General
- Java
- Mac
- Microsoft Office
- .NET
- Open Source
- PHP/MySQL
- SQL Server
- Visual Basic
- Web
- XML